When this book was first taught from, each student was asked to write a brief statement describing his or her understanding of what role electromagnetics plays in science, technology, and society. The following statement, submitted by Mr. Schaldenbrand, who has since graduated with a B.S.E. degree, was selected for inclusion here:

Electromagnetics has done more than just help science. Since we have such advanced communications, our understanding of other nations and nationalities has increased exponentially. This understanding has led and will lead the governments of the world to work towards global peace. The more knowledge we have about different cultures, the less foreign these cultures will seem. A global kinship will result, and the by-product will be harmony. Understanding is the first step, and communication is the means. Electromagnetics holds the key to this communication, and therefore is an important subject for not only science, but also the sake of humanity.

Mike Schaldenbrand

The University of Michigan

FUNDAMENTALS OF

APPLIED
ELECTROMAGNETICS
1999 Edition

FUNDAMENTALS OF

APPLIED ELECTROMAGNETICS
1999 Edition

Fawwaz T. Ulaby

The University of Michigan

PRENTICE HALL, Upper Saddle River, New Jersey 07458

Library of Congress Cataloging-in-Publication Data

Ulaby, Fawwaz T. (Fawwaz Tayssir)
 Fundamentals of applied electromagnetics / Fawwaz T. Ulaby
  p.   cm.
 Includes bibliographical references and index.
 ISBN 0-13-011554-1
  1. Electromagnetism. 2. Electromagnitism—Industrial applications
 I. Title.
 QC760.U49   1999
 537--dc21   98-41930
                                              CIP

Publisher: Tom Robbins
Acquisitions editor: Alice Dworkin
Editorial/production supervision: AnnMarie Kalajian
Cover photo: © Uniphoto Picture Agency
Cover designer: Joseph Sengotta/ Bruce Kensalaar
Creative director: Paula Maylahn
Managing editor: Eileen Clark
Editor-in-chief: Marcia Horton
Assistant vice president of production and manufacturing: David W. Riccardi
Manufacturing buyer: Pat Brown
Editorial assistant: Dan DePasquale

© 1997, 1999 Prentice-Hall, Inc.
Simon & Schuster / A Viacom Company
Upper Saddle River, New Jersey 07458

The author and publisher of this book have used their best efforts in preparing this book. These efforts
include the development, research, and testing of the theories and programs to determine their effectiveness. The author and publisher make no warranty of any kind, expressed or implied, with regard to these
programs or the documentation contained in this book. The author and publisher shall not be liable in any
event for incidental or consequential damages in connection with, or arising out of, the furnishing, performance, or use of these programs.

Printed in the United States of America

10 9 8 7 6 5 4 3

ISBN 0-13-011554-1

Prentice-Hall International (UK) Limited, *London*
Prentice-Hall of Australia Pty. Limited, *Sydney*
Prentice-Hall Canada Inc., *Toronto*
Prentice-Hall Hispanoamericana, S.A., *Mexico*
Prentice-Hall of India Private Limited, *New Delhi*
Prentice-Hall of Japan Inc., *Tokyo*
Simon & Schuster Asia Pte. Ltd., *Singapore*
Editora Prentice-Hall do Brasil, Ltda., *Rio de Janeiro*

Love is life's wonder!

I dedicate this book to Jean.

Children are life's joy.

I also dedicate this book to
Aziza, Gabriel, Laith, and Neda.

C O N T E N T S

P R E F A C E

Why Another EM Book?

Several textbooks are currently available for teaching electromagnetics to students majoring in electrical engineering. So, why do we need another one? The answer is simple: (1) the curriculum for a bachelor's degree in electrical engineering is undergoing a radical change, perhaps more radical in both outlook and content than any of the curricular changes we have witnessed over the past several decades, and (2) available textbooks are incompatible with the philosophy of the new curriculum proposed for the 21st century (see, for example, the article by Director et al. in the *Proceedings of the IEEE*, September 1995).

The Changing Curriculum

For a bachelor's degree in electrical engineering, the curriculum has undergone about one major change per decade. In the 1960s, courses concerning solid-state devices were introduced and those on vacuum-tube electronics were slowly phased out. In the 1970s, courses on electric machinery nearly disappeared from the curricula of most universities and were replaced with computer-programming courses. More computer-related and digital processing courses were added in the 1980s, mainly by increasing the volume of required courses and reducing the number of technical and free electives. Also, there has been a sustained effort to incorporate new knowledge and assimilate the rapidly evolving role of technology in the undergraduate curriculum. More material was added to courses, greater effort was expected from students, and the number of elective credit hours rapidly approached zero. By the early 1990s, the average student at a U.S. university needed closer to five years to complete what was originally designed as a four-year program.

This scenario was not limited to electrical engineering; indeed, in almost any engineering discipline, the curriculum had become too demanding in terms of time to complete the B.S. degree and too inflexible to accommodate changes. From these pressures was born a new philosophy embracing the following tenets: (1) the B.S. degree program should be scaled back to four years, (2) the required part of the B.S. degree program should focus on the teaching of fundamentals, but with greater exposure to engineering applications, and (3) the electives portion of the program should be increased substantially to allow the student to explore both engineering and nonengineering areas of interest. While this new philosophy has been adopted, as least in principle, by many engineering schools, the task of revising the curriculum has been a major challenge, particularly at the course level. The new electrical engineering curriculum calls for allocating fewer credit hours in many of the traditionally core areas—electromagnetics (EM) among them. At many universities the required EM content of the program has been reduced from two courses down to one. For some, the intent is to continue to offer a two-course sequence, but with only the first being part of the required core and the second being for students interested in deepening their knowledge in electromagnetics.

Course Contents

Given these objectives and associated boundary conditions, what then should be the contents of a one-semester or a two-course sequence in electromagnetics and what texts might one use for this purpose? To answer these questions, we should briefly review the traditional approaches that have been in use over the past two decades. Most EM books share a common template. They begin with one or more chapters on vector calculus and coordinate systems, followed by two or more chapters on statics (electrostatics and magnetostatics). These topics typically constitute half of the total material, and the second half usually covers dynamics (time-varying fields, wave propagation and reflection, waveguides and resonators, and antennas). Such an arrangement of topics presents two problems. First, starting a course with a heavy dose of mathematics tends to "turn the students off." And second, whereas electrostatics and magnetostatics are interesting subjects in themselves and have many practical applications, their importance pales in comparison with the many applications of time-varying fields to communication systems, radar, optics, computers, and solid-state electronics, among others. Thus, teaching statics only in the required portion of the B.S. degree curriculum would leave an electrical engineering student seriously lacking in ability to deal with most electromagnetic phenomena.

Because of the heavy emphasis on mathematical manipulations using vector calculus and the relatively dry character of electrostatics and magnetostatics, the student who completes such a required course is unlikely to select to take the second course in the sequence, which is where many of the topics relevant to engineering practice are covered. Clearly, teaching from a book based on this traditional sequence of topics is no longer compatible with the format of the new curriculum proposed for the 21st century.

In past years, attempts have been made by educators at various universities to teach the dynamics of electromagnetics without preceding it with a study of statics. These attempts have not been very successful, primarily because in dynamics students must deal with several fundamental parameters simultaneously. These include spatial coordinates denoting location, vectors denoting direction, time, and the interconnection between the electric and magnetic fields. In statics, the electric and magnetic fields are independent of one another and time is not a variable, thereby reducing the number of parameters from four to two. By studying electrostatics and magnetostatics first, the student can much more easily tackle situations involving time-varying fields.

This Book

In view of this background, how then does one design an appropriate one-semester or two-course sequence in electromagnetics? My answer to this question is characterized in this book by the following elements:

1. I wanted to begin by building a bridge between what was already familiar to a third-year electrical engineering student and the electromagnetics material. Prior to enrolling in an EM course, a typical student will have taken two or more courses in circuits. Thus, he or she is familiar with circuit analysis, Ohm's law, Kirchhoff's current and voltage laws, and related topics. Transmission lines constitute a *natural* bridge between electric circuits and electromagnetics. Without having to deal with vectors or fields, the student uses already familiar concepts to learn about wave motion, the reflection and transmission of power, phasors, impedance matching, and many of the properties of wave propagation in a guided structure. All of these newly learned concepts will prove invaluable later (in Chapters 7 through 9) and will facilitate the learning of how plane waves propagate in free space and in material media. Transmission lines are covered in Chapter 2, which is preceded in Chapter 1 with reviews of complex numbers and phasor analysis.

2. The next part of the book, contained in Chapters 3 through 5, covers vector analysis, electrostatics, and magnetostatics. Compared with most EM textbooks written for undergraduate instruction, the present book differs in terms of its presentation of these three topics in the following two ways: Of the total number of pages contained in the book, about 30% are allocated to these topics, compared with 50% or more in most EM textbooks. The electrostatics chapter begins with Maxwell's equations for the time-varying case, which are then specialized to electrostatics and magnetostatics, thereby providing the student with an overall framework for what is to come and showing him or her why electrostatics and magnetostatics are special cases of the more general time-varying case.

3. Unlike most EM textbooks, this book does not have a chapter on waveguides and resonators. Space limitations and the fact that waveguides are no longer used as widely as they had been prior to 1980 dictated this change.

4. Chapter 6 deals with time-varying fields and sets

Suggested Syllabi

Chapter	Two-semester Syllabus 6 credits (42 contact hours per semester)		One-semester Syllabus 4 credits (56 contact hours)	
	Sections	Hours	Sections	Hours
1 Introduction	All	4	All	4
2 Transmission Lines	All	12	2-1 to 2-8 and 2-11	8
3 Vector Analysis	All	8	All	8
4 Electrostatics	All	8	4-1 to 4-10	6
5 Magnetostatics	All	7	5-1 to 5-5 and 5-7 to 5-8	5
Exams		3		2
	Total for first semester	42		
6 Maxwell's Equations	All	6	6-1 to 6-3, and 6-6	3
7 Plane-wave Propagation	All	7	7-1 to 7-4, and 7-6	6
8 Wave Reflection and Transmission	All	9	8-1 to 8-3, and 8-6	7
9 Radiation and Antennas	All	10	9-1 to 9-6	6
10 Satellite Communication Systems and Radar Sensors	All	5	None	—
Exams		3		1
	Total for second semester	40	Total	56
Extra Hours	2	0		

the stage for the material in Chapters 7 through 9. Chapter 7 covers plane-wave propagation in dielectric and conducting media, and Chapter 8 covers reflection and transmission at discontinuous boundaries and introduces the student to fiber optics and the imaging properties of mirrors and lenses.

In Chapter 9, the student is introduced to the principles of radiation by currents flowing in wires, such as dipoles, as well as to radiation by apertures, such as a horn antenna or an opening in an opaque screen illuminated by a light source.

5. To give the student a taste of the wide-ranging applications of electromagnetics in today's technological society, Chapter 10 concludes the book with overview presentations of two system examples: satellite communication systems and radar sensors.

6. The material in this book was written for a two-semester sequence of six credits, but it is possible to trim it down to generate a syllabus for a one-semester

four-credit course. The above table provides syllabi for each of these two options.

In writing this book, I avoided lengthy derivations of theorems, particularly those involving extensive use of vector calculus. My goal has been to help the student to develop competence in applying vector calculus to solve electromagnetic problems of practical interest. I view vector calculus and mathematics in general as useful tools and not as ends in themselves. Throughout the material, emphasis is placed on using the mathematics to explain and clarify the physics, followed with practical examples intended to demonstrate the engineering relevance of physical concepts. I believe the combination of the approach used in presenting the material, the arrangement of topics covered in the book, and the relative emphasis in favor of dynamics constitutes an effective algorithm for equipping our future graduates with a relevant foundation in applied electromagnetics.

Acknowledgments

In writing this book I have been fortunate to have had valuable help from many talented individuals. First and foremost, I would like to acknowledge my most sincere gratitude to Roger DeRoo, Richard Carnes and Jim Ryan. I am indebted to Roger DeRoo, not only for the numerous discussions we have had on how to best present electromagnetic concepts and applications, without getting lost in the mathematics, but also for his painstaking review of several drafts of the manuscript. Richard Carnes is unquestionably the best technical typist I have ever worked with; his mastery of LaTeX, coupled with his attention to detail, made it possible to arrange the material in a clear and smooth format. The artwork was done by Jim Ryan, who skillfully transformed my rough sketches into drawings that are both professional looking and esthetically pleasing.

I would like to thank my colleagues at the Radiation Laboratory of the University of Michigan for their advice and support and, in particular, I gratefully acknowledge Linda Katehi, Chen-To Tai and Kamal Sarabandi for their generous counsel and constant encouragement. Thanks also are due to my students, who have suffered through two semesters of having a text in the form of notes and who have provided many suggestions for improving and clarifying the presentation. Moreover, I am grateful to the following graduate students for reading through parts or all of the manuscript and for helping me with the solutions manual: Bryan Hauck, Yanni Kouskoulas, and Paul Siqueira.

I am grateful to the reviewers for their valuable comments and suggestions. They include Constantine Balanis of Arizona State University, Harold Mott of the University of Alabama, David Pozar of the University of Massachusetts, S. N. Prasad of Bradley University, Robert Bond of New Mexico Institute of Technology, Mark Robinson of the University of Colorado at Colorado Springs, and Raj Mittra of the University of Illinois. I appreciate the dedicated efforts of the staff at Prentice Hall and I am grateful for their help in shepherding this project through the publication process in a very timely manner. I also would like to thank Mr. Ralph Pescatore for copyediting the manuscript.

Fawwaz T. Ulaby
Ann Arbor, Michigan

The 1999 Edition

This 1999 edition represents an intermediate step between the standard textbook format of the preceding edition and the CD-ROM interactive supplement to be introduced with the 2001 edition. The CD-ROM of the 2001 edition will include audio/video demonstrations, interactive "laboratory experiments," drill exercises, and complete solutions of a select subset of the end-of-chapter problems. The material developed under this project will be added to a web site at the Unversity of Michigan* on a continuing basis. Suggestions and ideas are very welcome.

The present edition differs from its predecessor in the following ways:

1. Typographical errors have been corrected.
2. The total number of end-of-chapter problems contained in the book has been increased from 238 problems to 373 problems, providing greater scope and a larger number of application examples.
3. Each copy of this book is accompanied with a CD-ROM, placed in the pocket attached to the inside back cover. The CD-ROM contains three folders:

   (a) Figures: This folder, containing copies of all figures appearing in the book, is made available to facilitate the process of reproducing the figures practicably by instructors who may want to generate viewgraphs of the figures electronically.

   (b) Smith Chart: The Smith-Chart file allows the student to generate a high quality print of a Smith chart for his or her use in Chapter 2.

   (c) Sample Solutions: Exposure to examples of problems and their solutions is an indespensable learning tool for the student. The book contains many such examples, but in an effort to further enhance this component of the book, complete solutions are provided for 45 problems, selected from among the end-of-chapter problems appearing in the book. Those selected for inclusion in the CD-ROM are indentified by the symbol ◉ next to the problem statement.

I would like to take this opportunity to thank my Prentice Hall editor, Mr. Tom Robbins, for his encouragement and enthusiastic support of the CD-ROM project

Fawwaz T. Ulaby
Ann Arbor, Michigan

*http://www.eecs.umich.edu/emag/

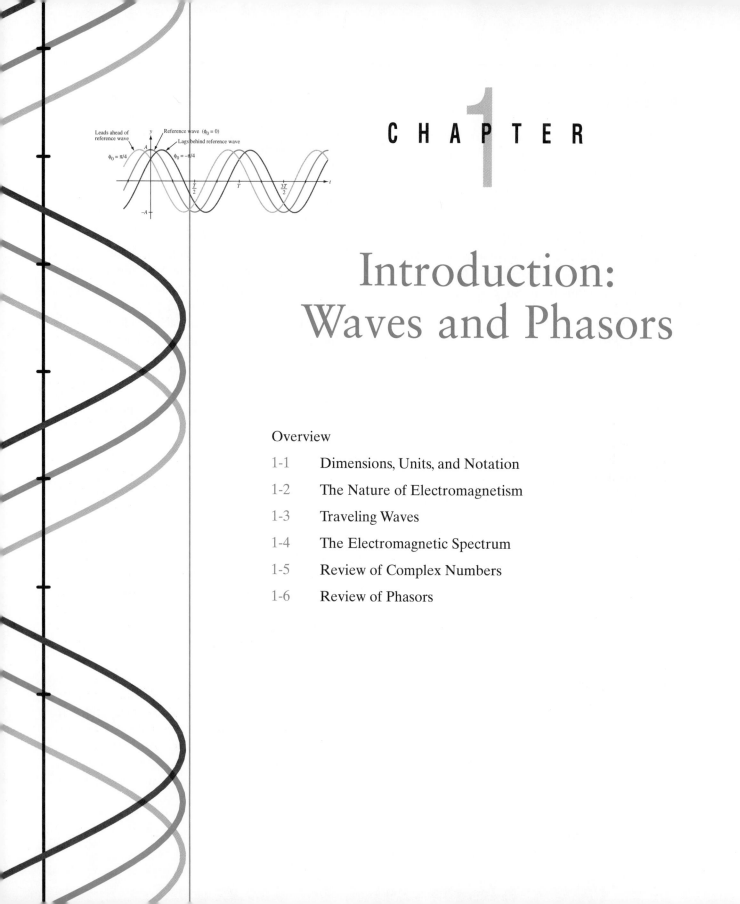

CHAPTER 1

Introduction: Waves and Phasors

Overview

OVERVIEW

The subject of this book is applied electromagnetics, which encompasses the study of electric and magnetic phenomena and their engineering applications, under both static and dynamic conditions. Primary emphasis is placed on the fundamental properties of time-varying (dynamic) electromagnetic fields because of their greater relevance to practical problems in many engineering disciplines, including microwave and optical communications, radar systems, bioelectromagnetics, and high-speed microelectronics, among others. We shall study wave propagation in guided media, such as coaxial transmission lines and optical fibers, wave reflection and transmission at the interface between dissimilar media, imaging properties of mirrors and lenses, radiation by antennas, and several other related topics. The concluding chapter is intended to illustrate a few aspects of applied electromagnetics through an examination of design considerations associated with the use and operation of radar sensors and satellite communication systems.

By way of an example, let us consider the communication network depicted in Fig. 1-1. The satellite segment uses a *multiple-beam* antenna so as to communicate with several ground stations simultaneously. The signal received from ground station 1 at frequency $f_1 = 6$ GHz* is amplified and then downconverted to a frequency $f_2 = 4$ GHz for retransmission to ground station 2. The frequency conversion is facilitated by a microwave solid-state device called a *mixer*, which in this application converts the input frequency f_1 at port M_1 to a frequency f_2 at port M_2 such that $f_2 = f_1 - f_0$, where f_0 is the satellite

---

*1 GHz = 10^9 Hz.

oscillator frequency. In this case $f_0 = 2$ GHz. Different frequencies are used for the uplink and downlink signals to avoid interference between them. The *circulator* connected to the satellite antenna allows the antenna to function as both a transmitter and a receiver. The circulator connects a signal received at port C_1 to the receiver at port C_2, and it can also connect a signal at port C_3 to port C_1 for transmission by the antenna. Although in the configuration shown in Fig. 1-1 one antenna beam is used for reception and another for transmission, in reality both beams can be used for both functions.

Ground station 1 is connected to multiple terminals through various types of *transmission lines*, including *coaxial cables* and *optical fibers*. The information carried on these lines includes voice for telephone communication, video and voice for TV, and digital data. One of the terminals is connected to a terrestrial microwave link.

In this book we will examine several elements of this communication network, including wave transmission on coaxial lines, the fundamentals of optical fiber, free-space propagation between antennas, antenna design, and many other topics. Our goal is to help the reader to develop an understanding of the fundamental physical laws of electromagnetism, as well as to gain insight into their practical applications.

We begin this chapter by describing the system of units and the symbolic notation used in the book. Next, we introduce the fundamental electric and magnetic field quantities we use in electromagnetics, their relationships to each other and to electric charges and electric currents, and the basic parameters by which we characterize the electromagnetic properties of materials. The

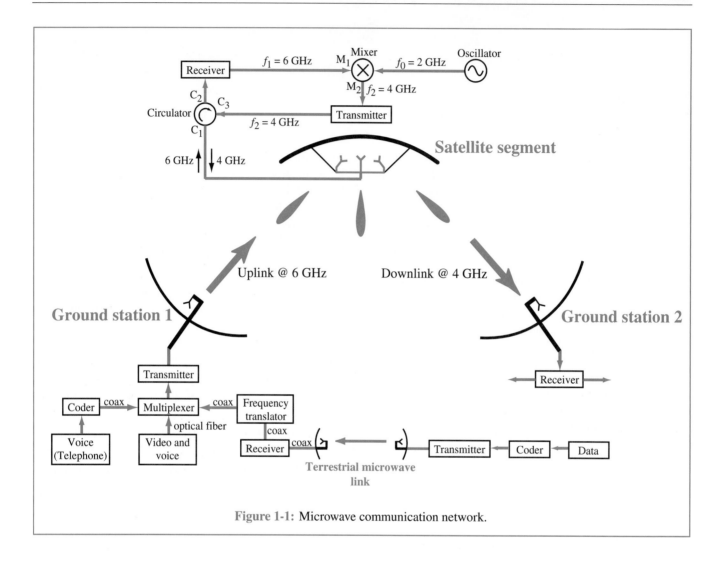

Figure 1-1: Microwave communication network.

laws governing these relationships constitute the basic infrastructure we use in the study of electromagnetic phenomena. Then, in preparation for the material presented in Chapter 2, we provide short reviews of three topics: traveling waves, complex numbers, and phasor analysis. Although the reader most likely has encountered these topics in circuit analysis or other engineering disciplines, short reviews of the properties of traveling waves and of the convenience phasor notation provides should prove useful in solving time-harmonic problems.

1-1 Dimensions, Units, and Notation

The *International System of Units*, abbreviated *SI* after its French name *Système Internationale*, is the standard system used in today's scientific literature for expressing

Table 1-1: Fundamental SI units.

Dimension	Unit	Symbol
Length	meter	m
Mass	kilogram	kg
Time	second	s
Electric Current	ampere	A
Temperature	kelvin	K
Amount of substance	mole	mol

Table 1-2: Multiple and submultiple prefixes.

Prefix	Symbol	Magnitude
exa	E	10^{18}
peta	P	10^{15}
tera	T	10^{12}
giga	G	10^{9}
mega	M	10^{6}
kilo	k	10^{3}
milli	m	10^{-3}
micro	μ	10^{-6}
nano	n	10^{-9}
pico	p	10^{-12}
femto	f	10^{-15}
atto	a	10^{-18}

the units of physical quantities. Length is a *dimension* and meter is the *unit* by which it is expressed relative to a reference standard. The SI system is based on the units for the six *fundamental dimensions* listed in Table 1-1. The units for all other dimensions are regarded as *secondary* because they are based on and can be expressed in terms of the six fundamental units. Appendix A contains a list of quantities used in this book, together with their symbols and units.

For quantities ranging in value between 10^{-18} and 10^{18}, a set of prefixes, arranged in steps of 10^{3}, are commonly used to denote multiples and submultiples of units. These prefixes, all of which were derived from Greek, Latin, Spanish, and Danish terms, are listed in Table 1-2. A length of 5×10^{-9} m, for example, may be written as 5 nm.

In electromagnetics we work with scalar and vector quantities. In this book we use a medium-weight italic font for symbols (other than Greek letters) denoting scalar quantities, such as R for resistance, whereas we use a boldface roman font for symbols denoting vectors, such as **E** for the electric field vector. A vector consists of a magnitude (scalar) and a direction, with the direction usually denoted by a unit vector. For example,

$$\mathbf{E} = \hat{\mathbf{x}}E, \qquad (1.1)$$

where E is the magnitude of **E** and $\hat{\mathbf{x}}$ is its direction. Unit vectors are printed in boldface with a circumflex (ˆ) above the letter.

Throughout this book, we make extensive use of *phasor representation* in solving problems involving electromagnetic quantities that vary sinusoidally in time. Letters denoting phasor quantities are printed with a tilde (∼) over the letter. Thus, $\widetilde{\mathbf{E}}$ is the phasor electric field vector corresponding to the instantaneous electric field vector $\mathbf{E}(t)$. This notation is discussed in more detail in Section 1-6.

1-2 The Nature of Electromagnetism

Our physical universe is governed by four fundamental forces of nature:

- The *nuclear force*, which is the strongest of the four, but its range is limited to *submicroscopic* systems, such as nuclei.

- The *weak-interaction force*, whose strength is only 10^{-14} that of the nuclear force. Its primary role is in interactions involving certain radioactive elementary particles.

- The *electromagnetic force*, which exists between all charged particles. It is the dominant force in *microscopic* systems, such as atoms and molecules, and its strength is on the order of 10^{-2} that of the nuclear force.
- The *gravitational force*, which is the weakest of all four forces, having a strength on the order of 10^{-41} that of the nuclear force. However, it is the dominant force in *macroscopic* systems, such as the solar system.

Our interest in this book is with the electromagnetic force and its consequences. Even though the electromagnetic force operates at the atomic scale, its effects can be transmitted in the form of electromagnetic waves that can propagate through both free space and material media. The purpose of this section is to provide an overview of the basic *framework of electromagnetism*, which consists of certain fundamental laws governing the electric and magnetic fields induced by static and moving electric charges, respectively, the relations between the electric and magnetic fields, and how these fields interact with matter. As a precursor, however, we will take advantage of our familiarity with the gravitational force by describing some of its properties because they provide a useful analogue to those of the electromagnetic force.

1-2.1   The Gravitational Force: A Useful Analogue

According to Newton's law of gravity, the gravitational force $\mathbf{F}_{g_{21}}$ acting on mass m_2 due to a mass m_1 at a distance R_{12} from m_2, as depicted in Fig. 1-2, is given by

$$\mathbf{F}_{g_{21}} = -\hat{\mathbf{R}}_{12} \frac{Gm_1m_2}{R_{12}^2} \quad \text{(N)}, \quad (1.2)$$

where G is the universal gravitational constant, $\hat{\mathbf{R}}_{12}$ is a unit vector that points from m_1 to m_2, and the unit for force is newton (N). The negative sign in Eq. (1.2) accounts for the fact that the gravitational force is attractive. Conversely, $\mathbf{F}_{g_{12}} = -\mathbf{F}_{g_{21}}$, where $\mathbf{F}_{g_{12}}$ is the force acting on mass m_1

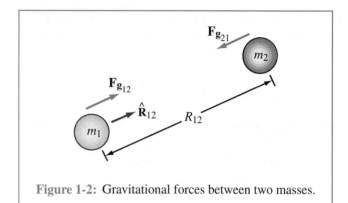

Figure 1-2: Gravitational forces between two masses.

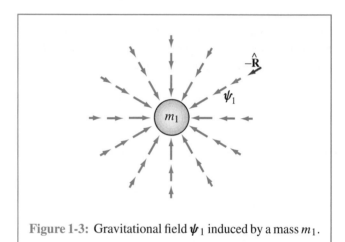

Figure 1-3: Gravitational field $\boldsymbol{\psi}_1$ induced by a mass m_1.

due to the gravitational pull of mass m_2. Note that the first subscript of $\mathbf{F}_g$ denotes the mass experiencing the force and the second subscript denotes the source of the force.

The force of gravitation acts at a distance; that is, the two objects do not have to be in direct contact for each to experience the pull by the other. This phenomenon of direct action at a distance has led to the concept of *fields*. An object of mass m_1 induces a gravitational field $\boldsymbol{\psi}_1$ (Fig. 1-3) that does not physically emanate from the object, but its influence exists at every point in space such that if another object of mass m_2 were to exist at a distance R_{12} from object m_1 then the second object m_2

would experience a force acting on it equal in strength to that given by Eq. (1.2). At a distance R from m_1, the field $\boldsymbol{\psi}_1$ is a vector defined as

$$\boldsymbol{\psi}_1 = -\hat{\mathbf{R}}\,\frac{Gm_1}{R^2} \quad \text{(N/kg)}, \qquad (1.3)$$

where $\hat{\mathbf{R}}$ is a unit vector that points in the radial direction away from object m_1, and therefore $-\hat{\mathbf{R}}$ points toward m_1. The force due to $\boldsymbol{\psi}_1$ acting on a mass m_2 at a distance $R = R_{12}$ along the direction $\hat{\mathbf{R}} = \hat{\mathbf{R}}_{12}$ is

$$\mathbf{F}_{g_{21}} = \boldsymbol{\psi}_1 m_2 = -\hat{\mathbf{R}}_{12}\,\frac{Gm_1 m_2}{R_{12}^2}\,. \qquad (1.4)$$

The field concept may be generalized by defining the gravitational field $\boldsymbol{\psi}$ at any point in space such that, when a test mass m is placed at that point, the force $\mathbf{F}_g$ acting on m is related to $\boldsymbol{\psi}$ by

$$\boldsymbol{\psi} = \frac{\mathbf{F}_g}{m}\,. \qquad (1.5)$$

The force $\mathbf{F}_g$ may be due to a single mass or a distribution of many masses.

1-2.2   Electric Fields

The electromagnetic force consists of an electrical force $\mathbf{F}_e$ and a magnetic force $\mathbf{F}_m$. The electrical force $\mathbf{F}_e$ is similar to the gravitational force, but with a major difference. *The source of the gravitational field is mass, and the source of the electrical field is electric charge,* and whereas both types of fields vary inversely as the square of the distance from their respective sources, electric charge may have positive or negative polarity, whereas mass does not exhibit such a property.

We know from atomic physics that all matter contains a mixture of neutrons, positively charged protons, and negatively charged electrons, with the fundamental quantity of charge being that of a single electron, usually denoted by the letter e. The unit by which electric charge is measured is the coulomb (C), named in honor of the eighteenth-century French scientist Charles Augustin de Coulomb (1736–1806). The magnitude of e is

$$\boxed{e = 1.6 \times 10^{-19} \quad \text{(C)}.} \qquad (1.6)$$

The charge of a single electron is $q_e = -e$ and that of a proton is equal in magnitude but opposite in polarity: $q_p = e$. Coulomb's experiments demonstrated that:

(1) *two like charges repel one another, whereas two charges of opposite polarity attract,*

(2) *the force acts along the line joining the charges, and*

(3) *its strength is proportional to the product of the magnitudes of the two charges and inversely proportional to the square of the distance between them.*

These properties constitute what today is called *Coulomb's law*, which can be expressed mathematically by the following equation:

$$\boxed{\mathbf{F}_{e_{21}} = \hat{\mathbf{R}}_{12}\,\frac{q_1 q_2}{4\pi\varepsilon_0 R_{12}^2} \quad \text{(N)} \quad \text{(in free space)},} \qquad (1.7)$$

where $\mathbf{F}_{e_{21}}$ is the electrical force acting on charge q_2 due to charge q_1, R_{12} is the distance between the two charges, $\hat{\mathbf{R}}_{12}$ is a unit vector pointing from charge q_1 to charge q_2 (Fig. 1-4), and ε_0 is a universal constant called the *electrical permittivity of free space* [$\varepsilon_0 = 8.854 \times 10^{-12}$ farad per meter (F/m)]. The two charges are assumed to be in *free space* (vacuum) and isolated from all other charges. The force $\mathbf{F}_{e_{12}}$ acting on charge q_1 due to charge q_2 is equal to force $\mathbf{F}_{e_{21}}$ in magnitude, but opposite in direction; $\mathbf{F}_{e_{12}} = -\mathbf{F}_{e_{21}}$.

The expression given by Eq. (1.7) for the electrical force is analogous to that given by Eq. (1.2) for the gravitational force, and we can extend the analogy further by defining

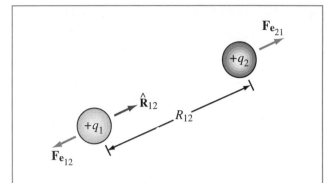

Figure 1-4: Electric forces on two positive point changes in free space.

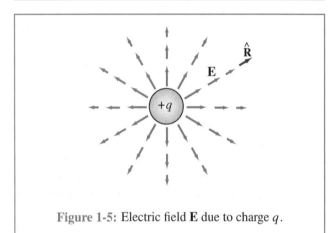

Figure 1-5: Electric field **E** due to charge q.

the existence of an *electric field intensity* **E** due to any charge q as follows:

$$\mathbf{E} = \hat{\mathbf{R}} \frac{q}{4\pi \varepsilon_0 R^2} \quad \text{(V/m)} \quad \text{(in free space)}, \quad (1.8)$$

where R is the distance between the charge and the observation point, and $\hat{\mathbf{R}}$ is the radial unit vector pointing away from the charge. Figure 1-5 depicts the electric-field lines due to a positive charge. For reasons that will become apparent in later chapters, the unit for **E** is volt per meter (V/m).

Electric charge exhibits two important properties. The first is the *law of conservation of electric charge*, which states that *the (net) electric charge can neither be created nor destroyed.* If a volume contains n_p protons and n_e electrons, then the total charge is

$$q = n_p e - n_e e = (n_p - n_e)e \quad \text{(C)}. \quad (1.9)$$

Even if some of the protons were to combine with an equal number of electrons to produce neutrons or other elementary particles, the net charge q remains unchanged. In matter, the quantum mechanical laws governing the behavior of the protons inside the atom's nucleus and the electrons outside it do not allow them to combine.

The second important property of electric charge is the *principle of linear superposition*, which states that *the total vector electric field at a point in space due to a system of point charges is equal to the vector sum of the electric fields at that point due to the individual charges.* This seemingly simple concept will allow us in future chapters to compute the electric field due to complex distributions of charge without having to be concerned with the forces acting on each individual charge due to the fields by all of the other charges.

The expression given by Eq. (1.8) describes the field induced by an electric charge when in free space. Let us now consider what happens when we place a positive point charge in a material composed of atoms. In the absence of the point charge, the material is electrically neutral, with each atom having a positively charged nucleus surrounded by a cloud of electrons of equal but opposite polarity. Hence, at any point in the material not occupied by an atom the electric field **E** is zero. Upon placing a point charge in the material, as shown in Fig. 1-6, the atoms experience forces that cause them to become distorted. The center of symmetry of the electron cloud is altered with respect to the nucleus, with one pole of the atom becoming more positively charged and the other pole becoming more negatively charged. Such a polarized atom is called an *electric dipole*, and

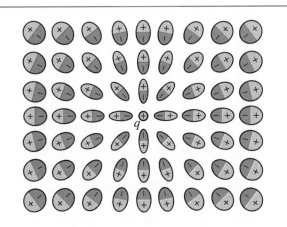

Figure 1-6: Polarization of the atoms of a dielectric material by a positive charge q.

where ε_r is a dimensionless quantity called the *relative permittivity* or *dielectric constant* of the material. For vacuum, $\varepsilon_r = 1$; for air near Earth's surface, $\varepsilon_r = 1.0006$; and for materials that we will have occasion to use in this book, their values of ε_r are tabulated in Appendix B.

In addition to the electric field intensity **E**, we will often find it convenient to also use a related quantity called the *electric flux density* **D**, given by

$$\mathbf{D} = \varepsilon \mathbf{E} \quad (\text{C/m}^2), \quad (1.12)$$

and its unit is coulomb per square meter (C/m^2). These two electrical quantities, **E** and **D**, constitute one of two fundamental pairs of electromagnetic fields. The second pair consists of the magnetic fields discussed next.

1-2.3 Magnetic Fields

As early as 800 B.C., the Greeks discovered that certain kinds of stones exhibit a force that attracts pieces of iron. These stones are now called *magnetite* (Fe_3O_4) and the phenomenon they exhibit is *magnetism*. In the thirteenth century, French scientists discovered that, when a needle was placed on the surface of a spherical natural magnet, the needle oriented itself along different directions for different locations on the magnet. By mapping the directions taken by the needle, it was determined that the magnetic force formed magnetic-field lines that encircled the sphere and appeared to pass through two points diametrically opposite each other. These points, called the *north and south poles* of the magnet, were found to exist for every magnet, regardless of its shape. The magnetic-field pattern of a bar magnet is displayed in Fig. 1-7. It was also observed that like poles of different magnets repel each other and unlike poles attract each other. This attraction–repulsion property is similar to the electric force between electric charges, except for one important difference: *electric charges can be isolated, but magnetic poles always exist in pairs.* If a permanent

the distortion process is called *polarization*. The degree of polarization depends on the distance between the atom and the isolated point charge, and the orientation of the dipole is such that the dipole axis connecting its two poles is directed toward the point charge, as illustrated schematically in Fig. 1-6. The net result of this polarization process is that the electric dipoles of the atoms (or molecules) tend to counteract the field due to the point charge. Consequently, the electric field at any point in the material would be different from the field that would have been induced by the point charge in the absence of the material. To extend Eq. (1.8) from the free-space case to any medium, we replace the permittivity of free space ε_0 with ε, where ε is now the permittivity of the material in which the electric field is measured and is therefore characteristic of that particular material. Thus,

$$\mathbf{E} = \hat{\mathbf{R}} \frac{q}{4\pi \varepsilon R^2} \quad (\text{V/m}). \quad (1.10)$$

Often, ε is expressed in the form

$$\varepsilon = \varepsilon_r \varepsilon_0 \quad (\text{F/m}), \quad (1.11)$$

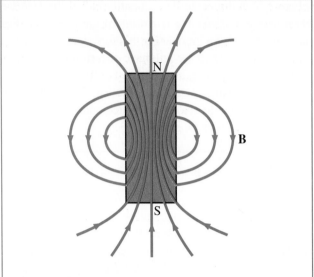

Figure 1-7: Pattern of magnetic field lines around a bar magnet.

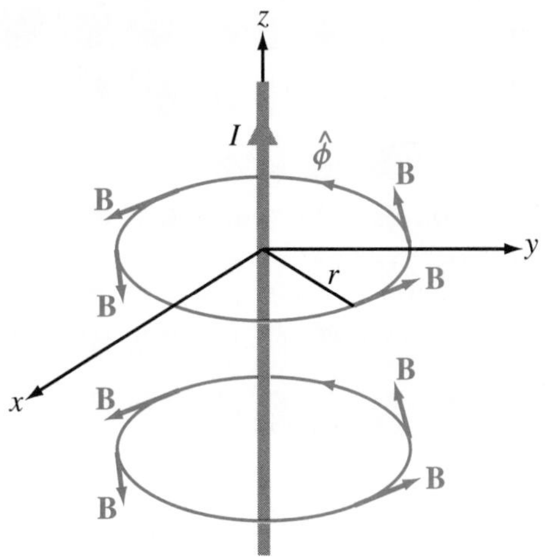

Figure 1-8: The magnetic field induced by a steady current flowing in the *z*-direction.

magnet is cut into small pieces, no matter how small each piece is, it will always have a north and a south pole.

The magnetic lines encircling a magnet are called *magnetic-field lines* and represent the existence of a magnetic field called the *magnetic flux density* **B**. A magnetic field not only exists around permanent magnets but can also be created by electric current. This connection between electricity and magnetism was discovered in 1819 by the Danish scientist Hans Oersted (1777–1851), who found that an electric current in a wire caused a compass needle placed in its vicinity to deflect and that the needle turned so that its direction was always perpendicular to the wire and to the radial line connecting the wire to the needle. From these observations, it was deduced that the current-carrying wire induced a magnetic field that formed closed circular loops around the wire, as illustrated in Fig. 1-8. Shortly after Oersted's discovery, French scientists Jean Baptiste Biot and Felix Savart developed an expression that relates the magnetic flux density **B** at a point in space to the current *I* in the conductor. Application

of their formulation, known today as the *Biot–Savart law*, to the situation depicted in Fig. 1-8 for a very long wire leads to the result that the *magnetic flux density* **B** induced by a constant current *I* flowing in the *z*-direction is given by

$$\mathbf{B} = \hat{\boldsymbol{\phi}}\,\frac{\mu_0 I}{2\pi r} \quad \text{(T)}, \qquad (1.13)$$

where *r* is the radial distance from the current and $\hat{\boldsymbol{\phi}}$ is an azimuthal unit vector denoting the fact that the magnetic field direction is tangential to the circle surrounding the current, as shown in Fig. 1-8. The magnetic field is measured in tesla (T), named in honor of Nikola Tesla (1856–1943), a Croatian-American electrical engineer whose work on transformers made it possible to transport electricity over long wires without too much loss. The quantity μ_0 is called the *magnetic permeability of free space* [$\mu_0 = 4\pi \times 10^{-7}$ henry per meter (H/m)], and it is analogous to the electric permittivity ε_0. In fact, as we will see in Chapter 2, the product of ε_0 and μ_0 specifies *c*, the *velocity of light in free space*, as follows:

$$c = \frac{1}{\sqrt{\mu_0 \varepsilon_0}} = 3 \times 10^8 \quad \text{(m/s)}. \quad (1.14)$$

The majority of natural materials are *nonmagnetic*, meaning that they exhibit a magnetic permeability $\mu = \mu_0$. For ferromagnetic materials, such as iron and nickel, μ can be much larger than μ_0. The magnetic permeability μ accounts for *magnetization* properties of a material. In analogy with Eq. (1.11), μ of a particular material can be defined as

$$\mu = \mu_r \mu_0 \quad \text{(H/m)}, \quad (1.15)$$

where μ_r is a dimensionless quantity called the *relative magnetic permeability* of the material. The values of μ_r for commonly used ferromagnetic materials are given in Appendix B.

We stated earlier that **E** and **D** constitute one of two pairs of electromagnetic field quantities. The second pair is **B** and the *magnetic field intensity* **H**, which are related to each other through μ:

$$\mathbf{B} = \mu \mathbf{H}. \quad (1.16)$$

1-2.4 Static and Dynamic Fields

Because the electric field **E** is governed by the charge q and the magnetic field **H** is governed by $I = dq/dt$, and since q and dq/dt are independent variables, the induced electric and magnetic fields are independent of one another as long as I remains constant. To demonstrate the validity of this statement, consider for example a small section of a beam of charged particles that are moving at a constant velocity. The moving charges constitute a d-c current. The electric field due to that section of the beam is determined by the total charge q contained in that

section. The magnetic field does not depend on q, but rather on the rate of charge (current) flowing through that section. Few charges moving very fast can constitute the same current as many charges moving slowly. In these two cases the induced magnetic field will be the same because the current I is the same, but the induced electric field will be quite different because the numbers of charges are not the same.

Electrostatics and *magnetostatics*, corresponding to stationary charges and steady currents, respectively, are special cases of electromagnetics. They represent two *independent* branches, so characterized because the induced electric and magnetic fields are uncoupled to each other. *Dynamics*, the third and more general branch of electromagnetics, involves *time-varying fields* induced by time-varying sources, that is, currents and charge densities. If the current associated with the beam of moving charged particles varies with time, then the amount of charge present in a given section of the beam also varies with time, and vice versa. As we will see in Chapter 6, the electric and magnetic fields become coupled to each other in that case. In fact, *a time-varying electric field will generate a time-varying magnetic field, and vice versa.* Table 1-3 provides a summary of the three branches of electromagnetics.

The electric and magnetic properties of materials are characterized by the two parameters ε and μ, respectively. A third fundamental parameter is also needed, the *conductivity* of a material σ, which is measured in siemens per meter (S/m). The conductivity characterizes the ease with which charges (electrons) can move freely in a material. If $\sigma = 0$, the charges do not move more than atomic distances and the material is said to be a *perfect dielectric*, and if $\sigma = \infty$, the charges can move very freely throughout the material, which is then called a *perfect conductor*. The material parameters ε, μ, and σ are often referred to as the *constitutive parameters* of a material (Table 1-4). A medium is said to be *homogeneous* if its constitutive parameters are constant throughout the medium.

Table 1-3: The three branches of electromagnetics.

Branch	Condition	Field Quantities (Units)
Electrostatics	Stationary charges $(\partial q/\partial t = 0)$	Electric field intensity $\mathbf{E}$ (V/m) Electric flux density $\mathbf{D}$ (C/m^2) $\mathbf{D} = \varepsilon\mathbf{E}$
Magnetostatics	Steady currents $(\partial I/\partial t = 0)$	Magnetic flux density $\mathbf{B}$ (T) Magnetic field intensity $\mathbf{H}$ (A/m) $\mathbf{B} = \mu\mathbf{H}$
Dynamics (Time-varying fields)	Time-varying currents $(\partial I/\partial t \neq 0)$	$\mathbf{E}, \mathbf{D}, \mathbf{B}$, and $\mathbf{H}$ $(\mathbf{E}, \mathbf{D})$ coupled to $(\mathbf{B}, \mathbf{H})$

Table 1-4: Constitutive parameters of materials.

Parameter	Units	Free-space Value
Electrical permittivity ε	F/m	$\varepsilon_0 = 8.854 \times 10^{-12}$ (F/m) $\simeq \dfrac{1}{36\pi} \times 10^{-9}$ (F/m)
Magnetic permeability μ	H/m	$\mu_0 = 4\pi \times 10^{-7}$ (H/m)
Conductivity σ	S/m	0

Q1.1   What are the four fundamental forces of nature and what are their relative strengths?

Q1.2   What is Coulomb's law? State its properties.

Q1.3   What are the two important properties of electric charge?

Q1.4   What do the electrical permittivity and magnetic permeability of a material account for?

Q1.5   What are the three branches and associated conditions of electromagnetics?

1-3 Traveling Waves

Waves are a natural consequence of many physical processes: waves and ripples on oceans and lakes; sound waves that travel through air; mechanical waves on stretched strings; electromagnetic waves that constitute light; earthquake waves; and many others.   All these various types of waves exhibit a number of common properties, including the following:

- *Moving waves carry energy from one point to another.*
- *Waves have velocity;* it takes time for a wave to travel from one point to another. In vacuum, light waves

travel at a speed of 3×10^8 m/s and sound waves in air travel at a speed approximately a million times slower, specifically 330 m/s.

- *Some waves exhibit a property called linearity.* Waves that do not affect the passage of other waves are called *linear* because they pass right through each other, and the total of two linear waves is simply the sum of the two waves as they would exist separately. Electromagnetic waves are linear, as are sound waves. When two people speak to one another, their sound waves do not reflect from one another, but simply pass through independently of each other. Water waves are approximately linear; the expanding circles of ripples caused by two pebbles thrown into two locations on a lake surface do not affect each other. Although the interaction of the two circles may exhibit a complicated pattern, it is simply the linear superposition of two independent expanding circles.

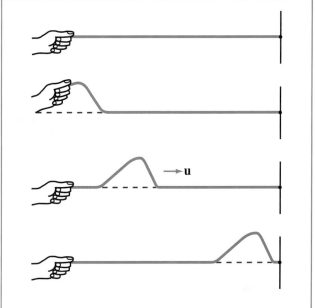

Figure 1-9: A one-dimensional wave traveling on a string.

Waves are of two types: *transient waves* caused by a short-duration disturbance and *continuous harmonic waves* generated by an oscillating source. We will encounter both types of waves in this book, but most of our discussion will deal with the propagation of continuous waves that vary sinusoidally with time.

An essential feature of a propagating wave is that it is a self-sustaining disturbance of the medium through which it travels. If this disturbance varies as a function of one space variable, such as the vertical displacement of the string shown in Fig. 1-9, we call the wave a *one-dimensional wave*. The vertical displacement varies with time and with the location along the length of the string. Even though the string rises up into a second dimension, the wave is only one-dimensional because the disturbance varies with only one space variable. A *two-dimensional wave* propagates out across a surface, like the ripples on a pond [Fig. 1-10(a)], and its disturbance can be described by two space variables. And by extension, a *three-dimensional wave* propagates

through a volume and its disturbance may be a function of all three space variables. Three-dimensional waves may take on many different shapes; they include *plane waves*, *cylindrical waves*, and *spherical waves*. A plane wave is characterized by a disturbance that at a given point in time has uniform properties across an infinite plane perpendicular to the direction of wave propagation [Fig. 1-10(b)] and, similarly, for cylindrical and spherical waves the disturbances are uniform across cylindrical and spherical surfaces, as shown in Figs. 1-10(b) and (c).

In the material that follows, we will examine some of the basic properties of waves by developing mathematical formulations that describe their functional dependence on time and space variables. To keep the presentation simple, we will limit our present discussion to sinusoidally varying waves whose disturbances are functions of only one space variable, and we will defer discussion of more complicated waves to later chapters.

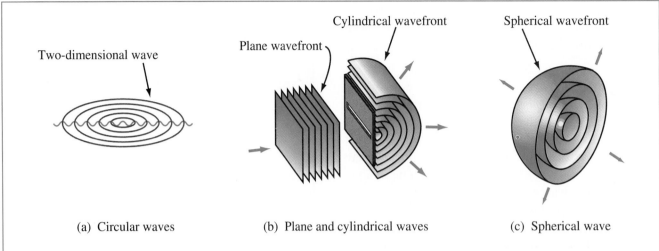

(a) Circular waves   (b) Plane and cylindrical waves   (c) Spherical wave

Figure 1-10: Examples of two-dimensional and three-dimensional waves: (a) circular waves on a pond, (b) a plane light wave exciting a cylindrical light wave through the use of a long narrow slit in an opaque screen, and (c) a sliced section of a spherical wave.

1-3.1 Sinusoidal Wave in a Lossless Medium

Regardless of the mechanism responsible for generating them, all waves can be described mathematically in common terms. By way of an example, let us consider a wave traveling on a lake surface. *A medium is said to be **lossless** if it does not attenuate the amplitude of the wave traveling within it or on its surface.* Let us assume for the time being that frictional forces can be ignored, thereby allowing a wave generated on the water surface to travel indefinitely with no loss in energy. If y denotes the height of the water surface relative to the mean height (undisturbed condition) and x denotes the distance of wave travel, the functional dependence of y on time t and the spatial coordinate x has the general form

$$y(x, t) = A \cos\left(\frac{2\pi t}{T} - \frac{2\pi x}{\lambda} + \phi_0\right) \quad \text{(m)}, \quad (1.17)$$

where A is the *amplitude* of the wave, T is its *time period*, λ is its *spatial wavelength*, and ϕ_0 is a *reference phase*.

The quantity $y(x, t)$ can also be expressed in the form

$$y(x, t) = A \cos\phi(x, t), \quad (1.18)$$

where

$$\phi(x, t) = \left(\frac{2\pi t}{T} - \frac{2\pi x}{\lambda} + \phi_0\right) \quad \text{(rad)}. \quad (1.19)$$

The angle $\phi(x, t)$ is called the *phase* of the wave, and it should not be confused with the reference phase ϕ_0, which is constant with respect to both time and space. Phase is measured by the same units as angles, that is, radians (rad) or degrees, with 2π radians $= 360°$.

Let us first analyze the simple case when $\phi_0 = 0$:

$$y(x, t) = A \cos\left(\frac{2\pi t}{T} - \frac{2\pi x}{\lambda}\right) \quad \text{(m)}. \quad (1.20)$$

The plots in Fig. 1-11 show the variation of $y(x, t)$ with x at $t = 0$ and with t at $x = 0$. The wave pattern repeats itself at a spatial period λ along x and at a temporal period T along t.

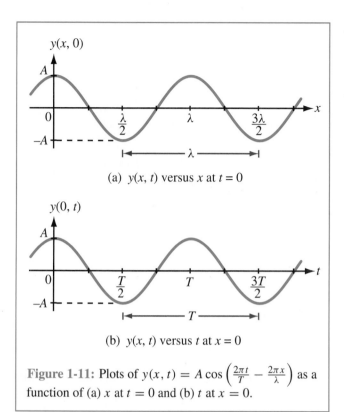

Figure 1-11: Plots of $y(x,t) = A \cos\left(\frac{2\pi t}{T} - \frac{2\pi x}{\lambda}\right)$ as a function of (a) x at $t=0$ and (b) t at $x=0$.

(a) $y(x,t)$ versus x at $t=0$

(b) $y(x,t)$ versus t at $x=0$

If we take time snapshots of the water surface, the height profile $y(x)$ would exhibit the sinusoidal patterns shown in Fig. 1-12. For each plot, corresponding to a specific value of t, the spacing between peaks is equal to the wavelength λ, but the patterns are shifted relative to one another because they correspond to different observation times. Because the pattern advances along the $+x$-direction at progressively increasing values of t, the height profile behaves like a wave traveling in that direction. If we choose any height level, such as the peak P, and follow it in time, we can measure the *phase velocity* of the wave. The peak corresponds to when the phase $\phi(x,t)$ of the wave is equal to zero or multiples of 2π radians. Thus,

$$\phi(x,t) = \frac{2\pi t}{T} - \frac{2\pi x}{\lambda} = 2n\pi, \qquad n = 0, 1, 2, \dots \quad (1.21)$$

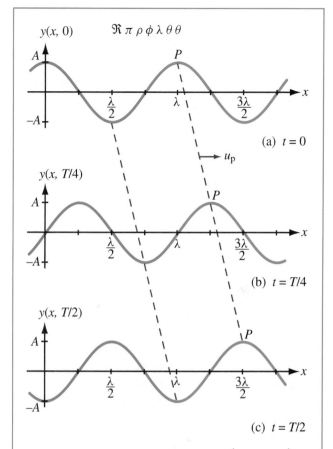

(a) $t=0$

(b) $t=T/4$

(c) $t=T/2$

Figure 1-12: Plots of $y(x,t) = A \cos\left(\frac{2\pi t}{T} - \frac{2\pi x}{\lambda}\right)$ as a function of x at (a) $t=0$, (b) $t=T/4$, and (c) $t=T/2$. Note that the wave moves in the $+x$-direction with a velocity $u_\text{p} = \lambda/T$.

Had we chosen any other fixed height of the wave, say y_0, and monitored its movement as a function of t and x, this again is equivalent to setting the phase $\phi(x,t)$ constant such that

$$y(x,t) = y_0 = A \cos\left(\frac{2\pi t}{T} - \frac{2\pi x}{\lambda}\right), \qquad (1.22)$$

or

$$\frac{2\pi t}{T} - \frac{2\pi x}{\lambda} = \cos^{-1}\left(\frac{y_0}{A}\right) = \text{constant}. \qquad (1.23)$$

The apparent velocity of that fixed height is obtained by taking the time derivative of Eq. (1.23),

$$\frac{2\pi}{T} - \frac{2\pi}{\lambda}\frac{dx}{dt} = 0, \qquad (1.24)$$

which gives the *phase velocity* u_p as

$$u_p = \frac{dx}{dt} = \frac{\lambda}{T} \quad \text{(m/s)}. \quad (1.25)$$

The phase velocity, also called the *propagation velocity*, is *the velocity of the wave pattern* as it moves across the water surface. The water itself mostly moves up and down; when the wave moves from one point to another, the water does not move physically along with it.

The direction of wave propagation is easily determined by inspecting the signs of the t and x terms in the expression for the phase $\phi(x, t)$ given by Eq. (1.19): *if one of the signs is positive and the other is negative, then the wave is traveling in the positive x-direction, and if both signs are positive or both are negative, then the wave is traveling in the negative x-direction.* The constant phase reference ϕ_0 has no influence on either the speed or the direction of wave propagation.

The *frequency* of a sinusoidal wave, f, is the reciprocal of its time period T:

$$f = \frac{1}{T} \quad \text{(Hz)}. \quad (1.26)$$

Combining the preceding two equations gives the relation

$$u_p = f\lambda \quad \text{(m/s)}. \quad (1.27)$$

The wave frequency f, which is measured in cycles per second, has been assigned the unit (Hz) (pronounced

"hertz"), named in honor of the German physicist Heinrich Hertz (1857–1894), who pioneered the development of radio waves.

Using Eq. (1.26), Eq. (1.20) can be rewritten in the shortened form as

$$y(x, t) = A\cos\left(2\pi f t - \frac{2\pi}{\lambda}x\right)$$
$$= A\cos(\omega t - \beta x), \qquad (1.28)$$

where ω is the *angular velocity* of the wave and β is its *phase constant* (or *wavenumber*), defined as

$$\omega = 2\pi f \quad \text{(rad/s)}, \qquad (1.29a)$$
$$\beta = \frac{2\pi}{\lambda} \quad \text{(rad/m)}. \qquad (1.29b)$$

In terms of these two quantities,

$$u_p = f\lambda = \frac{\omega}{\beta} . \qquad (1.30)$$

So far, we have examined the behavior of a wave traveling in the $+x$-direction. To describe a wave traveling in the $-x$-direction, we reverse the sign of x in Eq. (1.28):

$$y(x, t) = A\cos(\omega t + \beta x). \qquad (1.31)$$

We now examine the role of the phase reference ϕ_0 given previously in Eq. (1.17). If ϕ_0 is not zero, then Eq. (1.28) should be written as

$$y(x, t) = A\cos(\omega t - \beta x + \phi_0). \qquad (1.32)$$

A plot of $y(x, t)$ as a function of x at a specified t or as a function of t at a specified x will be shifted in space or time, respectively, relative to a plot with $\phi_0 = 0$ by an amount ϕ_0. This is illustrated by the plots shown in Fig. 1-13. We observe that when ϕ_0 is positive, $y(t)$ reaches its peak value, or any other specified value, sooner than when $\phi_0 = 0$. Thus, the wave with $\phi_0 = \pi/4$ is said to *lead* the wave with $\phi_0 = 0$ by a *phase lead* of $\pi/4$; and similarly, the wave with $\phi_0 = -\pi/4$ is said to *lag* the wave with $\phi_0 = 0$ by a *phase lag* of $\pi/4$. A wave function with a negative ϕ_0 takes longer to

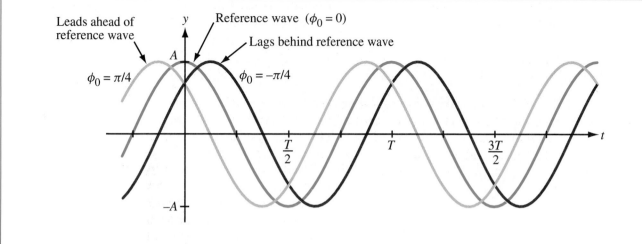

Figure 1-13: Plots of $y(0, t) = A \cos[(2\pi t/T) + \phi_0]$ for three different values of the reference phase ϕ_0.

reach a given value of $y(t)$ than the zero-phase reference function. When its value is positive, ϕ_0 signifies a phase lead in time, and when it is negative, it signifies a phase lag.

1-3.2 Sinusoidal Wave in a Lossy Medium

If a wave is traveling in the x-direction in a *lossy medium*, its amplitude will decrease as $e^{-\alpha x}$. This factor is called the *attenuation factor*, and α is called the *attenuation constant* of the medium and its unit is neper per meter (Np/m). Thus, in general,

$$y(x, t) = Ae^{-\alpha x}\cos(\omega t - \beta x + \phi_0). \qquad (1.33)$$

The wave amplitude is now $Ae^{-\alpha x}$, and not just A. Figure 1-14 shows a plot of $y(x, t)$ as a function of x at $t = 0$ for $A = 10$ m, $\lambda = 2$ m, $\alpha = 0.2$ Np/m, and $\phi_0 = 0$. Note that the envelope of the wave pattern decreases as $e^{-\alpha x}$.

The real unit of α is (1/m); the neper (Np) part is a dimensionless, artificial adjective traditionally used as a reminder that the unit (Np/m) refers to the attenuation constant of the medium, α. A similar practice is applied

to the phase constant β by assigning it the unit (rad/m) instead of just (1/m).

REVIEW QUESTIONS

Q1.6 How can you tell if a wave is traveling in the positive x-direction or the negative x-direction?

Q1.7 How does the envelope of the wave pattern vary with distance in (a) a lossless medium and (b) a lossy medium?

Q1.8 Why does a negative value of ϕ_0 signify a phase lag?

Example 1-1   Sound Wave in Water

An acoustic wave traveling in the x-direction in a fluid (liquid or gas) is characterized by a differential pressure $p(x, t)$. The unit for pressure is newton per square meter (N/m²). Find an expression for $p(x, t)$ for a sinusoidal sound wave traveling in the positive x-direction in water,

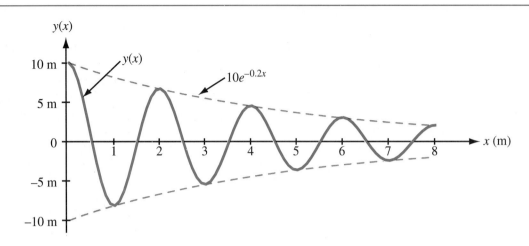

Figure 1-14: Plot of $y(x) = (10e^{-0.2x} \cos \pi x)$ meters. Note that the envelope is bounded between the curve given by $10e^{-0.2x}$ and its mirror image.

given that the wave frequency is 1 kHz, the velocity of sound in water is 1.5 km/s, the wave amplitude is $10\,\text{N/m}^2$, and $p(x,t)$ was observed to be at its maximum value at $t = 0$ and $x = 0.25$ m. Treat water as a lossless medium.

Solution: According to the general form given by Eq. (1.17) for a wave traveling in the positive x-direction,

$$p(x,t) = A \, \cos\left(\frac{2\pi}{T}t - \frac{2\pi}{\lambda}x + \phi_0\right) \quad (\text{N/m}^2).$$

The amplitude $A = 10\,\text{N/m}^2$, $T = 1/f = 10^{-3}$ s, and from $u_\text{p} = f\lambda$,

$$\lambda = \frac{u_\text{p}}{f} = \frac{1.5 \times 10^3}{10^3} = 1.5\,\text{m}.$$

Hence,

$$p(x,t) = 10 \cos\left(2\pi \times 10^3 t - \frac{4\pi}{3}x + \phi_0\right) \quad (\text{N/m}^2).$$

Since at $t = 0$ and $x = 0.25$ m, $p(0.25, 0) = 10\,\text{N/m}^2$, we have

$$10 = 10 \cos\left(\frac{-4\pi}{3}\,0.25 + \phi_0\right)$$

$$= 10 \cos\left(\frac{-\pi}{3} + \phi_0\right),$$

which yields the result $(\phi_0 - \pi/3) = \cos^{-1}(1)$, or $\phi_0 = \pi/3$. Hence,

$$p(x,t) = 10 \cos\left(2\pi \times 10^3 t - \frac{4\pi}{3}x + \frac{\pi}{3}\right) \quad (\text{N/m}^2).$$

∎

Example 1-2   Power Loss

A laser beam of light propagating through the atmosphere is characterized by an electric field intensity given by

$$E(x,t) =$$
$$150e^{-0.03x} \cos(3 \times 10^{15} t - 10^7 x) \quad (\text{V/m}),$$

where x is the distance from the source in meters. The attenuation is due to absorption by atmospheric gases. Determine (a) the direction of wave travel, (b) the wave velocity, and (c) the wave amplitude at a distance of 200 m.

Solution: (a) Since the coefficients of t and x in the argument of the cosine function have opposite signs, the wave must be traveling in the $+x$-direction.

(b)

$$u_p = \frac{\omega}{\beta} = \frac{3 \times 10^{15}}{10^7} = 3 \times 10^8 \text{ m/s},$$

which is equal to c, the velocity of light in free space.

(c) At $x = 200$ m, the amplitude of $E(x, t)$ is

$$150e^{-0.03 \times 200} = 0.37 \quad \text{(V/m)}. \quad \blacksquare$$

EXERCISE 1.1 The electric field of a traveling electromagnetic wave is given by

$$E(z, t) = 10 \cos(\pi \times 10^7 t + \pi z/15 + \pi/6) \quad \text{(V/m)}.$$

Determine (a) the direction of wave propagation, (b) the wave frequency f, (c) its wavelength λ, and (d) its phase velocity u_p.

Ans. (a) $-z$-direction, (b) $f = 5$ MHz, (c) $\lambda = 30$ m, (d) $u_p = 1.5 \times 10^8$ m/s.

EXERCISE 1.2 An electromagnetic wave is propagating in the z-direction in a lossy medium with attenuation constant $\alpha = 0.5$ Np/m. If the wave's electric-field amplitude is 100 V/m at $z = 0$, how far can the wave travel before its amplitude will have been reduced to (a) 10 V/m, (b) 1 V/m, (c) 1 μV/m?

Ans. (a) 4.6 m, (b) 9.2 m, (c) 37 m.

1-4 The Electromagnetic Spectrum

Visible light belongs to a family of waves called the *electromagnetic spectrum* (Fig. 1-15). Other members of this family include gamma rays, x rays, infrared waves, and radio waves. Generically, they all are called electromagnetic (EM) waves because they share the following fundamental properties:

- An EM wave consists of electric and magnetic field intensities that oscillate at the same frequency f.
- The phase velocity of an EM wave propagating in vacuum is a universal constant given by the velocity of light c, defined earlier by Eq. (1.14).
- In vacuum, the wavelength λ of an EM wave is related to its oscillation frequency f by

$$\lambda = \frac{c}{f}. \quad (1.34)$$

Whereas all EM waves share these properties, each is distinguished by its own wavelength λ, or equivalently by its own oscillation frequency f.

The visible part of the EM spectrum shown in Fig. 1-15 covers a very narrow wavelength range extending between $\lambda = 0.4$ μm (violet) and $\lambda = 0.7$ μm (red). As we move progressively toward shorter wavelengths, we encounter the ultraviolet, x-ray, and gamma-ray bands, each so named because of historical reasons associated with the discovery of waves with those wavelengths. On the other side of the visible spectrum lie the infrared band and then the radio region. Because of the link between λ and f given by Eq. (1.34), each of these spectral ranges may be specified in terms of its wavelength range or alternatively in terms of its frequency range. In practice, however, a wave is specified in terms of its wavelength λ if $\lambda < 1$ mm, which encompasses all parts of the EM spectrum except for the radio region, and the wave is specified in terms of its frequency f if $\lambda > 1$ mm (i.e., in the radio region). A wavelength of 1 mm corresponds to a frequency of 3×10^{11} Hz $= 300$ GHz in free space.

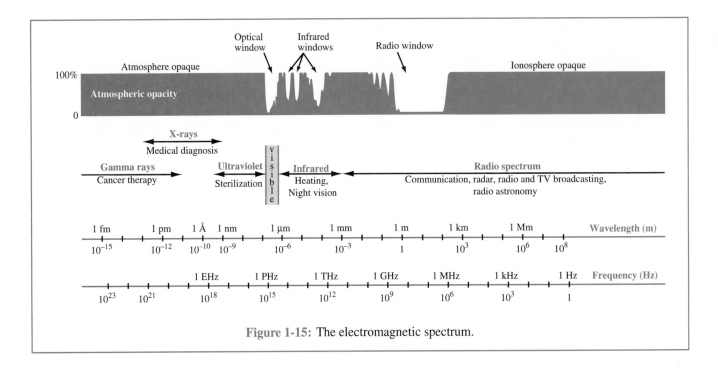

Figure 1-15: The electromagnetic spectrum.

The radio spectrum consists of several individual bands, as shown in the chart of Fig. 1-16. Each band covers one decade of the radio spectrum and has a letter designation based on a nomenclature defined by the International Telecommunication Union. Different frequencies have different applications because they are excited by different mechanisms, and the properties of an EM wave propagating in a material may vary considerably from one band to another. The extremely low frequency (ELF) band from 3 to 30 Hz is used primarily for the detection of buried metal objects. Lower frequencies down to 0.1 Hz are used in magnetotelluric sensing of the structure of the earth, and frequencies in the range from 1 Hz to 1 kHz sometimes are used for communications with submerged submarines and for certain kinds of sensing of Earth's ionosphere. The very low frequency (VLF) region from 3 to 30 kHz is used both for submarine communications and for position location by the Omega navigation system. The low-frequency (LF) band, from 30 to 300 kHz, is used for some forms of communication and for the Loran C position-location system. Some radio beacons and weather broadcast stations used in air navigation operate at frequencies in the higher end of the LF band. The medium-frequency (MF) region from 300 kHz to 3 MHz contains the standard AM broadcast band from 0.5 to 1.5 MHz.

Long-distance communications and short-wave broadcasting over long distances use frequencies in the high-frequency (HF) band from 3 to 30 MHz because waves in this band are strongly affected by reflections by the ionosphere and least affected by absorption in the ionosphere. The next frequency region, the very high frequency (VHF) band from 30 to 300 MHz, is used primarily for television and FM broadcasting over line-of-sight distances and also for communicating with aircraft and other vehicles. Some early radio-astronomy research was also conducted in this range. The ultrahigh frequency (UHF) region from 300 MHz to 3 GHz is

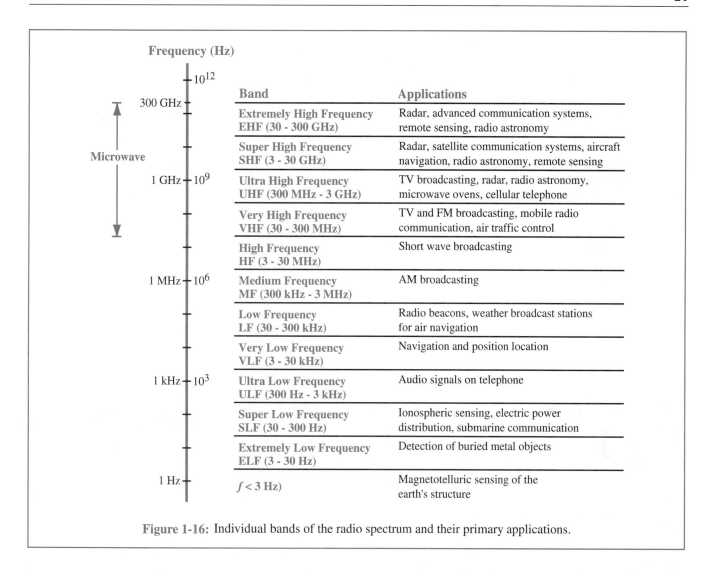

Figure 1-16: Individual bands of the radio spectrum and their primary applications.

extensively populated with radars, although part of this band also is used for television broadcasting and mobile communications with aircraft and surface vehicles. The radars in this region of the spectrum are normally used for aircraft detection and tracking. Some parts of this region have been reserved for radio astronomical observation.

Many point-to-point radio communication systems and various kinds of ground-based radars and ship radars operate at frequencies in the superhigh frequency (SHF) range from 3 to 30 GHz. Some aircraft navigation systems operate in this range as well.

Most of the extremely high frequency (EHF) band from 30 to 300 GHz is used less extensively, primarily because the technology is not as well developed and because of excessive absorption by the atmosphere in some parts of this band. Some advanced communication systems are being developed for operation at frequencies in the "atmospheric windows," where atmospheric absorption is not a serious problem, as are automobile collision-avoidance radars and some military imaging radar systems. These atmospheric windows include the ranges from 30 to 35 GHz, 70 to 75 GHz, 90 to 95 GHz, and 135 to 145 GHz.

Although no precise definition exists for the extent of the *microwave band*, it is conventionally regarded to cover the full ranges of the UHF, SHF, and EHF bands, with the EHF band sometimes referred to as the *millimeter-wave band*, because the wavelength range covered by this band extends from 1 mm (300 GHz) to 1 cm (30 GHz).

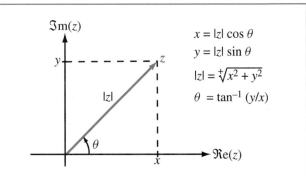

Figure 1-17: Relation between rectangular and polar representations of a complex number $z = x + jy = |z|e^{j\theta}$.

REVIEW QUESTIONS

Q1.9 What are the three fundamental properties of EM waves?

Q1.10 What is the range of frequencies covered by the microwave band?

Q1.11 What is the wavelength range of the visible spectrum? What are some of the applications of the infrared band?

1-5 Review of Complex Numbers

A *complex number z* is written in the form

$$z = x + jy, \qquad (1.35)$$

where x and y are the *real* ($\Re e$) and *imaginary* ($\Im m$) parts of z, respectively, and $j = \sqrt{-1}$. That is,

$$x = \Re e(z), \qquad y = \Im m(z). \qquad (1.36)$$

Alternatively, z may be written in *polar form* as

$$z = |z|e^{j\theta} = |z|\angle\theta \qquad (1.37)$$

where $|z|$ is the magnitude of z, θ is its phase angle, and the form $\angle\theta$ is a useful shorthand representation commonly used in numerical calculations. Applying *Euler's identity*,

$$e^{j\theta} = \cos\theta + j\sin\theta, \qquad (1.38)$$

we can convert z from polar form, as in Eq. (1.37), into rectangular form, as in Eq. (1.35),

$$z = |z|e^{j\theta} = |z|\cos\theta + j|z|\sin\theta, \qquad (1.39)$$

which leads to the relations

$$x = |z|\cos\theta, \qquad y = |z|\sin\theta, \qquad (1.40)$$

$$|z| = \sqrt[+]{x^2 + y^2}, \qquad \theta = \tan^{-1}(y/x). \qquad (1.41)$$

The two forms are illustrated graphically in Fig. 1-17. When using Eq. (1.41), care should be taken to ensure that θ is in the proper quadrant. Also note that, since $|z|$ is a positive quantity, only the positive root in Eq. (1.41) is applicable. This is denoted by the $+$ sign above the square-root sign.

The *complex conjugate* of z, denoted with a star superscript (or asterisk), is obtained by replacing j (wherever it appears) with $-j$, so that

$$z^* = (x + jy)^* = x - jy = |z|e^{-j\theta} = |z|\angle{-\theta}. \quad (1.42)$$

The magnitude $|z|$ is equal to the positive square root of the product of z and its complex conjugate:

$$|z| = \sqrt[+]{z\,z^*}. \quad (1.43)$$

We now highlight some of the properties of complex algebra that we will likely encounter in future chapters.

Equality: If two complex numbers z_1 and z_2 are given by

$$z_1 = x_1 + jy_1 = |z_1|e^{j\theta_1}, \quad (1.44)$$

$$z_2 = x_2 + jy_2 = |z_2|e^{j\theta_2}, \quad (1.45)$$

then $z_1 = z_2$ if and only if $x_1 = x_2$ and $y_1 = y_2$ or, equivalently, $|z_1| = |z_2|$ and $\theta_1 = \theta_2$.

Addition:

$$z_1 + z_2 = (x_1 + x_2) + j(y_1 + y_2). \quad (1.46)$$

Multiplication:

$$z_1 z_2 = (x_1 + jy_1)(x_2 + jy_2)$$
$$= (x_1 x_2 - y_1 y_2) + j(x_1 y_2 + x_2 y_1), \quad (1.47a)$$

or

$$z_1 z_2 = |z_1|e^{j\theta_1} \cdot |z_2|e^{j\theta_2}$$
$$= |z_1||z_2|e^{j(\theta_1+\theta_2)}$$
$$= |z_1||z_2|[\cos(\theta_1 + \theta_2) + j\sin(\theta_1 + \theta_2)]. \quad (1.47b)$$

Division: For $z_2 \neq 0$,

$$\frac{z_1}{z_2} = \frac{x_1 + jy_1}{x_2 + jy_2}$$
$$= \frac{(x_1 + jy_1)}{(x_2 + jy_2)} \cdot \frac{(x_2 - jy_2)}{(x_2 - jy_2)}$$
$$= \frac{(x_1 x_2 + y_1 y_2) + j(x_2 y_1 - x_1 y_2)}{x_2^2 + y_2^2}, \quad (1.48a)$$

or

$$\frac{z_1}{z_2} = \frac{|z_1|e^{j\theta_1}}{|z_2|e^{j\theta_2}}$$
$$= \frac{|z_1|}{|z_2|}e^{j(\theta_1-\theta_2)}$$
$$= \frac{|z_1|}{|z_2|}[\cos(\theta_1 - \theta_2) + j\sin(\theta_1 - \theta_2)]. \quad (1.48b)$$

Powers: For any positive integer n,

$$z^n = (|z|e^{j\theta})^n$$
$$= |z|^n e^{jn\theta} = |z|^n(\cos n\theta + j\sin n\theta), \quad (1.49)$$

$$z^{1/2} = \pm|z|^{1/2}e^{j\theta/2}$$
$$= \pm|z|^{1/2}[\cos(\theta/2) + j\sin(\theta/2)]. \quad (1.50)$$

Useful Relations:

$$-1 = e^{j\pi} = e^{-j\pi} = 1\angle{180°},$$
$$j = e^{j\pi/2} = 1\angle{90°}, \quad (1.51)$$
$$-j = -e^{j\pi/2} = e^{-j\pi/2} = 1\angle{-90°}, \quad (1.52)$$
$$\sqrt{j} = (e^{j\pi/2})^{1/2} = \pm e^{j\pi/4} = \frac{\pm(1 + j)}{\sqrt{2}}, \quad (1.53)$$
$$\sqrt{-j} = \pm e^{-j\pi/4} = \frac{\pm(1 - j)}{\sqrt{2}}. \quad (1.54)$$

Example 1-3   Working with Complex Numbers

Given two complex numbers

$$V = 3 - j4,$$

$$I = -(2 + j3).$$

(a) Express V and I in polar form, and find (b) VI, (c) VI^*, (d) V/I, and (e) $\sqrt{I}$.

Solution

(a) $|V| = \sqrt[+]{VV^*}$
$= \sqrt[+]{(3 - j4)(3 + j4)} = \sqrt[+]{9 + 16} = 5,$
$\theta_V = \tan^{-1}(-4/3) = -53.1°,$
$V = |V|e^{j\theta_V} = 5e^{-j53.1°} = 5\angle{-53.1°},$

$|I| = \sqrt[+]{2^2 + 3^2} = \sqrt[+]{13} = 3.61.$

Since $I = (-2 - j3)$ is in the third quadrant in the complex plane [Fig. 1-18],

$$\theta_I = 180° + \tan^{-1}\left(\tfrac{3}{2}\right) = 236.3°,$$

$$I = 3.61\angle{236.3°}.$$

(b) $VI = 5e^{-j53.1°} \times 3.61e^{j236.3°}$
$= 18.05e^{j(236.3° - 53.1°)} = 18.05e^{j183.2°}.$

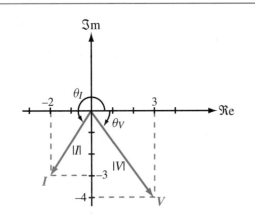

Figure 1-18: Complex numbers V and I in the complex plane (Example 1-3).

(c) $VI^* = 5e^{-j53.1°} \times 3.61e^{-j236.3°}$
$= 18.05e^{-j289.4°} = 18.05e^{j70.6°}.$

(d) $\dfrac{V}{I} = \dfrac{5e^{-j53.1°}}{3.61e^{j236.3°}}$
$= 1.39e^{-j289.4°} = 1.39e^{j70.6°}.$

(e) $\sqrt{I} = \sqrt{3.61e^{j236.3°}}$
$= \pm\sqrt{3.61}\ e^{j236.3°/2} = \pm1.90e^{j118.15°}.$ ∎

EXERCISE 1.3 Express the following complex functions in polar form:

$$z_1 = (4 - j3)^2,$$

$$z_2 = (4 - j3)^{1/2}.$$

Ans. $z_1 = 25\angle{-73.7°}, z_2 = \pm\sqrt{5}\,\angle{-18.4°}.$

EXERCISE 1.4 Show that $\sqrt{2j} = \pm(1 + j).$

1-6 Review of Phasors

Phasor analysis is a useful mathematical tool for solving problems involving linear systems in which the excitation is a periodic time function. Many engineering problems are cast in the form of linear integro-differential equations. If the excitation, more commonly known as the *forcing function*, varies sinusoidally with time, the use of phasor notation to represent time-dependent variables allows us to convert the integro-differential equation into a linear equation with no sinusoidal functions, thereby simplifying the method of solution. After solving for the desired variable, such as the voltage or current in a circuit, conversion from the phasor domain back to the time domain provides the desired result.

The phasor technique can also be used for analyzing linear systems when the forcing function is any arbitrary (nonsinusoidal) periodic time function, such as a square wave or a sequence of pulses. By expanding the forcing

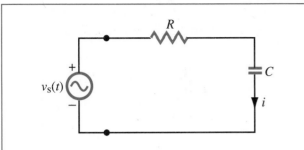

Figure 1-19: RC circuit connected to a voltage source $v_s(t)$.

function into a Fourier series of sinusoidal components, we can solve for the desired variable using phasor analysis for each Fourier component of the forcing function separately. According to the principle of superposition, the sum of the solutions due to all of the Fourier components gives the same result as one would obtain had the problem been solved entirely in the time domain without the aid of Fourier representation. The obvious advantage of the phasor–Fourier approach is simplicity. Moreover, in the case of nonperiodic source functions, such as a single pulse, the functions can be expressed as Fourier integrals, and a similar application of the principle of superposition can be used as well.

The simple RC circuit shown in Fig. 1-19 contains a sinusoidally time-varying voltage source given by

$$v_s(t) = V_0 \sin(\omega t + \phi_0), \tag{1.55}$$

where V_0 is the amplitude, ω is the angular frequency, and ϕ_0 is a reference phase. Application of Kirchhoff's voltage law gives the following loop equation:

$$R\, i(t) + \frac{1}{C} \int i(t)\, dt = v_s(t) \quad \text{(time domain).} \tag{1.56}$$

Our objective is to obtain an expression for the current $i(t)$. We can do this by solving Eq. (1.56) in the time domain, which is somewhat cumbersome because the

forcing function $v_s(t)$ is a sinusoid. Alternatively, we can take advantage of the phasor technique as follows.

Step 1: Adopt a cosine reference

This means that we should express the forcing function as a cosine, if not already in that form, and hence all time-varying functions, such as the current in the circuit and the voltages across R and C, will also have a cosine reference. Thus,

$$
\begin{aligned}
v_s(t) &= V_0 \sin(\omega t + \phi_0) \\
&= V_0 \cos\left(\frac{\pi}{2} - \omega t - \phi_0\right) \\
&= V_0 \cos\left(\omega t + \phi_0 - \frac{\pi}{2}\right), \tag{1.57}
\end{aligned}
$$

where we used the properties $\sin x = \cos(\pi/2 - x)$ and $\cos(-x) = \cos x$.

Step 2: Express time-dependent variables as phasors

Any cosinusoidally time-varying function $z(t)$ can be expressed in the form

$$z(t) = \Re\left[\widetilde{Z}\, e^{j\omega t}\right], \tag{1.58}$$

where $\widetilde{Z}$ is a time-independent function called the **phasor** of the **instantaneous** function $z(t)$. To distinguish instantaneous quantities from their phasor counterparts, a letter denoting a phasor is given a tilde ($\sim$) over the letter. The voltage $v_s(t)$ given by Eq. (1.57) can be cast in the form

$$
\begin{aligned}
v_s(t) &= \Re\left[V_0 e^{j(\omega t + \phi_0 - \pi/2)}\right] \\
&= \Re\left[V_0 e^{j(\phi_0 - \pi/2)} e^{j\omega t}\right] \\
&= \Re\left[\widetilde{V}_s e^{j\omega t}\right], \tag{1.59}
\end{aligned}
$$

where

$$\widetilde{V}_s = V_0 e^{j(\phi_0 - \pi/2)}. \tag{1.60}$$

The phasor $\widetilde{V}_s$, corresponding to the time function $v_s(t)$, contains amplitude and phase information but is independent of the time variable t. Next we define the unknown variable $i(t)$ in terms of a phasor $\tilde{I}$,

$$i(t) = \Re(\tilde{I}e^{j\omega t}), \qquad (1.61)$$

and if the equation we are trying to solve contains derivatives or integrals, we use the following two properties:

$$\frac{di}{dt} = \frac{d}{dt}\left[\Re(\tilde{I}e^{j\omega t})\right]$$

$$= \Re\left[\frac{d}{dt}(\tilde{I}e^{j\omega t})\right]$$

$$= \Re[j\omega\tilde{I}e^{j\omega t}], \qquad (1.62)$$

and

$$\int i \, dt = \int \Re(\tilde{I}e^{j\omega t}) \, dt$$

$$= \Re\left(\int \tilde{I}e^{j\omega t} \, dt\right)$$

$$= \Re\left(\frac{\tilde{I}}{j\omega}e^{j\omega t}\right). \qquad (1.63)$$

Thus, differentiation of the time function $i(t)$ is equivalent to multiplication of its phasor $\tilde{I}$ by $j\omega$, and integration is equivalent to division by $j\omega$.

Step 3: Recast the differential/integral equation in phasor form

Upon using Eqs. (1.59), (1.61), and (1.63) in Eq. (1.56), we have

$$R\,\Re(\tilde{I}e^{j\omega t}) + \frac{1}{C}\,\Re\left(\frac{\tilde{I}}{j\omega}e^{j\omega t}\right) = \Re(\tilde{V}_s e^{j\omega t}). \quad (1.64)$$

Since both R and C are real quantities and the $\Re(\ )$ operation is distributive, Eq. (1.64) simplifies to

$$\boxed{\tilde{I}\left(R + \frac{1}{j\omega C}\right) = \tilde{V}_s \qquad \text{(phasor domain).} \quad (1.65)}$$

The time factor $e^{j\omega t}$ has disappeared because it was contained in all three terms. Equation (1.65) is the phasor-domain equivalent of Eq. (1.56).

Step 4: Solve the phasor-domain equation

From Eq. (1.65) the phasor current $\tilde{I}$ is given by

$$\tilde{I} = \frac{\tilde{V}_s}{R + 1/(j\omega C)} \ . \qquad (1.66)$$

Before we apply the next step, we need to convert the right-hand side of Eq. (1.66) into the form $I_0 e^{j\theta}$ with I_0 being a real quantity. Thus,

$$\tilde{I} = V_0 e^{j(\phi_0 - \pi/2)}\left[\frac{j\omega C}{1 + j\omega RC}\right]$$

$$= V_0 e^{j(\phi_0 - \pi/2)}\left[\frac{\omega C e^{j\pi/2}}{\sqrt[+]{1 + \omega^2 R^2 C^2}\ e^{j\phi_1}}\right]$$

$$= \frac{V_0 \omega C}{\sqrt[+]{1 + \omega^2 R^2 C^2}}\ e^{j(\phi_0 - \phi_1)}, \qquad (1.67)$$

where we have used the identity $j = e^{j\pi/2}$. The phase angle $\phi_1 = \tan^{-1}(\omega RC)$ and lies in the first quadrant of the complex plane.

Step 5: Find the instantaneous value

To find $i(t)$, we simply apply Eq. (1.61). That is, we multiply the phasor $\tilde{I}$ given by Eq. (1.67) by $e^{j\omega t}$ and then take the real part:

$$i(t) = \mathfrak{Re}\left[\tilde{I}e^{j\omega t}\right]$$

$$= \mathfrak{Re}\left[\frac{V_0 \omega C}{\sqrt[+]{1 + \omega^2 R^2 C^2}}\, e^{j(\phi_0 - \phi_1)}e^{j\omega t}\right]$$

$$= \frac{V_0 \omega C}{\sqrt[+]{1 + \omega^2 R^2 C^2}}\cos(\omega t + \phi_0 - \phi_1). \quad (1.68)$$

In summary, we converted all time-varying quantities into the phasor domain, solved for the phasor $\tilde{I}$ of the desired instantaneous current $i(t)$, and then converted back to the time domain to obtain an expression for $i(t)$. Table 1-5 provides a summary of some time-domain functions and their phasor-domain equivalents.

Example 1-4   *RL* Circuit

The voltage source of the circuit shown in Fig. 1-20 is given by

$$v_s(t) = 5\sin(4 \times 10^4 t - 30°) \quad \text{(V)}. \quad (1.69)$$

Obtain an expression for the voltage across the inductor.

Solution: The voltage loop equation of the *RL* circuit is

$$Ri + L\frac{di}{dt} = v_s(t). \quad (1.70)$$

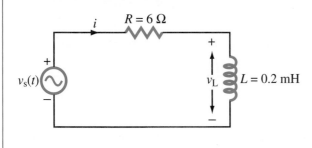

Figure 1-20: *RL* circuit (Example 1-4).

Table 1-5: Time-domain sinusoidal functions $z(t)$ and their cosine-reference phasor-domain equivalents $\tilde{Z}$, where $z(t) = \mathfrak{Re}[\tilde{Z}e^{j\omega t}]$.

$z(t)$		$\tilde{Z}$
$A\cos\omega t$	$\leftrightarrow$	A
$A\cos(\omega t + \phi_0)$	$\leftrightarrow$	$Ae^{j\phi_0}$
$A\cos(\omega t + \beta x + \phi_0)$	$\leftrightarrow$	$Ae^{j(\beta x + \phi_0)}$
$Ae^{-\alpha x}\cos(\omega t + \beta x + \phi_0)$	$\leftrightarrow$	$Ae^{-\alpha x}e^{j(\beta x + \phi_0)}$
$A\sin\omega t$	$\leftrightarrow$	$Ae^{-j\pi/2}$
$A\sin(\omega t + \phi_0)$	$\leftrightarrow$	$Ae^{j(\phi_0 - \pi/2)}$
$\dfrac{d}{dt}(z_1(t))$	$\leftrightarrow$	$j\omega\tilde{Z}_1$
$\dfrac{d}{dt}[A\cos(\omega t + \phi_0)]$	$\leftrightarrow$	$j\omega Ae^{j\phi_0}$
$\displaystyle\int z_1(t)\,dt$	$\leftrightarrow$	$\dfrac{1}{j\omega}\tilde{Z}_1$
$\displaystyle\int A\sin(\omega t + \phi_0)\,dt$	$\leftrightarrow$	$\dfrac{1}{j\omega}Ae^{j(\phi_0 - \pi/2)}$

Before converting Eq. (1.70) into the phasor domain, we need to express Eq. (1.69) in terms of a cosine reference:

$$v_s(t) = 5\sin(4 \times 10^4 t - 30°)$$

$$= 5\cos(4 \times 10^4 t - 120°) \quad \text{(V)}. \quad (1.71)$$

The coefficient of t specifies the angular frequency as $\omega = 4 \times 10^4$ (rad/s). The voltage phasor corresponding to $v_s(t)$ is

$$\tilde{V}_s = 5e^{-j120°} \quad \text{(V)},$$

and the phasor equation corresponding to Eq. (1.70) is

$$R\tilde{I} + j\omega L\tilde{I} = \tilde{V}_s. \quad (1.72)$$

Solving for the current phasor $\tilde{I}$, we have

$$\tilde{I} = \frac{\tilde{V}_s}{R + j\omega L}$$

$$= \frac{5e^{-j120°}}{6 + j4 \times 10^4 \times 2 \times 10^{-4}}$$

$$= \frac{5e^{-j120°}}{6 + j8} = \frac{5e^{-j120°}}{10e^{j53.1°}} = 0.5e^{-j173.1°} \quad (A).$$

The voltage phasor across the inductor is related to $\tilde{I}$ by

$$\tilde{V}_L = j\omega L \tilde{I}$$

$$= j4 \times 10^4 \times 2 \times 10^{-4} \times 0.5e^{-j173.1°}$$

$$= 4e^{j(90° - 173.1°)} = 4e^{-j83.1°} \quad (V),$$

and the corresponding instantaneous voltage $v_L(t)$ is therefore

$$v_L(t) = \Re\left[\tilde{V}_L e^{j\omega t}\right]$$

$$= \Re\left[4e^{-j83.1°} e^{j4 \times 10^4 t}\right]$$

$$= 4\cos(4 \times 10^4 t - 83.1°) \quad (V). \quad \blacksquare$$

REVIEW QUESTIONS

Q1.12 Why is the phasor technique useful? When is it used? Describe the process.

Q1.13 How is the phasor technique used when the forcing function is a non-sinusoidal periodic waveform, such as a train of pulses?

EXERCISE 1.5 A series RL circuit is connected to a voltage source given by $v_s(t) = 150\cos\omega t$ (V). Find (a) the phasor current $\tilde{I}$ and (b) the instantaneous current $i(t)$ for $R = 400\ \Omega$, $L = 3$ mH, and $\omega = 10^5$ rad/s.

Ans. (a) $\tilde{I} = 150/(R + j\omega L) = 0.3\angle{-36.9°}$ (A), (b) $i(t) = 0.3\cos(\omega t - 36.9°)$ (A).

EXERCISE 1.6 A phasor voltage is given by $\tilde{V} = j5$ V. Find $v(t)$.

Ans. $v(t) = 5\cos(\omega t + \pi/2) = -5\sin\omega t$ (V).

CHAPTER HIGHLIGHTS

- Electromagnetics is the study of electric and magnetic phenomena and their engineering applications.

- The International System of Units consists of the six fundamental dimensions listed in Table 1-1. The units of all other physical quantities can be expressed in terms of the six fundamental units.

- The four fundamental forces of nature are the nuclear, weak-interaction, electromagnetic and gravitational forces.

- The source of the electric field quantities $\mathbf{E}$ and $\mathbf{D}$ is the electric charge q. In a material, $\mathbf{E}$ and $\mathbf{D}$ are related by $\mathbf{D} = \varepsilon\mathbf{E}$, where ε is the electrical permittivity of the material. In free space, $\varepsilon = \varepsilon_0 \simeq (1/36\pi) \times 10^{-9}$ (F/m).

- The source of the magnetic field quantities $\mathbf{B}$ and $\mathbf{H}$ is the electric current I. In a material, $\mathbf{B}$ and $\mathbf{H}$ are related by $\mathbf{B} = \mu\mathbf{H}$, where μ is the magnetic permeability of the medium. In free space, $\mu = \mu_0 = 4\pi \times 10^{-7}$ (H/m).

- Electromagnetics consists of three branches: (1) electrostatics, which pertains to stationary charges, (2) magnetostatics, which pertains to steady currents, and (3) electrodynamics, which pertains to time-varying currents.

- A traveling wave is characterized by a spatial wavelength λ, a time period T, and a phase velocity $u_p = \lambda/T$.

- An electromagnetic (EM) wave consists of oscillating electric and magnetic field intensities and travels in free space at the velocity of light $c = 1/\sqrt{\varepsilon_0\mu_0}$. The EM spectrum encompasses gamma rays, x rays, visible light, infrared waves, and radio waves.
- Phasor analysis is a useful mathematical tool for solving problems involving time-periodic sources.

GLOSSARY OF IMPORTANT TERMS

Provide definitions or explain the meaning of the following terms:

SI system of units
fundamental dimensions
Coulomb's law
electric field intensity $\mathbf{E}$
law of conservation of electric charge
principle of linear superposition
electric dipole
electric polarization
electrical permittivity ε
relative permittivity or dielectric constant ε_r
electric flux density $\mathbf{D}$
electrostatics
magnetostatics
magnetic flux density $\mathbf{B}$
magnetic permeability μ
velocity of light c
nonmagnetic materials
magnetic field intensity $\mathbf{H}$
electrodynamics
conductivity σ
perfect dielectric
perfect conductor
constitutive parameters
transient wave
continuous harmonic wave
wave amplitude
wave period T

wavelength λ
reference phase ϕ_0
phase velocity (propagation velocity) u_p
wave frequency f
angular velocity ω
phase constant (wave number) β
phase lag and lead
attenuation factor
attenuation constant α
EM spectrum
microwave band
complex number
Euler's identity
complex conjugate
forcing function
phasor
instantaneous function

PROBLEMS

Section 1-3: Traveling Waves

1.1* A 4-kHz sound wave traveling in the x-direction in air was observed to have a differential pressure $p(x, t) = 5$ N/m^2 at $x = 0$ and $t = 25$ μs. If the reference phase of $p(x, t)$ is 42°, find a complete expression for $p(x, t)$. The velocity of sound in air is 330 m/s.

1.2 For the pressure wave described in Example 1-1, plot the following:

(a) $p(x, t)$ *versus* x at $t = 0$

(b) $p(x, t)$ *versus* t at $x = 0$

Be sure to use appropriate scales for x and t so that each of your plots covers at least two cycles.

1.3* A harmonic wave traveling along a string is generated by an oscillator that completes 120 vibrations per minute. If it is observed that a given crest, or maximum, travels 250 cm in 10 s, what is the wavelength?

---

*Answer(s) available in Appendix D.

⊚ Solution available in CD-ROM.

1.4 Two waves, $y_1(t)$ and $y_2(t)$, have identical amplitudes and oscillate at the same frequency, but $y_2(t)$ leads $y_1(t)$ by a phase angle of 60°. If

$$y_1(t) = 4\cos(2\pi \times 10^3 t)$$

write the expression appropriate for $y_2(t)$ and plot both functions over the time span from 0 to 2 ms.

1.5* The height of an ocean wave is described by the function

$$y(x, t) = 1.5\sin(0.5t - 0.6x) \quad \text{(m)}$$

Determine the phase velocity and wavelength, and then sketch $y(x, t)$ at $t = 2s$ over the range from $x = 0$ to $x = 2\lambda$.

1.6 A wave traveling along a string in the $+x$-direction is given by

$$y_1(x, t) = A\cos(\omega t - \beta x)$$

where $x = 0$ is the end of the string, which is tied rigidly to a wall, as shown in Fig. 1-21. When wave $y_1(x, t)$ arrives at the wall, a reflected wave $y_2(x, t)$ is generated. Hence, at any location on the string, the vertical displacement y_s is the sum of the incident and reflected waves:

$$y_s(x, t) = y_1(x, t) + y_2(x, t)$$

(a) Write an expression for $y_2(x, t)$, keeping in mind its direction of travel and the fact that the end of the string cannot move.

(b) Generate plots of $y_1(x, t)$, $y_2(x, t)$ and $y_s(x, t)$ versus x over the range $-2\lambda \le x \le 0$ at $\omega t = \pi/4$ and at $\omega t = \pi/2$.

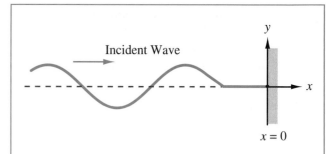

Figure 1-21: Wave on a string tied to a wall at $x = 0$ (Problem 1.6).

1.7* Two waves on a string are given by the following functions:

$$y_1(x, t) = 3\cos(20t - 30x) \quad \text{(cm)}$$
$$y_2(x, t) = -3\cos(20t + 30x) \quad \text{(cm)}$$

where x is in centimeters. The waves are said to interfere constructively when their superposition $|y_s| = |y_1 + y_2|$ is a maximum, and they interfere destructively when $|y_s|$ is a minimum.

(a) What are the directions of propagation of waves $y_1(x, t)$ and $y_2(x, t)$?

(b) At $t = (\pi/50)$ s, at what location x do the two waves interfere constructively, and what is the corresponding value of $|y_s|$?

(c) At $t = (\pi/50)$ s, at what location x do the two waves interfere destructively, and what is the corresponding value of $|y_s|$?

1.8 Give expressions for $y(x, t)$ for a sinusoidal wave traveling along a string in the negative x-direction, given that $y_{\max} = 20$ cm, $\lambda = 30$ cm, $f = 5$ Hz, and

(a) $y(x, 0) = 0$ at $x = 0$

(b) $y(x, 0) = 0$ at $x = 7.5$ cm

1.9* An oscillator that generates a sinusoidal wave on a string completes 20 vibrations in 30 s. The wave peak is observed to travel a distance of 2.8 m along the string in 5 s. What is the wavelength?

1.10 The vertical displacement of a string is given by the harmonic function:

$$y(x, t) = 5 \cos(12\pi t - 20\pi x) \quad \text{(m)}$$

where x is the horizontal distance along the string in meters. Suppose a tiny particle were attached to the string at $x = 5$ cm. Obtain an expression for the vertical velocity of the particle as a function of time.

1.11* Given two waves characterized by the following:

$$y_1(t) = 6 \cos \omega t$$
$$y_2(t) = 6 \sin(\omega t + 30°)$$

does $y_2(t)$ lead or lag $y_1(t)$ and by what phase angle?

1.12 The voltage of an electromagnetic wave traveling on a transmission line is given by

$$v(z, t) = 3e^{-\alpha z} \sin(2\pi \times 10^9 t - 10\pi z) \quad \text{(V)}$$

where z is the distance in meters from the generator.

(a) Find the frequency, wavelength, and phase velocity of the wave.

(b) At $z = 2$ m, the amplitude of the wave was measured to be 1 V. Find α.

1.13* A certain electromagnetic wave traveling in seawater was observed to have an amplitude of 19.025 (V/m) at a depth of 10 m, and an amplitude of 12.13 (V/m) at a depth of 100 m. What is the attenuation constant of seawater?

Section 1-5: Complex Numbers

1.14 Evaluate each of the following complex numbers and express the result in rectangular form:

(a) $z_1 = 3e^{j\pi/4}$

(b) $z_2 = \sqrt{3}\, e^{j3\pi/4}$

(c) $z_3 = 2\, e^{-j\pi/2}$

(d) $z_4 = j^3$

(e) $z_5 = j^{-4}$

(f) $z_6 = (1 - j)^3$

(g) $z_7 = (1 - j)^{1/2}$

1.15* Complex numbers z_1 and z_2 are given by the following:

$$z_1 = 3 - j2$$
$$z_2 = -4 + j2$$

(a) Express z_1 and z_2 in polar form.

(b) Find $|z_1|$ by first applying Eq. (1.41) and then by applying Eq. (1.43).

(c) Determine the product $z_1 z_2$ in polar form.

(d) Determine the ratio z_1/z_2 in polar form.

(e) Determine z_1^3 in polar form.

1.16 If $z = -2 + j3$, determine the following quantities in polar form:

(a) $1/z$

(b) z^3

(c) $|z|^2$

(d) $\Im\mathfrak{m}\{z\}$

(e) $\Im\mathfrak{m}\{z^*\}$

1.17* Find complex numbers $t = z_1 + z_2$ and $s = z_1 - z_2$, both in polar form, for each of the following pairs:

(a) $z_1 = 2 + j3$ and $z_2 = 1 - j2$

(b) $z_1 = 2$ and $z_2 = -j2$

(c) $z_1 = 3\angle 30°$ and $z_2 = 3\angle -30°$

(d) $z_1 = 3\angle 30°$ and $z_2 = 3\angle -150°$

1.18 Complex numbers z_1 and z_2 are given by the following:

$$z_1 = 5\angle -60°$$

$$z_2 = 2\angle 45°$$

(a) Determine the product $z_1 z_2$ in polar form.

(b) Determine the product $z_1 z_2^*$ in polar form.

(c) Determine the ratio z_1/z_2 in polar form.

(d) Determine the ratio z_1^*/z_2^* in polar form.

(e) Determine $\sqrt{z_1}$ in polar form.

1.19* If $z = 3 - j4$, find the value of $\ln(z)$.

1.20 If $z = 3 - j4$, find the value of e^z.

Section 1-6: Phasors

1.21* A voltage source given by

$$v_s(t) = 10\cos(2\pi \times 10^3 t - 30°) \quad \text{(V)}$$

is connected to a series RC load as shown in Fig. 1-19. If $R = 1\ \text{M}\Omega$ and $C = 100\ \text{pF}$, obtain an expression for $v_c(t)$, the voltage across the capacitor.

1.22 Find the phasors of the following time functions:

(a) $v(t) = 3\cos(\omega t - \pi/4)$ (V)

(b) $v(t) = 12\sin(\omega t + \pi/4)$ (V)

(c) $i(x, t) = 4e^{-3x}\sin(\omega t - \pi/6)$ (A)

(d) $i(t) = -2\cos(\omega t + 3\pi/4)$ (A)

(e) $i(t) = 2\sin(\omega t + \pi/3) + 3\cos(\omega t - \pi/6)$ (A)

1.23* Find the instantaneous time sinusoidal functions corresponding to the following phasors:

(a) $\tilde{V} = -3e^{j\pi/3}$ (V)

(b) $\tilde{V} = j6e^{j\pi/4}$ (V)

(c) $\tilde{I} = (3 + j4)$ (A)

(d) $\tilde{I} = -3 + j2$ (A)

(e) $\tilde{I} = j$ (A)

(f) $\tilde{I} = 2e^{j3\pi/4}$ (A)

1.24 A series RLC circuit is connected to a generator with a voltage $v_s(t) = V_0\cos(\omega t + \pi/3)$ (V).

(a) Write the voltage loop equation in terms of the current $i(t)$, R, L, C, and $v_s(t)$.

(b) Obtain the corresponding phasor-domain equation.

(c) Solve the equation to obtain an expression for the phasor current $\tilde{I}$.

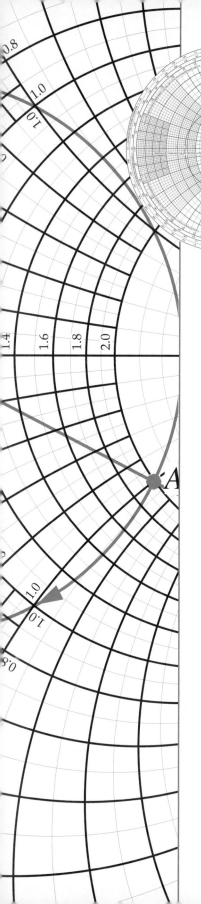

CHAPTER

2

Transmission Lines

2-1 General Considerations

In most electrical engineering curricula, the study of electromagnetics is preceded by one or more courses on electrical circuits. In this book, we use this background to build a bridge between circuit theory and electromagnetic theory. The bridge is provided by transmission lines, the topic of this chapter. By modeling the transmission line in the form of an equivalent circuit, we can use Kirchhoff's voltage and current laws to develop wave equations whose solutions provide an understanding of wave propagation, standing waves, and power transfer. Familiarity with these concepts facilitates the presentation of material in later chapters.

Although the family of *transmission lines* may encompass all structures and media that serve to transfer energy or information between two points, including nerve fibers in the human body, acoustic waves in fluids, and mechanical pressure waves in solids, we shall focus our treatment in this chapter on transmission lines used for guiding electromagnetic signals. Such transmission lines include telephone wires, coaxial cables carrying audio and video information to TV sets or digital data to computer monitors, and optical fibers carrying light waves for the transmission of data at very high rates. Fundamentally, a transmission line is a two-port network, with each port consisting of two terminals, as illustrated in Fig. 2-1. One of the ports is the sending end and the other is the receiving end. The source connected to its sending end may be any circuit with an output voltage, such as a radar transmitter, an amplifier, or a computer terminal operating in the transmission mode. From circuit theory, any such source can be represented by a Thévenin-equivalent *generator circuit* consisting of a generator voltage V_g in series with a generator resistance R_g, as shown in Fig. 2-1. The generator voltage may consist of digital pulses, a modulated time-varying sinusoidal signal, or any other signal waveform. In the case of a-c signals, the generator circuit is represented by a voltage phasor $\tilde{V}_g$ and an impedance Z_g.

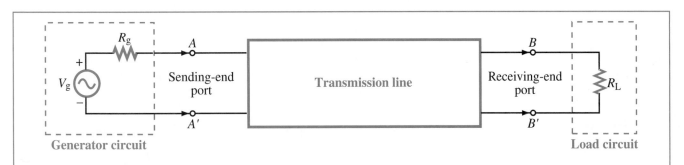

Figure 2-1: A transmission line is a two-port network connecting a generator circuit at the sending end to a load at the receiving end.

35

The circuit connected to the receiving end of the transmission line is called the load circuit, or simply the *load*. This may be an antenna in the case of a radar, a computer terminal operating in the receiving mode, the input terminals of an amplifier, or any output circuit whose input terminals can be represented by an equivalent load resistance R_L, or a load impedance Z_L in the a-c case.

2-1.1 The Role of Wavelength

In low-frequency electrical circuits, we usually use wires to connect the elements of the circuit in the desired configuration. In the circuit shown in Fig. 2-2, for example, the generator is connected to a simple RC load via a pair of wires. In view of our definition in the preceding paragraphs of what constitutes a transmission line, we pose the following question: Is the pair of wires between terminals AA' and terminals BB' a transmission line? If so, why is it important? After all, we usually solve for the current in the circuit and the voltage across its elements without regard for the wires connecting the elements. The answer to this question is yes; indeed the pair of wires constitutes a transmission line, but the impact of the line on the current and voltages in the circuit depends on the length of the line l and the frequency f of

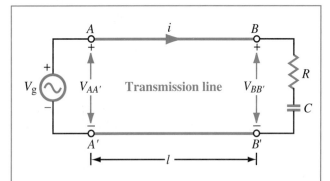

Figure 2-2: Generator connected to an RC circuit through a transmission line of length l.

the signal provided by the generator. (As we will see later, the determining factor is the ratio of the length l to the wavelength λ of the wave propagating on the transmission line between AA' and BB'.) If the generator voltage is cosinusoidal in time, then the voltage across the input terminals AA' is

$$V_{AA'} = V_g(t) = V_0 \cos \omega t \quad \text{(V)}, \quad (2.1)$$

where $\omega = 2\pi f$ is the angular frequency, and if we assume that the current flowing through the wires travels at the speed of light, $c = 3 \times 10^8$ m/s, then the voltage across the output terminals BB' will have to be delayed in time relative to that across AA' by the travel delay time l/c. Thus, assuming no significant ohmic losses in the transmission line,

$$V_{BB'}(t) = V_{AA'}(t - l/c)$$
$$= V_0 \cos [\omega(t - l/c)] \quad \text{(V)}. \quad (2.2)$$

Let us compare $V_{BB'}$ to $V_{AA'}$ at $t = 0$ for an ultralow-frequency electronic circuit operating at a frequency $f = 1$ kHz. For a typical wire length $l = 5$ cm, Eqs. (2.1) and (2.2) give $V_{AA'} = V_0$ and $V_{BB'} = V_0 \cos(2\pi f l/c) = 0.999999999998 \, V_0$. Thus, for all practical purposes, the length of the transmission line may be ignored and terminal AA' may be treated as identical with BB'. On the other hand, had the line been a 20-km long telephone cable carrying a 1-kHz voice signal, then the same calculation would have led to $V_{BB'} = 0.91 V_0$. The determining factor is the magnitude of $\omega l/c$. From Eq. (1.27), the velocity of propagation u_p of a traveling wave is related to the oscillation frequency f and the wavelength λ by

$$u_p = f\lambda \quad \text{(m/s)}. \quad (2.3)$$

In the present case, $u_p = c$. Hence, the phase factor

$$\frac{\omega l}{c} = \frac{2\pi f l}{c} = 2\pi \frac{l}{\lambda} \quad \text{radians}. \quad (2.4)$$

When l/λ is very small, transmission-line effects may be ignored, but when $l/\lambda \gtrsim 0.01$, it may be necessary to

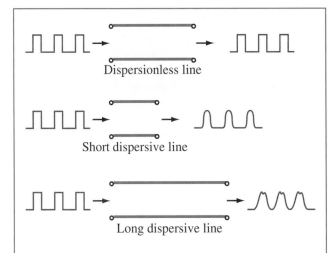

Dispersionless line

Short dispersive line

Long dispersive line

Figure 2-3: A dispersionless line does not distort signals passing through it regardless of its length, whereas a dispersive line distorts the shape of the input pulses because the different frequency components propagate at different velocities. The degree of distortion is proportional to the length of the dispersive line.

account not only for the phase shift associated with the time delay, but also for the presence of *reflected* signals that may have been bounced back by the load toward the generator. *Power loss* on the line and *dispersive* effects may need to be considered as well. A dispersive transmission line is one on which the wave velocity is not constant as a function of the frequency f. This means that the shape of a rectangular pulse, which through Fourier analysis is composed of many waves of different frequencies, will be distorted as it travels down the line because its different frequency components will not propagate at the same velocity (Fig. 2-3). Preservation of pulse shape is very important in high-speed data transmission, both between terminals as well as in high-speed integrated circuits in which transmission-line design and fabrication processes are an integral part of the IC design process. At 10 GHz, for example, the wavelength $\lambda = 3$ cm in air and is on the order of 1 cm in a semiconductor material. Hence, even connection lengths

between devices on the order of millimeters become significant, and their presence has to be incorporated in the overall design of the circuit.

2-1.2 Propagation Modes

A few examples of common types of transmission lines are shown in Fig. 2-4. Transmission lines may be classified into two basic types:

- *Transverse electromagnetic (TEM) transmission lines:* Waves propagating along these lines are characterized by electric and magnetic fields that are entirely *transverse* to the direction of propagation. This is called a TEM mode. A good example is the coaxial line shown in Fig. 2-5; the electric field lines are in the radial direction between the inner and outer conductors, the magnetic field forms circles around the inner conductor, and hence neither has any components along the length of the line (the direction of wave propagation). Other TEM transmission lines include the two-wire line and the parallel-plate line, both shown in Fig. 2-4. Although the fields present on a microstrip line do not adhere to the exact definition of a TEM mode, the nontransverse field components are sufficiently small in comparison to the transverse components to be ignored, thereby allowing the inclusion of microstrip lines in the TEM class. A common feature among TEM lines is that they consist of two parallel conducting surfaces.

- *Higher-order transmission lines:* Waves propagating along these lines have at least one significant field component in the direction of propagation. Hollow conducting waveguides, dielectric rods, and optical fibers belong to this class of lines.

Only TEM-mode transmission lines will be treated in this chapter. This is because less mathematical rigor is required for treating this class of lines than that required for treating waves characterized by higher-order modes

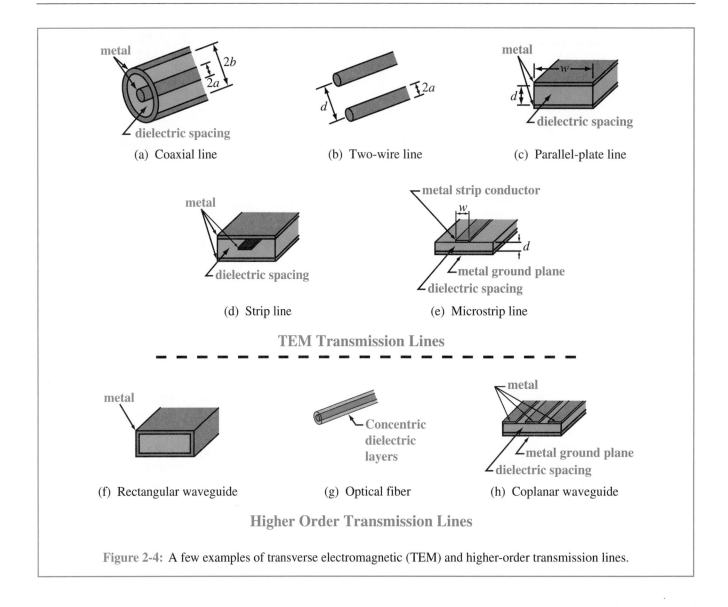

TEM Transmission Lines

Higher Order Transmission Lines

Figure 2-4: A few examples of transverse electromagnetic (TEM) and higher-order transmission lines.

and, in addition, TEM lines are more commonly used in practice. We start our treatment by representing the transmission line in terms of a lumped-element circuit model, and then we apply Kirchhoff's voltage and current laws to derive a set of two governing equations known as the *telegrapher's equations*. By combining these equations, we obtain wave equations for the voltage and current at any point on the line. Solution of the wave equations for the sinusoidal steady-state case leads to a set of formulas that can be used for solving a wide range of practical problems. In the latter part of this chapter we introduce a graphical technique known as

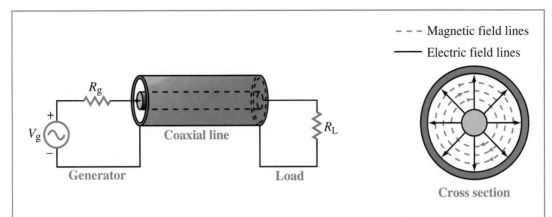

Figure 2-5: In a coaxial line, the electric field lines are in the radial direction between the inner and outer conductors, and the magnetic field forms circles around the inner conductor.

the *Smith chart*, which facilitates the solution of many transmission-line problems without having to perform laborious calculations involving complex numbers.

2-2 Lumped-Element Model

When we draw a schematic of an electronic circuit, we use specific symbols to represent resistors, capacitors, inductors, diodes, and the like. In each case, the symbol represents the functionality of the device, rather than its shape, size or other attributes. We shall do the same with regard to transmission lines; *we shall represent a transmission line by a parallel-wire configuration,* as shown in Fig. 2-6(a), *regardless of the specific shape of the line under consideration.* Thus, Fig. 2-6(a) may represent a coaxial line, a two-wire line, or any other TEM line.

Drawing again on our familiarity with electronic circuits, when we analyze a circuit containing a transistor, we represent the functionality of the transistor by an equivalent circuit composed of sources, resistors, and capacitors. We will apply the same approach to the transmission line by orienting the line along

the z-direction, subdividing it into differential sections each of length Δz [Fig. 2-6(b)] and then representing each section by an equivalent circuit, as illustrated in Fig. 2-6(c). This representation, which is called the *lumped-element circuit* model, consists of four basic elements, which henceforth will be called the *transmission line parameters*. These are

R': The combined *resistance* of both conductors per unit length, in Ω/m,

L': The combined *inductance* of both conductors per unit length, in H/m,

G': The *conductance* of the insulation medium per unit length, in S/m, and

C': The *capacitance* of the two conductors per unit length, in F/m.

Whereas the four line parameters have different expressions for different types and dimensions of transmission lines, the equivalent model represented by Fig. 2-6(c) is equally applicable to all transmission lines characterized by TEM-mode wave propagation. *The prime superscript is used as a reminder that the line parameters are differential quantities whose units are per unit length.*

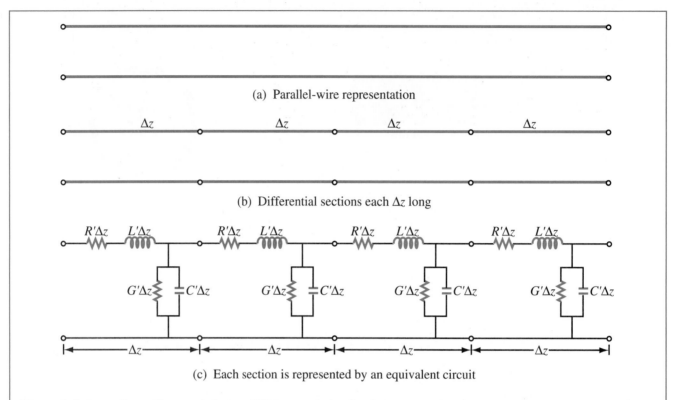

(a) Parallel-wire representation

(b) Differential sections each Δz long

(c) Each section is represented by an equivalent circuit

Figure 2-6: Regardless of its actual shape, a TEM transmission line is represented by the parallel-wire configuration shown in (a). To analyze the voltage and current relations, the line is subdivided into small differential sections (b), each of which is then represented by an equivalent circuit (c).

Expressions for the line parameters R', L', G', and C' are given in Table 2-1 for the three types of TEM transmission lines diagrammed in parts (a) through (c) of Fig. 2-4. For each of these lines, the expressions are functions of two sets of parameters: (1) geometric parameters defining the cross-sectional dimensions of the given line and (2) electromagnetic constitutive parameters characteristic of the materials of which the conductors and the insulating material between them are made. The pertinent geometric parameters are as follows:

Coaxial line [Fig. 2-4(a)]:

a = outer radius of inner conductor, m
b = inner radius of outer conductor, m

Two-wire line [Fig. 2-4(b)]:

a = radius of each wire, m
d = spacing between wires' centers, m

Parallel-plate line [Fig. 2-4(c)]:

w = width of each plate, m
d = thickness of insulation between plates, m

Table 2-1: Transmission-line parameters R', L', G', and C' for three types of lines.

Parameter	Coaxial	Two Wire	Parallel Plate	Unit
R'	$\dfrac{R_s}{2\pi}\left(\dfrac{1}{a}+\dfrac{1}{b}\right)$	$\dfrac{R_s}{\pi a}$	$\dfrac{2R_s}{w}$	Ω/m
L'	$\dfrac{\mu}{2\pi}\ln(b/a)$	$\dfrac{\mu}{\pi}\ln\left[(d/2a)+\sqrt{(d/2a)^2-1}\right]$	$\dfrac{\mu d}{w}$	H/m
G'	$\dfrac{2\pi\sigma}{\ln(b/a)}$	$\dfrac{\pi\sigma}{\ln\left[(d/2a)+\sqrt{(d/2a)^2-1}\right]}$	$\dfrac{\sigma w}{d}$	S/m
C'	$\dfrac{2\pi\varepsilon}{\ln(b/a)}$	$\dfrac{\pi\varepsilon}{\ln\left[(d/2a)+\sqrt{(d/2a)^2-1}\right]}$	$\dfrac{\varepsilon w}{d}$	F/m

Notes: (1) Refer to Fig. 2-4 for definitions of dimensions. (2) μ, ε, and σ pertain to the insulating material between the conductors. (3) $R_s = \sqrt{\pi f \mu_c/\sigma_c}$. (4) μ_c and σ_c pertain to the conductors. (5) If $(d/2a)^2 \gg 1$, then $\ln\left[(d/2a)+\sqrt{(d/2a)^2-1}\right] \simeq \ln(d/a)$.

The constitutive parameters apply to all three lines and consist of two groups: μ_c and σ_c are the magnetic permeability and electrical conductivity of the conductors, and ε, μ, and σ are the electrical permittivity, magnetic permeability, and electrical conductivity of the insulation material separating the conductors. Appendix B contains tabulated values for these constitutive parameters for various types of materials. For the purposes of the present chapter, we need not concern ourselves with the derivations responsible for the expressions given in Table 2-1. The formulations necessary for computing R', L', G', and C' will be made available in later chapters for the general case of any two-conductor configuration.

The lumped-element model shown in Fig. 2-6(c) represents the physical processes associated with the currents and voltages on any TEM transmission line. Other equivalent models are available also and are equally applicable as well. All these models, however, lead to exactly the same set of telegrapher's equations, from which all our future results will be derived. Hence, only the model described in Fig. 2-6(c) will be examined in the present treatment. It consists of two series elements, R' and L', and two shunt elements, G' and C'. By way of providing a physical explanation for the lumped-element model, let us consider a small section of a coaxial line, as shown in Fig. 2-7. The line consists of an inner conductor of radius a separated from an outer conducting cylinder of radius b by a material with permittivity ε, permeability μ, and conductivity σ. The two metal conductors are made of a material with conductivity σ_c and permeability μ_c. When a voltage source is connected across the two conductors at the sending end of the line, currents will flow through the conductors, primarily along the outer surface of the inner conductor and the inner surface of

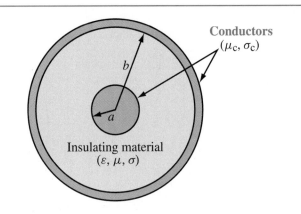

Figure 2-7: Cross section of a coaxial line with inner conductor of radius a and outer conductor of radius b. The conductors have magnetic permeability μ_c, and conductivity σ_c, and the spacing material between the conductors has permittivity ε, permeability μ, and conductivity σ.

the outer conductor. The line resistance R' accounts for the combined resistance per unit length of the inner and outer conductors. The expression for R' is derived in Chapter 7 and is given by Eq. (7.96) as

$$R' = \frac{R_s}{2\pi}\left(\frac{1}{a} + \frac{1}{b}\right) \qquad (\Omega/\text{m}), \qquad (2.5)$$

where R_s, which represents the surface resistance of the conductors, is called the *intrinsic resistance* and is given by Eq. (7.92a) as

$$R_s = \sqrt{\frac{\pi f \mu_c}{\sigma_c}} \qquad (\Omega). \qquad (2.6)$$

The intrinsic resistance depends not only on the material properties of the conductors (σ_c and μ_c), but on the frequency f of the wave traveling on the line as well. For a *perfect conductor* with $\sigma_c = \infty$ or a high-conductivity

material such that $(f\mu_c/\sigma_c) \ll 1$, R_s approaches zero, and so does R'.

Next, let us examine the inductance per unit length L'. Application of Ampère's law in Chapter 5 to the definition of inductance leads to the following expression [Eq. (5.99)] for the inductance per unit length of a coaxial line:

$$L' = \frac{\mu}{2\pi} \ln\left(\frac{b}{a}\right) \qquad (\text{H/m}). \qquad (2.7)$$

The shunt conductance per unit length G' accounts for current flow between the outer and inner conductors, made possible by the material conductivity σ of the insulator. It is precisely because the current flow is from one conductor to the other that G' is a shunt element in the lumped-element model. Its expression is given by Eq. (4.76) as

$$G' = \frac{2\pi\sigma}{\ln(b/a)} \qquad (\text{S/m}). \qquad (2.8)$$

If the material separating the inner and outer conductors is a *perfect dielectric* with $\sigma = 0$, then $G' = 0$.

The last line parameter on our list is the capacitance per unit length C'. When equal and opposite charges are placed on any two noncontacting conductors, a voltage difference will be induced between them. Capacitance is defined as the ratio of charge to voltage difference. For the coaxial line, C' is given by Eq. (4.117) as

$$C' = \frac{2\pi\varepsilon}{\ln(b/a)} \qquad (\text{F/m}). \qquad (2.9)$$

All TEM transmission lines share the following useful relations:

$$\boxed{L'C' = \mu\varepsilon, \qquad (2.10)}$$

and

$$\boxed{\frac{G'}{C'} = \frac{\sigma}{\varepsilon}. \qquad (2.11)}$$

If the insulating medium between the conductors is air, the transmission line is called an *air line* (e.g., coaxial air line or two-wire air line). For an air line, $\varepsilon = \varepsilon_0 = 8.854 \times 10^{-12}$ F/m, $\mu = \mu_0 = 4\pi \times 10^{-7}$ H/m, $\sigma = 0$, and $G' = 0$.

REVIEW QUESTIONS

Q2.1 What is a transmission line? When should transmission-line effects be considered?

Q2.2 What is the difference between dispersive and nondispersive transmission lines? What is the practical significance?

Q2.3 What constitutes a TEM transmission line?

Q2.4 What purpose does the lumped-element circuit model serve? How are the line parameters R', L', G', and C' related to the physical and electromagnetic constitutive properties of the transmission line?

EXERCISE 2.1 Use Table 2-1 to compute the line parameters of a two-wire air line whose wires are separated by a distance of 2 cm, and each is 1 mm in radius. The wires may be treated as perfect conductors with $\sigma_c = \infty$.

Ans. $R' = 0$, $L' = 1.20$ (μH/m), $G' = 0$, $C' = 9.29$ (pF/m).

EXERCISE 2.2 Calculate the transmission line parameters at 1 MHz for a rigid coaxial air line with an inner conductor diameter of 0.6 cm and an outer conductor diameter of 1.2 cm. The conductors are made of copper [see Appendix B for μ_c and σ_c of copper].

Ans. $R' = 2.08 \times 10^{-2}$ (Ω/m), $L' = 0.14$ (μH/m), $G' = 0$, $C' = 80.3$ (pF/m).

2-3 Transmission-Line Equations

A transmission line usually connects a source on one end to a load on the other end. Before we consider the complete circuit, however, we need to develop equations that describe the voltage across the transmission line and the current carried by the line as a function of time t and spatial position z. Using the lumped-element model described in Fig. 2-6(c), we begin by considering a differential length Δz as shown in Fig. 2-8. The quantities $v(z,t)$ and $i(z,t)$ denote the instantaneous voltage and current at the left end of the differential section (node N), and similarly $v(z + \Delta z, t)$ and $i(z + \Delta z, t)$ denote the same quantities at the right end (node $N + 1$). Application of Kirchhoff's voltage law accounts for the voltage drop across the series resistance $R' \Delta z$ and inductance $L' \Delta z$:

$$v(z,t) - R' \Delta z\, i(z,t)$$
$$- L' \Delta z \frac{\partial i(z,t)}{\partial t} - v(z + \Delta z, t) = 0. \quad (2.12)$$

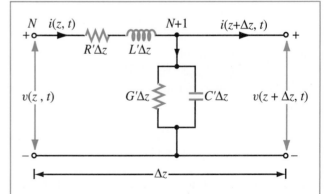

Figure 2-8: Equivalent circuit of a differential length Δz of a two-conductor transmission line.

Upon dividing all terms by Δz and rearranging terms, we obtain

$$-\left[\frac{v(z+\Delta z,t)-v(z,t)}{\Delta z}\right]$$
$$= R'\,i(z,t) + L'\frac{\partial\,i(z,t)}{\partial t}. \qquad (2.13)$$

In the limit as $\Delta z \to 0$, Eq. (2.13) becomes a differential equation:

$$-\frac{\partial v(z,t)}{\partial z} = R'\,i(z,t) + L'\frac{\partial i(z,t)}{\partial t}. \qquad (2.14)$$

Similarly, application of Kirchhoff's current law at node $N+1$ in Fig. 2-8 leads to

$$i(z,t) - G'\Delta z\,v(z+\Delta z,t)$$
$$- C'\Delta z\frac{\partial v(z+\Delta z,t)}{\partial t} - i(z+\Delta z,t) = 0. \qquad (2.15)$$

Upon dividing all terms by Δz and taking the limit as $\Delta z \to 0$, Eq. (2.15) provides a second differential equation,

$$-\frac{\partial i(z,t)}{\partial z} = G'\,v(z,t) + C'\frac{\partial v(z,t)}{\partial t}. \qquad (2.16)$$

The first-order differential equations given by Eqs. (2.14) and (2.16) are the time-domain form of the *transmission line equations*, otherwise called the *telegrapher's equations*.

Except for the last section, our primary interest in this chapter is in sinusoidal steady-state conditions. To this end, we shall make use of phasors with the cosine reference notation as outlined in Section 1-6. Thus, we define

$$v(z,t) = \mathfrak{Re}[\tilde{V}(z)\,e^{j\omega t}], \qquad (2.17a)$$
$$i(z,t) = \mathfrak{Re}[\tilde{I}(z)\,e^{j\omega t}], \qquad (2.17b)$$

where $\tilde{V}(z)$ and $\tilde{I}(z)$ are phasor quantities, each of which may be real or complex. Upon substituting Eqs. (2.17a) and (2.17b) into Eqs. (2.14) and (2.16) and utilizing the property given by Eq. (1.62) that $\partial/\partial t$ in the time domain

becomes equivalent to multiplication by $j\omega$ in the phasor domain, we obtain the following pair of equations:

$$-\frac{d\tilde{V}(z)}{dz} = (R'+j\omega L')\,\tilde{I}(z), \qquad (2.18a)$$
$$-\frac{d\tilde{I}(z)}{dz} = (G'+j\omega C')\,\tilde{V}(z). \qquad (2.18b)$$

These are the telegrapher's equations in phasor form.

2-4   Wave Propagation on a Transmission Line

The two first-order coupled equations given by Eqs. (2.18a) and (2.18b) can be combined to give two second-order uncoupled wave equations, one for $\tilde{V}(z)$ and another for $\tilde{I}(z)$. The wave equation for $\tilde{V}(z)$ is derived by differentiating both sides of Eq. (2.18a) with respect to z, giving

$$-\frac{d^2\tilde{V}(z)}{dz^2} = (R'+j\omega L')\frac{d\tilde{I}(z)}{dz}, \qquad (2.19)$$

and upon substituting Eq. (2.18b) into Eq. (2.19) for $d\tilde{I}(z)/dz$, Eq. (2.19) becomes

$$\frac{d^2\tilde{V}(z)}{dz^2} - (R'+j\omega L')(G'+j\omega C')\,\tilde{V}(z) = 0, \qquad (2.20)$$

or

$$\frac{d^2\tilde{V}(z)}{dz^2} - \gamma^2\,\tilde{V}(z) = 0, \qquad (2.21)$$

where

$$\gamma = \sqrt{(R'+j\omega L')(G'+j\omega C')}. \qquad (2.22)$$

Application of the same steps to Eqs. (2.18a) and (2.18b) but in reverse order, leads to

$$\frac{d^2 \tilde{I}(z)}{dz^2} - \gamma^2 \tilde{I}(z) = 0. \quad (2.23)$$

Equations (2.21) and (2.23) are called *wave equations* for $\tilde{V}(z)$ and $\tilde{I}(z)$, respectively, and γ is called the *complex propagation constant* of the transmission line. As such, γ consists of a real part α, called the *attenuation constant* of the line with units of Np/m, and an imaginary part β, called the *phase constant* of the line with units of rad/m. Thus,

$$\gamma = \alpha + j\beta \quad (2.24)$$

with

$$
\begin{aligned}
\alpha &= \mathfrak{Re}(\gamma) \\
&= \mathfrak{Re}\left(\sqrt{(R' + j\omega L')(G' + j\omega C')}\right) \quad \text{(Np/m)}, \\
& \hspace{10cm} (2.25a) \\
\beta &= \mathfrak{Im}(\gamma) \\
&= \mathfrak{Im}\left(\sqrt{(R' + j\omega L')(G' + j\omega C')}\right) \quad \text{(rad/m)}. \\
& \hspace{10cm} (2.25b)
\end{aligned}
$$

In Eqs. (2.25a) and (2.25b), we choose the square-root values that give positive values for α and β. For passive transmission lines, α is either zero or positive. Most transmission lines, and all those considered in this chapter, are of the passive type. The active region of a laser is an example of an active transmission line with a negative α.

The wave equations given by Eqs. (2.21) and (2.23) have traveling wave solutions of the following form:

$$\tilde{V}(z) = V_0^+ e^{-\gamma z} + V_0^- e^{\gamma z} \quad \text{(V)}, \quad (2.26a)$$

$$\tilde{I}(z) = I_0^+ e^{-\gamma z} + I_0^- e^{\gamma z} \quad \text{(A)}, \quad (2.26b)$$

where, as will be shown later, the $e^{-\gamma z}$ term represents wave propagation in the $+z$-direction and the $e^{\gamma z}$ term represents wave propagation in the $-z$-direction. Verification that these are indeed valid solutions is easily accomplished by substituting the expressions given by Eqs. (2.26a) and (2.26b), as well as their second derivatives, into Eqs. (2.21) and (2.23). In their present form, the solutions given by Eqs. (2.26a) and (2.26b) contain four unknowns, the wave amplitudes (V_0^+, I_0^+) of the $+z$ propagating wave and (V_0^-, I_0^-) of the $-z$ propagating wave. We can easily relate the current wave amplitudes, I_0^+ and I_0^-, to the voltage wave amplitudes, V_0^+ and V_0^-, respectively, by using Eq. (2.26a) in Eq. (2.18a) and then solving for the current $\tilde{I}(z)$ to get the result

$$\tilde{I}(z) = \frac{\gamma}{R' + j\omega L'}\left[V_0^+ e^{-\gamma z} - V_0^- e^{\gamma z}\right]. \quad (2.27)$$

Comparison of each term with the corresponding term in the expression given by Eq. (2.26b) leads to the conclusion that

$$\frac{V_0^+}{I_0^+} = Z_0 = \frac{-V_0^-}{I_0^-}, \quad (2.28)$$

where

$$Z_0 = \frac{R' + j\omega L'}{\gamma} = \sqrt{\frac{R' + j\omega L'}{G' + j\omega C'}} \quad (\Omega), \quad (2.29)$$

is defined as the *characteristic impedance* of the line. It should be noted that Z_0 *is equal to the ratio of the voltage amplitude to the current amplitude for each of the traveling waves individually (with an additional minus sign in the case of the $-z$ propagating wave), but it is not equal to the ratio of the total voltage $\tilde{V}(z)$ to the total current $\tilde{I}(z)$, unless one of the two waves is absent.* In terms of Z_0, Eq. (2.27) can be rewritten in the form

$$\tilde{I}(z) = \frac{V_0^+}{Z_0} e^{-\gamma z} - \frac{V_0^-}{Z_0} e^{\gamma z}. \quad (2.30)$$

In later sections, we will apply boundary conditions at the load and at the sending end of the transmission line to obtain expressions for the remaining wave amplitudes V_0^+ and V_0^-. In general, each will be a complex quantity composed of a magnitude and a phase angle. Thus,

$$V_0^+ = |V_0^+| e^{j\phi^+}, \qquad (2.31a)$$

$$V_0^- = |V_0^-| e^{j\phi^-}. \qquad (2.31b)$$

Upon substituting these definitions in Eq. (2.26a) and replacing γ with Eq. (2.24), we can convert back to the time domain to obtain an expression for $v(z, t)$, the instantaneous voltage on the line:

$$
\begin{aligned}
v(z, t) &= \mathfrak{Re}(\widetilde{V}(z) e^{j\omega t}) \\
&= \mathfrak{Re}\left[\left(V_0^+ e^{-\gamma z} + V_0^- e^{\gamma z}\right) e^{j\omega t}\right] \\
&= \mathfrak{Re}[|V_0^+| e^{j\phi^+} e^{j\omega t} e^{-(\alpha + j\beta)z} \\
&\quad + |V_0^-| e^{j\phi^-} e^{j\omega t} e^{(\alpha + j\beta)z}] \\
&= |V_0^+| e^{-\alpha z} \cos(\omega t - \beta z + \phi^+) \\
&\quad + |V_0^-| e^{\alpha z} \cos(\omega t + \beta z + \phi^-). \quad (2.32)
\end{aligned}
$$

From our review of traveling waves in Section 1-3, we recognize the first term in Eq. (2.32) as a wave traveling in the $+z$-direction (the coefficients of t and z have opposite signs) and the second term as a wave traveling in the $-z$-direction (the coefficients of t and z are both positive), both propagating with a phase velocity u_p given by Eq. (1.30):

$$u_\mathrm{p} = f\lambda = \frac{\omega}{\beta}. \qquad (2.33)$$

The factor $e^{-\alpha z}$ accounts for the attenuation of the $+z$ propagating wave, and the $e^{\alpha z}$ accounts for the attenuation of the $-z$ propagating wave. The presence of two waves on the line propagating in opposite directions produces a *standing wave*. To gain a physical understanding of what this means, we shall first examine the relatively simple but important case of a *lossless line* ($\alpha = 0$) and then extend the results to the more general case of *lossy transmission lines* ($\alpha \neq 0$). In fact, we shall devote the next several sections to the study of lossless transmission lines because in practice many lines can be designed to exhibit very low-loss characteristics.

Example 2-1   Air Line

An *air line* is a transmission line for which air is the dielectric material present between the two conductors, which renders $G' = 0$. In addition, the conductors are made of a material with high conductivity so that $R' \simeq 0$. For an air line with characteristic impedance of 50 Ω and phase constant of 20 rad/m at 700 MHz, find the inductance per meter and the capacitance per meter of the line.

Solution: The following quantities are given:

$$Z_0 = 50\ \Omega, \qquad \beta = 20\ \mathrm{rad/m},$$

$$f = 700\ \mathrm{MHz} = 7 \times 10^8\ \mathrm{Hz}.$$

With $R' = G' = 0$, Eqs. (2.25b) and (2.29) reduce to

$$
\begin{aligned}
\beta &= \mathfrak{Im}\left(\sqrt{(j\omega L')(j\omega C')}\right) \\
&= \mathfrak{Im}\left(j\omega\sqrt{L'C'}\right) = \omega\sqrt{L'C'},
\end{aligned}
$$

$$Z_0 = \sqrt{\frac{j\omega L'}{j\omega C'}} = \sqrt{\frac{L'}{C'}}.$$

The ratio is given by

$$\frac{\beta}{Z_0} = \omega C',$$

or

$$
\begin{aligned}
C' &= \frac{\beta}{\omega Z_0} \\
&= \frac{20}{2\pi \times 7 \times 10^8 \times 50} \\
&= 9.09 \times 10^{-11}\ \mathrm{(F/m)} = 90.9\ \mathrm{(pF/m)}.
\end{aligned}
$$

From $Z_0 = \sqrt{L'/C'}$,

$$L' = Z_0^2 C'$$
$$= (50)^2 \times 90.9 \times 10^{-12}$$
$$= 2.27 \times 10^{-7} \text{ (H/m)} = 227 \text{ (nH/m)}. \quad \blacksquare$$

EXERCISE 2.3   Verify that Eq. (2.26a) is indeed a solution of the wave equation given by Eq. (2.21).

EXERCISE 2.4   A two-wire air line has the following line parameters: $R' = 0.404$ (mΩ/m), $L' = 2.0$ (μH/m), $G' = 0$, and $C' = 5.56$ (pF/m). For operation at 5 kHz, determine (a) the attenuation constant α, (b) the phase constant β, (c) the phase velocity u_p, and (d) the characteristic impedance Z_0.

Ans. (a) $\alpha = 3.37 \times 10^{-7}$ (Np/m), (b) $\beta = 1.05 \times 10^{-4}$ (rad/m), (c) $u_p = 3.0 \times 10^8$ (m/s), (d) $Z_0 = (600 - j2.0)\ \Omega = 600\angle{-0.19°}\ \Omega$.

2-5   The Lossless Transmission Line

According to the preceding section, a transmission line is characterized by two fundamental properties, its propagation constant γ and characteristic impedance Z_0, both of which are specified by the angular frequency ω and the line parameters R', L', G', and C'. In many practical situations, the transmission line can be designed to minimize ohmic losses by selecting conductors with very high conductivities and dielectric materials (separating the conductors) with negligible conductivities. As a result, R' and G' assume very small values such that $R' \ll \omega L'$ and $G' \ll \omega C'$. These lossless-line conditions allow us to set $R' = G' = 0$ in Eq. (2.22), which then gives the result

$$\gamma = \alpha + j\beta = j\omega\sqrt{L'C'}, \qquad (2.34)$$

which means that

$$\boxed{\begin{aligned} \alpha &= 0 \qquad\qquad\text{(lossless line)}, \\ \beta &= \omega\sqrt{L'C'} \quad \text{(lossless line).} \end{aligned}} \qquad (2.35)$$

Application of the lossless-line conditions to Eq. (2.29) gives the characteristic impedance as

$$\boxed{Z_0 = \sqrt{\frac{L'}{C'}} \qquad \text{(lossless line)}, \qquad (2.36)}$$

which is now a real number. Using the lossless-line expression for β given by Eq. (2.35), we obtain the following relations for the wavelength λ and the phase velocity u_p:

$$\lambda = \frac{2\pi}{\beta} = \frac{2\pi}{\omega\sqrt{L'C'}}, \qquad (2.37)$$

$$u_p = \frac{\omega}{\beta} = \frac{1}{\sqrt{L'C'}}. \qquad (2.38)$$

Upon using the relation given by Eq. (2.10), which is shared by all TEM transmission lines, Eqs. (2.35) and (2.38) may be rewritten as

$$\boxed{\begin{aligned} \beta &= \omega\sqrt{\mu\varepsilon} \qquad \text{(rad/m)}, & (2.39) \\ u_p &= \frac{1}{\sqrt{\mu\varepsilon}} \qquad \text{(m/s)}, & (2.40) \end{aligned}}$$

where μ and ε are, respectively, the magnetic permeability and electrical permittivity of the insulating material separating the conductors. Materials used for this purpose are usually characterized by a permeability $\mu = \mu_0$, where $\mu_0 = 4\pi \times 10^{-7}$ H/m is the permeability of free space, and the permittivity is usually specified in terms of the relative permittivity ε_r defined as

$$\varepsilon_r = \varepsilon/\varepsilon_0, \qquad (2.41)$$

where $\varepsilon_0 = 8.854 \times 10^{-12}$ F/m $\simeq (1/36\pi) \times 10^{-9}$ F/m is the permittivity of free space. Hence, Eq. (2.40) becomes

$$u_p = \frac{1}{\sqrt{\mu_0 \varepsilon_r \varepsilon_0}} = \frac{1}{\sqrt{\mu_0 \varepsilon_0}} \cdot \frac{1}{\sqrt{\varepsilon_r}} = \frac{c}{\sqrt{\varepsilon_r}}, \qquad (2.42)$$

where $c = 1/\sqrt{\mu_0 \varepsilon_0} = 3 \times 10^8$ m/s is the velocity of light in a vacuum. If the insulating material between the conductors is air, then $\varepsilon_r = 1$ and $u_p = c$. In view of Eq. (2.41) and the relationship between λ and u_p given by Eq. (2.33), the wavelength is given by

$$\boxed{\lambda = \frac{u_p}{f} = \frac{c}{f} \frac{1}{\sqrt{\varepsilon_r}} = \frac{\lambda_0}{\sqrt{\varepsilon_r}},} \qquad (2.43)$$

where $\lambda_0 = c/f$ is the wavelength in air corresponding to a frequency f. Note that, because both u_p and λ depend on ε_r, the choice of the type of insulating material used in a transmission line is dictated not only by its mechanical properties, but by its electrical properties as well.

When the phase velocity of a medium is independent of frequency, the medium is called *nondispersive*, which clearly is the case for a lossless TEM transmission line. This is an important feature for the transmission of digital data in the form of pulses. A rectangular pulse or a series of pulses is composed of many Fourier components with different frequencies. If the phase velocity is the same for all frequency components (or at least for the dominant ones), the pulse shape will remain the same as the pulse travels on the line. In contrast, the shape of a pulse propagating in a dispersive medium gets progressively distorted, and the pulse length increases (stretches out) as a function of distance in the medium, thereby imposing a limitation on the maximum data rate (which is related to the length of the individual pulses and the spacing between adjacent pulses) that can be transmitted through the medium without loss of information.

Table 2-2 provides a list of the expressions for γ, Z_0, and u_p for the general case of a lossy line and for several types of lossless lines. The expressions for the lossless lines are based on the equations for L' and C' given in Table 2-1.

EXERCISE 2.5 For a lossless transmission line, $\lambda = 20.7$ cm at 1 GHz. Find ε_r of the insulating material.

Ans. $\varepsilon_r = 2.1$.

EXERCISE 2.6 A lossless transmission line uses a dielectric insulating material with $\varepsilon_r = 4$. If its line capacitance is $C' = 10$ (pF/m), find (a) the phase velocity u_p, (b) the line inductance L', and (c) the characteristic impedance Z_0.

Ans. (a) $u_p = 1.5 \times 10^8$ (m/s), (b) $L' = 4.45$ (μH/m), (c) $Z_0 = 667.1$ Ω.

2-5.1 Voltage Reflection Coefficient

With $\gamma = j\beta$ for the lossless line, the expressions given by Eqs. (2.26a) and (2.30) for the total voltage and current on the line become

$$\tilde{V}(z) = V_0^+ e^{-j\beta z} + V_0^- e^{j\beta z}, \qquad (2.44a)$$

$$\tilde{I}(z) = \frac{V_0^+}{Z_0} e^{-j\beta z} - \frac{V_0^-}{Z_0} e^{j\beta z}. \qquad (2.44b)$$

These expressions contain two unknowns, V_0^+ and V_0^-, the voltage amplitudes of the incident and reflected waves, respectively. To determine V_0^+ and V_0^-, we need to consider the lossless transmission line in the context of the complete circuit, including a generator circuit at its input terminals and a load at its output terminals, as shown in Fig. 2-9. The line, of length l, is terminated in an arbitrary *load impedance* Z_L. *For convenience, the reference of the spatial coordinate z is chosen such that $z = 0$ corresponds to the location of the load.* At the sending end at $z = -l$, the line is connected to a sinusoidal voltage source with phasor $\tilde{V}_g$ and an internal

Table 2-2: Characteristic parameters of transmission lines.

	Propagation Constant $\gamma = \alpha + j\beta$	Phase Velocity u_{p}	Characteristic Impedance Z_0
General case	$\gamma = \sqrt{(R' + j\omega L')(G' + j\omega C')}$	$u_{\mathrm{p}} = \omega/\beta$	$Z_0 = \sqrt{\dfrac{(R' + j\omega L')}{(G' + j\omega C')}}$
Lossless $(R' = G' = 0)$	$\alpha = 0, \ \beta = \omega\sqrt{\varepsilon_{\mathrm{r}}}/c$	$u_{\mathrm{p}} = c/\sqrt{\varepsilon_{\mathrm{r}}}$	$Z_0 = \sqrt{L'/C'}$
Lossless coaxial	$\alpha = 0, \ \beta = \omega\sqrt{\varepsilon_{\mathrm{r}}}/c$	$u_{\mathrm{p}} = c/\sqrt{\varepsilon_{\mathrm{r}}}$	$Z_0 = (60/\sqrt{\varepsilon_{\mathrm{r}}}) \ln(b/a)$
Lossless two wire	$\alpha = 0, \ \beta = \omega\sqrt{\varepsilon_{\mathrm{r}}}/c$	$u_{\mathrm{p}} = c/\sqrt{\varepsilon_{\mathrm{r}}}$	$Z_0 = (120/\sqrt{\varepsilon_{\mathrm{r}}})$ $\cdot \ln[(d/2a) + \sqrt{(d/2a)^2 - 1}]$ $Z_0 \simeq (120/\sqrt{\varepsilon_{\mathrm{r}}}) \ln(d/a),$ if $d \gg a$
Lossless parallel plate	$\alpha = 0, \ \beta = \omega\sqrt{\varepsilon_{\mathrm{r}}}/c$	$u_{\mathrm{p}} = c/\sqrt{\varepsilon_{\mathrm{r}}}$	$Z_0 = (120\pi/\sqrt{\varepsilon_{\mathrm{r}}}) (d/w)$

Notes: (1) $\mu = \mu_0$, $\varepsilon = \varepsilon_{\mathrm{r}}\varepsilon_0$, $c = 1/\sqrt{\mu_0\varepsilon_0}$, and $\sqrt{\mu_0/\varepsilon_0} \simeq (120\pi) \ \Omega$, where ε_{r} is the relative permittivity of insulating material. (2) For coaxial line, a and b are radii of inner and outer conductors. (3) For two-wire line, $a =$ wire radius and $d =$ separation between wire centers. (4) For parallel-plate line, $w =$ width of plate and $d =$ separation between the plates.

impedance Z_{g}. At the load, the phasor voltage across it, $\tilde{V}_{\mathrm{L}}$, and the phasor current through it, $\tilde{I}_{\mathrm{L}}$, are related by the load impedance Z_{L} as follows:

$$Z_{\mathrm{L}} = \frac{\tilde{V}_{\mathrm{L}}}{\tilde{I}_{\mathrm{L}}}. \qquad (2.45)$$

The voltage $\tilde{V}_{\mathrm{L}}$ is equal to the total voltage on the line $\tilde{V}(z)$ given by Eq. (2.44a), and $\tilde{I}_{\mathrm{L}}$ is equal to $\tilde{I}(z)$ given by Eq. (2.44b), both evaluated at $z = 0$:

$$\tilde{V}_{\mathrm{L}} = \tilde{V}(z{=}0) = V_0^+ + V_0^-, \qquad (2.46\mathrm{a})$$

$$\tilde{I}_{\mathrm{L}} = \tilde{I}(z{=}0) = \frac{V_0^+}{Z_0} - \frac{V_0^-}{Z_0}. \qquad (2.46\mathrm{b})$$

Upon using these expressions in Eq. (2.45), we obtain the result:

$$Z_{\mathrm{L}} = \left(\frac{V_0^+ + V_0^-}{V_0^+ - V_0^-}\right) Z_0. \qquad (2.47)$$

Solving for V_0^- gives

$$V_0^- = \left(\frac{Z_{\mathrm{L}} - Z_0}{Z_{\mathrm{L}} + Z_0}\right) V_0^+. \qquad (2.48)$$

The ratio of the amplitude of the reflected voltage wave to the amplitude of the incident voltage wave at the load

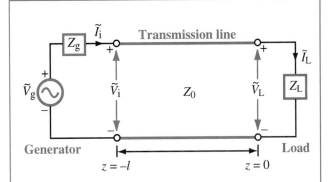

Figure 2-9: Transmission line of length l connected on one end to a generator circuit and on the other end to a load Z_L. The load is located at $z = 0$ and the generator terminals are at $z = -l$.

is known as the voltage reflection coefficient Γ. From Eq. (2.48), this definition gives the result

$$
\Gamma = \frac{V_0^-}{V_0^+} = \frac{Z_L - Z_0}{Z_L + Z_0}
$$
$$
= \frac{Z_L/Z_0 - 1}{Z_L/Z_0 + 1} \quad \text{(dimensionless)},\quad (2.49a)
$$

and in view of Eq. (2.28), the ratio of the current amplitudes is

$$
\frac{I_0^-}{I_0^+} = -\frac{V_0^-}{V_0^+} = -\Gamma. \quad (2.49b)
$$

We note that Γ is governed by a single parameter, the load impedance Z_L, normalized to the characteristic impedance of the line, Z_0. As indicated by Eq. (2.36), Z_0 of a lossless line is a real number. However, Z_L is in general a complex quantity, as in the case of a series RL

circuit, for example, for which $Z_L = R + j\omega L$. Hence, in general Γ may be complex also:

$$
\Gamma = |\Gamma|e^{j\theta_r}, \quad (2.50)
$$

where $|\Gamma|$ is the magnitude of Γ and θ_r is its phase angle. Note that $|\Gamma| \leq 1$.

A load is said to be matched to the line if $Z_L = Z_0$ *because then there will be no reflection by the load ($\Gamma = 0$ and $V_0^- = 0$).* On the other hand, when the load is an open circuit ($Z_L = \infty$), $\Gamma = 1$ and $V_0^- = V_0^+$, and when it is a short circuit ($Z_L = 0$), $\Gamma = -1$ and $V_0^- = -V_0^+$.

Example 2-2   Reflection Coefficient of a Series RC Load

A 100-Ω transmission line is connected to a load consisting of a 50-Ω resistor in series with a 10-pF capacitor. Find the reflection coefficient at the load for a 100-MHz signal.

Solution: The following quantities are given [Fig. 2-10]:

$$
R_L = 50\ \Omega, \qquad C_L = 10\ \text{pF} = 10^{-11}\ \text{F},
$$

$$
Z_0 = 100\ \Omega, \qquad f = 100\ \text{MHz} = 10^8\ \text{Hz}.
$$

The load impedance is

$$
Z_L = R_L - j/\omega C_L
$$
$$
= 50 - j\frac{1}{2\pi \times 10^8 \times 10^{-11}} = (50 - j159)\ \Omega.
$$

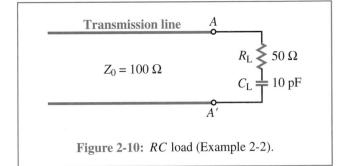

Figure 2-10: RC load (Example 2-2).

From Eq. (2.49a), the voltage reflection coefficient is given by

$$\Gamma = \frac{Z_L/Z_0 - 1}{Z_L/Z_0 + 1}$$

$$= \frac{0.5 - j1.59 - 1}{0.5 - j1.59 + 1}$$

$$= \frac{-0.5 - j1.59}{1.5 - j1.59} = \frac{-1.67e^{j72.6°}}{2.19e^{-j46.7°}} = -0.76e^{j119.3°}.$$

This result may be converted into a form with positive magnitude for Γ by replacing the minus sign with $e^{-j180°}$. Thus,

$$\Gamma = 0.76e^{j119.3°}e^{-j180°} = 0.76e^{-j60.7°} = 0.76\angle{-60.7°},$$

or

$$|\Gamma| = 0.76, \qquad \theta_r = -60.7°. \quad \blacksquare$$

Example 2-3 |Γ| for Purely Reactive Load

Show that $|\Gamma| = 1$ for a purely reactive load.

Solution: The load impedance of a purely reactive load is given by

$$Z_L = jX_L.$$

From Eq. (2.49a), the reflection coefficient is

$$\Gamma = \frac{Z_L - Z_0}{Z_L + Z_0}$$

$$= \frac{jX_L - Z_0}{jX_L + Z_0}$$

$$= \frac{-(Z_0 - jX_L)}{(Z_0 + jX_L)} = \frac{-\sqrt{Z_0^2 + X_L^2}\, e^{-j\theta}}{\sqrt{Z_0^2 + X_L^2}\, e^{j\theta}} = -e^{-j2\theta},$$

where $\theta = \tan^{-1} X_L/Z_0$. Hence

$$|\Gamma| = |-e^{-j2\theta}| = [(e^{-j2\theta})(e^{-j2\theta})^*]^{1/2} = 1. \quad \blacksquare$$

EXERCISE 2.7 A 50-Ω lossless transmission line is terminated in a load impedance $Z_L = (30 - j200)$ Ω. Calculate the voltage reflection coefficient at the load.

Ans. $\Gamma = 0.93\angle{-27.5°}$.

EXERCISE 2.8 A 150-Ω lossless line is terminated in a capacitor whose impedance is $Z_L = -j30$ Ω. Calculate Γ.

Ans. $\Gamma = 1\angle{-157.4°}$.

2-5.2 Standing Waves

Using the relation $V_0^- = \Gamma V_0^+$ in Eqs. (2.44a) and (2.44b) gives the expressions

$$\tilde{V}(z) = V_0^+(e^{-j\beta z} + \Gamma e^{j\beta z}), \qquad (2.51a)$$

$$\tilde{I}(z) = \frac{V_0^+}{Z_0}(e^{-j\beta z} - \Gamma e^{j\beta z}), \qquad (2.51b)$$

which now contain only one, yet to be determined, unknown, V_0^+. Before we proceed toward that goal, however, let us examine the physical meaning represented by these expressions. We begin by deriving an expression for $|\tilde{V}(z)|$, the magnitude of $\tilde{V}(z)$. Upon using Eq. (2.50) in (2.51a) and applying the relation $|\tilde{V}(z)| = [\tilde{V}(z)\,\tilde{V}^*(z)]^{1/2}$, where $\tilde{V}^*(z)$ is the complex conjugate of $\tilde{V}(z)$, we have

$$|\tilde{V}(z)| = \left\{ [V_0^+(e^{-j\beta z} + |\Gamma|e^{j\theta_r}e^{j\beta z})] \right.$$
$$\left. \cdot [(V_0^+)^*(e^{j\beta z} + |\Gamma|e^{-j\theta_r}e^{-j\beta z})] \right\}^{1/2}$$
$$= |V_0^+|\left[1 + |\Gamma|^2 \right.$$
$$\left. + |\Gamma|(e^{j(2\beta z + \theta_r)} + e^{-j(2\beta z + \theta_r)}) \right]^{1/2}$$
$$= |V_0^+|\left[1 + |\Gamma|^2 + 2|\Gamma|\cos(2\beta z + \theta_r) \right]^{1/2} (2.52)$$

where we have used the identity

$$e^{jx} + e^{-jx} = 2\cos x \qquad (2.53)$$

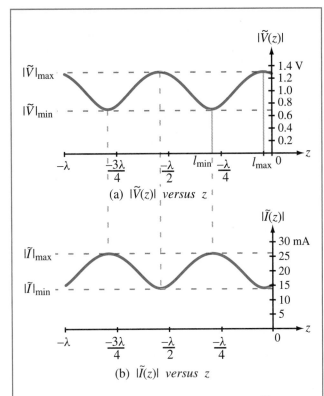

Figure 2-11: Standing-wave pattern for (a) $|\tilde{V}(z)|$ and (b) $|\tilde{I}(z)|$ for a lossless transmission line of characteristic impedance $Z_0 = 50 \ \Omega$, terminated in a load with a reflection coefficient $\Gamma = 0.3e^{j30°}$. The magnitude of the incident wave $|V_0^+| = 1$ V. The standing-wave ratio is $S = |\tilde{V}|_{max}/|\tilde{V}|_{min} = 1.3/0.7 = 1.86$.

for any real quantity x. By applying the same steps to Eq. (2.51b), a similar expression can be derived for $|\tilde{I}(z)|$, the magnitude of the current $\tilde{I}(z)$.

The variations of $|\tilde{V}(z)|$ and $|\tilde{I}(z)|$ as a function of z, the position on the line relative to the load at $z = 0$, are illustrated in Fig. 2-11 for a line with $|V_0^+| = 1$ V, $|\Gamma| = 0.3$, $\theta_r = 30°$, and $Z_0 = 50 \ \Omega$. The sinusoidal pattern is called a **standing wave**, and it is caused by the **interference** of the two waves. The maximum value of the **standing-wave pattern** of $|\tilde{V}(z)|$ corresponds

to the position on the line at which the incident and reflected waves are *in phase* $[2\beta z + \theta_r = -2n\pi$ in Eq. (2.52)] and therefore add constructively to give a value equal to $(1 + |\Gamma|)|V_0^+| = 1.3$ V. The minimum value of $|\tilde{V}(z)|$ corresponds to destructive interference, which occurs when the incident and reflected waves are in *phase opposition* $(2\beta z + \theta_r = -(2n + 1)\pi)$. In this case, $|\tilde{V}(z)| = (1 - |\Gamma|)|V_0^+| = 0.7$ V. Whereas the repetition period is λ for the incident and reflected waves individually, *the repetition period of the standing-wave pattern is $\lambda/2$.* The standing-wave pattern describes the spatial variation of the magnitude of $\tilde{V}(z)$ as a function of z. If one were to observe the variations of the instantaneous voltage as a function of time at any location z, corresponding to one of the maxima in the standing-wave pattern, for example, that variation would be as $\cos \omega t$ and would have an amplitude equal to 1.3 V [i.e., $v(t)$ would oscillate between -1.3 V and $+1.3$ V]. Similarly, the time oscillation of $v(z, t)$ at any location z will have an amplitude equal to $|\tilde{V}(z)|$ at that z.

Close inspection of the voltage and current standing-wave patterns shown in Fig. 2-11 reveals that the two patterns are in phase opposition (when one is at a maximum, the other is at a minimum, and vice versa). This is a consequence of the fact that the second term in Eq. (2.51a) is preceded by a plus sign, whereas the second term in Eq. (2.51b) is preceded by a negative sign.

The standing-wave patterns shown in Fig. 2-11 are for a typical situation with $\Gamma = 0.3 \, e^{j30°}$. The peak-to-peak variation of the pattern depends on $|\Gamma|$, which can vary between 0 and 1. For the special case of a matched line with $Z_L = Z_0$, we have $|\Gamma| = 0$ and $|\tilde{V}(z)| = |V_0^+|$ for all values of z, as shown in Fig. 2-12(a). *With no reflected wave present, there will be no interference and no standing waves.* The other end of the $|\Gamma|$ scale, at $|\Gamma| = 1$, corresponds to when the load is a short circuit $(\Gamma = -1)$ or an open circuit $(\Gamma = 1)$. The standing-wave patterns for these two cases are shown in Figs. 2-12(b) and (c), both of which have maxima equal to $2|V_0^+|$ and

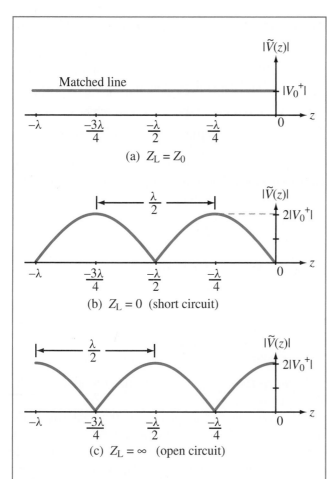

Figure 2-12: Voltage standing-wave patterns for (a) a matched load, (b) a short-circuited line, and (c) an open-circuited line.

then

$$|\widetilde{V}(z)| = |\widetilde{V}|_{\max} = |V_0^+|[1 + |\Gamma|], \qquad (2.54)$$

and this occurs when

$$2\beta z + \theta_r = -2\beta l_{\max} + \theta_r = -2n\pi, \qquad (2.55)$$

with $n = 0$ or a positive integer. Solving Eq. (2.55) for $l_{\max}$, we have

$$-z = l_{\max} = \frac{\theta_r + 2n\pi}{2\beta} = \frac{\theta_r \lambda}{4\pi} + \frac{n\lambda}{2},$$

$$\begin{cases} n = 1, 2, \ldots & \text{if } \theta_r < 0, \\ n = 0, 1, 2, \ldots & \text{if } \theta_r \ge 0, \end{cases} \qquad (2.56)$$

where we have used the relation $\beta = 2\pi/\lambda$. The phase angle of the voltage reflection coefficient, θ_r, is bounded between $-\pi$ and π radians. If $\theta_r \ge 0$, the *first voltage maximum* occurs at $l_{\max} = \theta_r \lambda/4\pi$, corresponding to $n = 0$, but if $\theta_r < 0$, the first physically meaningful maximum occurs at $l_{\max} = (\theta_r \lambda/4\pi) + \lambda/2$, corresponding to $n = 1$. Negative values of $l_{\max}$ correspond to locations "beyond" the load at the end of the line and therefore have no physical significance. As was mentioned earlier, *the locations on the line corresponding to voltage maxima also correspond to current minima, and vice versa.*

Similarly, the minimum values of $|\widetilde{V}(z)|$ occur at distances $l_{\min} = -z$ corresponding to when the argument of the cosine function in Eq. (2.52) is equal to $-(2n+1)\pi$, which gives the result

$$|\widetilde{V}|_{\min} = |V_0^+|[1 - |\Gamma|],$$

$$\text{when } (\theta_r - 2\beta l_{\min}) = -(2n + 1)\pi, \quad (2.57)$$

with $-\pi \le \theta_r \le \pi$. The first minimum corresponds to $n = 0$. The spacing between a maximum $l_{\max}$ and the

minima equal to zero, but the two patterns are shifted in z relative to each other by a distance of $\lambda/4$.

Now let us examine the maximum and minimum values of the voltage magnitude. From Eq. (2.52), $|\widetilde{V}(z)|$ is a maximum when the argument of the cosine function is equal to zero or multiples of 2π. Noting that the location on the line always corresponds to negative values of z (since the load is at $z = 0$), if we denote $l_{\max} = -z$ as the distance from the load at which $|\widetilde{V}(z)|$ is a maximum, adjacent minimum l_{min} is $\lambda/4$. Hence, the *first minimum* occurs at

$$l_{min} = \begin{cases} l_{max} + \lambda/4, & \text{if } l_{max} < \lambda/4, \\ l_{max} - \lambda/4, & \text{if } l_{max} \geq \lambda/4. \end{cases} \quad (2.58)$$

The ratio of $|\widetilde{V}|_{max}$ to $|\widetilde{V}|_{min}$ is called the *voltage standing-wave ratio S*, which from Eqs. (2.54) and (2.57) is given by

$$S = \frac{|\widetilde{V}|_{max}}{|\widetilde{V}|_{min}} = \frac{1 + |\Gamma|}{1 - |\Gamma|} \quad \text{(dimensionless).} \quad (2.59)$$

This quantity, which often is referred to by its acronym, VSWR, or the shorter acronym SWR, provides a measure of the mismatch between the load and the transmission line; for a matched load with $\Gamma = 0$, we get $S = 1$, and for a line with $|\Gamma| = 1$, $S = \infty$.

REVIEW QUESTIONS

Q2.5 The attenuation constant α represents ohmic losses. In view of the model given in Fig. 2-6(c), what should R' and G' be in order to have no losses? Verify your expectation through the expression for α given by Eq. (2.25a).

Q2.6 How is the wavelength λ of the wave traveling on the transmission line related to the free-space wavelength λ_0?

Q2.7 When is a load matched to the line? Why is it important?

Q2.8 What is a standing-wave pattern? Why is its period $\lambda/2$ and not λ?

Q2.9 What is the separation between the location of a voltage maximum and the adjacent current maximum on the line?

Example 2-4   Standing-wave Ratio

A 50-Ω transmission line is terminated in a load with $Z_L = (100 + j50)$ Ω. Find the voltage reflection coefficient and the voltage standing-wave ratio (SWR).

Solution: From Eq. (2.49a), Γ is given by

$$\Gamma = \frac{Z_L - Z_0}{Z_L + Z_0} = \frac{(100 + j50) - 50}{(100 + j50) + 50} = \frac{50 + j50}{150 + j50}.$$

Converting the numerator and denominator to polar form and then simplifying yields

$$\Gamma = \frac{70.7e^{j45°}}{158.1e^{j18.4°}} = 0.45e^{j26.6°}.$$

Using the definition for S given by Eq. (2.59), we have

$$S = \frac{1 + |\Gamma|}{1 - |\Gamma|} = \frac{1 + 0.45}{1 - 0.45} = 2.6. \quad \blacksquare$$

Example 2-5   Measuring Z_L

A *slotted-line* probe is an instrument used to measure the unknown impedance of a load, Z_L. A coaxial slotted line contains a narrow longitudinal slit in the outer conductor. A small probe inserted in the slit can be used to sample the magnitude of the electric field and, hence, the magnitude $|\widetilde{V}|$ of the voltage on the line (Fig. 2-13). By moving the probe along the length of the slotted line, it is possible to measure $|\widetilde{V}|_{max}$ and $|\widetilde{V}|_{min}$

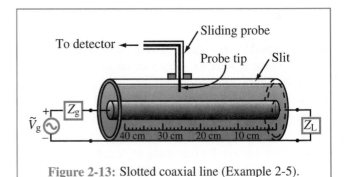

Figure 2-13: Slotted coaxial line (Example 2-5).

and the distances from the load at which they occur. Use of Eq. (2.59) then provides the voltage standing-wave ratio S. Measurements with a 50-Ω slotted line connected to an unknown load impedance determined that $S = 3$. The distance between successive voltage minima was found to be 30 cm, and the first voltage minimum was located at 12 cm from the load. Determine the load impedance Z_L.

Solution: The following quantities are given:

$$Z_0 = 50 \ \Omega, \quad S = 3, \quad l_{min} = 12 \text{ cm}.$$

Since the distance between successive voltage minima is equal to $\lambda/2$,

$$\lambda = 2 \times 0.3 = 0.6 \text{ m},$$

and

$$\beta = \frac{2\pi}{\lambda} = \frac{2\pi}{0.6} = \frac{10\pi}{3} \quad \text{(rad/m)}.$$

From Eq. (2.59), solving for $|\Gamma|$ in terms of S gives

$$|\Gamma| = \frac{S-1}{S+1} = \frac{3-1}{3+1} = 0.5.$$

Next, we use the condition given by Eq. (2.57) for the location of a voltage minimum to find θ_r:

$$\theta_r - 2\beta l_{min} = -\pi, \quad \text{for } n = 0 \text{ (first minimum)},$$

which gives

$$\theta_r = 2\beta l_{min} - \pi$$
$$= 2 \times \frac{10\pi}{3} \times 0.12 - \pi = -0.2\pi \text{ (rad)} = -36°.$$

Hence,

$$\Gamma = |\Gamma|e^{j\theta_r} = 0.5e^{-j36°} = 0.405 - j0.294.$$

Solving Eq. (2.49a) for Z_L, we have

$$Z_L = Z_0 \left[\frac{1+\Gamma}{1-\Gamma} \right]$$
$$= 50 \left[\frac{1 + 0.405 - j0.294}{1 - 0.405 + j0.294} \right] = (85 - j67) \ \Omega. \quad \blacksquare$$

EXERCISE 2.9 If $\Gamma = 0.5\angle{-60°}$ and $\lambda = 24$ cm, find the locations of the voltage maximum and minimum nearest to the load.

Ans. $l_{max} = 10$ cm, $l_{min} = 4$ cm.

EXERCISE 2.10 A 140-Ω lossless line is terminated in a load impedance $Z_L = (280 + j182) \ \Omega$. If $\lambda = 72$ cm, find (a) the reflection coefficient Γ, (b) the voltage standing-wave ratio S, (c) the locations of voltage maxima, and (d) the locations of voltage minima.

Ans. (a) $\Gamma = 0.5\angle{29°}$, (b) $S = 3.0$, (c) $l_{max} = 2.9$ cm$+$ $n\lambda/2$, (d) $l_{min} = 20.9$ cm$+n\lambda/2$, where $n = 0, 1, 2, \ldots$.

2-6 Input Impedance of the Lossless Line

The standing-wave patterns indicate that for a mismatched line the voltage and current magnitudes are oscillatory with position on the line and in phase opposition with each other. Hence, the voltage to current ratio, called the *input impedance* Z_{in}, must vary with position also. Using Eqs. (2.51a) and (2.51b), Z_{in} is given by

$$Z_{in}(z) = \frac{\tilde{V}(z)}{\tilde{I}(z)}$$
$$= \frac{V_0^+[e^{-j\beta z} + \Gamma e^{j\beta z}]}{V_0^+[e^{-j\beta z} - \Gamma e^{j\beta z}]} Z_0$$
$$= Z_0 \left[\frac{1 + \Gamma e^{j2\beta z}}{1 - \Gamma e^{j2\beta z}} \right] \quad (\Omega). \quad (2.60)$$

Note that $Z_{in}(z)$ *is the ratio of the total voltage (incident- and reflected-wave voltages) to the total current at any point z on the line, in contrast with the characteristic impedance of the line* Z_0*, which relates the voltage and current of each of the two waves individually* ($Z_0 = V_0^+/I_0^+ = -V_0^-/I_0^-$).

Of particular interest in many transmission-line problems is the input impedance at the input of the line at $z = -l$, which is given by

$$Z_{in}(-l) = Z_0 \left[\frac{e^{j\beta l} + \Gamma e^{-j\beta l}}{e^{j\beta l} - \Gamma e^{-j\beta l}} \right]$$

$$= Z_0 \left[\frac{1 + \Gamma e^{-j2\beta l}}{1 - \Gamma e^{-j2\beta l}} \right]. \qquad (2.61)$$

By replacing Γ with Eq. (2.49a) and using the relations

$$e^{j\beta l} = \cos\beta l + j\sin\beta l, \qquad (2.62a)$$

$$e^{-j\beta l} = \cos\beta l - j\sin\beta l, \qquad (2.62b)$$

Eq. (2.61) can be rewritten in terms of Z_L as

$$Z_{in}(-l) = Z_0 \left(\frac{Z_L \cos\beta l + j Z_0 \sin\beta l}{Z_0 \cos\beta l + j Z_L \sin\beta l} \right)$$

$$= Z_0 \left(\frac{Z_L + j Z_0 \tan\beta l}{Z_0 + j Z_L \tan\beta l} \right). \qquad (2.63)$$

From the standpoint of the generator circuit, the transmission line can be replaced with an impedance Z_{in}, as shown in Fig. 2-14. The phasor voltage across Z_{in} is given by

$$\tilde{V}_i = \tilde{I}_i Z_{in} = \frac{\tilde{V}_g Z_{in}}{Z_g + Z_{in}}, \qquad (2.64)$$

but from the standpoint of the transmission line, the voltage across it at the input of the line is given by Eq. (2.51a) with $z = -l$:

$$\tilde{V}_i = \tilde{V}(-l) = V_0^+[e^{j\beta l} + \Gamma e^{-j\beta l}]. \qquad (2.65)$$

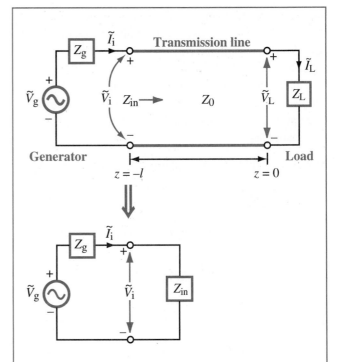

Figure 2-14: At the generator end, the terminated transmission line can be replaced with the input impedance of the line Z_{in}.

Equating Eq. (2.64) to Eq. (2.65) and then solving for V_0^+ leads to the result

$$V_0^+ = \left(\frac{\tilde{V}_g Z_{in}}{Z_g + Z_{in}} \right) \left(\frac{1}{e^{j\beta l} + \Gamma e^{-j\beta l}} \right). \qquad (2.66)$$

This completes the solution of the transmission-line wave equations, given by Eqs. (2.21) and (2.23), for the special case of a lossless transmission line. We started out with the general solutions given by Eqs. (2.26a) and (2.26b), which included four unknown amplitudes, V_0^+, V_0^-, I_0^+, and I_0^-. We then found out that $Z_0 = V_0^+/I_0^+ = -V_0^-/I_0^-$, thereby reducing the unknowns to the two voltage amplitudes only. Upon applying the boundary condition at the load, we were able to relate V_0^- to V_0^+ through Γ, and, finally, by applying the boundary condition at the sending end of the line, we obtained an expression for V_0^+.

Example 2-6 **Complete Solution for $v(z,t)$ and $i(z,t)$**

A 1.05-GHz generator circuit with series impedance $Z_g = 10\ \Omega$ and voltage source given by

$$v_g(t) = 10\sin(\omega t + 30°) \quad (V)$$

is connected to a load $Z_L = (100 + j50)\ \Omega$ through a 50-Ω, 67-cm-long lossless transmission line. The phase velocity of the line is $0.7c$, where c is the velocity of light in a vacuum. Find $v(z,t)$ and $i(z,t)$ on the line.

Solution: From the relationship $u_p = \lambda f$, we find the wavelength:

$$\lambda = \frac{u_p}{f} = \frac{0.7 \times 3 \times 10^8}{1.05 \times 10^9} = 0.2\ \text{m},$$

and

$$\tan(\beta l) = \tan\left(\frac{2\pi}{\lambda}l\right)$$

$$= \tan\left(\frac{2\pi}{0.2} \times 0.67\right)$$

$$= \tan 6.7\pi = \tan 0.7\pi = \tan 126°,$$

where we have subtracted multiples of 2π. The voltage reflection coefficient at the load is

$$\Gamma = \frac{Z_L - Z_0}{Z_L + Z_0} = \frac{(100 + j50) - 50}{(100 + j50) + 50} = 0.45e^{j26.6°}.$$

With reference to Fig. 2-14, the input impedance of the line, given by Eq. (2.63), is

$$Z_{\text{in}} = Z_0\left[\frac{Z_L + jZ_0\tan\beta l}{Z_0 + jZ_L\tan\beta l}\right]$$

$$= Z_0\left[\frac{Z_L/Z_0 + j\tan\beta l}{1 + j(Z_L/Z_0)\tan\beta l}\right]$$

$$= 50\left[\frac{(2 + j1) + j\tan 126°}{1 + j(2 + j1)\tan 126°}\right] = (21.9 + j17.4)\ \Omega.$$

Rewriting the expression for the generator voltage with the cosine reference, we have

$$v_g(t) = 10\sin(\omega t + 30°)$$

$$= 10\cos(\pi/2 - \omega t - 30°)$$

$$= 10\cos(\omega t - 60°)$$

$$= \Re[10e^{-j60°}e^{j\omega t}] = \Re[\tilde{V}_g e^{j\omega t}] \quad (V).$$

Hence, the phasor voltage $\tilde{V}_g$ is given by

$$\tilde{V}_g = 10\,e^{-j60°}\ (V) = 10\angle{-60°} \quad (V).$$

Application of Eq. (2.66) gives

$$V_0^+ = \left(\frac{\tilde{V}_g Z_{\text{in}}}{Z_g + Z_{\text{in}}}\right)\left(\frac{1}{e^{j\beta l} + \Gamma e^{-j\beta l}}\right)$$

$$= \left[\frac{10e^{-j60°}(21.9 + j17.4)}{10 + 21.9 + j17.4}\right]$$

$$\cdot \left[e^{j126°} + 0.45e^{j26.6°}e^{-j126°}\right]^{-1}$$

$$= 10.2e^{j159°}\ (V) = 10.2\angle{159°} \quad (V).$$

The phasor voltage on the line is then

$$\tilde{V}(z) = V_0^+(e^{-j\beta z} + \Gamma e^{j\beta z})$$

$$= 10.2e^{j159°}(e^{-j\beta z} + 0.45e^{j26.6°}e^{j\beta z}),$$

and the instantaneous voltage $v(z, t)$ is

$$v(z, t) = \Re[\tilde{V}(z)e^{j\omega t}]$$
$$= 10.2 \cos(\omega t - \beta z + 159°)$$
$$+ 4.55 \cos(\omega t + \beta z + 185.6°) \quad (V).$$

Similarly, use of V_0^+ in Eq. (2.51b) leads to

$$\tilde{I}(z) = 0.20 e^{j159°}(e^{-j\beta z} - 0.45 e^{j26.6°} e^{j\beta z}),$$
$$i(z, t) = 0.20 \cos(\omega t - \beta z + 159°)$$
$$+ 0.091 \cos(\omega t + \beta z + 5.6°) \quad (A). \quad \blacksquare$$

2-7 Special Cases of the Lossless Line

We often encounter situations involving lossless transmission lines with particular terminations or lines whose lengths exhibit particularly useful properties. We shall now consider some of these special cases.

2-7.1 Short-Circuited Line

The transmission line shown in Fig. 2-15(a) is terminated in a short circuit, $Z_L = 0$. Consequently, the voltage reflection coefficient defined by Eq. (2.49a) is $\Gamma = -1$, and the voltage standing-wave ratio given by Eq. (2.59) is $S = \infty$. From Eqs. (2.51a) and (2.51b), the voltage and current on a short-circuited lossless transmission line are given by

$$\tilde{V}_{sc}(z) = V_0^+[e^{-j\beta z} - e^{j\beta z}] = -2jV_0^+ \sin \beta z, (2.67a)$$
$$\tilde{I}_{sc}(z) = \frac{V_0^+}{Z_0}[e^{-j\beta z} + e^{j\beta z}] = \frac{2V_0^+}{Z_0} \cos \beta z. \quad (2.67b)$$

The voltage $\tilde{V}_{sc}(z)$ is zero at the load ($z = 0$), as it should be for a short circuit, and its amplitude varies as $\sin \beta z$, whereas the current $\tilde{I}_{sc}(z)$ is a maximum at the load and it varies as $\cos \beta z$. Both quantities are displayed in Fig. 2-15 as a function of negative z.

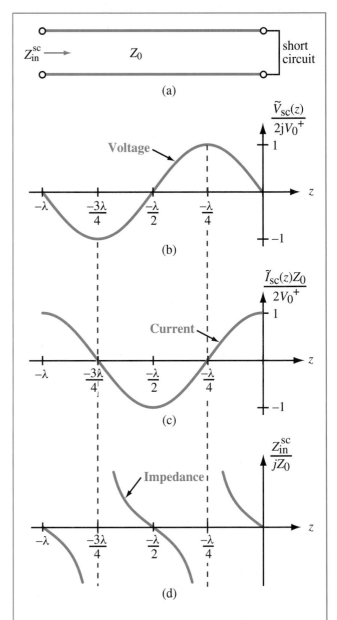

Figure 2-15: Transmission line terminated in a short circuit: (a) schematic representation, (b) normalized voltage on the line, (c) normalized current, and (d) normalized input impedance.

The input impedance of the line at $z = -l$ is given by the ratio of $\tilde{V}_{sc}(-l)$ to $\tilde{I}_{sc}(-l)$. Denoting Z_{in}^{sc} as the input impedance for a short-circuited line, we have

$$Z_{in}^{sc} = \frac{\tilde{V}_{sc}(-l)}{\tilde{I}_{sc}(-l)} = jZ_0 \tan\beta l. \quad (2.68)$$

A plot of Z_{in}^{sc}/jZ_0 versus negative z is shown in Fig. 2-15(d).

In general, the input impedance Z_{in} may consist of a real part, or input resistance R_{in}, and an imaginary part, or input reactance X_{in}:

$$Z_{in} = R_{in} + jX_{in}. \quad (2.69)$$

In the case of the short-circuited lossless line, the input impedance is purely reactive ($R_{in} = 0$). If $\tan\beta l \geq 0$, the line appears inductive, acting like an equivalent inductor L_{eq} whose impedance is equal to Z_{in}^{sc}. Thus,

$$j\omega L_{eq} = jZ_0 \tan\beta l, \quad \text{if } \tan\beta l \geq 0, \quad (2.70a)$$

or

$$L_{eq} = \frac{Z_0 \tan\beta l}{\omega} \quad \text{(H)}. \quad (2.70b)$$

The minimum line length l that would result in an input impedance Z_{in}^{sc} equivalent to that of an inductor of inductance L_{eq} is

$$l = \frac{1}{\beta} \tan^{-1}\left(\frac{\omega L_{eq}}{Z_0}\right) \quad \text{(m)}. \quad (2.70c)$$

Similarly, if $\tan\beta l \leq 0$, the input impedance is capacitive, in which case the line acts like an equivalent capacitor C_{eq} such that

$$\frac{1}{j\omega C_{eq}} = jZ_0 \tan\beta l, \quad \text{if } \tan\beta l \leq 0, \quad (2.71a)$$

or

$$C_{eq} = -\frac{1}{Z_0 \omega \tan\beta l} \quad \text{(F)}. \quad (2.71b)$$

Since l is a positive number, the shortest length l for which $\tan\beta l \leq 0$ corresponds to the range $\pi/2 \leq \beta l \leq \pi$. Hence, the minimum line length l that would result in an input impedance Z_{in}^{sc} equivalent to that of a capacitor of capacitance C_{eq} is

$$l = \frac{1}{\beta}\left[\pi - \tan^{-1}\left(\frac{1}{\omega C_{eq} Z_0}\right)\right] \quad \text{(m)}. \quad (2.71c)$$

These results mean that, through proper choice of the length of a short-circuited line, we can make substitutes for capacitors and inductors with any desired reactance. Such a practice is indeed common in the design of microwave circuits and high-speed integrated circuits, because making an actual capacitor or inductor is often more difficult than making a shorted transmission line.

Example 2-7   Equivalent Reactive Elements

Choose the length of a shorted 50-Ω lossless transmission line (Fig. 2-16) such that its input impedance at 2.25 GHz is equivalent to the reactance of a capacitor with capacitance $C_{eq} = 4$ pF. The wave velocity on the line is $0.75c$.

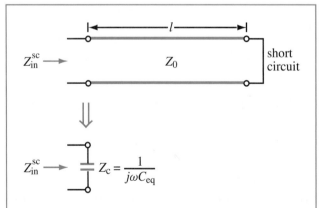

Figure 2-16: Shorted line as equivalent capacitor (Example 2-7).

Solution: We are given

$$u_p = 0.75c = 0.75 \times 3 \times 10^8 = 2.25 \times 10^8 \text{ m/s},$$

$$Z_0 = 50 \text{ }\Omega,$$

$$f = 2.25 \text{ GHz} = 2.25 \times 10^9 \text{ Hz},$$

$$C_{eq} = 4 \text{ pF} = 4 \times 10^{-12} \text{ F}.$$

The phase constant is

$$\beta = \frac{2\pi}{\lambda} = \frac{2\pi f}{u_p} = \frac{2\pi \times 2.25 \times 10^9}{2.25 \times 10^8} = 62.8 \quad \text{(rad/m)}.$$

From Eq. (2.71a),

$$\tan \beta l = -\frac{1}{Z_0 \omega C_{eq}}$$

$$= -\frac{1}{50 \times 2\pi \times 2.25 \times 10^9 \times 4 \times 10^{-12}} = -0.354.$$

The tangent function is negative when its argument is in the second or fourth quadrants. The solution for the second quadrant is

$$\beta l_1 = 2.8 \text{ rad} \quad \text{or} \quad l_1 = \frac{2.8}{\beta} = \frac{2.8}{62.8} = 4.46 \text{ cm},$$

and the solution for the fourth quadrant is

$$\beta l_2 = 5.94 \text{ rad} \quad \text{or} \quad l_2 = \frac{5.94}{62.8} = 9.46 \text{ cm}.$$

We also could have obtained the value of l_1 by applying Eq. (2.71c). The length l_2 is greater than l_1 by exactly $\lambda/2$. In fact, any length $l = 4.46$ cm $+ n\lambda/2$, where n is a positive integer, is also a solution. ∎

2-7.2   Open-Circuited Line

With $Z_L = \infty$, as illustrated in Fig. 2-17(a), we have $\Gamma = 1$, $S = \infty$, and the voltage, current, and input impedance are given by

$$\tilde{V}_{oc}(z) = V_0^+ [e^{-j\beta z} + e^{j\beta z}] = 2V_0^+ \cos \beta z, \qquad (2.72a)$$

$$\tilde{I}_{oc}(z) = \frac{V_0^+}{Z_0}[e^{-j\beta z} - e^{j\beta z}] = \frac{-2jV_0^+}{Z_0} \sin \beta z, (2.72b)$$

$$\boxed{Z_{in}^{oc} = \frac{\tilde{V}_{oc}(-l)}{\tilde{I}_{oc}(-l)} = -jZ_0 \cot \beta l. \quad (2.73)}$$

Plots of these quantities are displayed in Fig. 2-17 as a function of negative z.

2-7.3   Application of Short-Circuit and Open-Circuit Measurements

A network analyzer is a radio-frequency (RF) instrument capable of measuring the impedance of any load connected to its input terminal. When used to measure Z_{in}^{sc}, the input impedance of a lossless line terminated in a short circuit, and again Z_{in}^{oc}, the input impedance of the line when terminated in an open circuit, the combination of the two measurements can be used to determine the characteristic impedance of the line Z_0 and its phase constant β. The product of Eqs. (2.68) and (2.73) gives the result

$$\boxed{Z_0 = +\sqrt{Z_{in}^{sc} \, Z_{in}^{oc}}, \quad (2.74)}$$

and the ratio of the same equations leads to

$$\boxed{\tan \beta l = \sqrt{\frac{-Z_{in}^{sc}}{Z_{in}^{oc}}}. \quad (2.75)}$$

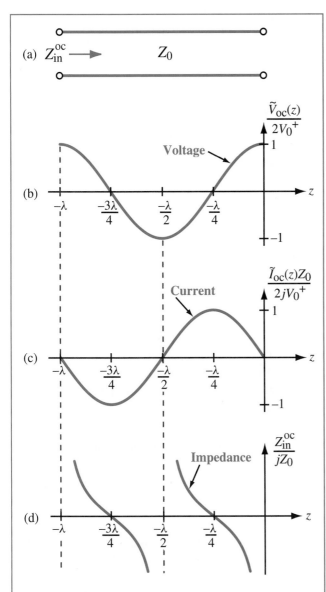

Figure 2-17: Transmission line terminated in an open circuit: (a) schematic representation, (b) normalized voltage on the line, (c) normalized current, and (d) normalized input impedance.

Because of the π phase ambiguity associated with the tangent function, the length l should be less than or equal to $\lambda/2$ to provide an unambiguous result.

Example 2-8 **Measuring Z_0 and β**

Find Z_0 and β of a 57-cm-long lossless transmission line whose input impedance was measured as $Z_{in}^{sc} = j40.42 \ \Omega$ when terminated in a short circuit and as $Z_{in}^{oc} = -j121.24 \ \Omega$ when terminated in an open circuit. From other measurements, we know that the line is between 3 and 3.25 wavelengths long.

Solution: From Eqs. (2.74) and (2.75),

$$Z_0 = \sqrt[+]{Z_{in}^{sc} \ Z_{in}^{oc}} = \sqrt{(j40.42)(-j121.24)} = 70 \ \Omega,$$

$$\tan \beta l = \sqrt{\frac{-Z_{in}^{sc}}{Z_{in}^{oc}}} = \sqrt{\frac{1}{3}}.$$

Since l is between 3λ and 3.25λ, $\beta l = (2\pi l/\lambda)$ is between 6π radians and $(13\pi/2)$ radians. This places βl in the first quadrant (0 to $\pi/2$) in a polar coordinate system. Hence, the only acceptable solution for the above equation is $\beta l = \pi/6$ radians. This value, however, does not include the 2π multiples associated with the integer λ multiples of l. Hence, the true value of βl is

$$\beta l = 6\pi + \frac{\pi}{6} = 19.4 \quad \text{(rad)},$$

in which case

$$\beta = \frac{19.4}{0.57} = 34 \quad \text{(rad/m).} \ \blacksquare$$

2-7.4 Lines of Length $l = n\lambda/2$

If $l = n\lambda/2$, where n is an integer,

$$\tan \beta l = \tan [(2\pi/\lambda)(n\lambda/2)] = \tan n\pi = 0.$$

Consequently, Eq. (2.63) reduces to

$$Z_{\text{in}} = Z_{\text{L}}, \quad \text{for } l = n\lambda/2, \quad (2.76)$$

which means that a half-wavelength line (or any integer multiple of $\lambda/2$) does not modify the load impedance. Thus, *a generator connected to a load through a half-wavelength lossless line would induce the same voltage across the load and current through it as when the line is not there.*

2-7.5 Quarter-Wave Transformer

Another case of interest is when the length of the line is a quarter-wavelength (or $\lambda/4 + n\lambda/2$, where $n = 0$ or a positive integer), corresponding to $\beta l = (2\pi/\lambda)(\lambda/4) = \pi/2$. From Eq. (2.63), the input impedance becomes

$$Z_{\text{in}} = \frac{Z_0^2}{Z_{\text{L}}}, \quad \text{for } l = \lambda/4 + n\lambda/2. \quad (2.77)$$

The utility of such a quarter-wave transformer is illustrated by the problem in Example 2-9.

Example 2-9   Quarter-Wave Transformer

A 50-Ω lossless transmission line is to be matched to a resistive load impedance with $Z_{\text{L}} = 100\ \Omega$ via a quarter-wave section as shown in Fig. 2-18, thereby eliminating reflections along the feedline. Find the characteristic impedance of the quarter-wave transformer.

Solution: To eliminate reflections at terminal AA', the input impedance Z_{in} looking into the quarter-wave line

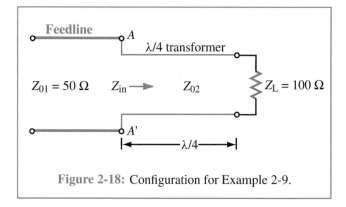

Figure 2-18: Configuration for Example 2-9.

should be equal to Z_{01}, the characteristic impedance of the feedline. Thus, $Z_{\text{in}} = 50\ \Omega$. From Eq. (2.77),

$$Z_{\text{in}} = \frac{Z_{02}^2}{Z_{\text{L}}},$$

or

$$Z_{02} = \sqrt{Z_{\text{in}} Z_{\text{L}}} = \sqrt{50 \times 100} = 70.7\ \Omega.$$

Whereas this eliminates reflections on the feedline, it does not eliminate them on the $\lambda/4$ line. However, since the lines are lossless, all the incident power will end up getting transferred into the load Z_{L}. ∎

2-7.6 Matched Transmission Line: $Z_{\text{L}} = Z_0$

For a matched lossless transmission line with $Z_{\text{L}} = Z_0$, (1) the input impedance $Z_{\text{in}} = Z_0$ for all locations z on the line, (2) $\Gamma = 0$, and (3) all the incident power is delivered to the load, regardless of the line length l. A summary of the properties of standing waves is given in Table 2-3.

REVIEW QUESTIONS

Q2.10 What is the difference between the characteristic impedance Z_0 and the input impedance Z_{in}? When are they the same?

Table 2-3: Properties of standing waves on a lossless transmission line.

Voltage maximum	$|\tilde{V}|_{\max} = |V_0^+|[1 + |\Gamma|]$
Voltage minimum	$|\tilde{V}|_{\min} = |V_0^+|[1 - |\Gamma|]$
Positions of voltage maxima (also positions of current minima)	$l_{\max} = \dfrac{\theta_r \lambda}{4\pi} + \dfrac{n\lambda}{2}, \quad n = 0, 1, 2, \ldots$
Position of first maximum (also position of first current minimum)	$l_{\max} = \begin{cases} \dfrac{\theta_r \lambda}{4\pi}, & \text{if } 0 \leq \theta_r \leq \pi \\ \dfrac{\theta_r \lambda}{4\pi} + \dfrac{\lambda}{2}, & \text{if } -\pi \leq \theta_r \leq 0 \end{cases}$
Positions of voltage minima (also positions of first current maxima)	$l_{\min} = \dfrac{\theta_r \lambda}{4\pi} + \dfrac{(2n+1)\lambda}{4}, \quad n = 0, 1, 2, \ldots$
Position of first minimum (also position of first current maximum)	$l_{\min} = \dfrac{\lambda}{4}\left(1 + \dfrac{\theta_r}{\pi}\right)$
Input impedance	$Z_{\text{in}} = Z_0 \left(\dfrac{Z_L + j Z_0 \tan \beta l}{Z_0 + j Z_L \tan \beta l} \right)$
Positions at which Z_{in} is real	at voltage maxima and minima
Z_{in} at voltage maxima	$Z_{\text{in}} = Z_0 \left(\dfrac{1 + |\Gamma|}{1 - |\Gamma|} \right)$
Z_{in} at voltage minima	$Z_{\text{in}} = Z_0 \left(\dfrac{1 - |\Gamma|}{1 + |\Gamma|} \right)$
Z_{in} of short-circuited line	$Z_{\text{in}}^{\text{sc}} = j Z_0 \tan \beta l$
Z_{in} of open-circuited line	$Z_{\text{in}}^{\text{oc}} = -j Z_0 \cot \beta l$
Z_{in} of line of length $l = n\lambda/2$	$Z_{\text{in}} = Z_L, \quad n = 0, 1, 2, \ldots$
Z_{in} of line of length $l = \lambda/4 + n\lambda/2$	$Z_{\text{in}} = Z_0^2/Z_L, \quad n = 0, 1, 2, \ldots$
Z_{in} of matched line	$Z_{\text{in}} = Z_0$
$|V_0^+|$ = amplitude of incident wave, $\Gamma = |\Gamma| e^{j\theta_r}$ with $-\pi < \theta_r < \pi$; θ_r in radians.	

Q2.11 What is a quarter-wave transformer? How is it used?

Q2.12 A lossless transmission line of length l is terminated in a short circuit. If $l < \lambda/4$, is the input impedance inductive or capacitive?

Q2.13 What is the input impedance of an infinitely long line?

Q2.14 If the input impedance of a lossless line is inductive when terminated in a short circuit, will it be inductive or capacitive when the line is terminated in an open circuit?

EXERCISE 2.11 A 50-Ω lossless transmission line uses an insulating material with $\varepsilon_r = 2.25$. When terminated in an open circuit, how long should the line be for its input impedance to be equivalent to a 10-pF capacitor at 50 MHz?

Ans. $l = 9.92$ cm.

EXERCISE 2.12 A 300-Ω feedline is to be connected to a 3-m long, 150-Ω line terminated in a 150-Ω resistor. Both lines are lossless and use air as the insulating material, and the operating frequency is 50 MHz. Determine (a) the input impedance of the 3-m long line, (b) the voltage standing-wave ratio on the feedline, and (c) the characteristic impedance of a quarter-wave transformer were it to be used between the two lines in order to achieve $S = 1$ on the feedline.

Ans. (a) $Z_{in} = 150$ Ω, (b) $S = 2$, (c) $Z_0 = 212.1$ Ω.

2-8   Power Flow on a Lossless Transmission Line

Our discussion thus far has focused on the voltage and current aspects of wave propagation on a transmission line. Now we shall examine the flow of power carried by the incident and reflected waves. We begin by

reintroducing Eqs. (2.51a) and (2.51b), the general expressions for the voltage and current phasors on a lossless transmission line:

$$\tilde{V}(z) = V_0^+ (e^{-j\beta z} + \Gamma e^{j\beta z}), \qquad (2.78a)$$

$$\tilde{I}(z) = \frac{V_0^+}{Z_0} (e^{-j\beta z} - \Gamma e^{j\beta z}). \qquad (2.78b)$$

In these expressions, the first terms represent the incident-wave voltage and current, and the terms involving Γ represent the reflected-wave voltage and current. At the load ($z = 0$), the incident and reflected voltages and currents are

$$\tilde{V}^i = V_0^+, \qquad \tilde{I}^i = \frac{V_0^+}{Z_0}, \qquad \text{(at } z = 0), \quad (2.79)$$

$$\tilde{V}^r = \Gamma V_0^+, \qquad \tilde{I}^r = -\Gamma \frac{V_0^+}{Z_0}, \qquad \text{(at } z = 0). \quad (2.80)$$

2-8.1   Instantaneous Power

The instantaneous power carried by the incident wave, *as it arrives at the load*, is equal to the product of the instantaneous voltage $v^i(t)$ and the instantaneous current $i^i(t)$,

$$\begin{aligned}
P^i(t) &= v^i(t) \cdot i^i(t) \\
&= \mathfrak{Re}[\tilde{V}^i e^{j\omega t}] \cdot \mathfrak{Re}\left[\tilde{I}^i e^{j\omega t} \right] \\
&= \mathfrak{Re}[|V_0^+| e^{j\phi^+} e^{j\omega t}] \cdot \mathfrak{Re}\left[\frac{|V_0^+|}{Z_0} e^{j\phi^+} e^{j\omega t} \right] \\
&= |V_0^+| \cos(\omega t + \phi^+) \cdot \frac{|V_0^+|}{Z_0} \cos(\omega t + \phi^+) \\
&= \frac{|V_0^+|^2}{Z_0} \cos^2(\omega t + \phi^+) \qquad \text{(W)}, \qquad (2.81)
\end{aligned}$$

where use was made of Eq. (2.31a) to express V_0^+ in terms of its magnitude $|V_0^+|$ and phase angle ϕ^+.

Similarly, upon replacing Γ in Eq. (2.80) with $|\Gamma| e^{j\theta_r}$ and then following the same steps, we obtain the following expression for the instantaneous power reflected by the load:

$$P^r(t) = v^r(t) \cdot i^r(t)$$

$$= -|\Gamma|^2 \frac{|V_0^+|^2}{Z_0} \cos^2(\omega t + \phi^+ + \theta_r) \quad (W). \quad (2.82)$$

The negative sign in Eq. (2.82) signifies the fact that the reflected power flows in the $-z$-direction.

2-8.2 Time-Average Power

From a practical standpoint, we usually are more interested in the time-average power flow along the transmission line, P_{av}, than in the instantaneous power $P(t)$. To compute P_{av}, we can use a time-domain approach or a computationally simpler phasor-domain approach. For completeness, we will consider both.

Time-Domain Approach

The time-average power flow is equal to the instantaneous power averaged over one time period $T = 1/f = 2\pi/\omega$. For the incident wave, the time-average power is

$$P^i_{av} = \frac{1}{T} \int_0^T P^i(t)\, dt = \frac{\omega}{2\pi} \int_0^{2\pi/\omega} P^i(t)\, dt. \quad (2.83)$$

Upon inserting Eq. (2.81) for the incident power $P^i(t)$ into Eq. (2.83) and performing the integration, we obtain the result

$$P^i_{av} = \frac{|V_0^+|^2}{2Z_0} \quad (W). \quad (2.84)$$

The factor of $1/2$ is a consequence of the integration of $\cos^2(\omega t + \phi^+)$ over one period. A similar treatment for the reflected wave gives

$$P^r_{av} = -|\Gamma|^2 \frac{|V_0^+|^2}{2Z_0} = -|\Gamma|^2 P^i_{av}. \quad (2.85)$$

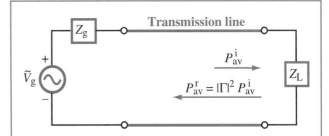

Figure 2-19: The time-average power reflected by a load connected to a lossless transmission line is equal to the incident power multiplied by $|\Gamma|^2$.

Thus, the magnitude of the average reflected power is equal to the average incident power, reduced by a multiplicative factor of $|\Gamma|^2$.

The *net average power delivered to the load* shown in Fig. 2-19 is

$$P_{av} = P^i_{av} + P^r_{av} = \frac{|V_0^+|^2}{2Z_0}[1 - |\Gamma|^2] \quad (W). \quad (2.86)$$

Phasor-Domain Approach

For any propagating wave with voltage and current phasors $\widetilde{V}$ and $\widetilde{I}$, a useful formula for computing the time-average power is

$$P_{av} = \tfrac{1}{2} \mathfrak{Re}[\widetilde{V} \cdot \widetilde{I}^*], \quad (2.87)$$

where $\widetilde{I}^*$ is the complex conjugate of $\widetilde{I}$. Application of this formula to Eqs. (2.79) and (2.80) gives

$$P^i_{av} = \tfrac{1}{2} \mathfrak{Re}\left[V_0^+ \cdot \frac{V_0^{+*}}{Z_0} \right] = \frac{|V_0^+|^2}{2Z_0}, \quad (2.88)$$

$$P^r_{av} = \tfrac{1}{2} \mathfrak{Re}\left[\Gamma V_0^+ \cdot \left(\frac{-\Gamma^* V_0^{+*}}{Z_0} \right) \right]$$

$$= -|\Gamma|^2 \frac{|V_0^+|^2}{2Z_0} , \qquad (2.89)$$

which are, respectively, identical to the expressions given by Eqs. (2.84) and (2.85).

EXERCISE 2.13 For a 50-Ω lossless transmission line terminated in a load impedance $Z_L = (100 + j50)$ Ω, determine the fraction of the average incident power reflected by the load.

Ans. 20%.

EXERCISE 2.14 For the line of Exercise 2.13, what is the magnitude of the average reflected power if $|V_0^+| = 1$ V?

Ans. $P_{av}^r = 2$ (mW).

REVIEW QUESTIONS

Q2.15 According to Eq. (2.82), the instantaneous value of the reflected power depends on the phase of the reflection coefficient θ_r, but the average reflected power given by Eq. (2.85) does not. Explain.

Q2.16 What is the average power delivered by a lossless transmission line to a reactive load?

Q2.17 What fraction of the incident power is delivered to a matched load?

Q2.18 Verify that

$$\frac{1}{T} \int_0^T \cos^2 \left(\frac{2\pi t}{T} + \phi \right) dt = \frac{1}{2} ,$$

regardless of the value of ϕ.

2-9 The Smith Chart

Prior to the age of computers and programmable calculators, several types of charts were developed to assist in the solution of transmission-line problems. The *Smith chart*, which was developed by P. H. Smith in 1939, has been and continues to be the most widely used graphical technique for analyzing and designing transmission-line circuits. Even though the original intent of its inventor was to provide a useful graphical tool for performing calculations involving complex impedances, the Smith chart has become a principal presentation medium in computer-aided design (CAD) software for displaying the performance of microwave circuits. As the material in this and the next section will demonstrate, use of the Smith chart not only avoids tedious manipulations of complex numbers, but it also allows an engineer to design impedance-matching circuits with relative ease. The Smith chart can be used for both lossy and lossless transmission lines. In the present treatment, however, we will confine our discussion to the lossless case.

2-9.1 Parametric Equations

The reflection coefficient Γ is, in general, a complex quantity composed of a magnitude $|\Gamma|$ and a phase angle θ_r or, equivalently, a real part Γ_r and an imaginary part Γ_i,

$$\Gamma = |\Gamma|e^{j\theta_r} = \Gamma_r + j\Gamma_i , \qquad (2.90)$$

where

$$\Gamma_r = |\Gamma| \cos \theta_r, \qquad (2.91a)$$

$$\Gamma_i = |\Gamma| \sin \theta_r. \qquad (2.91b)$$

The Smith chart lies in the complex plane of Γ. In Fig. 2-20, point A represents a reflection coefficient $\Gamma_A = 0.3 + j0.4$ or, equivalently,

$$|\Gamma_A| = [(0.3)^2 + (0.4)^2]^{1/2} = 0.5$$

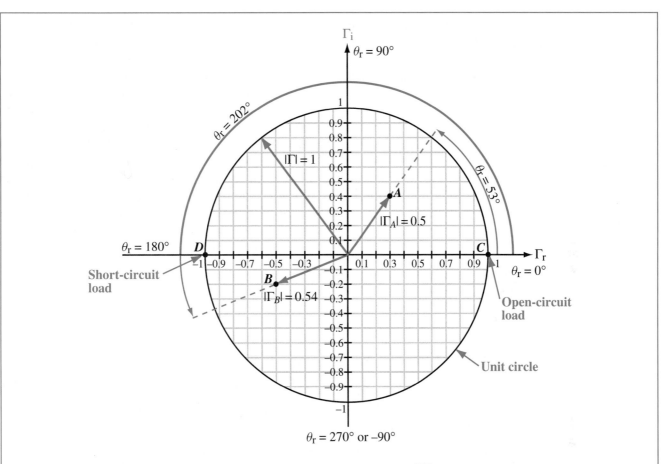

Figure 2-20: The complex Γ plane. Point A is at $\Gamma_A = 0.3 + j0.4 = 0.5e^{j53°}$, and point B is at $\Gamma_B = -0.5 - j0.2 = |0.54|e^{j202°}$. The unit circle corresponds to $|\Gamma| = 1$. At point C, $\Gamma = 1$, corresponding to an open-circuit load, and at point D, $\Gamma = -1$, corresponding to a short circuit.

and

$$\theta_r = \tan^{-1}(0.4/0.3) = 53°.$$

Similarly, point B represents $\Gamma_B = -0.5 - j0.2$, or $|\Gamma_B| = 0.54$ and $\theta_r = 202°$ [or, equivalently, $\theta_r = (360° - 202°) = -158°$]. Note that when both Γ_r and Γ_i are negative numbers θ_r is in the third quadrant in the Γ_r–Γ_i plane. Thus, when using $\theta = \tan^{-1}(\Gamma_i/\Gamma_r)$ to

compute θ_r, it may be necessary to add or subtract 180° to obtain the correct value of θ_r.

The *unit circle* shown in Fig. 2-20 corresponds to $|\Gamma| = 1$. Because $|\Gamma| \leq 1$ for a transmission line, only that part of the Γ_r–Γ_i plane that lies within the unit circle has physical meaning; hence, future drawings will be limited to the domain contained within the unit circle.

Impedances on a Smith chart are represented by

normalized values, with Z_0, the characteristic impedance of the line, serving as the normalization constant. *Normalized impedances are denoted by lowercase letters,* as in $z = Z/Z_0$. The *normalized load impedance* is then given by

$$z_L = Z_L/Z_0 \quad \text{(dimensionless)}, \qquad (2.92)$$

and the reflection coefficient Γ, defined by Eq. (2.49a), can be written as

$$\Gamma = \frac{Z_L/Z_0 - 1}{Z_L/Z_0 + 1} = \frac{z_L - 1}{z_L + 1}. \qquad (2.93)$$

The inverse relation of Eq. (2.93) is

$$\boxed{z_L = \frac{1 + \Gamma}{1 - \Gamma}. \qquad (2.94)}$$

The normalized load impedance z_L is, in general, a complex quantity composed of a *normalized load resistance* r_L and a *normalized load reactance* x_L:

$$z_L = r_L + jx_L. \qquad (2.95)$$

Using Eqs. (2.90) and (2.95) in Eq. (2.94), we have

$$r_L + jx_L = \frac{(1 + \Gamma_r) + j\Gamma_i}{(1 - \Gamma_r) - j\Gamma_i}, \qquad (2.96)$$

which can be solved to obtain explicit expressions for r_L and x_L in terms of Γ_r and Γ_i. This is accomplished by multiplying the numerator and denominator of the right-hand side of Eq. (2.96) by the complex conjugate of the denominator and then separating the result into real and imaginary parts. These steps lead to

$$r_L = \frac{1 - \Gamma_r^2 - \Gamma_i^2}{(1 - \Gamma_r)^2 + \Gamma_i^2}, \qquad (2.97a)$$

$$x_L = \frac{2\Gamma_i}{(1 - \Gamma_r)^2 + \Gamma_i^2}. \qquad (2.97b)$$

These expressions state that for a given set of values for Γ_r and Γ_i there corresponds a unique set of values for r_L

and x_L. However, if we fix the value of r_L, say at 2, many possible combinations of values can be assigned to Γ_r and Γ_i, each of which can give the same value of r_L. For example, $(\Gamma_r, \Gamma_i) = (0.33, 0)$ gives $r_L = 2$, as does the combination $(\Gamma_r, \Gamma_i) = (0.5, 0.29)$, as well as an infinite number of other combinations. In fact, if we were to plot in the Γ_r–Γ_i plane all the possible combinations of Γ_r and Γ_i corresponding to $r_L = 2$, we would obtain the circle denoted by $r_L = 2$ in Fig. 2-21. Similar circles apply to other values of r_L, and within the $|\Gamma| = 1$ domain all these circles pass through the point $(\Gamma_r, \Gamma_i) = (1, 0)$. After some algebraic manipulations, Eq. (2.97a) can be rearranged to give the following *parametric equation* for the circle in the Γ_r–Γ_i plane corresponding to a given value of r_L:

$$\left(\Gamma_r - \frac{r_L}{1 + r_L}\right)^2 + \Gamma_i^2 = \left(\frac{1}{1 + r_L}\right)^2. \qquad (2.98)$$

The standard equation for a circle in the x–y plane with center at (x_0, y_0) and radius a is given by

$$(x - x_0)^2 + (y - y_0)^2 = a^2. \qquad (2.99)$$

Comparison of Eq. (2.98) with Eq. (2.99) shows that the r_L circle is centered at $\Gamma_r = r_L/(1 + r_L)$ and $\Gamma_i = 0$, and its radius is $1/(1 + r_L)$. The largest circle shown in Fig. 2-21 corresponds to $r_L = 0$, which is also the unit circle corresponding to $|\Gamma| = 1$. This is to be expected, because when $r_L = 0$, $|\Gamma| = 1$ regardless of the magnitude of x_L.

A similar examination of the expression for x_L given by Eq. (2.97b) also leads to an equation for a circle given by

$$(\Gamma_r - 1)^2 + \left(\Gamma_i - \frac{1}{x_L}\right)^2 = \left(\frac{1}{x_L}\right)^2, \qquad (2.100)$$

but the x_L circles in the Γ_r–Γ_i plane exhibit a different character from that for r_L. To start with, the normalized reactance x_L may assume both positive and negative values, whereas the normalized resistance cannot be negative (negative resistances are physically

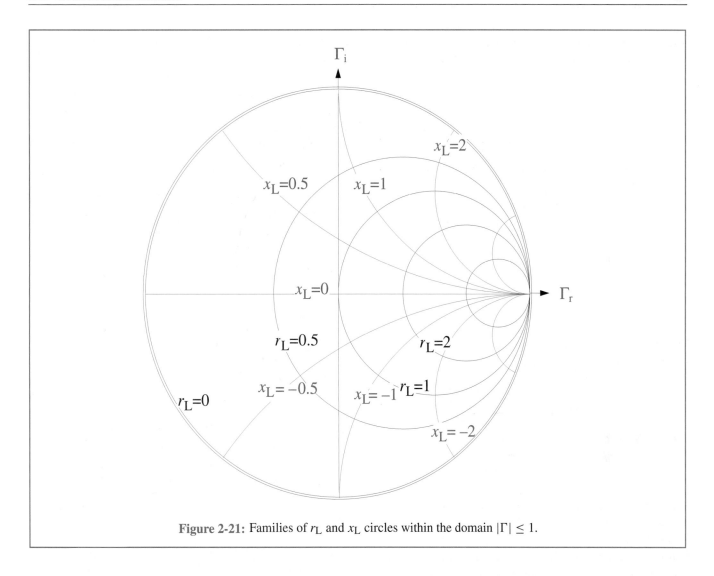

Figure 2-21: Families of r_L and x_L circles within the domain $|\Gamma| \leq 1$.

meaningless). Hence, Eq. (2.100) can generate two families of circles, one family corresponding to positive values of x_L and another corresponding to negative values of x_L. Furthermore, as shown in Fig. 2-21, only part of a given circle falls within the bounds of the unit circle. The families of circles of the two parametric equations given by Eqs. (2.98) and (2.100) plotted for selected values of r_L and x_L constitute the Smith chart shown in Fig. 2-22. A given point on the Smith chart, such as point P in Fig. 2-22, represents a normalized load impedance $z_L = 2 - j1$, with a corresponding voltage reflection coefficient $\Gamma = 0.45 \exp(-j26.6°)$. The magnitude $|\Gamma| = 0.45$ is obtained by dividing the length of the line between the center of the Smith chart and the point P by

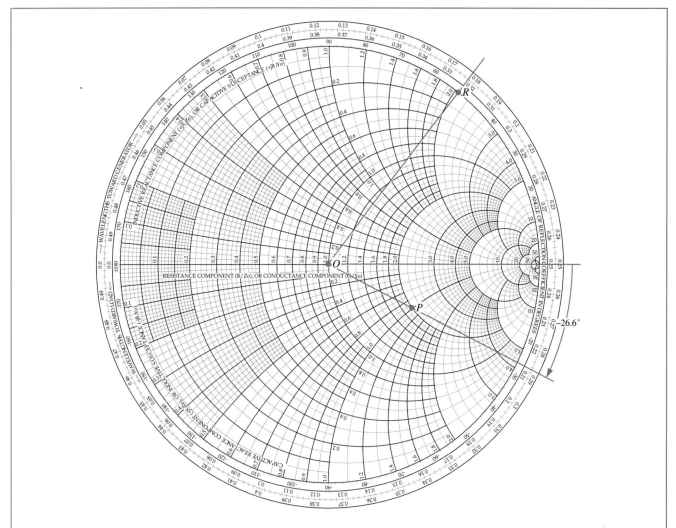

Figure 2-22: Point P represents a normalized load impedance $z_L = 2 - j1$. The reflection coefficient has a magnitude $|\Gamma| = \overline{OP}/\overline{OR} = 0.45$ and an angle $\theta_r = -26.6°$. Point R is an arbitrary point on the $r_L = 0$ circle (which also is the $|\Gamma| = 1$ circle).

the length of the line between the center of the Smith chart and the edge of the unit circle (the radius of the unit circle corresponds to $|\Gamma| = 1$). The perimeter of the Smith chart contains three concentric scales. The innermost scale is labeled *angle of reflection coefficient in degrees*. This is the scale for θ_r. As indicated in Fig. 2-22, $\theta_r = -26.6°$ for point P. The meanings and uses of the other two scales are discussed next.

EXERCISE 2.15 Use the Smith chart to find the values of Γ corresponding to the following normalized load impedances: (a) $z_L = 2 + j0$, (b) $z_L = 1 - j1$, (c) $z_L = 0.5 - j2$, (d) $z_L = -j3$, (e) $z_L = 0$ (short circuit), (f) $z_L = \infty$ (open circuit), (g) $z_L = 1$ (matched load).

Ans. (a) $\Gamma = 0.33$, (b) $\Gamma = 0.45\angle{-63.4°}$, (c) $\Gamma = 0.83\angle{-50.9°}$, (d) $\Gamma = 1\angle{-36.9°}$, (e) $\Gamma = -1$, (f) $\Gamma = 1$, (g) $\Gamma = 0$.

2-9.2 Input Impedance

From Eq. (2.61), the input impedance looking toward the load at a distance l from the load is given by

$$Z_{in} = Z_0 \left[\frac{1 + \Gamma e^{-j2\beta l}}{1 - \Gamma e^{-j2\beta l}} \right] \quad (\Omega). \quad (2.101)$$

To use the Smith chart, we always normalize impedances to the characteristic impedance Z_0. Hence, the *normalized input impedance* is

$$z_{in} = \frac{Z_{in}}{Z_0} = \frac{1 + \Gamma e^{-j2\beta l}}{1 - \Gamma e^{-j2\beta l}} \quad \text{(dimensionless)}. \quad (2.102)$$

The quantity $\Gamma = |\Gamma| e^{j\theta_r}$ is the voltage reflection coefficient at the load. Let us define

$$\Gamma_l = \Gamma e^{-j2\beta l} = |\Gamma| e^{j\theta_r} e^{-j2\beta l} = |\Gamma| e^{j(\theta_r - 2\beta l)} \quad (2.103)$$

as the *phase-shifted* voltage reflection coefficient, meaning that Γ_l has the same magnitude as Γ, but the phase of Γ_l is shifted by $2\beta l$ relative to that of Γ. In terms of Γ_l, Eq. (2.102) can be rewritten as

$$z_{in} = \frac{1 + \Gamma_l}{1 - \Gamma_l}. \quad (2.104)$$

The form of Eq. (2.104) is identical with that for z_L given by Eq. (2.94):

$$z_L = \frac{1 + \Gamma}{1 - \Gamma}. \quad (2.105)$$

The similarity in form suggests that, if Γ is transformed into Γ_l, z_L gets transformed into z_{in}. On the Smith chart, transforming Γ into Γ_l means maintaining $|\Gamma|$ constant and decreasing the phase θ_r by $2\beta l$, which corresponds to rotation in a clockwise direction on the Smith chart. Noting that a complete rotation around the Smith chart is equal to a phase change of 2π, the length l corresponding to such a change is obtained from

$$2\beta l = 2 \frac{2\pi}{\lambda} l = 2\pi, \quad (2.106)$$

or $l = \lambda/2$. The outermost scale around the perimeter of the Smith chart (Fig. 2-22), called the *wavelengths toward generator* (WTG) scale, has been constructed to denote movement on the transmission line toward the generator, in units of the wavelength λ. That is, l is measured in wavelengths, and one complete rotation corresponds to $l = \lambda/2$. In some transmission-line problems, it may be necessary to move from some point on the transmission line toward another point closer to the load, in which case the phase is increased, which corresponds to rotation in a counterclockwise direction. For convenience, the Smith chart contains a third scale around its perimeter (in between the θ_r scale and the WTG scale) for accommodating such a need. It is called the *wavelengths toward load* (WTL) scale.

To illustrate how the Smith chart is used to find Z_{in}, let us consider a 50-Ω lossless transmission line terminated in a load impedance $Z_L = (100 - j50)$ Ω. Our objective is to find Z_{in} at a distance $l = 0.1\lambda$ from the load. The normalized load impedance is $z_L = Z_L/Z_0 = 2 - j1$, and it is denoted by point A on the Smith chart shown in Fig. 2-23. On the WTG scale, the location of point A is at 0.287λ. Using a compass, a circle is drawn through point A, with the center of the circle being at the center of the Smith chart. Since the center of the Smith chart is the intersection point of the Γ_r and Γ_i axes, all points on the drawn circle have the same value of $|\Gamma|$. This is called the *constant-$|\Gamma|$ circle*, or more commonly the *SWR circle*. The reason for this second name is that the

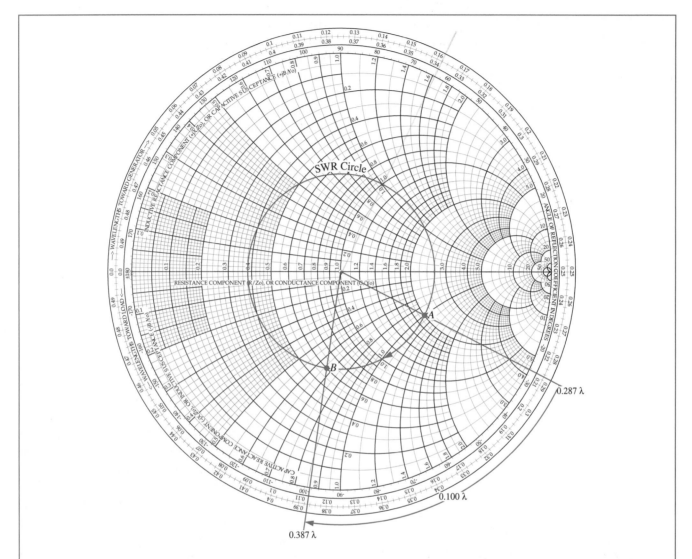

Figure 2-23: Point A represents a normalized load $z_L = 2 - j1$ at 0.287λ on the WTG scale. Point B represents the line input at 0.1λ from the load. At B, $z_{in} = 0.6 - j0.66$.

voltage standing-wave ratio (SWR) is related to $|\Gamma|$ by Eq. (2.59) as

$$S = \frac{1 + |\Gamma|}{1 - |\Gamma|}. \qquad (2.107)$$

Thus, a constant value of $|\Gamma|$ corresponds to a specific value for S. As was stated earlier, to transform z_L to z_{in}, we need to maintain $|\Gamma|$ constant, which means staying on the SWR circle, and to decrease the phase of Γ by $2\beta l$. This is equivalent to moving a distance $l = 0.1\lambda$ toward the generator on the WTG scale. Since the location of point A is at 0.287λ, we need to move to location $0.287\lambda + 0.1\lambda = 0.387\lambda$ on the WTG scale. A radial line through this new position on the WTG scale intersects the SWR circle at point B. This point represents z_{in}, and its value is $z_{in} = 0.6 - j0.66$. Finally, we unnormalize z_{in} by multiplying it by $Z_0 = 50\ \Omega$ to get $Z_{in} = (30 - j33)\ \Omega$. This result can be verified analytically using Eq. (2.101). The points between points A and B on the SWR circle represent different points along the transmission line.

EXERCISE 2.16 Use the Smith chart to find the normalized input impedance of a lossless line of length l terminated in a normalized load impedance z_L for each of the following combinations: (a) $l = 0.25\lambda$, $z_L = 1 + j0$, (b) $l = 0.5\lambda$, $z_L = 1 + j1$, (c) $l = 0.3\lambda$, $z_L = 1 - j1$, (d) $l = 1.2\lambda$, $z_L = 0.5 - j0.5$, (e) $l = 0.1\lambda$, $z_L = 0$ (short circuit), (f) $l = 0.4\lambda$, $z_L = j3$, (g) $l = 0.2\lambda$, $z_L = \infty$ (open circuit).

Ans. (a) $z_{in} = 1 + j0$, (b) $z_{in} = 1 + j1$, (c) $z_{in} = 0.76 + j0.84$, (d) $z_{in} = 0.59 + j0.66$, (e) $z_{in} = 0 + j0.73$, (f) $z_{in} = 0 + j0.72$, (g) $z_{in} = 0 - j0.32$.

2-9.3 SWR, Voltage Maxima and Minima

Consider a load with $z_L = 2 + j1$. Figure 2-24 shows a Smith chart with a SWR circle drawn through z_L (point A). The SWR circle intersects the real axis (Γ_r) at two points, designated P_{max} and P_{min}. Thus, at both points $\Gamma_i = 0$ and $\Gamma = \Gamma_r$. Also, on the real axis, the imaginary part of the load impedance $x_L = 0$. From the definition of Γ,

$$\Gamma = \frac{z_L - 1}{z_L + 1}, \qquad (2.108)$$

points P_{max} and P_{min} correspond to the special case

$$\Gamma = \Gamma_r = \frac{r_L - 1}{r_L + 1} \quad (\text{for } \Gamma_i = 0), \qquad (2.109)$$

with P_{min} corresponding to the condition when $r_L < 1$ and P_{max} corresponding to the condition when $r_L > 1$. Rewriting Eq. (2.107) for $|\Gamma|$ in terms of S, we have

$$|\Gamma| = \frac{S - 1}{S + 1} . \qquad (2.110)$$

For points P_{max} and P_{min}, $|\Gamma| = \Gamma_r$; hence

$$\Gamma_r = \frac{S - 1}{S + 1} . \qquad (2.111)$$

The similarity in form of Eqs. (2.109) and (2.111) suggests that $S = r_L$. However, since by definition $S \geq 1$, only point P_{max} (for which $r_L > 1$) satisfies the similarity condition. In Fig. 2-24, $r_L = 2.6$ at P_{max}; hence $S = 2.6$. In other words, S is *numerically equal to the value of* r_L *at* P_{max}, *the point at which the SWR circle intersects the real* Γ *axis on the right-hand side of the chart's center.*

The points P_{min} and P_{max} also represent the distances from the load at which the magnitude of the voltage on the line, $|\widetilde{V}|$, is a minimum and a maximum, respectively. This statement is easily demonstrated by considering the definition of Γ_l given by Eq. (2.103). At point P_{max}, the total phase of Γ_l, that is, $(\theta_r - 2\beta l)$, is equal to zero (if $\theta_r > 0$) or 2π (if $\theta_r < 0$), which is the condition corresponding to $|\widetilde{V}|_{max}$, as indicated by Eq. (2.55). Similarly, at P_{min} the total phase of Γ_l is equal to π, which is the condition for $|\widetilde{V}|_{min}$. Thus, for the transmission line represented by the SWR circle shown in Fig. 2-24, the distance between the load and the nearest voltage maximum is $l_{max} = 0.037\lambda$, obtained by moving clockwise from the load at point A to point P_{max}, and the distance to the nearest voltage minimum is $l_{min} = 0.287\lambda$, corresponding to the clockwise rotation from A to P_{min}. Since the location of $|\widetilde{V}|_{max}$ is also the location of $|\widetilde{I}|_{min}$ and the location of $|\widetilde{V}|_{min}$ is also the location of $|\widetilde{I}|_{max}$, the Smith chart provides a convenient way to determine the distances to all maxima and minima on the line (the standing-wave pattern has a repetition period of $\lambda/2$).

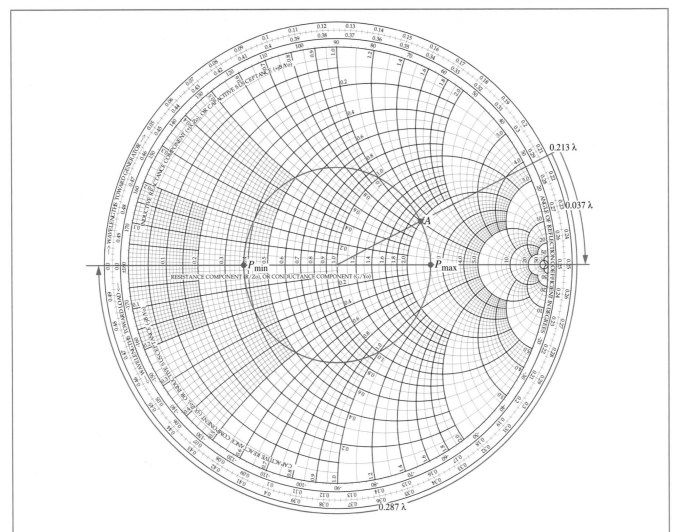

Figure 2-24: Point A represents a normalized load with $z_{\mathrm{L}} = 2 + j1$. The standing wave ratio is $S = 2.6$ (at $P_{\max}$), the distance between the load and the first voltage maximum is $l_{\max} = (0.25 - 0.213)\lambda = 0.037\lambda$, and the distance between the load and the first voltage minimum is $l_{\min} = (0.037 + 0.25)\lambda = 0.287\lambda$.

2-9.4 Impedance to Admittance Transformations

In solving certain types of transmission line problems, it is often more convenient to work with admittances than with impedances. Any impedance Z is in general a complex quantity consisting of a resistance R and a reactance X:

$$Z = R + jX \quad (\Omega). \qquad (2.112)$$

The *admittance* Y corresponding to Z is the reciprocal of Z:

$$Y = \frac{1}{Z} = \frac{1}{R + jX} = \frac{R - jX}{R^2 + X^2} \quad \text{(S)}. \qquad (2.113)$$

The real part of Y is called the *conductance* G, and the imaginary part of Y is called the *susceptance* B. That is,

$$Y = G + jB \quad \text{(S)} \qquad (2.114)$$

Comparison of Eq. (2.114) with Eq. (2.113) leads to

$$G = \frac{R}{R^2 + X^2} \quad \text{(S)}, \qquad (2.115a)$$

$$B = \frac{-X}{R^2 + X^2} \quad \text{(S)}. \qquad (2.115b)$$

A normalized impedance z is defined as the ratio of Z to Z_0, the characteristic impedance of the line. The same concept applies to the definition of the *normalized admittance* y; that is,

$$y = \frac{Y}{Y_0} = \frac{G}{Y_0} + j\frac{B}{Y_0} = g + jb \quad \text{(dimensionless)},$$
$$\qquad (2.116)$$

where $Y_0 = 1/Z_0$ is the *characteristic admittance of the line* and

$$g = \frac{G}{Y_0} = GZ_0 \quad \text{(dimensionless)}, \quad (2.117a)$$

$$b = \frac{B}{Y_0} = BZ_0 \quad \text{(dimensionless)}. \quad (2.117b)$$

The lowercase quantities g and b represent the *normalized conductance* and *normalized susceptance* of y, respectively. Of course, the normalized admittance y is the reciprocal of the normalized impedance z,

$$y = \frac{Y}{Y_0} = \frac{Z_0}{Z} = \frac{1}{z}. \qquad (2.118)$$

Accordingly, using Eq. (2.105), the normalized load admittance y_L is given by

$$y_L = \frac{1}{z_L} = \frac{1 - \Gamma}{1 + \Gamma} \quad \text{(dimensionless)}. \quad (2.119)$$

Now let us consider the normalized input impedance z_{in} at a distance $l = \lambda/4$ from the load. Using Eq. (2.102) with $2\beta l = 4\pi l/\lambda = 4\pi\lambda/4\lambda = \pi$ gives

$$z_{in}(l = \lambda/4) = \frac{1 + \Gamma e^{-j\pi}}{1 - \Gamma e^{-j\pi}} = \frac{1 - \Gamma}{1 + \Gamma} = y_L. \quad (2.120)$$

Thus, *rotation by $\lambda/4$ on the Smith chart transforms z_L into y_L.* In Fig. 2-25, the points representing z_L and y_L are diametrically opposite to each other on the SWR circle. In fact, such a transformation on the Smith chart can be used to determine any normalized admittance from its corresponding normalized impedance, and vice versa.

The Smith chart can be used with normalized impedances or with normalized admittances. As an impedance chart, the Smith chart consists of r_L and x_L circles, representing the normalized resistance and reactance of a normalized load impedance z_L. When used as an admittance chart, the r_L circles become g_L circles and the x_L circles become b_L circles, where g_L and b_L are the normalized conductance and susceptance of the normalized load admittance y_L, respectively.

Example 2-10   Smith Chart Calculations

A 50-Ω lossless transmission line is terminated in a load impedance $Z_L = (25 + j50)$ Ω. Use the Smith chart to find (a) the voltage reflection coefficient, (b) the voltage standing-wave ratio, (c) the distances of the first voltage maximum and first voltage minimum from the load, (d) the input impedance of the line, given that the line is 3.3λ long, and (e) the input admittance of the line.

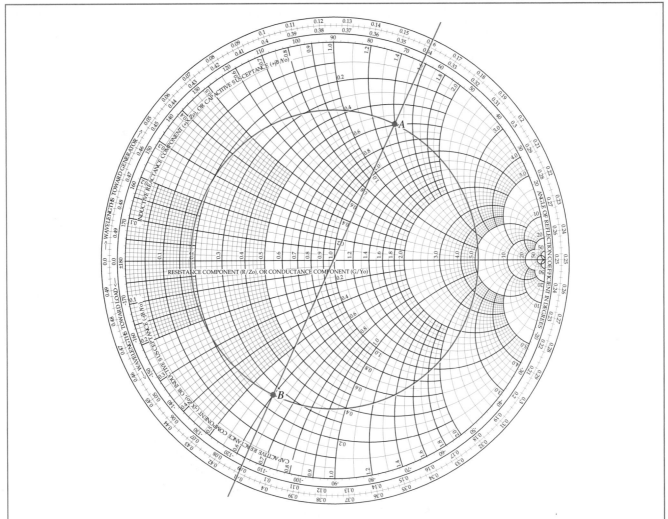

Figure 2-25: Point A represents a normalized load $z_L = 0.6 + j1.4$. Its corresponding normalized admittance is $y_L = 0.25 - j0.6$, and it is at point B.

Solution: (a) The normalized load impedance is

$$z_L = \frac{Z_L}{Z_0} = \frac{25 + j50}{50} = 0.5 + j1,$$

which is marked as point A on the Smith chart in Fig. 2-26. Using a ruler, a radial line is drawn from the center of the chart at point O through point A, outward to the outer perimeter of the chart. The line crosses the scale labeled "angle of reflection coefficient in degrees" at $\theta_r = 83°$. Next, a ruler is used to measure the length d_A

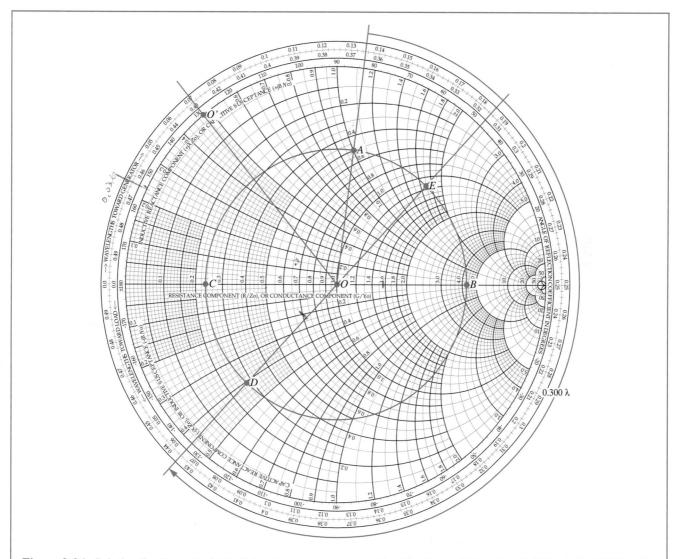

Figure 2-26: Solution for Example 2-10. Point A represents a normalized load $z_L = 0.5 + j1$ at 0.135λ on the WTG scale. At A, $\theta_r = 83°$ and $|\Gamma| = d_A/d_{O'} = \overline{OA}/\overline{OO'} = 0.62$. At B, the standing-wave ratio is $S = 4.26$. The distance from A to B gives $l_{max} = 0.115\lambda$ and from A to C gives $l_{min} = 0.365\lambda$. Point D represents the normalized input impedance $z_{in} = 0.28 - j0.40$, and point E represents the normalized input admittance $y_{in} = 1.15 + j1.7$.

of the line between points O and A and the length $d_{O'}$ of the line between points O and O', where O' is an arbitrary point on the $r_L = 0$ circle. The length $d_{O'}$ is equal to the radius of the $|\Gamma| = 1$ circle. The magnitude

of Γ is then obtained from $|\Gamma| = d_A/d_{O'} = 0.62$. Hence,

$$\Gamma = 0.62e^{j83°} = 0.62\underline{/83°}. \qquad (2.121)$$

(b) Using a compass, the SWR circle with center at

point O is drawn through point A. The circle crosses the Γ_r axis at points B and C. The value of r_L at point B is 4.26, which also is equal to S. Thus,

$$S = 4.26.$$

(c) The first voltage maximum is at point B on the SWR circle, which is at location 0.25λ on the WTG scale. The load, represented by point A, is at 0.135λ on the WTG scale. Hence, the distance between the load and the first voltage maximum is

$$l_{max} = (0.25 - 0.135)\lambda = 0.115\lambda.$$

The first voltage minimum is at point C. Moving on the WTG scale between A and C gives

$$l_{min} = (0.5 - 0.135)\lambda = 0.365\lambda,$$

which is 0.25λ past l_{max}.

(d) The line is 3.3λ long; subtracting multiples of 0.5λ leaves 0.3λ. From the load at 0.135λ on the WTG scale, the input of the line is at $(0.135+0.3)\lambda = 0.435\lambda$. This is labeled as point D on the SWR circle, and the normalized impedance is read as

$$z_{in} = 0.28 - j0.40,$$

and therefore

$$Z_{in} = z_{in}Z_0 = (0.28 - j0.40)50 = (14 - j20) \ \Omega.$$

(e) The normalized input admittance y_{in} is found by moving 0.25λ on the Smith chart to the image point of z_{in} across the circle, labeled point E on the SWR circle. The coordinates of point E give

$$y_{in} = 1.15 + j1.7,$$

and the corresponding input admittance is

$$Y_{in} = y_{in}Y_0 = \frac{y_{in}}{Z_0} = \frac{1.15 + j1.7}{50} = (0.023+j0.034) \text{ S.} \quad \blacksquare$$

Example 2-11   Determining Z_L Using the Smith Chart

This problem is similar to Example 2-5, except now we demonstrate the solution using the Smith chart.

Given that the voltage standing-wave ratio is $S = 3$ on a 50-Ω line, that the first voltage minimum occurs at 5 cm from the load, and that the distance between successive minima is 20 cm, find the load impedance.

Solution: The distance between successive minima is equal to $\lambda/2$. Hence, $\lambda = 40$ cm. In wavelength units, the first voltage minimum is at

$$l_{min} = \frac{5}{40} = 0.125\lambda.$$

Point A on the Smith chart in Fig. 2-27 corresponds to $r_L = S = 3$. Using a compass, the constant S circle is drawn through point A. Point B corresponds to the locations of voltage minima. Upon moving 0.125λ from point B toward the load on the WTL scale (counterclockwise), we arrive at point C, which represents the location of the load. The normalized load impedance at point C is

$$z_L = 0.6 - j0.8.$$

Multiplying by $Z_0 = 50 \ \Omega$, we obtain

$$Z_L = 50(0.6 - j0.8) = (30 - j40) \ \Omega. \quad \blacksquare$$

REVIEW QUESTIONS

Q2.19 The outer perimeter of the Smith chart represents what value of $|\Gamma|$? What point on the Smith chart represents a matched load?

Q2.20 What is the SWR circle? What quantity is constant at all points on the SWR circle?

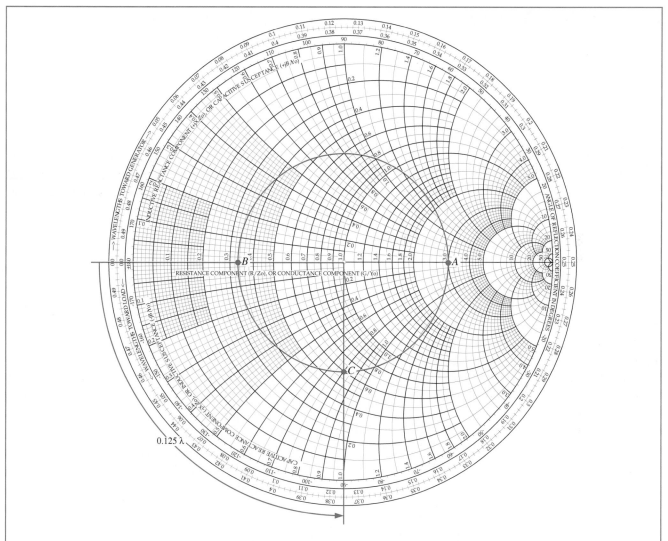

Figure 2-27: Solution for Example 2-11. Point *A* denotes that *S* = 3, point *B* represents the location of the voltage minimum, and point *C* represents the load at 0.125λ on the WTL scale from point *B*. At *C*, $z_L = 0.6 - j0.8$.

Q2.21 What line length corresponds to one complete rotation around the Smith chart? Why?

Q2.22 What points on the SWR circle correspond to the locations of the voltage maxima and minima on the line and why?

Q2.23 Given a normalized impedance z_L, how do you use the Smith chart to find the corresponding normalized admittance $y_L = 1/z_L$?

2-10 Impedance Matching

A transmission line usually connects a generator circuit at one end to a load at the other end. The load may be an antenna or any circuit with an equivalent input impedance Z_L. *The transmission line is said to be matched to the load when its characteristic impedance $Z_0 = Z_L$*, in which case no reflection occurs at the load end of the line. Since the primary uses of transmission lines are to transfer power and to transmit coded signals (such as digital data), a matched load ensures that the power delivered to the load is a maximum.

The simplest solution to matching a load to a transmission line is to design the load circuit such that its impedance $Z_L = Z_0$. Unfortunately, this may not be possible in practice because the load circuit may have to satisfy other requirements. An alternative solution is to place an *impedance-matching network* between the load and the transmission line as illustrated in Fig. 2-28. The purpose of the matching network is to eliminate reflections at the juncture MM' between the transmission line and the network. This is achieved by designing the matching network to exhibit an impedance equal to Z_0 at MM' when looking into the network from the transmission line side. If the network is lossless, then all the power going into it will end up in the load. Matching networks may consist of lumped elements (and in order to avoid ohmic losses only capacitors and inductors are used) or of sections of transmission lines with appropriate lengths and terminations. We will demonstrate the latter approach using a *single-stub matching network*.

The matching network is intended to match a load impedance $Z_L = R_L + jX_L$ to a lossless transmission line with characteristic impedance Z_0. This means that the network has to transform the real part of the load impedance from R_L at the load to Z_0 at MM' in Fig. 2-28 and to transform the reactive part from X_L at the load to zero at MM'. To achieve these two transformations, the matching network has to have at least two degrees of freedom; that is, at least two adjustable parameters. The single-stub matching network shown in Fig. 2-29 consists of two transmission line sections, one of length d connecting the load to the feedline at MM' and another of length l connected in parallel with the other two lines at MM'. This second line is called a *stub*, and it is usually terminated in either a short circuit or open circuit. The stub shown in Fig. 2-29 has a short-circuit termination.

The required two degrees of freedom are provided by the length l of the stub and the distance d from the load to the stub position. Because at MM' the stub is added in "parallel" to the line (and hence called a *shunt* stub), it is easier to work with admittances than with impedances. The matching procedure consists of two basic steps. In the first step, the distance d is selected so as to transform the load admittance $Y_L = 1/Z_L$ into an admittance of the form $Y_d = Y_0 + jB$, when looking toward the load at MM'. Then, in the second step, the length l of the stub line is selected so that its input admittance Y_s at MM' is equal to $-jB$. The parallel sum of the two admittances at MM' yields Y_0, the characteristic admittance of the line. The procedure is illustrated by Example 2-12.

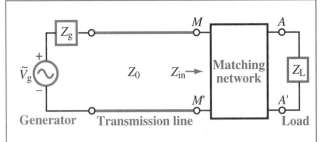

Figure 2-28: The function of a matching network is to transform the load impedance Z_L such that the input impedance Z_{in} looking into the network is equal to Z_0 of the transmission line.

Example 2-12   Single-Stub Matching

A 50-Ω transmission line is connected to an antenna with load impedance $Z_L = (25 - j50)$ Ω. Find the

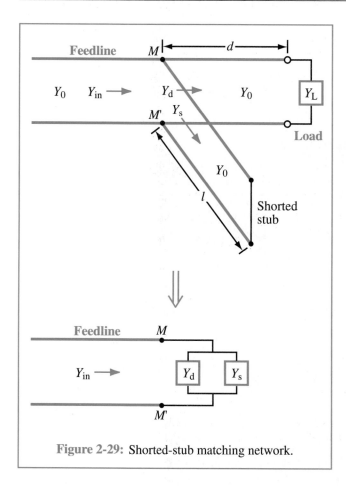

Figure 2-29: Shorted-stub matching network.

position and length of the short-circuited stub required to match the line.

Solution:

The normalized load impedance is

$$z_L = \frac{Z_L}{Z_0} = \frac{25 - j50}{50} = 0.5 - j1,$$

which is located at point A in the Smith chart of Fig. 2-30. Next, we draw the constant S circle through point A. To perform the matching task, it is easier to work with admittances than with impedances. The normalized load

admittance y_L is at point B, obtained by rotating 0.25λ, or equivalently by drawing a line from point A through the center to the image point on the S circle. The value of y_L at B is

$$y_L = 0.4 + j0.8,$$

and it is located at position 0.115λ on the WTG scale. In the admittance domain, the r_L circles become g_L circles, and the x_L circles become b_L circles. To achieve matching, we need to move from the load toward the generator a distance d such that the normalized input admittance y_d of the line terminated in the load (Fig. 2-29) has a real part equal to 1. This condition is satisfied by either of the two *matching points* C and D on the Smith charts of Figs. 2-30 and 2-31, respectively, corresponding to the intersections of the S circle with the $g_L = 1$ circle. Points C and D represent two possible solutions for the distance d in Fig. 2-29.

Solution for Point C [Fig. 2-30]: At C,

$$y_d = 1 + j1.58,$$

and it is located at 0.178λ on the WTG scale. The distance between points B and C is

$$d_1 = (0.178 - 0.115)\lambda = 0.063\lambda.$$

Looking from the generator toward the parallel combination of the line connected to the load and the short-circuited stub, the normalized input admittance at the juncture is

$$y_{in} = y_s + y_d,$$

where y_s is the normalized input admittance of the stub line. To match the feed line to the parallel combinations, we need to have $y_{in} = 1 + j0$. Thus,

$$1 + j0 = y_s + 1 + j1.58,$$

or

$$y_s = -j1.58.$$

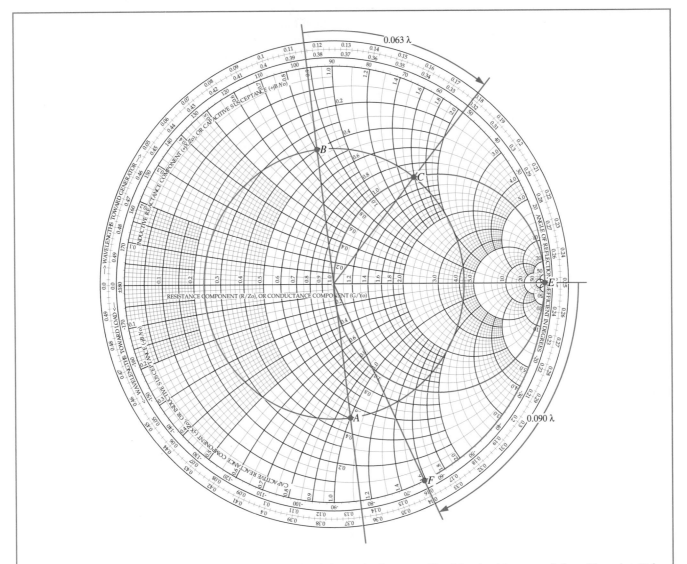

Figure 2-30: Solution for point C of Example 2-12. Point A is the normalized load with $z_L = 0.5 - j1$; point B is $y_L = 0.4 + j0.8$. Point C is the intersection of the SWR circle with the $g_L = 1$ circle. The distance from B to C is $d_1 = 0.063\lambda$. The length of the shorted stub (E to F) is $l_1 = 0.09\lambda$.

The normalized admittance of a short circuit is $-j\infty$ and is located at point E on the Smith chart, whose position is 0.25λ on the WTG scale. An input normalized admittance of $-j1.58$ is located at point F and is at position 0.34λ on the WTG scale. Hence,

$$l_1 = (0.34 - 0.25)\lambda = 0.09\lambda.$$

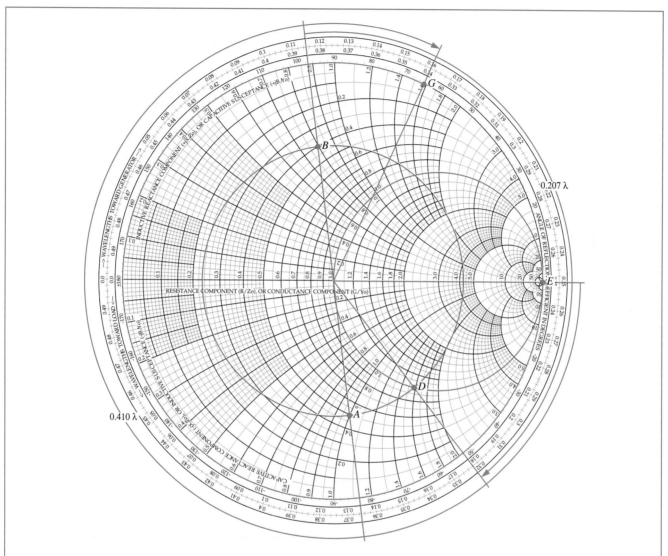

Figure 2-31: Solution for point D of Example 2-12. Point D is the second point of intersection of the SWR circle and the $y_L = 1$ circle. The distance B to D gives $d_2 = 0.207\lambda$, and the distance E to G gives $l_2 = 0.410\lambda$.

Solution for Point D (Fig. 2-31): At point D,

$$y_d = 1 - j1.58,$$

and the distance between points B and D is

$$d_2 = (0.322 - 0.115)\lambda = 0.207\lambda.$$

The needed normalized input admittance of the stub is $y_s = +j1.58$, which is located at point G at position

0.16λ on the WTG scale. Rotating from point E to point G involves a rotation of 0.25λ plus an additional rotation of 0.16λ, or

$$l_2 = (0.25 + 0.16)\lambda = 0.41\lambda. \quad \blacksquare$$

REVIEW QUESTIONS

Q2.24 To match an arbitrary load impedance to a lossless transmission line through a matching network, what is the required minimum number of degrees of freedom that the network should provide?

Q2.25 In the case of the single-stub matching network, what are the two degrees of freedom?

Q2.26 When a transmission line is matched to a load through a single-stub matching network, no waves will be reflected toward the generator. What happens to the waves reflected by the load and by the shorted stub when they arrive at terminals MM' in Fig. 2-29?

2-11   Transients on Transmission Lines

Thus far, our treatment of wave propagation on transmission lines has focused on the analysis of single-frequency, time-harmonic signals under steady-state conditions. The tools we developed—including the impedance-matching techniques and the use of the Smith chart— are useful for a wide range of applications, but they are inappropriate for dealing with digital or wideband signals on telephone lines or in a computer network. For such signals, we need to examine their transient behavior as a function of time. The *transient response* of a voltage pulse on a transmission line is a time record of its back and forth travel between the sending and receiving ends of the line, taking into account all the multiple reflections (echoes) at both ends.

Let us start by considering the simple case of a single rectangular pulse of amplitude V_0 and duration τ, as shown in Fig. 2-32(a). The amplitude of the pulse is zero prior to $t = 0$, V_0 over the duration $0 \leq t \leq \tau$, and again zero afterward. The pulse can be described mathematically as the sum of two unit step functions:

$$V(t) = V_1(t) + V_2(t)$$
$$= V_0\, U(t) - V_0\, U(t - \tau), \quad (2.122)$$

where the unit step function $U(x)$ is defined in terms of its argument x as

$$U(x) = \begin{cases} 1 & \text{for } x \geq 0, \\ 0 & \text{for } x < 0. \end{cases} \quad (2.123)$$

The first component, $V_1(t) = V_0\, U(t)$, represents a d-c voltage of amplitude V_0 that gets switched on at $t = 0$ and remains that way indefinitely, and the second component, $V_2(t) = -V_0\, U(t - \tau)$, represents a d-c voltage of amplitude $-V_0$ that gets switched on at $t = \tau$ and then remains that way indefinitely. As can be seen from Fig. 2-32(b), the sum of the two components is equal to V_0 for $0 \leq t \leq \tau$ and equal to zero for $t > \tau$. This representation of a pulse in terms of two step functions allows us to analyze the transient behavior of the pulse on a transmission line as the superposition of two d-c signals. Hence, if we can develop basic tools for describing the transient behavior of a single step function, we can apply the same tools for each of the two components of the pulse and then add the results appropriately.

2-11.1   Transient Response

The circuit shown in Fig. 2-33(a) consists of a generator composed of a d-c voltage source V_g and a series resistance R_g connected to a lossless transmission line of length l and characteristic impedance Z_0. The line is terminated in a purely resistive load Z_L at $z = l$. Hence, all impedances in the circuit are real.

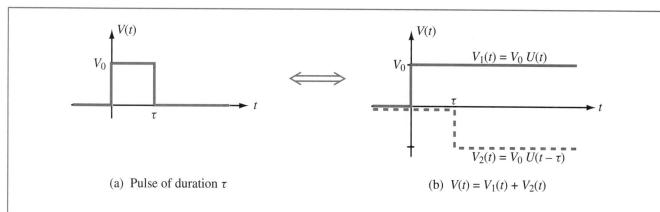

(a) Pulse of duration τ (b) $V(t) = V_1(t) + V_2(t)$

Figure 2-32: A rectangular pulse $V(t)$ of duration τ can be represented as the sum of two step functions of opposite polarities displaced by τ relative to each other.

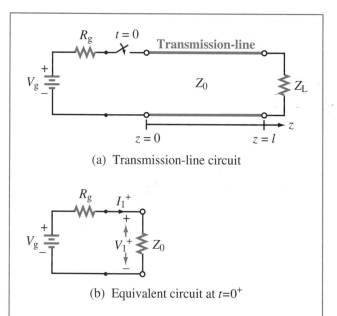

(a) Transmission-line circuit

(b) Equivalent circuit at $t=0^+$

Figure 2-33: At $t = 0^+$, immediately after closing the switch in the circuit in (a), the circuit can be represented by the equivalent circuit in (b).

The switch between the generator circuit and the transmission line is switched on at $t = 0$. At the instant

the switch is closed, the transmission line appears to the generator circuit as a load with impedance Z_0, the characteristic impedance of the line. This is because, in the absence of a signal on the line, the input impedance of the line is unaffected by the load impedance Z_L. The circuit representing the *initial condition* is shown in Fig. 2-33(b). The *initial current* I_1^+ and corresponding *initial voltage* V_1^+ at the input end of the transmission line are given by

$$I_1^+ = \frac{V_g}{R_g + Z_0}, \qquad (2.124\text{a})$$

$$V_1^+ = I_1^+ Z_0 = \frac{V_g Z_0}{R_g + Z_0}. \qquad (2.124\text{b})$$

The combination of V_1^+ and I_1^+ constitutes a wave that starts to travel along the line with a velocity $u_p = 1/\sqrt{\mu\varepsilon}$, immediately after the instant at which the switch is closed. The plus-sign superscript denotes the fact that the wave is traveling in the $+z$-direction. The transient response of the wave is shown in Fig. 2-34 at each of three instances in time for a circuit with

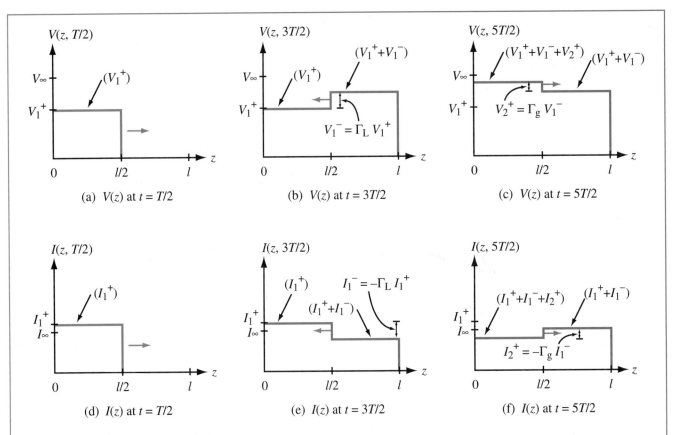

Figure 2-34: Voltage and current distributions on a lossless transmission line at $t = T/2$, $t = 3T/2$, and $t = 5T/2$, due to a unit step voltage applied to a circuit with $R_g = 4Z_0$ and $Z_L = 2Z_0$. The corresponding reflection coefficients are $\Gamma_L = 1/3$ and $\Gamma_g = 3/5$.

$R_g = 4Z_0$ and $Z_L = 2Z_0$. The first response is at time $t_1 = T/2$, where $T = l/u_p$ is the time it takes the wave to travel the full length of the line. By time t_1, the wave has traveled half-way down the line; consequently, the voltage on the first half of the line is equal to V_1^+, and the voltage on the second half is still zero [Fig. 2-34(a)]. At $t = T$, the wave reaches the load at $z = l$, and because $Z_L \neq Z_0$, the mismatch generates a reflected wave with amplitude

$$V_1^- = \Gamma_L V_1^+, \qquad (2.125)$$

where

$$\Gamma_L = \frac{Z_L - Z_0}{Z_L + Z_0} \qquad (2.126)$$

is the reflection coefficient of the load. For the specific case illustrated in Fig. 2-34, $Z_L = 2Z_0$, which results in $\Gamma_L = 1/3$. After this first reflection, the voltage on the line consists of the sum of two waves, the initial wave V_1^+ and the reflected wave V_1^-. The voltage on the transmission line at $t_2 = 3T/2$ is shown in Fig. 2-34(b); $V(z, 3T/2)$ is equal to V_1^+ on the first half of the line

$(0 \le z < l/2)$, and it is equal to $(V_1^+ + V_1^-)$ on the second half $(l/2 \le z \le l)$.

At $t = 2T$, the reflected wave V_1^- arrives at the sending end of the line. If $R_g \ne Z_0$, the mismatch at the sending end generates a reflection at $z = 0$ in the form of a wave with voltage amplitude V_2^+ given by

$$V_2^+ = \Gamma_g V_1^- = \Gamma_g \Gamma_L V_1^+, \qquad (2.127)$$

where

$$\Gamma_g = \frac{R_g - Z_0}{R_g + Z_0} \qquad (2.128)$$

is the reflection coefficient of the generator resistance R_g. For $R_g = 4Z_0$, we have $\Gamma_g = 0.6$. As time progresses after $t = 2T$, the wave V_2^+ travels down the line toward the load and, as it does that, it adds to the previous voltage condition on the line. Hence, at $t = 5T/2$, the total voltage on the first half of the line is

$$V(z, 5T/2) = V_1^+ + V_1^- + V_2^+$$
$$= (1 + \Gamma_L + \Gamma_L \Gamma_g) V_1^+$$
$$(0 \le z < l/2), \qquad (2.129a)$$

and on the second half of the line the voltage is

$$V(z, 5T/2) = V_1^+ + V_1^-$$
$$= (1 + \Gamma_L) V_1^+ \qquad (l/2 \le z \le l). \quad (2.129b)$$

The voltage distribution is shown in Fig. 2-34(c).

So far, we have examined the transient response of only the voltage wave $V(z, t)$. The associated transient response of the current $I(z, t)$ is shown in Figs. 2-34(d)–(f). The process is similar to that we described for the voltage $V(z, t)$, except for one important difference. Whereas at either end of the line the reflected voltage is related to the incident voltage by the reflection coefficient at that end, the reflected current is related to the incident current by the negative of the reflection coefficient. This property of wave reflection is expressed by Eq. (2.49b). Accordingly,

$$I_1^- = -\Gamma_L I_1^+, \qquad (2.130a)$$
$$I_2^+ = -\Gamma_g I_1^- = \Gamma_g \Gamma_L I_1^+, \qquad (2.130b)$$

and so on.

The multiple-reflection process continues indefinitely, and the ultimate value that $V(z, t)$ reaches as t approaches ∞ is the same at all locations on the transmission line and is given by

$$V_\infty = V_1^+ + V_1^- + V_2^+ + V_2^- + V_3^+ + V_3^- + \cdots$$
$$= V_1^+ [1 + \Gamma_L + \Gamma_L \Gamma_g + \Gamma_L^2 \Gamma_g + \Gamma_L^2 \Gamma_g^2 + \Gamma_L^3 \Gamma_g^2 + \cdots]$$
$$= V_1^+ [(1 + \Gamma_L)(1 + \Gamma_L \Gamma_g + \Gamma_L^2 \Gamma_g^2 + \cdots)]$$
$$= V_1^+ (1 + \Gamma_L)[1 + x + x^2 + \cdots], \qquad (2.131)$$

where $x = \Gamma_L \Gamma_g$. The series inside the square bracket is the binomial series of the function

$$\frac{1}{1 - x} = 1 + x + x^2 + \cdots \qquad \text{for } |x| < 1. \quad (2.132)$$

Hence, Eq. (2.131) can be rewritten in the compact form

$$V_\infty = V_1^+ \frac{1 + \Gamma_L}{1 - \Gamma_L \Gamma_g}. \qquad (2.133)$$

Upon replacing V_1^+, Γ_L, and Γ_g with the expressions given by Eqs. (2.124b), (2.126), and (2.128), respectively, and then simplifying the resulting expression, we obtain

$$\boxed{V_\infty = \frac{V_g Z_L}{R_g + Z_L}. \qquad (2.134)}$$

The voltage V_∞ is called the *steady-state voltage* on the line, and its expression is exactly what we should expect on the basis of d-c analysis of the circuit in Fig. 2-33(a) if we were to treat the transmission line as simply a connecting wire between the generator circuit and the load. The corresponding *steady-state current* is

$$I_\infty = \frac{V_\infty}{Z_L} = \frac{V_g}{R_g + Z_L} . \qquad (2.135)$$

2-11.2 Bounce Diagrams

Keeping track of the voltage and current waves as they bounce back and forth on the line is a rather tedious process. The *bounce diagram* is a graphical presentation that allows us to accomplish the same goal, but with relative ease. The horizontal axis in Figs. 2-35(a) and (b) represents position along the transmission line, and the vertical axis denotes time. Figure 2-35(a) pertains to $V(z,t)$ and part (b) pertains to $I(z,t)$. The bounce diagram in Fig. 2-35(a) consists of a zigzag line indicating the progress of the voltage wave on the line. The incident wave V_1^+ starts at $z = t = 0$ and travels in the $+z$-direction until it reaches the load at $z = l$ at time $t = T$. At the very top of the bounce diagram, the reflection coefficients are indicated by $\Gamma = \Gamma_g$ at the generator end and by $\Gamma = \Gamma_L$ at the load end. At the end of the first straight-line segment of the zigzag line, a second line is drawn to represent the reflected voltage wave $V_1^- = \Gamma_L V_1^+$. The amplitude of each new straight-line segment is equal to the product of the amplitude of the preceding straight-line segment and the reflection coefficient at that end of the line. The bounce diagram for the current $I(z,t)$ in Fig. 2-35(b) follows the same procedure as for the voltage except for the reversal of the signs of Γ_L and Γ_g at the top of the bounce diagram.

Using the bounce diagram, the total voltage (or current) at any point z_1 and time t_1 can be determined by drawing a vertical line through the point z_1, then adding the voltages (or currents) of all the zigzag segments intersected by that line between $t = 0$ and $t = t_1$. To find the voltage

at $z = l/4$ and $T = 4T$, for example, we draw a dashed vertical line in Fig. 2-35(a) through $z = l/4$ and we extend it from $t = 0$ to $t = 4T$. The dashed line intersects four line segments. The total voltage at $z = l/4$ and $t = 4T$ is therefore given by

$$V(l/4, 4T) = V_1^+ + \Gamma_L V_1^+ + \Gamma_g \Gamma_L V_1^+ + \Gamma_g \Gamma_L^2 V_1^+$$
$$= V_1^+(1 + \Gamma_L + \Gamma_g \Gamma_L + \Gamma_g \Gamma_L^2).$$

The time variation of V at a specific location z can be obtained by plotting the values of V along the (dashed) vertical line passing through z. Figure 2-35(c) shows the variation of V as a function of time at $z = l/4$ for a circuit with $\Gamma_g = 3/5$ and $\Gamma_L = 1/3$.

Example 2-13   Time-Domain Reflectometer

A time-domain reflectometer (TDR) is an instrument used to locate faults on a transmission line. Consider, for example, a long underground or undersea cable that gets damaged at some distance d from the sending end of the line. The damage may alter the electrical properties or the shape of the cable, causing it to exhibit an impedance Z_{Lf} at the fault location that is different from its characteristic impedance, Z_0. A TDR sends a step voltage down the line, and by observing the voltage at the sending end as a function of time, it is possible to determine the location of the fault and its severity.

If the voltage waveform shown in Fig. 2-36(a) is seen on an oscilloscope connected to the input of a 75-Ω matched transmission line, determine (a) the generator voltage, (b) the location of the fault, and (c) the fault shunt resistance. The line's insulating material is Teflon with $\varepsilon_r = 2.1$.

Solution: (a) Since the line is properly matched, it means that $R_g = Z_L = Z_0$. In Fig. 2-36(b), the fault located at distance d from the sending end is represented by a shunt resistance R_f. For a matched line, Eq. (2.124b) gives

$$V_1^+ = \frac{V_g Z_0}{R_g + Z_0} = \frac{V_g Z_0}{2Z_0} = \frac{V_g}{2} .$$

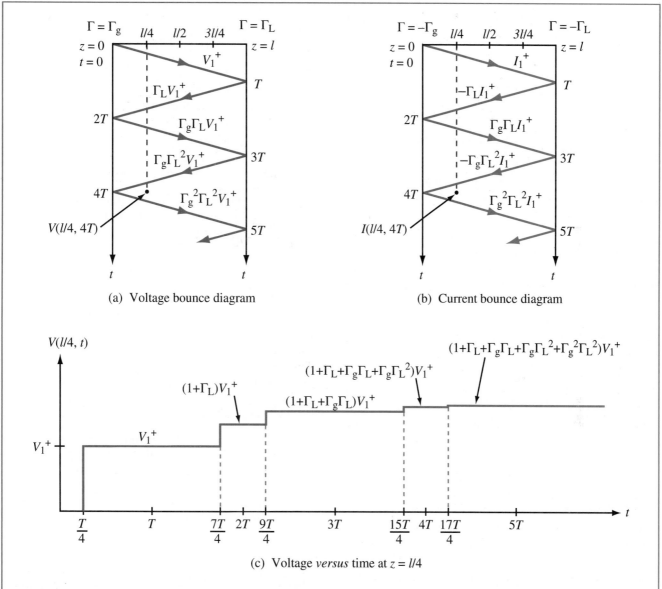

(a) Voltage bounce diagram

(b) Current bounce diagram

(c) Voltage *versus* time at $z = l/4$

Figure 2-35: Bounce diagrams for (a) voltage and (b) current. In (c), the voltage variation with time at $z = l/4$ for a circuit with $\Gamma_g = 3/5$ and $\Gamma_L = 1/3$ is deduced from the vertical dashed line at $l/4$ in (a).

According to Fig. 2-36(a), $V_1^+ = 6$ V. Hence,

$$V_g = 2V_1^+ = 12 \text{ V.}$$

(b) The propagation velocity on the line is

$$u_p = \frac{c}{\sqrt{\varepsilon_r}} = \frac{3 \times 10^8}{\sqrt{2.1}} = 2.07 \times 10^8 \text{ m/s.}$$

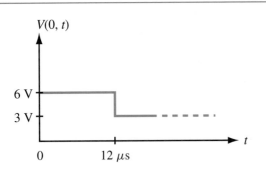

(a) Observed voltage at the sending end

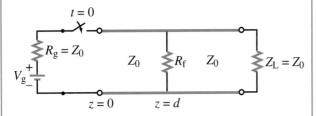

(b) The fault at $z = d$ is represented by a
fault resistance R_f

Figure 2-36: Time-domain reflectometer of Example 2-13.

For a fault at a distance d, the round-trip time delay of the echo is

$$\Delta t = \frac{2d}{u_p}.$$

From Fig. 2-36(a), $\Delta t = 12 \ \mu s$. Hence,

$$d = \frac{\Delta t}{2} u_p = \frac{12 \times 10^{-6}}{2} \times 2.07 \times 10^8 = 1,242 \ \text{m}.$$

(c) The change in level of $V(0, t)$ shown in Fig. 2-36(a) represents V_1^-. Thus,

$$V_1^- = \Gamma_f V_1^+ = -3 \ \text{V},$$

or

$$\Gamma_f = \frac{-3}{6} = -0.5,$$

where Γ_f is the reflection coefficient due to the fault load Z_{Lf} that appears at $z = d$.

From Eq. (2.49a),

$$\Gamma_f = \frac{Z_{Lf} - Z_0}{Z_{Lf} + Z_0},$$

and it follows that $Z_{Lf} = 25 \ \Omega$. This fault load is composed of the fault shunt resistance R_f and the characteristic impedance Z_0 of the line to the right of the fault:

$$\frac{1}{Z_{Lf}} = \frac{1}{R_f} + \frac{1}{Z_0},$$

so that the shunt resistance is 37.5 Ω. ∎

REVIEW QUESTIONS

Q2.27 What is transient analysis used for?

Q2.28 The transient analysis presented in this section was for a step voltage. How does one use it for analyzing the response to a pulse?

Q2.29 What is the difference between the bounce diagram for voltage and the bounce diagram for current?

CHAPTER HIGHLIGHTS

- A transmission line is a two-port network connecting a generator to a load. EM waves traveling on the line may experience ohmic power losses, dispersive effects, and reflections at the generator and load ends of the line. These transmission-line effects may be ignored if the line length is much shorter than λ.

- TEM transmission lines consist of two conductors that can support the propagation of transverse electromagnetic waves characterized by electric and magnetic fields that are transverse to the direction of propagation. TEM lines may be represented by a lumped-element model consisting of four line parameters (R', L', G', and C') whose values are specified by the specific line geometry, the constitutive parameters of the conductors and of the insulating material between them, and the angular frequency ω.

- Wave propagation on a transmission line, which is represented by the phasor voltage $\widetilde{V}(z)$ and associated current $\widetilde{I}(z)$, is governed by the propagation constant of the line, $\gamma = \alpha + j\beta$, and its characteristic impedance Z_0. Both γ and Z_0 are specified by ω and the four line parameters.

- If $R' = G' = 0$, the line becomes lossless ($\alpha = 0$). A lossless line is nondispersive, meaning that the phase velocity of a wave is independent of its oscillation frequency.

- In general, a line supports two waves, an incident wave supplied by the generator and another wave reflected by the load. The sum of the two waves generates a standing-wave pattern with a period of $\lambda/2$. The voltage standing-wave ratio S, which is equal to the ratio of the maximum to minimum voltage magnitude on the line, varies between 1 for a matched load ($Z_L = Z_0$) to ∞ for a line terminated in an open circuit, a short circuit, or a purely reactive load.

- The input impedance of a line terminated in a short circuit or open circuit is purely reactive. This property can be used to design equivalent inductors and capacitors.

- The fraction of the incident power delivered to the load by a lossless line is equal to $(1 - |\Gamma|^2)$.

- The Smith chart is a useful graphical technique for analyzing transmission-line problems and for designing impedance-matching networks.

- Matching networks are placed between the load and the feed transmission line for the purpose of eliminating reflections toward the generator. A matching network may consist of lumped elements in the form of capacitors or inductors, or it may consist of sections of transmission lines with appropriate lengths and terminations.

- Transient analysis of pulses can be performed using a bounce-diagram graphical technique that tracks reflections at both the load and generator ends of the transmission line.

GLOSSARY OF IMPORTANT TERMS

Provide definitions or explain the meaning of the following terms:

dispersive transmission line
TEM transmission lines
higher-order transmission lines
lumped-element model
transmission-line parameters
intrinsic resistance R_s
perfect conductor
perfect dielectric
air line
telegrapher's equations
complex propagation constant γ
attenuation constant α
phase constant β
characteristic impedance Z_0
lossless line
standing wave
distortionless line
voltage reflection coefficient Γ
matched transmission line
standing-wave pattern
in-phase
phase opposition
voltage maxima and minima

current maxima and minima

voltage standing-wave ratio (VSWR or SWR) S

slotted line

input impedance Z_{in}

short-circuited line

open-circuited line

quarter-wave transformer

Smith chart

unit circle

normalized impedance

normalized load resistance r_L

normalized load reactance x_L

WTG and WTL

SWR circle

admittance Y

conductance G and susceptance B

impedance matching

matching network

single-stub matching

transient response

bounce diagram

PROBLEMS

Sections 2-1 to 2-4: Transmission-Line Model

2.1* A transmission line of length l connects a load to a sinusoidal voltage source with an oscillation frequency f. Assuming that the velocity of wave propagation on the line is c, for which of the following situations is it reasonable to ignore the presence of the transmission line in the solution of the circuit:

(a) $l = 20$ cm, $f = 10$ kHz

---

*Answer(s) available in Appendix D.

⦿ Solution available in CD-ROM.

(b) $l = 50$ km, $f = 60$ Hz

(c) $l = 20$ cm, $f = 300$ MHz

(d) $l = 1$ mm, $f = 100$ GHz

2.2 Calculate the line parameters R', L', G', and C' for a coaxial line with an inner conductor diameter of 0.5 cm and an outer conductor diameter of 1 cm, filled with an insulating material where $\mu = \mu_0$, $\varepsilon_r = 2.25$, and $\sigma = 10^{-3}$ S/m. The conductors are made of copper with $\mu_c = \mu_0$ and $\sigma_c = 5.8 \times 10^7$ S/m. The operating frequency is 1 GHz.

2.3* A 1-GHz parallel-plate transmission line consists of 1.5-cm-wide copper strips separated by a 0.2-cm-thick layer of polystyrene. Appendix B gives $\mu_c = \mu_0 = 4\pi \times 10^{-7}$ (H/m) and $\sigma_c = 5.8 \times 10^7$ (S/m) for copper, and $\varepsilon_r = 2.6$ for polystyrene. Use Table 2-1 to determine the line parameters of the transmission line. Assume that $\mu = \mu_0$ and $\sigma \simeq 0$ for polystyrene.

2.4 Show that the transmission-line model shown in Fig. 2-37 yields the same telegrapher's equations given by Eqs. (2.14) and (2.16).

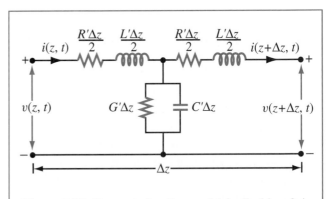

Figure 2-37: Transmission-line model for Problem 2.4.

2.5* Find α, β, u_p, and Z_0 for the coaxial line of Problem 2.2.

Section 2-5: The Lossless Line

2.6 In addition to not dissipating power, a lossless line has two important features: (1) it is dispersionless (u_p is independent of frequency); and (2) its characteristic impedance Z_0 is purely real. Sometimes, it is not possible to design a transmission line such that $R' \ll \omega L'$ and $G' \ll \omega C'$, but it is possible to choose the dimensions of the line and its material properties so as to satisfy the condition

$$R'C' = L'G' \quad \text{(distortionless line)}$$

Such a line is called a *distortionless* line, because despite the fact that it is not lossless, it nonetheless possesses the previously mentioned features of the lossless line. Show that for a distortionless line,

$$\alpha = R'\sqrt{\frac{C'}{L'}} = \sqrt{R'G'}$$

$$\beta = \omega\sqrt{L'C'}$$

$$Z_0 = \sqrt{\frac{L'}{C'}}$$

2.7* For a distortionless line [see Problem 2.6] with $Z_0 = 50\ \Omega$, $\alpha = 40$ (mNp/m), and $u_p = 2.5 \times 10^8$ (m/s), find the line parameters and λ at 250 MHz.

2.8 Find α and Z_0 of a distortionless line whose $R' = 4\ \Omega/\text{m}$ and $G' = 4 \times 10^{-4}$ S/m.

2.9* A transmission line operating at 125 MHz has $Z_0 = 40\ \Omega$, $\alpha = 0.02$ (Np/m), and $\beta = 0.75$ rad/m. Find the line parameters R', L', G', and C'.

2.10 Using a slotted line, the voltage on a lossless transmission line was found to have a maximum magnitude of 1.5 V and a minimum magnitude of 0.8 V. Find the magnitude of the load's reflection coefficient.

2.11* Polyethylene with $\varepsilon_r = 2.25$ is used as the insulating material in a lossless coaxial line with a characteristic impedance of 50 Ω. The radius of the inner conductor is 1 mm.

(a) What is the radius of the outer conductor?

(b) What is the phase velocity of the line?

2.12 A 50-Ω lossless transmission line is terminated in a load with impedance $Z_L = (30 - j60)\ \Omega$. The wavelength is 5 cm. Find the following:

(a) The reflection coefficient at the load.

(b) The standing-wave ratio on the line.

(c) The position of the voltage maximum nearest the load.

(d) The position of the current maximum nearest the load.

2.13* On a 150-Ω lossless transmission line, the following observations were noted: distance of first voltage minimum from the load = 3 cm; distance of first voltage maximum from the load = 9 cm; $S = 3$. Find Z_L.

2.14 Using a slotted line, the following results were obtained: distance of first minimum from the load = 4 cm; distance of second minimum from the load = 14 cm; voltage standing-wave ratio = 2.5. If the line is lossless and $Z_0 = 50\ \Omega$, find the load impedance.

2.15* A load with impedance $Z_L = (25 - j50)\ \Omega$ is to be connected to a lossless transmission line with characteristic impedance Z_0, with Z_0 chosen such that the standing-wave ratio is the smallest possible. What should Z_0 be?

2.16 A 50-Ω lossless line terminated in a purely resistive load has a voltage standing-wave ratio of 4. Find all possible values of Z_L.

Section 2-6: Input Impedance

2.17* At an operating frequency of 300 MHz, a lossless 50-Ω air-spaced transmission line 2.5 m in length is terminated with an impedance $Z_L = (60 + j20)$ Ω. Find the input impedance.

2.18 A lossless transmission line of electrical length $l = 0.35\lambda$ is terminated in a load impedance as shown in Fig. 2-38. Find Γ, S, and Z_{in}.

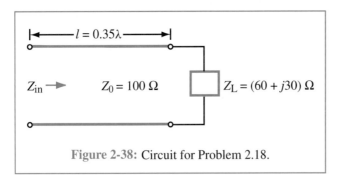

Figure 2-38: Circuit for Problem 2.18.

2.19 Show that the input impedance of a quarter-wavelength–long lossless line terminated in a short circuit appears as an open circuit.

2.20 Show that at the position where the magnitude of the voltage on the line is a maximum, the input impedance is purely real.

2.21* A voltage generator with

$$v_g(t) = 5\cos(2\pi \times 10^9 t) \text{ V}$$

and internal impedance $Z_g = 50$ Ω is connected to a 50-Ω lossless air-spaced transmission line. The line length is 5 cm and it is terminated in a load with impedance $Z_L = (100 - j100)$ Ω. Find the following:

(a) Γ at the load.

(b) Z_{in} at the input to the transmission line.

(c) The input voltage $\tilde{V}_i$ and input current $\tilde{I}_i$.

2.22 A 6-m section of 150-Ω lossless line is driven by a source with

$$v_g(t) = 5\cos(8\pi \times 10^7 t - 30°) \quad \text{(V)}$$

and $Z_g = 150$ Ω. If the line, which has a relative permittivity $\epsilon_r = 2.25$, is terminated in a load $Z_L = (150 - j50)$ Ω, find the following:

(a) λ on the line.

(b) The reflection coefficient at the load.

(c) The input impedance.

(d) The input voltage $\tilde{V}_i$.

(e) The time-domain input voltage $v_i(t)$.

2.23* Two half-wave dipole antennas, each with an impedance of 75 Ω, are connected in parallel through a pair of transmission lines, and the combination is connected to a feed transmission line, as shown in Fig. 2-39. All lines are 50 Ω and lossless.

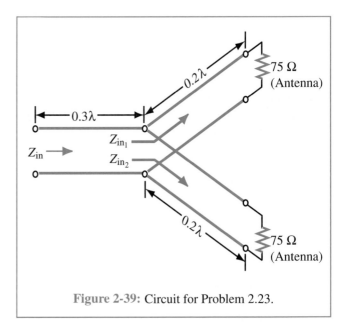

Figure 2-39: Circuit for Problem 2.23.

(a) Calculate Z_{in_1}, the input impedance of the antenna-terminated line, at the parallel juncture.

(b) Combine Z_{in_1} and Z_{in_2} in parallel to obtain Z'_L, the effective load impedance of the feedline.

(c) Calculate Z_{in} of the feedline.

Section 2-7: Special Cases

2.24 At an operating frequency of 200 MHz, it is desired to use a section of a lossless 50-Ω transmission line terminated in a short circuit to construct an equivalent load with reactance $X = 25\ \Omega$. If the phase velocity of the line is $0.75c$, what is the shortest possible line length that would exhibit the desired reactance at its input?

2.25* A lossless transmission line is terminated in a short circuit. How long (in wavelengths) should the line be for it to appear as an open circuit at its input terminals?

2.26 The input impedance of a 31-cm-long lossless transmission line of unknown characteristic impedance was measured at 1 MHz. With the line terminated in a short circuit, the measurement yielded an input impedance equivalent to an inductor with inductance of $0.128\ \mu$H, and when the line was open-circuited, the measurement yielded an input impedance equivalent to a capacitor with capacitance of 20 pF. Find Z_0 of the line, the phase velocity, and the relative permittivity of the insulating material.

2.27* A 60-Ω resistive load is preceded by a $\lambda/4$ section of a 50-Ω lossless line, which itself is preceded by another $\lambda/4$ section of a 100-Ω line. What is the input impedance?

2.28 A 100-MHz FM broadcast station uses a 300-Ω transmission line between the transmitter and a tower-mounted half-wave dipole antenna. The antenna impedance is 73 Ω. You are asked to design a quarter-wave transformer to match the antenna to the line.

(a) Determine the electrical length and characteristic impedance of the quarter-wave section.

(b) If the quarter-wave section is a two-wire line with $d = 2.5$ cm, and the spacing between the wires is made of polystyrene with $\varepsilon_r = 2.6$, determine the physical length of the quarter-wave section and the radius of the two wire conductors.

2.29* A 50-MHz generator with $Z_g = 50\ \Omega$ is connected to a load $Z_L = (50 - j25)\ \Omega$. The time-average power transferred from the generator into the load is maximum when $Z_g = Z_L^*$, where Z_L^* is the complex conjugate of Z_L. To achieve this condition without changing Z_g, the effective load impedance can be modified by adding an open-circuited line in series with Z_L, as shown in Fig. 2-40. If the line's $Z_0 = 100\ \Omega$,

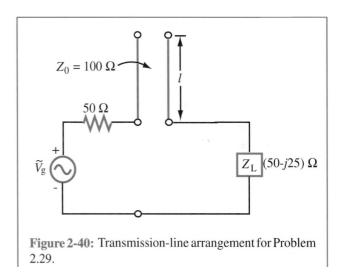

Figure 2-40: Transmission-line arrangement for Problem 2.29.

determine the shortest length of line (in wavelengths) necessary for satisfying the maximum-power-transfer condition.

2.30 A 50-Ω lossless line of length $l = 0.375\lambda$ connects a 200-MHz generator with $\tilde{V}_g = 150$ V and $Z_g = 50\ \Omega$ to a load Z_L. Determine the time-domain current through the load for:

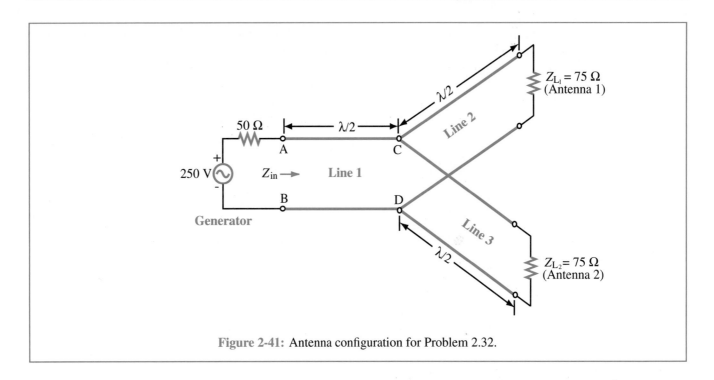

Figure 2-41: Antenna configuration for Problem 2.32.

(a) $Z_L = (50 - j50) \ \Omega$

(b) $Z_L = 50 \ \Omega$

(c) $Z_L = 0$ (short circuit)

Section 2-8: Power Flow on Lossless Line

2.31* A generator with $\tilde{V}_g = 100$ V and $Z_g = 50 \ \Omega$ is connected to a load $Z_L = 75 \ \Omega$ through a 50-Ω lossless line of length $l = 0.15\lambda$.

(a) Compute Z_{in}, the input impedance of the line at the generator end.

(b) Compute $\tilde{I}_i$ and $\tilde{V}_i$.

(c) Compute the time-average power delivered to the line, $P_{in} = \frac{1}{2}\Re[\tilde{V}_i \tilde{I}_i^*]$.

(d) Compute $\tilde{V}_L$, $\tilde{I}_L$, and the time-average power delivered to the load, $P_L = \frac{1}{2}\Re[\tilde{V}_L \tilde{I}_L^*]$. How does P_{in} compare to P_L? Explain.

(e) Compute the time-average power delivered by the generator, P_g, and the time-average power dissipated in Z_g. Is conservation of power satisfied?

2.32 If the two-antenna configuration shown in Fig. 2-41 is connected to a generator with $\tilde{V}_g = 250$ V and $Z_g = 50 \ \Omega$, how much average power is delivered to each antenna?

2.33* For the circuit shown in Fig. 2-42, calculate the average incident power, the average reflected power, and the average power transmitted into the infinite 100-Ω line. The $\lambda/2$ line is lossless and the infinitely long line is slightly lossy. (Hint: The input impedance of an infinitely long line is equal to its characteristic impedance so long as $\alpha \neq 0$.)

2.34 An antenna with a load impedance

$$Z_L = (75 + j25) \ \Omega$$

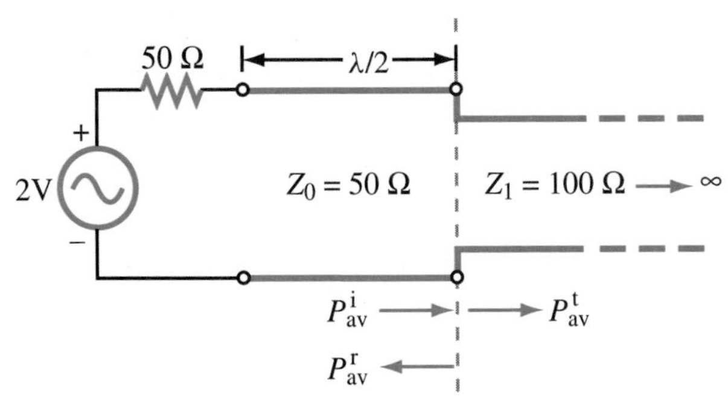

Figure 2-42: Circuit for Problem 2.33.

is connected to a transmitter through a 50-Ω lossless transmission line. If under matched conditions (50-Ω load) the transmitter can deliver 10 W to the load, how much power can it deliver to the antenna? Assume that $Z_g = Z_0$.

Section 2-9: Smith Chart

2.35* Use the Smith chart to find the reflection coefficient corresponding to a load impedance of

(a) $Z_L = 3Z_0$

(b) $Z_L = (2 - 2j)Z_0$

(c) $Z_L = -2jZ_0$

(d) $Z_L = 0$ (short circuit)

2.36 Use the Smith chart to find the normalized load impedance corresponding to a reflection coefficient of

(a) $\Gamma = 0.5$

(b) $\Gamma = 0.5\angle 60°$

(c) $\Gamma = -1$

(d) $\Gamma = 0.3\angle -30°$

(e) $\Gamma = 0$

(f) $\Gamma = j$

2.37* On a lossless transmission line terminated in a load $Z_L = 100\ \Omega$, the standing-wave ratio was measured to be 2.5. Use the Smith chart to find the two possible values of Z_0.

2.38 A lossless 50-Ω transmission line is terminated in a load with $Z_L = (50 + j25)\ \Omega$. Use the Smith chart to find the following:

(a) The reflection coefficient Γ.

(b) The standing-wave ratio.

(c) The input impedance at 0.35λ from the load.

(d) The input admittance at 0.35λ from the load.

(e) The shortest line length for which the input impedance is purely resistive.

(f) The position of the first voltage maximum from the load.

2.39* A lossless 50-Ω transmission line is terminated in a short circuit. Use the Smith chart to find the following:

(a) The input impedance at a distance 2.3λ from the load.

(b) The distance from the load at which the input admittance is $Y_{in} = -j0.04$ S.

2.40 Use the Smith chart to find y_L if $z_L = 1.5 - j0.7$.

2.41* A lossless 100-Ω transmission line $3\lambda/8$ in length is terminated in an unknown impedance. If the input impedance is $Z_{in} = -j2.5\ \Omega$,

(a) Use the Smith chart to find Z_L.

(b) What length of open-circuit line could be used to replace Z_L?

2.42 A 75-Ω lossless line is 0.6λ long. If $S = 1.8$ and $\theta_r = -60°$, use the Smith chart to find $|\Gamma|$, Z_L, and Z_{in}.

2.43* Using a slotted line on a 50-Ω air-spaced lossless line, the following measurements were obtained:

$S = 1.6$ and $|\widetilde{V}|_{\max}$ occurred only at 10 cm and 24 cm from the load. Use the Smith chart to find Z_L.

2.44   At an operating frequency of 5 GHz, a 50-Ω lossless coaxial line with insulating material having a relative permittivity $\varepsilon_r = 2.25$ is terminated in an antenna with an impedance $Z_L = 75 \ \Omega$. Use the Smith chart to find Z_{in}. The line length is 30 cm.

Section 2-10: Impedance Matching

2.45*   A 50-Ω lossless line 0.6λ long is terminated in a load with $Z_L = (50 + j25) \ \Omega$. At 0.3$\lambda$ from the load, a resistor with resistance $R = 30 \ \Omega$ is connected as shown in Fig. 2-43. Use the Smith chart to find Z_{in}.

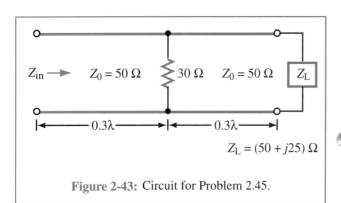

Figure 2-43: Circuit for Problem 2.45.

2.46   A 50-Ω lossless line is to be matched to an antenna with $Z_L = (75 - j20) \ \Omega$ using a shorted stub. Use the Smith chart to determine the stub length and distance between the antenna and stub.

2.47*   Repeat Problem 2.46 for a load with $Z_L = (100 + j50) \ \Omega$.

2.48   Use the Smith chart to find Z_{in} of the feed line shown in Fig. 2-44. All lines are lossless with $Z_0 = 50 \ \Omega$.

2.49*   Repeat Problem 2.48 for the case where all three transmission lines are $\lambda/4$ in length.

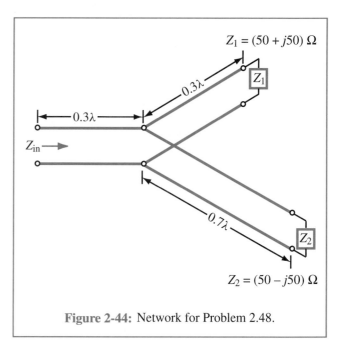

Figure 2-44: Network for Problem 2.48.

Section 2-11: Transients on Transmission Lines

2.50   Generate a bounce diagram for the voltage $V(z, t)$ for a 1-m-long lossless line characterized by $Z_0 = 50 \ \Omega$ and $u_p = 2c/3$ (where c is the velocity of light) if the line is fed by a step voltage applied at $t = 0$ by a generator circuit with $V_g = 60 \ V$ and $R_g = 100 \ \Omega$. The line is terminated in a load $Z_L = 25 \ \Omega$. Use the bounce diagram to plot $V(t)$ at a point midway along the length of the line from $t = 0$ to $t = 25$ ns.

2.51   Repeat Problem 2.50 for the current I on the line.

2.52   In response to a step voltage, the voltage waveform shown in Fig. 2-45 was observed at the sending end of a lossless transmission line with $R_g = 50 \ \Omega$, $Z_0 = 50 \ \Omega$, and $\varepsilon_r = 2.25$. Determine the following:

(a) The generator voltage.

(b) The length of the line.

(c) The load impedance.

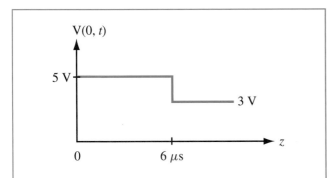

Figure 2-45: Voltage waveform for Problems 2.52 and 2.54.

2.53* In response to a step voltage, the voltage waveform shown in Fig. 2-46 was observed at the sending end of a shorted line with $Z_0 = 50 \ \Omega$ and $\varepsilon_r = 4$. Determine V_g, R_g, and the line length.

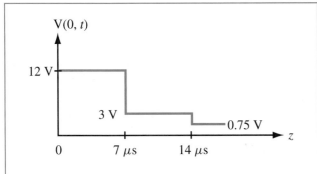

Figure 2-46: Voltage waveform of Problem 2.53.

2.54 Suppose the voltage waveform shown in Fig. 2-45 was observed at the sending end of a 50-Ω transmission line in response to a step voltage introduced by a generator with $V_g = 15$ V and an unknown series resistance R_g. The line is 1 km in length, its velocity of propagation is 1×10^8 m/s, and it is terminated in a load $Z_L = 100 \ \Omega$.

(a) Determine R_g.

(b) Explain why the drop in level of $V(0, t)$ at $t = 6 \ \mu$s cannot be due to reflection from the load.

(c) Determine the shunt resistance R_f and location of the fault responsible for the observed waveform.

2.55 A generator circuit with $V_g = 200$ V and $R_g = 25 \ \Omega$ was used to excite a 75-Ω lossless line with a rectangular pulse of duration $\tau = 0.4 \ \mu$s. The line is 200 m long, its $u_p = 2 \times 10^8$ m/s, and it is terminated in a load $Z_L = 125 \ \Omega$.

(a) Synthesize the voltage pulse exciting the line as the sum of two step functions, $V_{g_1}(t)$ and $V_{g_2}(t)$.

(b) For each voltage step function, generate a bounce diagram for the voltage on the line.

(c) Use the bounce diagrams to plot the total voltage at the sending end of the line.

2.56 For the circuit of Problem 2.55, generate a bounce diagram for the current and plot its time history at the middle of the line.

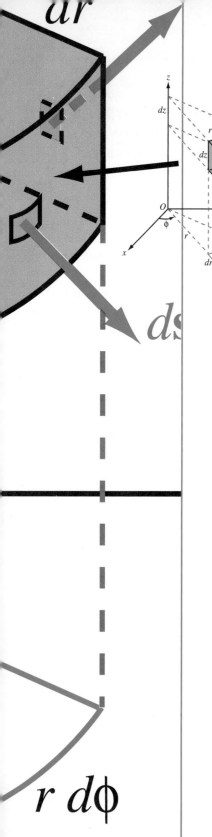

CHAPTER 3

Vector Analysis

Overview

OVERVIEW

In our examination of wave propagation on a transmission line in Chapter 2, the primary quantities we worked with were voltage, current, impedance, and power. Each of these is a *scalar* quantity, meaning that it can be completely specified by its magnitude, if it is a positive real number, or by its magnitude and phase angle if it is a negative or a complex number (a negative number has a positive magnitude and a phase angle of π (rad)). A *vector* specifies both the magnitude and direction of a quantity. The speed of an object is a scalar, whereas its velocity is a vector.

Starting in the next chapter and throughout the succeeding chapters in the book, the electromagnetic quantities we will deal with the most are the electric and magnetic fields, **E** and **H**. These, and many other related quantities, are vectors. Vector analysis provides the mathematical tools necessary for expressing and manipulating vector quantities in an efficient and convenient manner. To specify a vector in three-dimensional space, it is necessary to specify its components along each of the three dimensions. Several types of coordinate systems are used in the study of vector quantities, the most common being the Cartesian (or rectangular), cylindrical, and spherical systems. A particular coordinate system is usually chosen to best suit the geometry of the particular problem under consideration.

Vector algebra governs the laws of addition, subtraction, and multiplication of vectors in any given coordinate system. The rules of vector algebra and vector representation in each of the aforementioned orthogonal coordinate systems (including vector transformation between them) are two of the three major topics treated in this chapter. The third topic is *vector calculus*, which

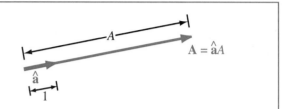

Figure 3-1: Vector $\mathbf{A} = \hat{\mathbf{a}}A$ has a magnitude $A = |\mathbf{A}|$ and unit vector $\hat{\mathbf{a}} = \mathbf{A}/A$.

encompasses the laws of differentiation and integration of vectors, the use of special vector operators (gradient, divergence, and curl), and the application of certain particularly useful theorems, most notably the divergence and Stokes's theorems.

3-1 Basic Laws of Vector Algebra

A vector **A** has a *magnitude* $A = |\mathbf{A}|$ and a direction specified by a unit vector $\hat{\mathbf{a}}$:

$$\mathbf{A} = \hat{\mathbf{a}}|\mathbf{A}| = \hat{\mathbf{a}}A. \qquad (3.1)$$

The *unit vector* $\hat{\mathbf{a}}$ has a magnitude of unity ($|\hat{\mathbf{a}}| = 1$), and its direction is given by

$$\hat{\mathbf{a}} = \frac{\mathbf{A}}{|\mathbf{A}|} = \frac{\mathbf{A}}{A}. \qquad (3.2)$$

Figure 3-1 shows a graphical representation of the vector **A** as a straight line of length A with its tip pointing in the direction of $\hat{\mathbf{a}}$.

In the Cartesian (or rectangular) coordinate system shown in Fig. 3-2(a), the directions of the x, y,

101

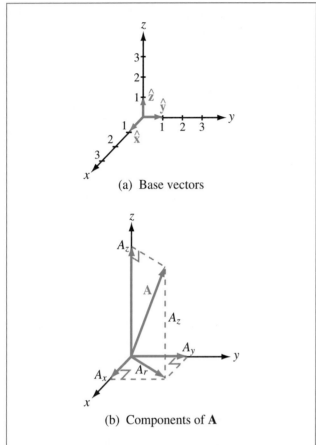

(a) Base vectors

(b) Components of **A**

Figure 3-2: Cartesian coordinate system: (a) base vectors $\hat{\mathbf{x}}$, $\hat{\mathbf{y}}$, and $\hat{\mathbf{z}}$, and (b) components of vector **A**.

and z coordinates are denoted by the three mutually perpendicular unit vectors $\hat{\mathbf{x}}$, $\hat{\mathbf{y}}$, and $\hat{\mathbf{z}}$, which are called the *base vectors*. The vector **A** in Fig. 3-2(b) may be represented as

$$\mathbf{A} = \hat{\mathbf{x}}A_x + \hat{\mathbf{y}}A_y + \hat{\mathbf{z}}A_z, \qquad (3.3)$$

where A_x, A_y, and A_z are the components of **A** along the x-, y-, and z-directions, respectively. The component A_z is equal to the perpendicular projection of **A** onto the

z-axis, and similar definitions apply to A_x and A_y. Application of the Pythagorean theorem, first to the right triangle in the x–y plane to express the hypotenuse A_r in terms of A_x and A_y, and then again to the vertical right triangle with sides A_r and A_z and hypotenuse A, gives the following expression for the magnitude of **A**:

$$A = |\mathbf{A}| = \sqrt[+]{A_x^2 + A_y^2 + A_z^2}. \qquad (3.4)$$

Since A is a nonnegative scalar, only the positive root applies. From Eq. (3.2), the unit vector $\hat{\mathbf{a}}$ is given by

$$\hat{\mathbf{a}} = \frac{\mathbf{A}}{A} = \frac{\hat{\mathbf{x}}A_x + \hat{\mathbf{y}}A_y + \hat{\mathbf{z}}A_z}{\sqrt[+]{A_x^2 + A_y^2 + A_z^2}}. \qquad (3.5)$$

Occasionally, we shall use the shorthand notation $\mathbf{A} = (A_x, A_y, A_z)$ to denote a vector with components A_x, A_y, and A_z in a Cartesian coordinate system.

3-1.1 Equality of Two Vectors

Two vectors **A** and **B** are said to be equal if they have equal magnitudes and identical unit vectors. Thus, if

$$\mathbf{A} = \hat{\mathbf{a}}A = \hat{\mathbf{x}}A_x + \hat{\mathbf{y}}A_y + \hat{\mathbf{z}}A_z, \qquad (3.6a)$$

$$\mathbf{B} = \hat{\mathbf{b}}B = \hat{\mathbf{x}}B_x + \hat{\mathbf{y}}B_y + \hat{\mathbf{z}}B_z, \qquad (3.6b)$$

then $\mathbf{A} = \mathbf{B}$ if and only if $A = B$ and $\hat{\mathbf{a}} = \hat{\mathbf{b}}$, which requires that $A_x = B_x$, $A_y = B_y$, and $A_z = B_z$. *Equality of two vectors does not necessarily imply that they are identical;* in Cartesian coordinates, two displaced parallel vectors of equal magnitude and pointing in the same direction are equal, but they are identical only if they lie on top of one another.

3-1.2 Vector Addition and Subtraction

The sum of two vectors $\mathbf{A}$ and $\mathbf{B}$ is a vector $\mathbf{C}$ given by

$$\mathbf{C} = \mathbf{A} + \mathbf{B} = \mathbf{B} + \mathbf{A}. \tag{3.7}$$

Graphically, vector addition is obtained by either the parallelogram rule or the head-to-tail rule, as illustrated in Fig. 3-3. With $\mathbf{A}$ and $\mathbf{B}$ drawn from the same point, while keeping their magnitudes and directions unchanged, the vector $\mathbf{C}$ is the diagonal of the parallelogram found by $\mathbf{A}$ and $\mathbf{B}$. With the head-to-tail rule, we may either add $\mathbf{A}$ to $\mathbf{B}$ or $\mathbf{B}$ to $\mathbf{A}$. When $\mathbf{A}$ is added to $\mathbf{B}$, it is positioned so that its tail starts at the tip of $\mathbf{B}$, again keeping its length and direction unchanged. The sum vector $\mathbf{C}$ starts at the tail of $\mathbf{B}$ and ends at the tip of $\mathbf{A}$.

If $\mathbf{A}$ and $\mathbf{B}$ are given in a rectangular coordinate system by Eqs. (3.6a) and (3.6b), vector addition gives

$$\mathbf{C} = \mathbf{A} + \mathbf{B}$$
$$= (\hat{\mathbf{x}}A_x + \hat{\mathbf{y}}A_y + \hat{\mathbf{z}}A_z) + (\hat{\mathbf{x}}B_x + \hat{\mathbf{y}}B_y + \hat{\mathbf{z}}B_z)$$
$$= \hat{\mathbf{x}}(A_x + B_x) + \hat{\mathbf{y}}(A_y + B_y) + \hat{\mathbf{z}}(A_z + B_z). \tag{3.8}$$

Subtraction of vector $\mathbf{B}$ from vector $\mathbf{A}$ is equivalent to the addition of $\mathbf{A}$ to negative $\mathbf{B}$. Thus,

$$\mathbf{D} = \mathbf{A} - \mathbf{B}$$
$$= \mathbf{A} + (-\mathbf{B})$$
$$= \hat{\mathbf{x}}(A_x - B_x) + \hat{\mathbf{y}}(A_y - B_y) + \hat{\mathbf{z}}(A_z - B_z). \tag{3.9}$$

Graphically, the same rules used for vector addition are also applicable to vector subtraction; the only difference is that the arrowhead of $(-\mathbf{B})$ is drawn on the opposite end of the line segment representing the vector $\mathbf{B}$ (i.e., the tail and head are interchanged).

3-1.3 Position and Distance Vectors

In a given coordinate system, the *position vector* of a point P in space is the vector from the origin to P. Points P_1 and P_2 in Fig. 3-4 are located at (x_1, y_1, z_1)

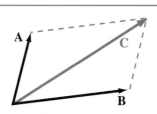

(a) Parallelogram rule

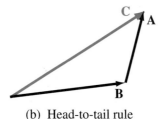

(b) Head-to-tail rule

Figure 3-3: Vector addition by (a) the parallelogram rule and (b) the head-to-tail rule.

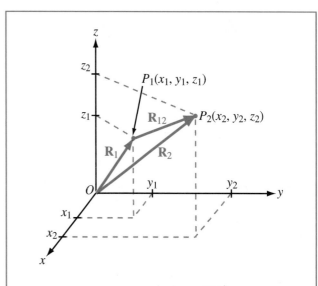

Figure 3-4: Position vector $\mathbf{R}_{12} = \overrightarrow{P_1 P_2} = \mathbf{R}_2 - \mathbf{R}_1$.

and (x_2, y_2, z_2), respectively. Their position vectors are

$$\mathbf{R}_1 = \overrightarrow{OP_1} = \hat{\mathbf{x}}x_1 + \hat{\mathbf{y}}y_1 + \hat{\mathbf{z}}z_1, \qquad (3.10a)$$

$$\mathbf{R}_2 = \overrightarrow{OP_2} = \hat{\mathbf{x}}x_2 + \hat{\mathbf{y}}y_2 + \hat{\mathbf{z}}z_2, \qquad (3.10b)$$

where point O is the origin. The *distance vector* from P_1 to P_2 is defined as

$$
\begin{aligned}
\mathbf{R}_{12} &= \overrightarrow{P_1 P_2} \\
&= \mathbf{R}_2 - \mathbf{R}_1 \\
&= \hat{\mathbf{x}}(x_2 - x_1) + \hat{\mathbf{y}}(y_2 - y_1) + \hat{\mathbf{z}}(z_2 - z_1), \quad (3.11)
\end{aligned}
$$

and the distance d between P_1 and P_2 is equal to the magnitude of $\mathbf{R}_{12}$:

$$
\begin{aligned}
d &= |\mathbf{R}_{12}| \\
&= [(x_2 - x_1)^2 + (y_2 - y_1)^2 + (z_2 - z_1)^2]^{1/2}. \quad (3.12)
\end{aligned}
$$

Note that the first subscript of $\mathbf{R}_{12}$ denotes the location of the tail of vector $\mathbf{R}_{12}$ and the second subscript denotes the location of its head, as shown in Fig. 3-4.

3-1.4 Vector Multiplication

Three types of products can occur in vector calculus. These are the simple, scalar (or dot), and vector (or cross) products.

Simple Product

Multiplication of a vector by a scalar is called a *simple product*. The product of the vector $\mathbf{A} = \hat{\mathbf{a}}A$ by a scalar k results in a vector $\mathbf{B}$ whose magnitude is kA and whose direction is the same as that of $\mathbf{A}$. That is,

$$\mathbf{B} = k\mathbf{A} = \hat{\mathbf{a}}kA = \hat{\mathbf{x}}(kA_x) + \hat{\mathbf{y}}(kA_y) + \hat{\mathbf{z}}(kA_z). \quad (3.13)$$

Scalar or Dot Product

The *scalar* (or *dot*) *product* of two vectors $\mathbf{A}$ and $\mathbf{B}$, denoted by $\mathbf{A} \cdot \mathbf{B}$ and pronounced "A dot B," is defined geometrically as the product of the magnitude of one of the vectors and the projection of the other vector onto the first one, or vice versa. Thus,

$$\boxed{\mathbf{A} \cdot \mathbf{B} = AB \cos\theta_{AB}, \quad (3.14)}$$

where θ_{AB} is the angle between $\mathbf{A}$ and $\mathbf{B}$, as shown in Fig. 3-5. The scalar product of two vectors yields a scalar whose magnitude is less than or equal to the products of the magnitudes of the two vectors (equality holds when $\theta_{AB} = 0$) and whose sign is positive if $0 < \theta_{AB} < 90°$ and negative if $90° < \theta_{AB} < 180°$. When $\theta_{AB} = 90°$, the two vectors are orthogonal. The dot product of two orthogonal vectors is zero. The quantity $A\cos\theta_{AB}$ is the component of $\mathbf{A}$ along $\mathbf{B}$ and

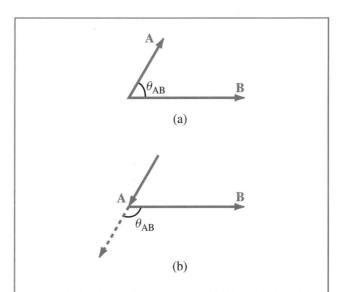

(a)

(b)

Figure 3-5: The angle θ_{AB} is the angle between $\mathbf{A}$ and $\mathbf{B}$, measured from $\mathbf{A}$ to $\mathbf{B}$ between vector tails. The dot product is positive if $0 \le \theta_{AB} < 90°$, as in (a), and it is negative if $90° < \theta_{AB} \le 180°$, as in (b).

is equal to the *projection* of vector **A** along the direction of vector **B**, and similarly $B \cos \theta_{AB}$ is the component of **B** along **A**.

If $\mathbf{A} = (A_x, A_y, A_z)$ and $\mathbf{B} = (B_x, B_y, B_z)$, then

$$\mathbf{A} \cdot \mathbf{B} = (\hat{\mathbf{x}} A_x + \hat{\mathbf{y}} A_y + \hat{\mathbf{z}} A_z) \cdot (\hat{\mathbf{x}} B_x + \hat{\mathbf{y}} B_y + \hat{\mathbf{z}} B_z). \quad (3.15)$$

Since the base vectors $\hat{\mathbf{x}}$, $\hat{\mathbf{y}}$, and $\hat{\mathbf{z}}$ are each orthogonal to the other two, it follows that

$$\hat{\mathbf{x}} \cdot \hat{\mathbf{x}} = \hat{\mathbf{y}} \cdot \hat{\mathbf{y}} = \hat{\mathbf{z}} \cdot \hat{\mathbf{z}} = 1, \quad (3.16a)$$
$$\hat{\mathbf{x}} \cdot \hat{\mathbf{y}} = \hat{\mathbf{y}} \cdot \hat{\mathbf{z}} = \hat{\mathbf{z}} \cdot \hat{\mathbf{x}} = 0. \quad (3.16b)$$

Use of Eqs. (3.16a) and (3.16b) in Eq. (3.15) leads to

$$\mathbf{A} \cdot \mathbf{B} = A_x B_x + A_y B_y + A_z B_z. \quad (3.17)$$

The dot product obeys both the commutative and distributive properties of multiplication; that is,

$$\mathbf{A} \cdot \mathbf{B} = \mathbf{B} \cdot \mathbf{A} \quad \text{(commutative property)}, \quad (3.18a)$$
$$\mathbf{A} \cdot (\mathbf{B} + \mathbf{C}) = \mathbf{A} \cdot \mathbf{B} + \mathbf{A} \cdot \mathbf{C} \quad \text{(distributive property)}. \quad (3.18b)$$

The dot product of a vector with itself gives

$$\mathbf{A} \cdot \mathbf{A} = |\mathbf{A}|^2 = A^2. \quad (3.19)$$

If the vector **A** is defined in a given coordinate system, its magnitude A can be determined from

$$A = |\mathbf{A}| = \sqrt[+]{\mathbf{A} \cdot \mathbf{A}}. \quad (3.20)$$

Also, if vectors **A** and **B** are specified in a given coordinate system, then the smaller angle between them, θ_{AB}, can be determined from

$$\theta_{AB} = \cos^{-1}\left[\frac{\mathbf{A} \cdot \mathbf{B}}{\sqrt[+]{\mathbf{A} \cdot \mathbf{A}} \sqrt[+]{\mathbf{B} \cdot \mathbf{B}}}\right]. \quad (3.21)$$

Vector or Cross Product

The *vector* (or *cross*) *product* of two vectors **A** and **B**, denoted by $\mathbf{A} \times \mathbf{B}$ and pronounced "*A* cross *B*," yields a vector defined as

$$\mathbf{A} \times \mathbf{B} = \hat{\mathbf{n}} \, AB \sin \theta_{AB}, \quad (3.22)$$

where θ_{AB} is the angle between **A** and **B**, measured *from* the tail of **A** *to* the tail of **B**, and $\hat{\mathbf{n}}$ is a unit vector normal to the plane containing **A** and **B**. The magnitude of the cross product is equal to the area of the parallelogram defined by the two vectors, as illustrated in Fig. 3-6(a), and its direction is specified by $\hat{\mathbf{n}}$ in accordance with the

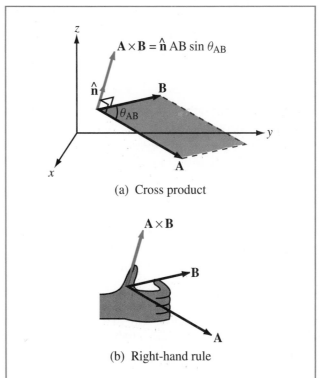

(a) Cross product

(b) Right-hand rule

Figure 3-6: Cross product $\mathbf{A} \times \mathbf{B}$ points in the direction $\hat{\mathbf{n}}$, which is perpendicular to the plane containing **A** and **B** and defined by the right-hand rule.

following *right-hand rule* [Fig. 3-6(b)]: the direction of $\hat{\mathbf{n}}$ points along the right thumb when the fingers rotate from **A** to **B** through the angle θ_{AB}. We note that, since $\hat{\mathbf{n}}$ is perpendicular to the plane containing **A** and **B**, it is also perpendicular to the vectors **A** and **B**.

The cross product is anticommutative, meaning that

$$\mathbf{A} \times \mathbf{B} = -\mathbf{B} \times \mathbf{A} \quad \text{(anticommutative)}. \quad (3.23)$$

This property can be verified by rotating the fingers of the right hand from **B** to **A** through the angle θ_{AB}. Other properties of the cross product include

$$\mathbf{A} \times (\mathbf{B} + \mathbf{C}) = \mathbf{A} \times \mathbf{B} + \mathbf{A} \times \mathbf{C} \quad \text{(distributive)}, (3.24\text{a})$$

$$\mathbf{A} \times \mathbf{A} = 0. \quad (3.24\text{b})$$

From the definition of the cross product given by Eq. (3.22), it is easy to verify that the base vectors $\hat{\mathbf{x}}$, $\hat{\mathbf{y}}$, and $\hat{\mathbf{z}}$ of the Cartesian coordinate system obey the following right-hand cyclic relations:

$$\hat{\mathbf{x}} \times \hat{\mathbf{y}} = \hat{\mathbf{z}}, \quad \hat{\mathbf{y}} \times \hat{\mathbf{z}} = \hat{\mathbf{x}}, \quad \hat{\mathbf{z}} \times \hat{\mathbf{x}} = \hat{\mathbf{y}}. \quad (3.25)$$

Note the cyclic order ($xyzxyz\ldots$). Also,

$$\hat{\mathbf{x}} \times \hat{\mathbf{x}} = \hat{\mathbf{y}} \times \hat{\mathbf{y}} = \hat{\mathbf{z}} \times \hat{\mathbf{z}} = 0. \quad (3.26)$$

If $\mathbf{A} = (A_x, A_y, A_z)$ and $\mathbf{B} = (B_x, B_y, B_z)$, use of Eqs. (3.25) and (3.26) leads to

$$\mathbf{A} \times \mathbf{B} = (\hat{\mathbf{x}}A_x + \hat{\mathbf{y}}A_y + \hat{\mathbf{z}}A_z) \times (\hat{\mathbf{x}}B_x + \hat{\mathbf{y}}B_y + \hat{\mathbf{z}}B_z)$$
$$= \hat{\mathbf{x}}(A_y B_z - A_z B_y) + \hat{\mathbf{y}}(A_z B_x - A_x B_z)$$
$$+ \hat{\mathbf{z}}(A_x B_y - A_y B_x). \quad (3.27)$$

The cyclical form of the result given by Eq. (3.27) allows us to express the cross product in the form of a determinant:

$$\mathbf{A} \times \mathbf{B} = \begin{vmatrix} \hat{\mathbf{x}} & \hat{\mathbf{y}} & \hat{\mathbf{z}} \\ A_x & A_y & A_z \\ B_x & B_y & B_z \end{vmatrix}. \quad (3.28)$$

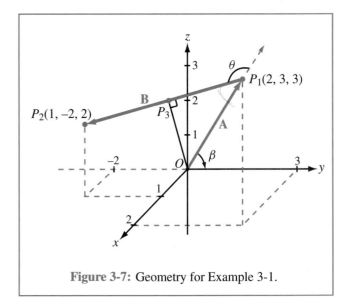

Figure 3-7: Geometry for Example 3-1.

Example 3-1   Vectors and Angles

In Cartesian coordinates, vector **A** is directed from the origin to point $P_1(2, 3, 3)$, and vector **B** is directed from P_1 to point $P_2(1, -2, 2)$. Find

(a) vector **A**, its magnitude A, and unit vector $\hat{\mathbf{a}}$,
(b) the angle that **A** makes with the y-axis,
(c) vector **B**,
(d) the angle between **A** and **B**, and
(e) the perpendicular distance from the origin to vector **B**.

Solution: (a) Vector **A** is given by the distance vector from $O(0, 0, 0)$ to $P_1(2, 3, 3)$ as shown in Fig. 3-7. Thus,

$$\mathbf{A} = \hat{\mathbf{x}}2 + \hat{\mathbf{y}}3 + \hat{\mathbf{z}}3,$$

$$A = |\mathbf{A}| = \sqrt{2^2 + 3^2 + 3^2} = \sqrt{22},$$

$$\hat{\mathbf{a}} = \frac{\mathbf{A}}{A} = (\hat{\mathbf{x}}2 + \hat{\mathbf{y}}3 + \hat{\mathbf{z}}3)/\sqrt{22}.$$

(b) The angle β between **A** and the y-axis is obtained from

$$\mathbf{A} \cdot \hat{\mathbf{y}} = |\mathbf{A}| \cos \beta,$$

or

$$\beta = \cos^{-1}\left(\frac{\mathbf{A}\cdot\hat{\mathbf{y}}}{A}\right) = \cos^{-1}\left(\frac{3}{\sqrt{22}}\right) = 50.2°.$$

(c)

$$\mathbf{B} = \hat{\mathbf{x}}(1-2) + \hat{\mathbf{y}}(-2-3) + \hat{\mathbf{z}}(2-3) = -\hat{\mathbf{x}} - \hat{\mathbf{y}}5 - \hat{\mathbf{z}}.$$

(d)

$$\theta = \cos^{-1}\left[\frac{\mathbf{A}\cdot\mathbf{B}}{|\mathbf{A}||\mathbf{B}|}\right] = \cos^{-1}\left[\frac{(-2-15-3)}{\sqrt{22}\,\sqrt{27}}\right] = 145.1°.$$

(e) The perpendicular distance between the origin and vector $\mathbf{B}$ is the distance $|\,\overrightarrow{OP_3}\,|$ shown in Fig. 3-7. From right triangle OP_1P_3,

$$|\,\overrightarrow{OP_3}\,| = |\mathbf{A}|\sin(180° - \theta)$$
$$= \sqrt{22}\,\sin(180° - 145.1°) = 2.68. \quad \blacksquare$$

EXERCISE 3.1 Find the distance vector between $P_1(1, 2, 3)$ and $P_2(-1, -2, 3)$ in Cartesian coordinates.

Ans. $\overrightarrow{P_1P_2} = -\hat{\mathbf{x}}2 - \hat{\mathbf{y}}4.$

EXERCISE 3.2 Find the angle θ between vectors $\mathbf{A}$ and $\mathbf{B}$ of Example 3-1 using the cross product between them.

Ans. $\theta = 145.1°.$

EXERCISE 3.3 Find the angle that vector $\mathbf{B}$ of Example 3-1 makes with the z-axis.

Ans. $101.1°.$

3-1.5   Scalar and Vector Triple Products

When three vectors are multiplied, not all combinations of dot and cross products are meaningful. For example, the product

$$\mathbf{A} \times (\mathbf{B}\cdot\mathbf{C})$$

does not make sense because $\mathbf{B}\cdot\mathbf{C}$ gives a scalar, and the cross product of the vector $\mathbf{A}$ with a scalar is not defined under the rules of vector algebra. Other than the product of the form $\mathbf{A}(\mathbf{B}\cdot\mathbf{C})$, the only two meaningful products of three vectors are the scalar triple product and the vector triple product.

Scalar Triple Product

The dot product of a vector with the cross product of two other vectors is called a scalar triple product, so named because the result is a scalar. A scalar triple product obeys the following cyclic order:

$$\boxed{\mathbf{A}\cdot(\mathbf{B}\times\mathbf{C}) = \mathbf{B}\cdot(\mathbf{C}\times\mathbf{A}) = \mathbf{C}\cdot(\mathbf{A}\times\mathbf{B}).} \quad (3.29)$$

The equalities hold as long as the cyclic order $(ABCABC\ldots)$ is preserved. The scalar triple product of vectors $\mathbf{A} = (A_x, A_y, A_z)$, $\mathbf{B} = (B_x, B_y, B_z)$, and $\mathbf{C} = (C_x, C_y, C_z)$ can be written in the form of a 3×3 determinant:

$$\mathbf{A}\cdot(\mathbf{B}\times\mathbf{C}) = \begin{vmatrix} A_x & A_y & A_z \\ B_x & B_y & B_z \\ C_x & C_y & C_z \end{vmatrix}. \quad (3.30)$$

The validity of Eqs. (3.29) and (3.30) can be verified by expanding $\mathbf{A}$, $\mathbf{B}$, and $\mathbf{C}$ in component form and carrying out the multiplications.

Vector Triple Product

The vector triple product involves the cross product of a vector with the cross product of two others, such as

$$\mathbf{A} \times (\mathbf{B} \times \mathbf{C}). \tag{3.31}$$

Since each cross product yields a vector, the result of a vector triple product is also a vector. The vector triple product does not, in general, obey the associative law. That is,

$$\mathbf{A} \times (\mathbf{B} \times \mathbf{C}) \neq (\mathbf{A} \times \mathbf{B}) \times \mathbf{C}, \tag{3.32}$$

which means that it is important to specify which cross multiplication is to be performed first. By expanding the vectors $\mathbf{A}$, $\mathbf{B}$, and $\mathbf{C}$ in component form, it can be shown that

$$\boxed{\mathbf{A} \times (\mathbf{B} \times \mathbf{C}) = \mathbf{B}(\mathbf{A} \cdot \mathbf{C}) - \mathbf{C}(\mathbf{A} \cdot \mathbf{B}),} \tag{3.33}$$

which sometimes is known as the "bac-cab" rule.

Example 3-2   Vector Triple Product

Given $\mathbf{A} = \hat{\mathbf{x}} - \hat{\mathbf{y}} + \hat{\mathbf{z}}2$, $\mathbf{B} = \hat{\mathbf{y}} + \hat{\mathbf{z}}$, and $\mathbf{C} = -\hat{\mathbf{x}}2 + \hat{\mathbf{z}}3$, find $(\mathbf{A} \times \mathbf{B}) \times \mathbf{C}$ and compare it with $\mathbf{A} \times (\mathbf{B} \times \mathbf{C})$.

Solution

$$\mathbf{A} \times \mathbf{B} = \begin{vmatrix} \hat{\mathbf{x}} & \hat{\mathbf{y}} & \hat{\mathbf{z}} \\ 1 & -1 & 2 \\ 0 & 1 & 1 \end{vmatrix} = -\hat{\mathbf{x}}3 - \hat{\mathbf{y}} + \hat{\mathbf{z}}$$

and

$$(\mathbf{A} \times \mathbf{B}) \times \mathbf{C} = \begin{vmatrix} \hat{\mathbf{x}} & \hat{\mathbf{y}} & \hat{\mathbf{z}} \\ -3 & -1 & 1 \\ -2 & 0 & 3 \end{vmatrix} = -\hat{\mathbf{x}}3 + \hat{\mathbf{y}}7 - \hat{\mathbf{z}}2.$$

A similar procedure gives $\mathbf{A} \times (\mathbf{B} \times \mathbf{C}) = \hat{\mathbf{x}}2 + \hat{\mathbf{y}}4 + \hat{\mathbf{z}}$. The fact that the results of two vector triple products are different is a demonstration of the inequality stated in Eq. (3.32). ∎

REVIEW QUESTIONS

Q3.1   When are two vectors equal and when are they identical?

Q3.2   When is the position vector of a point identical to the distance vector between two points?

Q3.3   If $\mathbf{A} \cdot \mathbf{B} = 0$, what is θ_{AB}?

Q3.4   If $\mathbf{A} \times \mathbf{B} = 0$, what is θ_{AB}?

Q3.5   Is $\mathbf{A}(\mathbf{B} \cdot \mathbf{C})$ a vector triple product?

Q3.6   If $\mathbf{A} \cdot \mathbf{B} = \mathbf{A} \cdot \mathbf{C}$, does it follow that $\mathbf{B} = \mathbf{C}$?

3-2 Orthogonal Coordinate Systems

In electromagnetics, the physical quantities we deal with are, in general, functions of space and time. A three-dimensional coordinate system allows us to uniquely specify the location of a point in space or the direction of a vector quantity. Coordinate systems may be orthogonal or nonorthogonal. An *orthogonal coordinate system* is one whose coordinates are mutually perpendicular, whereas in a nonorthogonal system not all three coordinates are mutually perpendicular. Nonorthogonal systems are very specialized and seldom used in solving practical problems. Many orthogonal coordinate systems have been devised, but the most standard and commonly used are

- the Cartesian (otherwise called the rectangular) coordinate system,
- the cylindrical coordinate system, and
- the spherical coordinate system.

Why do we need more than one coordinate system? Whereas a point in space has the same location and an object has the same shape regardless of which specific coordinate system is used to describe them, the solution of a given practical problem can be greatly facilitated by the proper choice of a coordinate system that best fits the geometry of the problem. Hence, in the following subsections we shall examine the properties of each of the aforementioned orthogonal systems, and in Section 3-3 we shall describe how a point or vector may be transformed from one coordinate system to another.

3-2.1 Cartesian Coordinates

The Cartesian coordinate system was introduced in Section 3-1, where we used it to illustrate many of the laws of vector algebra. In lieu of repeating these laws for the Cartesian system, we have summarized them for easy access in Table 3-1. In differential calculus, we often work with differential quantities. Differential length in Cartesian coordinates is a vector (Fig. 3-8) defined as

$$d\mathbf{l} = \hat{\mathbf{x}}\,dl_x + \hat{\mathbf{y}}\,dl_y + \hat{\mathbf{z}}\,dl_z = \hat{\mathbf{x}}\,dx + \hat{\mathbf{y}}\,dy + \hat{\mathbf{z}}\,dz, \quad (3.34)$$

where $dl_x = dx$ is a differential length along $\hat{\mathbf{x}}$, and similar definitions apply to $dl_y = dy$ and $dl_z = dz$.

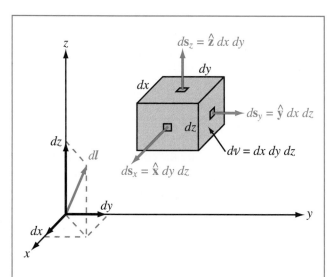

Figure 3-8: Differential length, area, and volume in Cartesian coordinates.

A differential surface area $d\mathbf{s}$ is a vector quantity with a magnitude ds equal to the product of two differential lengths (such as dl_y and dl_z), and its direction is denoted by a unit vector along the third direction (such as $\hat{\mathbf{x}}$). Thus, for a differential area in the y–z plane,

$$d\mathbf{s}_x = \hat{\mathbf{x}}\,dl_y\,dl_z = \hat{\mathbf{x}}\,dy\,dz \quad (y\text{–}z \text{ plane}), \quad (3.35\text{a})$$

with the subscript of $d\mathbf{s}$ denoting its direction. Similarly,

$$d\mathbf{s}_y = \hat{\mathbf{y}}\,dx\,dz \quad (x\text{–}z \text{ plane}), \quad (3.35\text{b})$$

$$d\mathbf{s}_z = \hat{\mathbf{z}}\,dx\,dy \quad (x\text{–}y \text{ plane}). \quad (3.35\text{c})$$

A differential volume is equal to the product of all three differential lengths:

$$dv = dx\,dy\,dz. \quad (3.36)$$

3-2.2 Cylindrical Coordinates

A cylindrical coordinate system is useful for solving problems having cylindrical symmetry, such as calculating the capacitance per unit length of a coaxial transmission line. The location of a point in space is uniquely defined by three variables, r, ϕ, and z, as shown in Fig. 3-9. The coordinate r is the *radial distance* in the x–y plane, ϕ is the *azimuth angle* measured from the positive x-axis, and z is as previously defined in the Cartesian coordinate system. Their ranges are $0 \leq r < \infty$, $0 \leq \phi < 2\pi$, and $-\infty < z < \infty$. Point $P(r_1, \phi_1, z_1)$ in Fig. 3-9 is located at the intersection of three surfaces. These are the cylindrical surface defined by $r = r_1$, the vertical half-plane defined by $\phi = \phi_1$ (which extends outwardly from the z-axis), and the horizontal plane defined by $z = z_1$. The mutually perpendicular base vectors are $\hat{\mathbf{r}}, \hat{\boldsymbol{\phi}}$, and $\hat{\mathbf{z}}$, with $\hat{\mathbf{r}}$ pointing away from the origin along r, $\hat{\boldsymbol{\phi}}$ pointing in a direction tangential to the cylindrical surface, and $\hat{\mathbf{z}}$ pointing along the vertical. Unlike the Cartesian system, in which the base vectors $\hat{\mathbf{x}}, \hat{\mathbf{y}}$, and $\hat{\mathbf{z}}$ are independent of the location of P, in the cylindrical system both $\hat{\mathbf{r}}$ and $\hat{\boldsymbol{\phi}}$ are functions of ϕ.

Table 3-1: Summary of vector relations.

	Cartesian Coordinates	Cylindrical Coordinates	Spherical Coordinates
Coordinate variables	x, y, z	r, ϕ, z	R, θ, ϕ
Vector representation, A =	$\hat{\mathbf{x}}A_x + \hat{\mathbf{y}}A_y + \hat{\mathbf{z}}A_z$	$\hat{\mathbf{r}}A_r + \hat{\boldsymbol{\phi}}A_\phi + \hat{\mathbf{z}}A_z$	$\hat{\mathbf{R}}A_R + \hat{\boldsymbol{\theta}}A_\theta + \hat{\boldsymbol{\phi}}A_\phi$
Magnitude of A, $\lvert A \rvert$ =	$\sqrt[+]{A_x^2 + A_y^2 + A_z^2}$	$\sqrt[+]{A_r^2 + A_\phi^2 + A_z^2}$	$\sqrt[+]{A_R^2 + A_\theta^2 + A_\phi^2}$
Position vector $\overrightarrow{OP_1}$ =	$\hat{\mathbf{x}}x_1 + \hat{\mathbf{y}}y_1 + \hat{\mathbf{z}}z_1,$ for $P(x_1, y_1, z_1)$	$\hat{\mathbf{r}}r_1 + \hat{\mathbf{z}}z_1,$ for $P(r_1, \phi_1, z_1)$	$\hat{\mathbf{R}}R_1,$ for $P(R_1, \theta_1, \phi_1)$
Base vectors properties	$\hat{\mathbf{x}}\cdot\hat{\mathbf{x}} = \hat{\mathbf{y}}\cdot\hat{\mathbf{y}} = \hat{\mathbf{z}}\cdot\hat{\mathbf{z}} = 1$ $\hat{\mathbf{x}}\cdot\hat{\mathbf{y}} = \hat{\mathbf{y}}\cdot\hat{\mathbf{z}} = \hat{\mathbf{z}}\cdot\hat{\mathbf{x}} = 0$ $\hat{\mathbf{x}}\times\hat{\mathbf{y}} = \hat{\mathbf{z}}$ $\hat{\mathbf{y}}\times\hat{\mathbf{z}} = \hat{\mathbf{x}}$ $\hat{\mathbf{z}}\times\hat{\mathbf{x}} = \hat{\mathbf{y}}$	$\hat{\mathbf{r}}\cdot\hat{\mathbf{r}} = \hat{\boldsymbol{\phi}}\cdot\hat{\boldsymbol{\phi}} = \hat{\mathbf{z}}\cdot\hat{\mathbf{z}} = 1$ $\hat{\mathbf{r}}\cdot\hat{\boldsymbol{\phi}} = \hat{\boldsymbol{\phi}}\cdot\hat{\mathbf{z}} = \hat{\mathbf{z}}\cdot\hat{\mathbf{r}} = 0$ $\hat{\mathbf{r}}\times\hat{\boldsymbol{\phi}} = \hat{\mathbf{z}}$ $\hat{\boldsymbol{\phi}}\times\hat{\mathbf{z}} = \hat{\mathbf{r}}$ $\hat{\mathbf{z}}\times\hat{\mathbf{r}} = \hat{\boldsymbol{\phi}}$	$\hat{\mathbf{R}}\cdot\hat{\mathbf{R}} = \hat{\boldsymbol{\theta}}\cdot\hat{\boldsymbol{\theta}} = \hat{\boldsymbol{\phi}}\cdot\hat{\boldsymbol{\phi}} = 1$ $\hat{\mathbf{R}}\cdot\hat{\boldsymbol{\theta}} = \hat{\boldsymbol{\theta}}\cdot\hat{\boldsymbol{\phi}} = \hat{\boldsymbol{\phi}}\cdot\hat{\mathbf{R}} = 0$ $\hat{\mathbf{R}}\times\hat{\boldsymbol{\theta}} = \hat{\boldsymbol{\phi}}$ $\hat{\boldsymbol{\theta}}\times\hat{\boldsymbol{\phi}} = \hat{\mathbf{R}}$ $\hat{\boldsymbol{\phi}}\times\hat{\mathbf{R}} = \hat{\boldsymbol{\theta}}$
Dot product, A · B =	$A_x B_x + A_y B_y + A_z B_z$	$A_r B_r + A_\phi B_\phi + A_z B_z$	$A_R B_R + A_\theta B_\theta + A_\phi B_\phi$
Cross product, A × B =	$\begin{vmatrix} \hat{\mathbf{x}} & \hat{\mathbf{y}} & \hat{\mathbf{z}} \\ A_x & A_y & A_z \\ B_x & B_y & B_z \end{vmatrix}$	$\begin{vmatrix} \hat{\mathbf{r}} & \hat{\boldsymbol{\phi}} & \hat{\mathbf{z}} \\ A_r & A_\phi & A_z \\ B_r & B_\phi & B_z \end{vmatrix}$	$\begin{vmatrix} \hat{\mathbf{R}} & \hat{\boldsymbol{\theta}} & \hat{\boldsymbol{\phi}} \\ A_R & A_\theta & A_\phi \\ B_R & B_\theta & B_\phi \end{vmatrix}$
Differential length, $d\mathbf{l}$ =	$\hat{\mathbf{x}}\,dx + \hat{\mathbf{y}}\,dy + \hat{\mathbf{z}}\,dz$	$\hat{\mathbf{r}}\,dr + \hat{\boldsymbol{\phi}}r\,d\phi + \hat{\mathbf{z}}\,dz$	$\hat{\mathbf{R}}\,dR + \hat{\boldsymbol{\theta}}R\,d\theta + \hat{\boldsymbol{\phi}}R\sin\theta\,d\phi$
Differential surface areas	$d\mathbf{s}_x = \hat{\mathbf{x}}\,dy\,dz$ $d\mathbf{s}_y = \hat{\mathbf{y}}\,dx\,dz$ $d\mathbf{s}_z = \hat{\mathbf{z}}\,dx\,dy$	$d\mathbf{s}_r = \hat{\mathbf{r}}r\,d\phi\,dz$ $d\mathbf{s}_\phi = \hat{\boldsymbol{\phi}}\,dr\,dz$ $d\mathbf{s}_z = \hat{\mathbf{z}}r\,dr\,d\phi$	$d\mathbf{s}_R = \hat{\mathbf{R}}R^2\sin\theta\,d\theta\,d\phi$ $d\mathbf{s}_\theta = \hat{\boldsymbol{\theta}}R\sin\theta\,dR\,d\phi$ $d\mathbf{s}_\phi = \hat{\boldsymbol{\phi}}R\,dR\,d\theta$
Differential volume, $d\mathcal{V}$ =	$dx\,dy\,dz$	$r\,dr\,d\phi\,dz$	$R^2\sin\theta\,dR\,d\theta\,d\phi$

The base unit vectors obey the following right-hand cyclic relations:

$$\hat{\mathbf{r}}\times\hat{\boldsymbol{\phi}} = \hat{\mathbf{z}}, \qquad \hat{\boldsymbol{\phi}}\times\hat{\mathbf{z}} = \hat{\mathbf{r}}, \qquad \hat{\mathbf{z}}\times\hat{\mathbf{r}} = \hat{\boldsymbol{\phi}}, \qquad (3.37)$$

and like all unit vectors, $\hat{\mathbf{r}}\cdot\hat{\mathbf{r}} = \hat{\boldsymbol{\phi}}\cdot\hat{\boldsymbol{\phi}} = \hat{\mathbf{z}}\cdot\hat{\mathbf{z}} = 1$, and $\hat{\mathbf{r}}\times\hat{\mathbf{r}} = \hat{\boldsymbol{\phi}}\times\hat{\boldsymbol{\phi}} = \hat{\mathbf{z}}\times\hat{\mathbf{z}} = 0$.

In cylindrical coordinates, a vector is expressed as

$$\mathbf{A} = \hat{\mathbf{a}}\lvert\mathbf{A}\rvert = \hat{\mathbf{r}}A_r + \hat{\boldsymbol{\phi}}A_\phi + \hat{\mathbf{z}}A_z, \qquad (3.38)$$

where A_r, A_ϕ, and A_z are the components of **A** along the $\hat{\mathbf{r}}$-, $\hat{\boldsymbol{\phi}}$-, and $\hat{\mathbf{z}}$-directions. The magnitude of **A** is obtained by applying Eq. (3.20), which gives

$$\lvert\mathbf{A}\rvert = \sqrt[+]{\mathbf{A}\cdot\mathbf{A}} = \sqrt[+]{A_r^2 + A_\phi^2 + A_z^2}. \qquad (3.39)$$

The position vector $\overrightarrow{OP}$ shown in Fig. 3-9 has components along r and z only. Thus,

$$\mathbf{R}_1 = \overrightarrow{OP} = \hat{\mathbf{r}}r_1 + \hat{\mathbf{z}}z_1. \qquad (3.40)$$

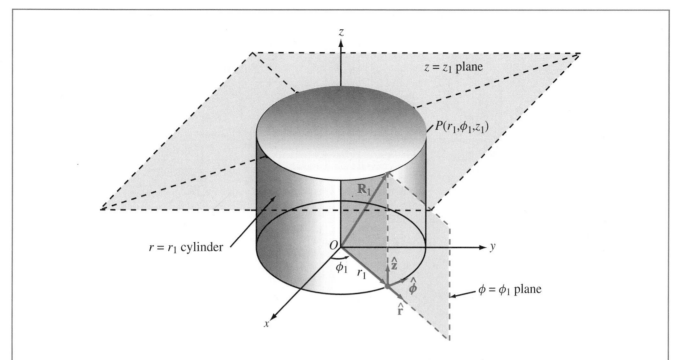

Figure 3-9: Point $P(r_1, \phi_1, z_1)$ in cylindrical coordinates; r_1 is the radial distance from the origin in the x–y plane, ϕ_1 is the angle in the x–y plane measured from the x-axis toward the y-axis, and z_1 is the vertical distance from the x–y plane.

The dependence of $\mathbf{R}_1$ on ϕ_1 is implicit through the dependence of $\hat{\mathbf{r}}$ on ϕ_1. Hence, when using Eq. (3.40) to denote the position vector of point $P(r_1, \phi_1, z_1)$, it is necessary to specify that $\hat{\mathbf{r}}$ is at ϕ_1.

Figure 3-10 shows a differential volume element in cylindrical coordinates. The differential lengths along $\hat{\mathbf{r}}$, $\hat{\boldsymbol{\phi}}$, and $\hat{\mathbf{z}}$ are

$$dl_r = dr, \quad dl_\phi = r\,d\phi, \quad dl_z = dz. \quad (3.41)$$

Note that the differential length along $\hat{\boldsymbol{\phi}}$ is $r\,d\phi$, not just $d\phi$. The differential length $d\mathbf{l}$ in cylindrical coordinates is given by

$$d\mathbf{l} = \hat{\mathbf{r}}\,dl_r + \hat{\boldsymbol{\phi}}\,dl_\phi + \hat{\mathbf{z}}\,dl_z = \hat{\mathbf{r}}\,dr + \hat{\boldsymbol{\phi}}r\,d\phi + \hat{\mathbf{z}}\,dz. \quad (3.42)$$

As was stated previously for the Cartesian coordinate system, the product of any pair of differential lengths is equal to the magnitude of a vector differential surface area with a surface normal pointing along the direction of the third coordinate. Thus,

$$d\mathbf{s}_r = \hat{\mathbf{r}}\,dl_\phi\,dl_z = \hat{\mathbf{r}}r\,d\phi\,dz \quad (\phi\text{–}z \text{ cylindrical surface}),$$
$$(3.43a)$$
$$d\mathbf{s}_\phi = \hat{\boldsymbol{\phi}}\,dl_r\,dl_z = \hat{\boldsymbol{\phi}}\,dr\,dz \quad (r\text{–}z \text{ plane}), \quad (3.43b)$$
$$d\mathbf{s}_z = \hat{\mathbf{z}}\,dl_r\,dl_\phi = \hat{\mathbf{z}}r\,dr\,d\phi \quad (r\text{–}\phi \text{ plane}). \quad (3.43c)$$

The differential volume is the product of the three differential lengths,

$$dv = dl_r\,dl_\phi\,dl_z = r\,dr\,d\phi\,dz. \quad (3.44)$$

The preceding properties of the cylindrical coordinate system are summarized in Table 3-1.

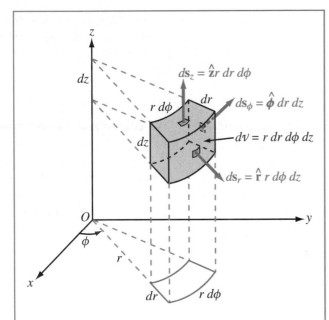

Figure 3-10: Differential areas and volume in cylindrical coordinates.

Example 3-3 **Distance Vector in Cylindrical Coordinates**

Find an expression for the unit vector of vector **A** shown in Fig. 3-11 in cylindrical coordinates.

Solution: In triangle OP_1P_2,

$$\overrightarrow{OP_2} = \overrightarrow{OP_1} + \mathbf{A}.$$

Hence,

$$\mathbf{A} = \overrightarrow{OP_2} - \overrightarrow{OP_1} = \hat{\mathbf{r}}r_0 - \hat{\mathbf{z}}h,$$

and

$$\hat{\mathbf{a}} = \frac{\mathbf{A}}{|\mathbf{A}|} = \frac{\hat{\mathbf{r}}r_0 - \hat{\mathbf{z}}h}{\sqrt{r_0^2 + h^2}}.$$

We note that the expression for **A** is independent of ϕ_0. That is, all vectors from point P_1 to any point on the

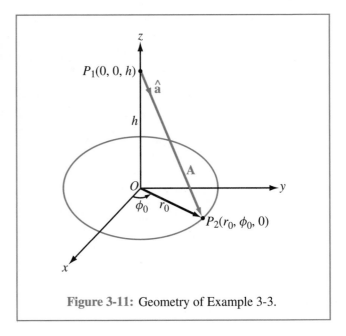

Figure 3-11: Geometry of Example 3-3.

circle defined by $r = r_0$ in the x–y plane are equal in the cylindrical coordinate system. The ambiguity can be eliminated by specifying that **A** passes through a point whose $\phi = \phi_0$. ∎

Example 3-4 **Cylindrical Area**

Find the area of a cylindrical surface described by $r = 5$, $30° \leq \phi \leq 60°$, and $0 \leq z \leq 3$ (Fig. 3-12).

Solution: The prescribed surface is shown in Fig. 3-12. Use of Eq. (3.43a) for a surface element with constant r gives

$$S = r \int_{\phi=30°}^{60°} d\phi \int_{z=0}^{3} dz = 5\phi \Big|_{\pi/6}^{\pi/3} z \Big|_{0}^{3} = \frac{5\pi}{2}.$$

Note that ϕ had to be converted to radians before evaluating the integration limits. ∎

EXERCISE 3.4 A circular cylinder of radius $r = 5$ cm is concentric with the z-axis and extends between

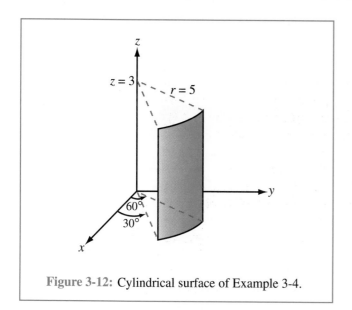

Figure 3-12: Cylindrical surface of Example 3-4.

$z = -3$ cm and $z = 3$ cm. Use Eq. (3.44) to find the cylinder's volume.

Ans. 471.2 cm^3.

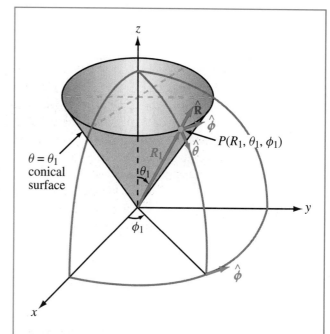

Figure 3-13: Point $P(R_1, \theta_1, \phi_1)$ in spherical coordinates.

3-2.3 Spherical Coordinates

In the spherical coordinate system, the location of a point in space is uniquely specified by the variables R, θ, and ϕ, as shown in Fig. 3-13. The coordinate R, which sometimes is called the *range* coordinate, describes a sphere of radius R centered at the origin. The *zenith* angle θ is measured from the positive z-axis and it describes a conical surface with its apex at the origin, and the azimuth angle ϕ is the same as in the cylindrical coordinate system. The ranges of R, θ, and ϕ are $0 \leq R < \infty$, $0 \leq \theta \leq \pi$, and $0 \leq \phi < 2\pi$. The base vectors $\hat{\mathbf{R}}$, $\hat{\boldsymbol{\theta}}$, and $\hat{\boldsymbol{\phi}}$ obey the following right-hand cyclic relations:

$$\hat{\mathbf{R}} \times \hat{\boldsymbol{\theta}} = \hat{\boldsymbol{\phi}}, \quad \hat{\boldsymbol{\theta}} \times \hat{\boldsymbol{\phi}} = \hat{\mathbf{R}}, \quad \hat{\boldsymbol{\phi}} \times \hat{\mathbf{R}} = \hat{\boldsymbol{\theta}}. \quad (3.45)$$

A vector with components A_R, A_θ, and A_ϕ is written as

$$\mathbf{A} = \hat{\mathbf{a}}|\mathbf{A}| = \hat{\mathbf{R}}A_R + \hat{\boldsymbol{\theta}}A_\theta + \hat{\boldsymbol{\phi}}A_\phi, \quad (3.46)$$

and its magnitude is given by

$$|\mathbf{A}| = \sqrt[+]{\mathbf{A} \cdot \mathbf{A}} = \sqrt[+]{A_R^2 + A_\theta^2 + A_\phi^2}. \quad (3.47)$$

The position vector of point $P(R_1, \theta_1, \phi_1)$ is simply

$$\mathbf{R}_1 = \overrightarrow{OP} = \hat{\mathbf{R}}R_1, \quad (3.48)$$

while keeping in mind that $\hat{\mathbf{R}}$ is implicitly dependent on θ_1 and ϕ_1.

As shown in Fig. 3-14, the differential lengths along $\hat{\mathbf{R}}$, $\hat{\boldsymbol{\theta}}$, and $\hat{\boldsymbol{\phi}}$ are

$$dl_R = dR, \quad dl_\theta = R \, d\theta, \quad dl_\phi = R \sin\theta \, d\phi. \quad (3.49)$$

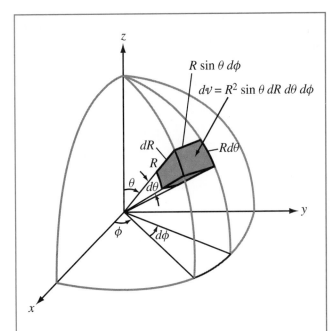

Figure 3-14: Differential volume in spherical coordinates.

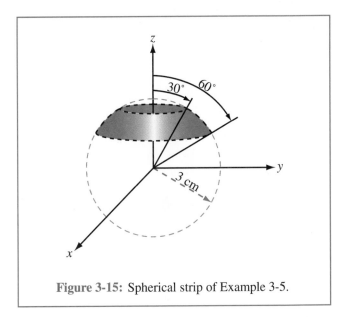

Figure 3-15: Spherical strip of Example 3-5.

Hence, the expressions for the vector differential length $d\mathbf{l}$, the vector differential surface $d\mathbf{s}$, and the differential volume dv are

$$d\mathbf{l} = \hat{\mathbf{R}}\, dl_R + \hat{\boldsymbol{\theta}}\, dl_\theta + \hat{\boldsymbol{\phi}}\, dl_\phi$$

$$= \hat{\mathbf{R}}\, dR + \hat{\boldsymbol{\theta}} R\, d\theta + \hat{\boldsymbol{\phi}} R \sin\theta\, d\phi, \qquad (3.50\text{a})$$

$$d\mathbf{s}_R = \hat{\mathbf{R}}\, dl_\theta\, dl_\phi = \hat{\mathbf{R}} R^2 \sin\theta\, d\theta\, d\phi$$

$$(\theta{-}\phi \text{ spherical surface}), \qquad (3.50\text{b})$$

$$d\mathbf{s}_\theta = \hat{\boldsymbol{\theta}}\, dl_R\, dl_\phi = \hat{\boldsymbol{\theta}} R \sin\theta\, dR\, d\phi$$

$$(R{-}\phi \text{ conical surface}), \qquad (3.50\text{c})$$

$$d\mathbf{s}_\phi = \hat{\boldsymbol{\phi}}\, dl_R\, dl_\theta = \hat{\boldsymbol{\phi}} R\, dR\, d\theta \quad (R{-}\theta \text{ plane}), \quad (3.50\text{d})$$

$$dv = dl_R\, dl_\theta\, dl_\phi = R^2 \sin\theta\, dR\, d\theta\, d\phi. \qquad (3.50\text{e})$$

These relations are summarized in Table 3-1.

<hr>

Example 3-5   Surface Area in Spherical Coordinates

The spherical strip shown in Fig. 3-15 is a section of a sphere of radius 3 cm. Find the area of the strip.

Solution: Use of Eq. (3.50b) for the area of an elemental spherical area with constant radius R gives

$$S = R^2 \int_{\theta=30°}^{60°} \sin\theta\, d\theta \int_{\phi=0}^{2\pi} d\phi$$

$$= 9(-\cos\theta)\Big|_{30°}^{60°}\ \phi\Big|_0^{2\pi} \quad (\text{cm}^2)$$

$$= 18\pi(\cos 30° - \cos 60°) = 20.7 \text{ cm}^2. \ \blacksquare$$

Example 3-6   Charge in a Sphere

A sphere of radius 2 cm contains a volume charge density ρ_v given by

$$\rho_v = 4\cos^2\theta \quad (\text{C/m}^3).$$

Find the total charge Q contained in the sphere.

Solution

$$Q = \int_v \rho_v \, dv$$

$$= \int_{\phi=0}^{2\pi} \int_{\theta=0}^{\pi} \int_{R=0}^{2\times10^{-2}} (4\cos^2\theta) R^2 \sin\theta \, dR \, d\theta \, d\phi$$

$$= 4 \int_0^{2\pi} \int_0^{\pi} \left(\frac{R^3}{3}\right)\Big|_0^{2\times10^{-2}} \sin\theta \cos^2\theta \, d\theta \, d\phi$$

$$= \frac{32}{3} \times 10^{-6} \int_0^{2\pi} \left(-\frac{\cos^3\theta}{3}\right)\Big|_0^{\pi} d\phi$$

$$= \frac{64}{9} \times 10^{-6} \int_0^{2\pi} d\phi$$

$$= \frac{128\pi}{9} \times 10^{-6} = 44.68 \quad (\mu C).$$

Note that the limits on R were converted to meters prior to evaluating the integral on R. ■

3-3 Transformations between Coordinate Systems

The position of a given point in space is invariant with respect to the choice of coordinate system. That is, its location is the same irrespective of which specific coordinate system is used to represent it. The same is true for vectors. In this section, we shall establish the relations between the variables (x, y, z) of the Cartesian system, (r, ϕ, z) of the cylindrical system, and (R, θ, ϕ) of the spherical system. These relations will then be used to transform vectors expressed in any one of the three systems into vectors expressed in any of the other two.

3-3.1 Cartesian to Cylindrical Transformations

Point P in Fig. 3-16 has Cartesian coordinates (x, y, z) and cylindrical coordinates (r, ϕ, z). Both systems share the coordinate z, and the relations between the other two

pairs of coordinates can be obtained from the geometry in Fig. 3-16. They are

$$r = \sqrt[+]{x^2 + y^2}, \qquad \phi = \tan^{-1}\left(\frac{y}{x}\right), \qquad (3.51)$$

and the inverse relations are

$$x = r\cos\phi, \qquad y = r\sin\phi. \qquad (3.52)$$

Next, with the help of Fig. 3-17, which shows the

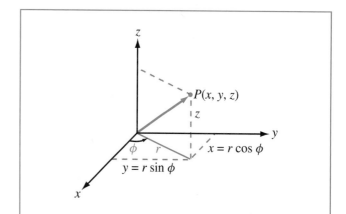

Figure 3-16: Interrelationships between Cartesian coordinates (x, y, z) and cylindrical coordinates (r, ϕ, z).

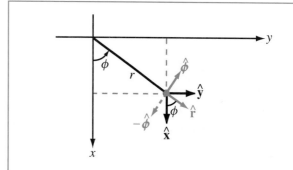

Figure 3-17: Interrelationships between base vectors $(\hat{x}, \hat{y})$ and $(\hat{r}, \hat{\phi})$.

directions of the unit vectors $\hat{\mathbf{x}}$, $\hat{\mathbf{y}}$, $\hat{\mathbf{r}}$, and $\hat{\boldsymbol{\phi}}$ in the x–y plane, we obtain the following relations:

$$\hat{\mathbf{r}} \cdot \hat{\mathbf{x}} = \cos \phi, \qquad \hat{\mathbf{r}} \cdot \hat{\mathbf{y}} = \sin \phi, \qquad (3.53a)$$

$$\hat{\boldsymbol{\phi}} \cdot \hat{\mathbf{x}} = -\sin \phi, \qquad \hat{\boldsymbol{\phi}} \cdot \hat{\mathbf{y}} = \cos \phi, \qquad (3.53b)$$

To express $\hat{\mathbf{r}}$ in terms of $\hat{\mathbf{x}}$ and $\hat{\mathbf{y}}$, let us write $\hat{\mathbf{r}}$ as

$$\hat{\mathbf{r}} = \hat{\mathbf{x}}a + \hat{\mathbf{y}}b, \qquad (3.54)$$

where a and b are unknown transformation coefficients. The dot product $\hat{\mathbf{r}} \cdot \hat{\mathbf{x}}$ gives

$$\hat{\mathbf{r}} \cdot \hat{\mathbf{x}} = \hat{\mathbf{x}} \cdot \hat{\mathbf{x}}a + \hat{\mathbf{y}} \cdot \hat{\mathbf{x}}b = a. \qquad (3.55)$$

Comparison of Eq. (3.55) with Eq. (3.53a) leads to the conclusion that $a = \cos \phi$. Similarly, application of the dot product $\hat{\mathbf{r}} \cdot \hat{\mathbf{y}}$ to Eq. (3.54) gives $b = \sin \phi$. Hence,

$$\boxed{\hat{\mathbf{r}} = \hat{\mathbf{x}} \cos \phi + \hat{\mathbf{y}} \sin \phi. \qquad (3.56a)}$$

Repetition of the procedure for $\hat{\boldsymbol{\phi}}$ leads to

$$\boxed{\hat{\boldsymbol{\phi}} = -\hat{\mathbf{x}} \sin \phi + \hat{\mathbf{y}} \cos \phi. \qquad (3.56b)}$$

The third base vector $\hat{\mathbf{z}}$ is the same in both coordinate systems. By solving Eqs. (3.56a) and (3.56b) simultaneously for $\hat{\mathbf{x}}$ and $\hat{\mathbf{y}}$, we obtain the following inverse relations:

$$\boxed{\begin{aligned} \hat{\mathbf{x}} &= \hat{\mathbf{r}} \cos \phi - \hat{\boldsymbol{\phi}} \sin \phi, & (3.57a) \\ \hat{\mathbf{y}} &= \hat{\mathbf{r}} \sin \phi + \hat{\boldsymbol{\phi}} \cos \phi. & (3.57b) \end{aligned}}$$

The relations given by Eqs. (3.56a) to (3.57b) are not only useful for transforming the base vectors $(\hat{\mathbf{x}}, \hat{\mathbf{y}})$ into $(\hat{\mathbf{r}}, \hat{\boldsymbol{\phi}})$, and vice versa, they can also be used to transform the components of a vector expressed in either coordinate system into its corresponding components expressed in the other system. For example, a vector $\mathbf{A} = \hat{\mathbf{x}}A_x + \hat{\mathbf{y}}A_y + \hat{\mathbf{z}}A_z$ in Cartesian coordinates can be

transformed into $\mathbf{A} = \hat{\mathbf{r}}A_r + \hat{\boldsymbol{\phi}}A_\phi + \hat{\mathbf{z}}A_z$ in cylindrical coordinates by applying Eqs. (3.56a) and (3.56b). That is,

$$\boxed{\begin{aligned} A_r &= A_x \cos \phi + A_y \sin \phi, & (3.58a) \\ A_\phi &= -A_x \sin \phi + A_y \cos \phi, & (3.58b) \end{aligned}}$$

and, conversely,

$$\boxed{\begin{aligned} A_x &= A_r \cos \phi - A_\phi \sin \phi, & (3.59a) \\ A_y &= A_r \sin \phi + A_\phi \cos \phi. & (3.59b) \end{aligned}}$$

The transformation relations given in this and the following two subsections are summarized in Table 3-2.

Example 3-7   Cartesian to Cylindrical Transformations

Given point $P_1(3, -4, 3)$ and vector $\mathbf{A} = \hat{\mathbf{x}}2 - \hat{\mathbf{y}}3 + \hat{\mathbf{z}}4$, defined in Cartesian coordinates, express P_1 and $\mathbf{A}$ in cylindrical coordinates and evaluate $\mathbf{A}$ at P_1.

Solution: For point P_1, $x = 3$, $y = -4$, and $z = 3$. Using Eq. (3.51), we have

$$r = \sqrt{x^2 + y^2} = 5, \qquad \phi = \tan^{-1}\frac{y}{x} = -53.1° = 306.9°,$$

and z remains unchanged. Hence, $P_1 = P_1(5, 306.9°, 3)$ in cylindrical coordinates.

For vector $\mathbf{A} = \hat{\mathbf{r}}A_r + \hat{\boldsymbol{\phi}}A_\phi + \hat{\mathbf{z}}A_z$ in cylindrical coordinates, its components can be determined by applying Eqs. (3.58a) and (3.58b):

$$A_r = A_x \cos \phi + A_y \sin \phi = 2 \cos \phi - 3 \sin \phi,$$

$$A_\phi = -A_x \sin \phi + A_y \cos \phi = -2 \sin \phi - 3 \cos \phi,$$

$$A_z = 4.$$

Hence,

$$\mathbf{A} = \hat{\mathbf{r}}(2 \cos \phi - 3 \sin \phi) - \hat{\boldsymbol{\phi}}(2 \sin \phi + 3 \cos \phi) + \hat{\mathbf{z}}4.$$

At point P, $\phi = 306.9°$, which gives

$$\mathbf{A} = \hat{\mathbf{r}}3.60 - \hat{\boldsymbol{\phi}}0.20 + \hat{\mathbf{z}}4. \qquad \blacksquare$$

Table 3-2: Coordinate transformation relations.

Transformation	Coordinate Variables	Unit Vectors	Vector Components
Cartesian to cylindrical	$r = \sqrt[+]{x^2 + y^2}$ $\phi = \tan^{-1}(y/x)$ $z = z$	$\hat{\mathbf{r}} = \hat{\mathbf{x}}\cos\phi + \hat{\mathbf{y}}\sin\phi$ $\hat{\boldsymbol{\phi}} = -\hat{\mathbf{x}}\sin\phi + \hat{\mathbf{y}}\cos\phi$ $\hat{\mathbf{z}} = \hat{\mathbf{z}}$	$A_r = A_x\cos\phi + A_y\sin\phi$ $A_\phi = -A_x\sin\phi + A_y\cos\phi$ $A_z = A_z$
Cylindrical to Cartesian	$x = r\cos\phi$ $y = r\sin\phi$ $z = z$	$\hat{\mathbf{x}} = \hat{\mathbf{r}}\cos\phi - \hat{\boldsymbol{\phi}}\sin\phi$ $\hat{\mathbf{y}} = \hat{\mathbf{r}}\sin\phi + \hat{\boldsymbol{\phi}}\cos\phi$ $\hat{\mathbf{z}} = \hat{\mathbf{z}}$	$A_x = A_r\cos\phi - A_\phi\sin\phi$ $A_y = A_r\sin\phi + A_\phi\cos\phi$ $A_z = A_z$
Cartesian to spherical	$R = \sqrt[+]{x^2 + y^2 + z^2}$ $\theta = \tan^{-1}[\sqrt[+]{x^2 + y^2}/z]$ $\phi = \tan^{-1}(y/x)$	$\hat{\mathbf{R}} = \hat{\mathbf{x}}\sin\theta\cos\phi$ $\qquad + \hat{\mathbf{y}}\sin\theta\sin\phi + \hat{\mathbf{z}}\cos\theta$ $\hat{\boldsymbol{\theta}} = \hat{\mathbf{x}}\cos\theta\cos\phi$ $\qquad + \hat{\mathbf{y}}\cos\theta\sin\phi - \hat{\mathbf{z}}\sin\theta$ $\hat{\boldsymbol{\phi}} = -\hat{\mathbf{x}}\sin\phi + \hat{\mathbf{y}}\cos\phi$	$A_R = A_x\sin\theta\cos\phi$ $\qquad + A_y\sin\theta\sin\phi + A_z\cos\theta$ $A_\theta = A_x\cos\theta\cos\phi$ $\qquad + A_y\cos\theta\sin\phi - A_z\sin\theta$ $A_\phi = -A_x\sin\phi + A_y\cos\phi$
Spherical to Cartesian	$x = R\sin\theta\cos\phi$ $y = R\sin\theta\sin\phi$ $z = R\cos\theta$	$\hat{\mathbf{x}} = \hat{\mathbf{R}}\sin\theta\cos\phi$ $\qquad + \hat{\boldsymbol{\theta}}\cos\theta\cos\phi - \hat{\boldsymbol{\phi}}\sin\phi$ $\hat{\mathbf{y}} = \hat{\mathbf{R}}\sin\theta\sin\phi$ $\qquad + \hat{\boldsymbol{\theta}}\cos\theta\sin\phi + \hat{\boldsymbol{\phi}}\cos\phi$ $\hat{\mathbf{z}} = \hat{\mathbf{R}}\cos\theta - \hat{\boldsymbol{\theta}}\sin\theta$	$A_x = A_R\sin\theta\cos\phi$ $\qquad + A_\theta\cos\theta\cos\phi - A_\phi'\sin\phi$ $A_y = A_R\sin\theta\sin\phi$ $\qquad + A_\theta\cos\theta\sin\phi + A_\phi\cos\phi$ $A_z = A_R\cos\theta - A_\theta\sin\theta$
Cylindrical to spherical	$R = \sqrt[+]{r^2 + z^2}$ $\theta = \tan^{-1}(r/z)$ $\phi = \phi$	$\hat{\mathbf{R}} = \hat{\mathbf{r}}\sin\theta + \hat{\mathbf{z}}\cos\theta$ $\hat{\boldsymbol{\theta}} = \hat{\mathbf{r}}\cos\theta - \hat{\mathbf{z}}\sin\theta$ $\hat{\boldsymbol{\phi}} = \hat{\boldsymbol{\phi}}$	$A_R = A_r\sin\theta + A_z\cos\theta$ $A_\theta = A_r\cos\theta - A_z\sin\theta$ $A_\phi = A_\phi$
Spherical to cylindrical	$r = R\sin\theta$ $\phi = \phi$ $z = R\cos\theta$	$\hat{\mathbf{r}} = \hat{\mathbf{R}}\sin\theta + \hat{\boldsymbol{\theta}}\cos\theta$ $\hat{\boldsymbol{\phi}} = \hat{\boldsymbol{\phi}}$ $\hat{\mathbf{z}} = \hat{\mathbf{R}}\cos\theta - \hat{\boldsymbol{\theta}}\sin\theta$	$A_r = A_R\sin\theta + A_\theta\cos\theta$ $A_\phi = A_\phi$ $A_z = A_R\cos\theta - A_\theta\sin\theta$

3-3.2 Cartesian to Spherical Transformations

From Fig. 3-18, we obtain the following relations between the Cartesian coordinates (x, y, z) and the spherical coordinates (R, θ, ϕ):

$$R = \sqrt[+]{x^2 + y^2 + z^2}, \qquad (3.60a)$$

$$\theta = \tan^{-1}\left[\frac{\sqrt[+]{x^2 + y^2}}{z}\right], \qquad (3.60b)$$

$$\phi = \tan^{-1}\left(\frac{y}{x}\right), \qquad (3.60c)$$

and the converse relations are

$$x = R\sin\theta\cos\phi, \qquad (3.61a)$$

$$y = R\sin\theta\sin\phi, \qquad (3.61b)$$

$$z = R\cos\theta. \qquad (3.61c)$$

The unit vector $\hat{\mathbf{R}}$ lies in the $\hat{\mathbf{r}}$–$\hat{\mathbf{z}}$ plane. Hence, it can be expressed as a linear combination of $\hat{\mathbf{r}}$ and $\hat{\mathbf{z}}$ as follows:

$$\hat{\mathbf{R}} = \hat{\mathbf{r}}a + \hat{\mathbf{z}}b, \qquad (3.62)$$

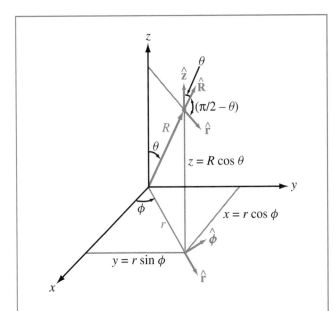

Figure 3-18: Interrelationships between (x, y, z) and (R, θ, ϕ).

where a and b are transformation coefficients. Since $\hat{\mathbf{r}}$ and $\hat{\mathbf{z}}$ are mutually perpendicular,

$$\hat{\mathbf{R}} \cdot \hat{\mathbf{r}} = a, \tag{3.63a}$$

$$\hat{\mathbf{R}} \cdot \hat{\mathbf{z}} = b. \tag{3.63b}$$

From Fig. 3-18, the angle between $\hat{\mathbf{R}}$ and $\hat{\mathbf{r}}$ is the complement of θ and that between $\hat{\mathbf{R}}$ and $\hat{\mathbf{z}}$ is θ. Hence, $a = \hat{\mathbf{R}} \cdot \hat{\mathbf{r}} = \sin \theta$ and $b = \hat{\mathbf{R}} \cdot \hat{\mathbf{z}} = \cos \theta$. Upon inserting these expressions for a and b in Eq. (3.62) and replacing $\hat{\mathbf{r}}$ with Eq. (3.56a), we have

$$\hat{\mathbf{R}} = \hat{\mathbf{x}} \sin \theta \cos \phi + \hat{\mathbf{y}} \sin \theta \sin \phi + \hat{\mathbf{z}} \cos \theta. \tag{3.64a}$$

A similar procedure can be followed to obtain the following expression for $\hat{\boldsymbol{\theta}}$:

$$\hat{\boldsymbol{\theta}} = \hat{\mathbf{x}} \cos \theta \cos \phi + \hat{\mathbf{y}} \cos \theta \sin \phi - \hat{\mathbf{z}} \sin \theta, \tag{3.64b}$$

and the expression for $\hat{\boldsymbol{\phi}}$ is given by Eq. (3.56b) as

$$\hat{\boldsymbol{\phi}} = -\hat{\mathbf{x}} \sin \phi + \hat{\mathbf{y}} \cos \phi. \tag{3.64c}$$

Equations (3.64a) through (3.64c) can be solved simultaneously to give the following expressions for $(\hat{\mathbf{x}}, \hat{\mathbf{y}}, \hat{\mathbf{z}})$ in terms of $(\hat{\mathbf{R}}, \hat{\boldsymbol{\theta}}, \hat{\boldsymbol{\phi}})$:

$$\hat{\mathbf{x}} = \hat{\mathbf{R}} \sin \theta \cos \phi + \hat{\boldsymbol{\theta}} \cos \theta \cos \phi - \hat{\boldsymbol{\phi}} \sin \phi, \tag{3.65a}$$

$$\hat{\mathbf{y}} = \hat{\mathbf{R}} \sin \theta \sin \phi + \hat{\boldsymbol{\theta}} \cos \theta \sin \phi + \hat{\boldsymbol{\phi}} \cos \phi, \tag{3.65b}$$

$$\hat{\mathbf{z}} = \hat{\mathbf{R}} \cos \theta - \hat{\boldsymbol{\theta}} \sin \theta. \tag{3.65c}$$

Equations (3.64a) to (3.65c) can also be used to transform (A_x, A_y, A_z) of vector $\mathbf{A}$ into its spherical components (A_R, A_θ, A_ϕ), and vice versa, by replacing $(\hat{\mathbf{x}}, \hat{\mathbf{y}}, \hat{\mathbf{z}}, \hat{\mathbf{R}}, \hat{\boldsymbol{\theta}}, \hat{\boldsymbol{\phi}})$ with $(A_x, A_y, A_z, A_R, A_\theta, A_\phi)$, respectively.

Example 3-8   Cartesian to Spherical Transformation

Express vector $\mathbf{A} = \hat{\mathbf{x}}(x+y) + \hat{\mathbf{y}}(y-x) + \hat{\mathbf{z}}z$ in spherical coordinates.

Solution: Using the transformation relation for A_R given in Table 3-2, we have

$$A_R = A_x \sin \theta \cos \phi + A_y \sin \theta \sin \phi + A_z \cos \theta$$
$$= (x + y) \sin \theta \cos \phi + (y - x) \sin \theta \sin \phi + z \cos \theta.$$

Using the expressions for x, y, and z given by Eq. (3.61c), we have

$$A_R = (R \sin \theta \cos \phi + R \sin \theta \sin \phi) \sin \theta \cos \phi$$
$$\qquad + (R \sin \theta \sin \phi - R \sin \theta \cos \phi) \sin \theta \sin \phi + R \cos^2 \theta$$
$$= R \sin^2 \theta (\cos^2 \phi + \sin^2 \phi) + R \cos^2 \theta$$
$$= R \sin^2 \theta + R \cos^2 \theta = R.$$

Similarly,

$$A_\theta = (x + y) \cos\theta \cos\phi + (y - x) \cos\theta \sin\phi - z\sin\theta,$$

$$A_\phi = -(x + y)\sin\phi + (y - x)\cos\phi,$$

and following the procedure used with A_R, we obtain the results

$$A_\theta = 0,$$

$$A_\phi = -R\sin\theta.$$

Hence,

$$\mathbf{A} = \hat{\mathbf{R}}A_R + \hat{\boldsymbol{\theta}}A_\theta + \hat{\boldsymbol{\phi}}A_\phi = \hat{\mathbf{R}}R - \hat{\boldsymbol{\phi}}R\sin\theta. \quad \blacksquare$$

3-3.3 Cylindrical to Spherical Transformations

Transformations between cylindrical and spherical coordinates can be realized by combining the transformation relations of the preceding two subsections. The results are given in Table 3-2.

3-3.4 Distance between Two Points

In Cartesian coordinates, the distance d between two points $P_1(x_1, y_1, z_1)$ and $P_2(x_2, y_2, z_2)$ is given by Eq. (3.12) as

$$d = |\mathbf{R}_{12}|$$
$$= [(x_2 - x_1)^2 + (y_2 - y_1)^2 + (z_2 - z_1)^2]^{1/2}. \quad (3.66)$$

Upon using Eq. (3.52) to convert the Cartesian coordinates of P_1 and P_2 into their cylindrical equivalents, we have

$$d = \big[(r_2 \cos\phi_2 - r_1 \cos\phi_1)^2$$
$$+ (r_2 \sin\phi_2 - r_1 \sin\phi_1)^2 + (z_2 - z_1)^2\big]^{1/2}$$
$$= \big[r_2^2 + r_1^2 - 2r_1 r_2 \cos(\phi_2 - \phi_1) + (z_2 - z_1)^2\big]^{1/2}$$
$$\text{(cylindrical).} \quad (3.67)$$

A similar transformation using Eqs. (3.61a-c) leads to an expression for d in terms of the spherical coordinates of P_1 and P_2:

$$d = \big\{R_2^2 + R_1^2 - 2R_1 R_2[\cos\theta_2 \cos\theta_1$$
$$+ \sin\theta_1 \sin\theta_2 \cos(\phi_2 - \phi_1)]\big\}^{1/2}$$
$$\text{(spherical).} \quad (3.68)$$

REVIEW QUESTIONS

Q3.7 Why do we use more than one coordinate system?

Q3.8 Why is it that the base vectors $(\hat{\mathbf{x}}, \hat{\mathbf{y}}, \hat{\mathbf{z}})$ are independent of the location of a point, but $\hat{\mathbf{r}}$ and $\hat{\boldsymbol{\phi}}$ are not?

Q3.9 What are the cyclic relations for the base vectors in (a) Cartesian coordinates, (b) cylindrical coordinates, and (c) spherical coordinates?

Q3.10 How is the position vector of a point in cylindrical coordinates related to its position vector in spherical coordinates?

EXERCISE 3.5 Point $P(2\sqrt{3}, \pi/3, -2)$ is given in cylindrical coordinates. Express P in spherical coordinates.

Ans. $P(4, 2\pi/3, \pi/3)$.

EXERCISE 3.6 Transform vector

$$\mathbf{A} = \hat{\mathbf{x}}(x + y) + \hat{\mathbf{y}}(y - x) + \hat{\mathbf{z}}z$$

from Cartesian to cylindrical coordinates.

Ans. $\mathbf{A} = \hat{\mathbf{r}}r - \hat{\boldsymbol{\phi}}r + \hat{\mathbf{z}}z$.

3-4   Gradient of a Scalar Field

When dealing with a scalar physical quantity whose magnitude depends on a single variable, such as the temperature T as a function of height z, the rate of change of T with height can be described by the derivative dT/dz. However, if T is also a function of x and y in a Cartesian coordinate system, its space rate of change becomes more difficult to describe because we now have to deal not only with three separate variables, but also do it in a unified arrangement. The differential change in T along x, y, and z can be described in terms of the partial derivatives of T with respect to the three coordinate variables, but it is not immediately obvious as to how we should combine the three partial derivatives so as to describe the space rate of change of T along a specified direction. Furthermore, many of the quantities we deal with in electromagnetics are vectors, and therefore both their magnitudes and directions may vary with spatial position. In vector calculus, we use three fundamental operators to describe the differential spatial variations of scalars and vectors; these are the *gradient*, *divergence*, and *curl* operators. The gradient operator applies to scalar fields and is the subject of the present section. The other two operators, which apply to vector fields, are discussed in succeeding sections.

Suppose that $T_1(x, y, z)$ is the temperature at point $P_1(x, y, z)$ in some region of space, and $T_2(x + dx, y + dy, z + dz)$ is the temperature at a nearby point P_2, as shown in Fig. 3-19. The differential distances dx, dy, and dz are the components of the differential distance vector $d\mathbf{l}$. That is,

$$d\mathbf{l} = \hat{\mathbf{x}}\, dx + \hat{\mathbf{y}}\, dy + \hat{\mathbf{z}}\, dz. \qquad (3.69)$$

From differential calculus, the differential temperature $dT = T_2 - T_1$ is given by

$$dT = \frac{\partial T}{\partial x}\, dx + \frac{\partial T}{\partial y}\, dy + \frac{\partial T}{\partial z}\, dz, \qquad (3.70)$$

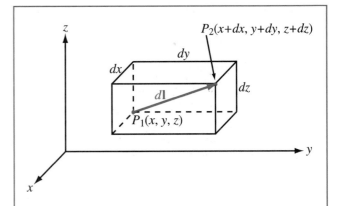

Figure 3-19: Differential distance vector $d\mathbf{l}$ between points P_1 and P_2.

and since by definition $dx = \hat{\mathbf{x}} \cdot d\mathbf{l}$, $dy = \hat{\mathbf{y}} \cdot d\mathbf{l}$, and $dz = \hat{\mathbf{z}} \cdot d\mathbf{l}$, Eq. (3.70) can be rewritten as

$$dT = \hat{\mathbf{x}}\frac{\partial T}{\partial x} \cdot d\mathbf{l} + \hat{\mathbf{y}}\frac{\partial T}{\partial y} \cdot d\mathbf{l} + \hat{\mathbf{z}}\frac{\partial T}{\partial z} \cdot d\mathbf{l}$$

$$= \left[\hat{\mathbf{x}}\frac{\partial T}{\partial x} + \hat{\mathbf{y}}\frac{\partial T}{\partial y} + \hat{\mathbf{z}}\frac{\partial T}{\partial z} \right] \cdot d\mathbf{l}. \qquad (3.71)$$

The vector inside the square brackets in Eq. (3.71) defines the change in temperature dT corresponding to a vector change in position $d\mathbf{l}$. This vector is called the *gradient* of T, or *grad T* for short, and it is usually written symbolically as ∇T. That is,

$$\nabla T = \text{grad } T \triangleq \hat{\mathbf{x}}\frac{\partial T}{\partial x} + \hat{\mathbf{y}}\frac{\partial T}{\partial y} + \hat{\mathbf{z}}\frac{\partial T}{\partial z}, \qquad (3.72)$$

and Eq. (3.71) can be expressed in the form

$$dT = \nabla T \cdot d\mathbf{l}. \qquad (3.73)$$

The symbol ∇ is called the *del* or *gradient operator* and is defined as

$$\nabla \triangleq \hat{\mathbf{x}}\frac{\partial}{\partial x} + \hat{\mathbf{y}}\frac{\partial}{\partial y} + \hat{\mathbf{z}}\frac{\partial}{\partial z} \qquad \text{(Cartesian)}. \qquad (3.74)$$

We note that, *whereas the gradient operator has no physical meaning by itself, it attains a physical meaning once it operates on a scalar physical quantity, and the result of the operation is a vector whose magnitude is equal to the maximum rate of change of the physical quantity per unit distance and whose direction is along the direction of maximum increase.* With $d\mathbf{l} = \hat{\mathbf{a}}_l dl$, where $\hat{\mathbf{a}}_l$ is the unit vector of $d\mathbf{l}$, the *directional derivative* of T along the direction $\hat{\mathbf{a}}_l$ is given by

$$\frac{dT}{dl} = \nabla T \cdot \hat{\mathbf{a}}_l. \quad (3.75)$$

If ∇T is a known function of the coordinate variables of a given coordinate system, we can find the difference $(T_2 - T_1)$, where T_1 and T_2 are the values of T at points P_1 and P_2, respectively, by integrating both sides of Eq. (3.73). Thus,

$$T_2 - T_1 = \int_{P_1}^{P_2} \nabla T \cdot d\mathbf{l}. \quad (3.76)$$

Example 3-9   Directional Derivative

Find the directional derivative of $T = x^2 + y^2 z$ along the direction $\hat{\mathbf{x}}2 + \hat{\mathbf{y}}3 - \hat{\mathbf{z}}2$ and evaluate it at $(1, -1, 2)$.

Solution: The directional derivative dT/dl is given by Eq. (3.75). First, we find the gradient of T:

$$\nabla T = \left(\hat{\mathbf{x}} \frac{\partial}{\partial x} + \hat{\mathbf{y}} \frac{\partial}{\partial y} + \hat{\mathbf{z}} \frac{\partial}{\partial z} \right) (x^2 + y^2 z)$$

$$= \hat{\mathbf{x}}2x + \hat{\mathbf{y}}2yz + \hat{\mathbf{z}}y^2.$$

We denote $\mathbf{l}$ as the given direction,

$$\mathbf{l} = \hat{\mathbf{x}}2 + \hat{\mathbf{y}}3 - \hat{\mathbf{z}}2.$$

Its unit vector is

$$\hat{\mathbf{a}}_l = \frac{\mathbf{l}}{|\mathbf{l}|} = \frac{\hat{\mathbf{x}}2 + \hat{\mathbf{y}}3 - \hat{\mathbf{z}}2}{\sqrt{2^2 + 3^2 + 2^2}} = \frac{\hat{\mathbf{x}}2 + \hat{\mathbf{y}}3 - \hat{\mathbf{z}}2}{\sqrt{17}}.$$

Application of Eq. (3.75) gives

$$\frac{dT}{dl} = \nabla T \cdot \hat{\mathbf{a}}_l = (\hat{\mathbf{x}}2x + \hat{\mathbf{y}}2yz + \hat{\mathbf{z}}y^2) \cdot \left(\frac{\hat{\mathbf{x}}2 + \hat{\mathbf{y}}3 - \hat{\mathbf{z}}2}{\sqrt{17}} \right)$$

$$= \frac{4x + 6yz - 2y^2}{\sqrt{17}}.$$

At $(1, -1, 2)$,

$$\left. \frac{dT}{dl} \right|_{(1,-1,2)} = \frac{4 - 12 - 2}{\sqrt{17}} = \frac{-10}{\sqrt{17}}. \quad \blacksquare$$

3-4.1 Gradient Operator in Cylindrical and Spherical Coordinates

Even though Eq. (3.73) was derived using Cartesian coordinates, it should be equally valid in any orthogonal coordinate system. In order to apply the gradient operator to a scalar quantity expressed in cylindrical or spherical coordinates, we need expressions for ∇ in those coordinate systems. To convert Eq. (3.72) into the cylindrical coordinates (r, ϕ, z), we start by restating the coordinate relations

$$r = \sqrt{x^2 + y^2}, \qquad \tan \phi = \frac{y}{x}. \quad (3.77)$$

From differential calculus,

$$\frac{\partial T}{\partial x} = \frac{\partial T}{\partial r} \frac{\partial r}{\partial x} + \frac{\partial T}{\partial \phi} \frac{\partial \phi}{\partial x} + \frac{\partial T}{\partial z} \frac{\partial z}{\partial x}. \quad (3.78)$$

Since z is orthogonal to x, the last term is equal to zero because $\partial z / \partial x = 0$. Using the coordinate relations given by Eq. (3.77), it is easy to show that

$$\frac{\partial r}{\partial x} = \frac{x}{\sqrt{x^2 + y^2}} = \cos \phi, \quad (3.79\text{a})$$

$$\frac{\partial \phi}{\partial x} = -\frac{1}{r} \sin \phi. \quad (3.79\text{b})$$

Hence,

$$\frac{\partial T}{\partial x} = \cos \phi \frac{\partial T}{\partial r} - \frac{\sin \phi}{r} \frac{\partial T}{\partial \phi}. \quad (3.80)$$

This expression can be used to replace the coefficient of $\hat{\mathbf{x}}$ in Eq. (3.72), and a similar procedure can be followed to obtain an expression for $\partial T/\partial y$ in terms of r and ϕ. If, in addition, we use the relations $\hat{\mathbf{x}} = \hat{\mathbf{r}}\cos\phi - \hat{\boldsymbol{\phi}}\sin\phi$ and $\hat{\mathbf{y}} = \hat{\mathbf{r}}\sin\phi + \hat{\boldsymbol{\phi}}\cos\phi$ [from Eqs. (3.57a) and (3.57b)], Eq. (3.72) becomes

$$\nabla T = \hat{\mathbf{r}}\frac{\partial T}{\partial r} + \hat{\boldsymbol{\phi}}\frac{1}{r}\frac{\partial T}{\partial \phi} + \hat{\mathbf{z}}\frac{\partial T}{\partial z}, \qquad (3.81)$$

and therefore the gradient operator in cylindrical coordinates can be defined as

$$\nabla = \hat{\mathbf{r}}\frac{\partial}{\partial r} + \hat{\boldsymbol{\phi}}\frac{1}{r}\frac{\partial}{\partial \phi} + \hat{\mathbf{z}}\frac{\partial}{\partial z} \qquad \text{(cylindrical)}. \quad (3.82)$$

A similar procedure leads to the following expression for the gradient in spherical coordinates:

$$\nabla = \hat{\mathbf{R}}\frac{\partial}{\partial R} + \hat{\boldsymbol{\theta}}\frac{1}{R}\frac{\partial}{\partial \theta} + \hat{\boldsymbol{\phi}}\frac{1}{R\sin\theta}\frac{\partial}{\partial \phi} \qquad \text{(spherical)}.$$
$$\qquad (3.83)$$

3-4.2   Properties of the Gradient Operator

For any two scalar functions U and V, the following relations apply:

(1) $\quad \nabla(U + V) = \nabla U + \nabla V,$ (3.84a)

(2) $\quad \nabla(UV) = U\,\nabla V + V\,\nabla U,$ (3.84b)

(3) $\quad \nabla V^n = nV^{n-1}\,\nabla V,$ for any n.   (3.84c)

The gradient of a vector is meaningless under the rules of vector calculus.

Example 3-10   **Calculating the Gradient**

Find the gradient of each of the following scalar functions and then evaluate it at the given point.

(a) $V_1 = 24V_0 \cos(\pi y/3)\sin(2\pi z/3)$ at $(3, 2, 1)$ in Cartesian coordinates,

(b) $V_2 = V_0 e^{-2r}\sin 3\phi$ at $(1, \pi/2, 3)$ in cylindrical coordinates,

(c) $V_3 = V_0(a/R)\cos 2\theta$ at $(2a, 0, \pi)$ in spherical coordinates.

Solution: (a) In Cartesian coordinates,

$$\nabla V_1 = \hat{\mathbf{x}}\frac{\partial V_1}{\partial x} + \hat{\mathbf{y}}\frac{\partial V_1}{\partial y} + \hat{\mathbf{z}}\frac{\partial V_1}{\partial z}$$

$$= -\hat{\mathbf{y}}8\pi V_0 \sin\frac{\pi y}{3}\sin\frac{2\pi z}{3} + \hat{\mathbf{z}}16\pi V_0 \cos\frac{\pi y}{3}\cos\frac{2\pi z}{3}$$

$$= 8\pi V_0 \left[-\hat{\mathbf{y}}\sin\frac{\pi y}{3}\sin\frac{2\pi z}{3} + \hat{\mathbf{z}}2\cos\frac{\pi y}{3}\cos\frac{2\pi z}{3}\right].$$

At $(3, 2, 1)$,

$$\nabla V_1 = 8\pi V_0 \left[-\hat{\mathbf{y}}\sin^2\frac{2\pi}{3} + \hat{\mathbf{z}}2\cos^2\frac{2\pi}{3}\right]$$

$$= \pi V_0 \left[-\hat{\mathbf{y}}6 + \hat{\mathbf{z}}4\right].$$

(b) The function V_2 is expressed in terms of cylindrical variables. Hence, we need to use Eq. (3.82) for ∇:

$$\nabla V_2 = \left(\hat{\mathbf{r}}\frac{\partial}{\partial r} + \hat{\boldsymbol{\phi}}\frac{1}{r}\frac{\partial}{\partial \phi} + \hat{\mathbf{z}}\frac{\partial}{\partial z}\right) V_0 e^{-2r}\sin 3\phi$$

$$= -\hat{\mathbf{r}}2V_0 e^{-2r}\sin 3\phi + \hat{\boldsymbol{\phi}}(3V_0 e^{-2r}\cos 3\phi)/r$$

$$= \left[-\hat{\mathbf{r}}2\sin 3\phi + \hat{\boldsymbol{\phi}}\frac{3\cos 3\phi}{r}\right] V_0 e^{-2r}.$$

At $(1, \pi/2, 3)$, $r = 1$ and $\phi = \pi/2$. Hence,

$$\nabla V_2 = \left[-\hat{\mathbf{r}}2\sin\frac{3\pi}{2} + \hat{\boldsymbol{\phi}}3\cos\frac{3\pi}{2}\right] V_0 e^{-2}$$

$$= \hat{\mathbf{r}}2V_0 e^{-2} = \hat{\mathbf{r}}0.27V_0.$$

(c) As V_3 is expressed in spherical coordinates, we apply Eq. (3.83) to V_3:

$$\nabla V_3 = \left(\hat{\mathbf{R}}\frac{\partial}{\partial R} + \hat{\boldsymbol{\theta}}\frac{1}{R}\frac{\partial}{\partial \theta} + \hat{\boldsymbol{\phi}}\frac{1}{R\sin\theta}\frac{\partial}{\partial \phi}\right) V_0\left(\frac{a}{R}\right)\cos 2\theta$$

$$= -\hat{\mathbf{R}}\frac{V_0 a}{R^2}\cos 2\theta - \hat{\boldsymbol{\theta}}\frac{2V_0 a}{R^2}\sin 2\theta$$

$$= -[\hat{\mathbf{R}}\cos 2\theta + \hat{\boldsymbol{\theta}}2\sin 2\theta]\frac{V_0 a}{R^2}.$$

At $(2a, 0, \pi)$, $R = 2a$ and $\theta = 0$, which gives

$$\nabla V_3 = -\hat{\mathbf{R}}\frac{V_0}{4a} . \quad \blacksquare$$

EXERCISE 3.7 Given $V = x^2 y + xy^2 + xz^2$, (a) find the gradient of V, and (b) evaluate it at $(1, -1, 2)$.

Ans. (a) $\nabla V = \hat{\mathbf{x}}(2xy + y^2 + z^2) + \hat{\mathbf{y}}(x^2 + 2xy) + \hat{\mathbf{z}}2xz$,
(b) $\nabla V\big|_{(1,-1,2)} = \hat{\mathbf{x}}3 - \hat{\mathbf{y}} + \hat{\mathbf{z}}4$.

EXERCISE 3.8 Find the directional derivative of $V = rz^2 \cos 2\phi$ along the direction $\mathbf{A} = \hat{\mathbf{r}}2 - \hat{\mathbf{z}}$ and evaluate it at $(1, \pi/2, 2)$.

Ans. $(dV/dl)\big|_{(1,\pi/2,2)} = -4/\sqrt{5}$.

3-5 Divergence of a Vector Field

From our brief introduction of Coulomb's law in Chapter 1, we know that an isolated, positive point charge q induces an electric field $\mathbf{E}$ in the space around it, with the direction of $\mathbf{E}$ being along the outward direction away from the charge. Also, the strength (magnitude) of $\mathbf{E}$ is proportional to q and decreases with distance R from the charge as $1/R^2$. In a graphical presentation, a vector field is usually represented by *field lines*, as shown in Fig. 3-20. The arrowhead denotes the direction of the field at the point where the field line is drawn, and the length of the line provides a qualitative depiction of the field's magnitude.

Even though the electric field vector does not actually move, we regard its presence as a flux flowing through space, and we refer to its field lines as *flux lines*. At a surface boundary, *flux density* is defined as the amount of outward flux crossing a unit surface ds:

$$\text{Flux density of } \mathbf{E} = \frac{\mathbf{E} \cdot d\mathbf{s}}{|d\mathbf{s}|} = \frac{\mathbf{E} \cdot \hat{\mathbf{n}}\, ds}{ds}, \quad (3.85)$$

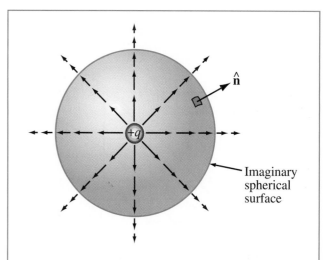

Figure 3-20: Flux lines of the electric field $\mathbf{E}$ due to a positive charge q.

where $\hat{\mathbf{n}}$ is the outward surface normal of $d\mathbf{s}$. The *total flux* crossing a closed surface S, such as the enclosed surface of the imaginary sphere outlined in Fig. 3-20, is

$$\text{Total flux} = \oint_S \mathbf{E} \cdot d\mathbf{s}. \quad (3.86)$$

Let us now consider the case of a differential rectangular parallelepiped, such as a cube, whose edges are lined up with the axes of a Cartesian coordinate system as shown in Fig. 3-21. The lengths of the edges are Δx along x, Δy along y, and Δz along z. A vector field $\mathbf{E}(x, y, z)$ exists in the region of space containing the parallelepiped, and we wish to determine the flux of $\mathbf{E}$ through its total surface S. Since S includes six faces, we need to sum up the fluxes through all of them, and by definition the flux through any face is the *outward* flux from the volume Δv through that face.

Let $\mathbf{E}$ be defined as

$$\mathbf{E} = \hat{\mathbf{x}}E_x + \hat{\mathbf{y}}E_y + \hat{\mathbf{z}}E_z. \quad (3.87)$$

The area of the face marked 1 in Fig. 3-21 is $\Delta y\, \Delta z$, and its unit vector $\hat{\mathbf{n}}_1 = -\hat{\mathbf{x}}$. Hence, the outward flux F_1

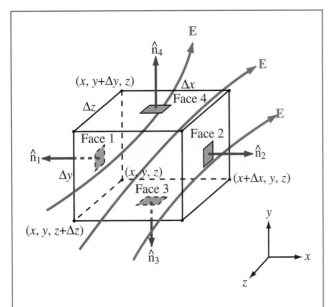

Figure 3-21: Flux lines of a vector field **E** passing through a differential rectangular parallelepiped of volume $\Delta v = \Delta x\, \Delta y\, \Delta z$.

through face 1 is

$$
\begin{aligned}
F_1 &= \int_{\text{Face 1}} \mathbf{E} \cdot \hat{\mathbf{n}}_1 \, ds \\
&= \int_{\text{Face 1}} (\hat{\mathbf{x}} E_x + \hat{\mathbf{y}} E_y + \hat{\mathbf{z}} E_z) \cdot (-\hat{\mathbf{x}}) \, dy \, dz \\
&= -E_x(1) \, \Delta y \, \Delta z, \qquad (3.88)
\end{aligned}
$$

where $E_x(1)$ is the value of E_x at the center of face 1. Approximating E_x over face 1 by its value at the center is justified by the assumption that the differential volume under consideration is very small.

Similarly, the flux out of face 2 (with $\hat{\mathbf{n}}_2 = \hat{\mathbf{x}}$) is

$$
F_2 = E_x(2) \, \Delta y \, \Delta z, \qquad (3.89)
$$

where $E_x(2)$ is the value of E_x at the center of face 2. Over a differential separation Δx, $E_x(2)$ is related to

$E_x(1)$ by

$$
E_x(2) = E_x(1) + \frac{\partial E_x}{\partial x} \, \Delta x, \qquad (3.90)
$$

where we have ignored higher-order terms involving $(\Delta x)^2$ and higher powers because their contributions are negligibly small when Δx is very small. Substituting Eq. (3.90) into Eq. (3.89) gives

$$
F_2 = \left[E_x(1) + \frac{\partial E_x}{\partial x} \, \Delta x \right] \Delta y \, \Delta z. \qquad (3.91)
$$

The sum of the fluxes out of faces 1 and 2 is obtained by adding Eqs. (3.88) and (3.91),

$$
F_1 + F_2 = \frac{\partial E_x}{\partial x} \, \Delta x \, \Delta y \, \Delta z. \qquad (3.92a)
$$

Repeating the same procedure to each of the other pairs of faces leads to

$$
F_3 + F_4 = \frac{\partial E_y}{\partial y} \, \Delta x \, \Delta y \, \Delta z, \qquad (3.92b)
$$

$$
F_5 + F_6 = \frac{\partial E_z}{\partial z} \, \Delta x \, \Delta y \, \Delta z. \qquad (3.92c)
$$

The sum of fluxes F_1 through F_6 gives the total flux through surface S of the parallelepiped:

$$
\begin{aligned}
\oint_S \mathbf{E} \cdot d\mathbf{s} &= \left(\frac{\partial E_x}{\partial x} + \frac{\partial E_y}{\partial y} + \frac{\partial E_z}{\partial z} \right) \Delta x \, \Delta y \, \Delta z \\
&= (\text{div } \mathbf{E}) \, \Delta v, \qquad (3.93)
\end{aligned}
$$

where $\Delta v = \Delta x \, \Delta y \, \Delta z$ and div **E** is a differential function called the **divergence** of **E** and is defined in Cartesian coordinates as

$$
\text{div } \mathbf{E} = \frac{\partial E_x}{\partial x} + \frac{\partial E_y}{\partial y} + \frac{\partial E_z}{\partial z}. \qquad (3.94)
$$

By shrinking the volume Δv to zero, we define the divergence of **E** at a point as the net outward flux per unit volume over a closed incremental surface. Thus, from Eq. (3.93), we have

$$\text{div } \mathbf{E} \triangleq \lim_{\Delta v \to 0} \frac{\oint_S \mathbf{E} \cdot d\mathbf{s}}{\Delta v}, \qquad (3.95)$$

where S encloses the elemental volume Δv. Instead of denoting the divergence of $\mathbf{E}$ by div $\mathbf{E}$, it is common practice to denote it as $\nabla \cdot \mathbf{E}$. That is,

$$\nabla \cdot \mathbf{E} \triangleq \text{div } \mathbf{E} = \frac{\partial E_x}{\partial x} + \frac{\partial E_y}{\partial y} + \frac{\partial E_z}{\partial z} \qquad (3.96)$$

for a vector $\mathbf{E}$ in Cartesian coordinates. Further discussion of these two forms of notation is given in Section 3-5.2.

From the definition of the divergence of $\mathbf{E}$ given by Eq. (3.95), the field $\mathbf{E}$ has positive divergence if the net flux out of surface S is positive, which may be viewed as if the volume Δv contains a *source* of flux. If the divergence is negative, Δv may be viewed as a *sink* because the net flux is inward into Δv. For a uniform field $\mathbf{E}$, the same amount of flux enters Δv as leaves from it; hence, its divergence is zero and the field is said to be *divergenceless*. The divergence is a differential operator, it can operate only on vectors, and the result of its operation is a scalar. This is in contrast with the gradient operator, which can operate only on scalars and the result is a vector. Expressions for the divergence of a vector in cylindrical and spherical coordinates are given on the inside back cover of this book.

The divergence operator is distributive. That is, for any pair of vectors $\mathbf{E}_1$ and $\mathbf{E}_2$,

$$\nabla \cdot (\mathbf{E}_1 + \mathbf{E}_2) = \nabla \cdot \mathbf{E}_1 + \nabla \cdot \mathbf{E}_2. \qquad (3.97)$$

If $\nabla \cdot \mathbf{E} = 0$, the vector field $\mathbf{E}$ is called *solenoidal*.

3-5.1 Divergence Theorem

The result given by Eq. (3.93) for a differential volume Δv can be extended to relate the volume integral of $\nabla \cdot \mathbf{E}$ over any volume v to the flux of $\mathbf{E}$ through the closed surface S that bounds v. That is,

$$\int_v \nabla \cdot \mathbf{E} \, dv = \oint_S \mathbf{E} \cdot d\mathbf{s} \qquad \text{(divergence theorem)}. \qquad (3.98)$$

This relationship, known as the *divergence theorem*, is used extensively in electromagnetics.

3-5.2 Remarks on Notation

Books on vector analysis use one or both of the following two notations for denoting the gradient, divergence, and curl operators. In the linguistic notation, the gradient of a scalar T is written as "grad T," and the divergence and curl of a vector $\mathbf{E}$ are written as "div $\mathbf{E}$" and "curl $\mathbf{E}$." A few years after James Clerk Maxwell published his classic treatise on electricity and magnetism in 1873, at which time vector analysis was not yet available, the American mathematician J. Willard Gibbs published a pair of pamphlets entitled *Elements of Vector Analysis*, in which he established the field of vector analysis as we know it today. In his presentation, he introduced a new notation for the divergence and curl operations: "$\nabla \cdot \mathbf{E}$" for div $\mathbf{E}$ and "$\nabla \times \mathbf{E}$" for curl $\mathbf{E}$. Because of its symbolic simplicity, Gibbs's notation was quickly adopted and is now used in the overwhelming majority of books on electromagnetics. The problem with Gibbs's notation, however, is that it is prone to misinterpretation. In Gibbs's notation, $\nabla \cdot \mathbf{E}$ stands for the divergence of vector $\mathbf{E}$. This is commonly misinterpreted as the dot product of the operator ∇ with the vector $\mathbf{E}$, although Gibbs himself did not call it that. The reason for the misinterpretation is that algebraically the dot product of ∇, as given by Eq. (3.74), with $\mathbf{E}$, as given by

Eq. (3.87), gives the same result as the expression for the divergence of **E** given by Eq. (3.96). However, the "dot product" interpretation is incorrect on two grounds. First, the operator ∇ is treated as if it were a vector quantity, which it is not. As we mentioned earlier, a differential operator has no physical meaning in itself and only attains a meaning after it operates on a physical quantity. This may be a subtle point, but it is nonetheless fundamental. The second criticism of the dot-product interpretation is that it can lead to invalid results under coordinate transformations, which is unacceptable because the divergence of a physical quantity should be the same regardless of the choice of coordinate system. Similar criticisms apply to the interpretation of $\nabla \times \mathbf{E}$ as the cross product of ∇ and **E**.

To avoid the misinterpretation problems associated with Gibbs's notation, Tai* introduced the notation $\nabla\, \mathbf{E}$ for the divergence of **E** and $\nabla\mathbf{E}$ for the curl of **E**. Tai's proposed notation, however, is relatively new, and it will take some time before its logical foundation is properly appreciated. Hence, in keeping with the established tradition, we will still adopt Gibbs's notation in this book, but we reiterate that "$\nabla\, \cdot$" and "$\nabla\times$" should not be regarded as the del operator followed with a dot or cross product.

Example 3-11 Calculating the Divergence

Determine the divergence of each of the following vector fields and then evaluate it at the indicated point:
(a) $\mathbf{E} = \hat{\mathbf{x}}3x^2 + \hat{\mathbf{y}}2z + \hat{\mathbf{z}}x^2z$ at $(2, -2, 0)$;
(b) $\mathbf{E} = \hat{\mathbf{R}}(a^3\cos\theta/R^2) - \hat{\boldsymbol{\theta}}(a^3\sin\theta/R^2)$ at $(a/2, 0, \pi)$.

Solution

(a) $\nabla \cdot \mathbf{E} = \dfrac{\partial E_x}{\partial x} + \dfrac{\partial E_y}{\partial y} + \dfrac{\partial E_z}{\partial z}$

$= \dfrac{\partial}{\partial x}(3x^2) + \dfrac{\partial}{\partial y}(2z) + \dfrac{\partial}{\partial z}(x^2z)$

$= 6x + 0 + x^2 = x^2 + 6x.$

---

*C. T. Tai, *Generalized Vector and Dyadic Analysis*, IEEE Press, New York, 1992, pp. 41–44.

At $(2, -2, 0)$, $\nabla \cdot \mathbf{E}\big|_{(2,-2,0)} = 16.$

(b) Using the expression given on the inside of the back cover of the book for the divergence of a vector in spherical coordinates, we have

$$\nabla \cdot \mathbf{E} = \frac{1}{R^2}\frac{\partial}{\partial R}(R^2 E_R) + \frac{1}{R\sin\theta}\frac{\partial}{\partial\theta}(E_\theta\sin\theta)$$

$$+ \frac{1}{R\sin\theta}\frac{\partial E_\phi}{\partial\phi}$$

$$= \frac{1}{R^2}\frac{\partial}{\partial R}(a^3\cos\theta) + \frac{1}{R\sin\theta}\frac{\partial}{\partial\theta}\left(-\frac{a^3\sin^2\theta}{R^2}\right)$$

$$= 0 - \frac{2a^3\cos\theta}{R^3} = -\frac{2a^3\cos\theta}{R^3}.$$

At $R = a/2$ and $\theta = 0$, $\nabla \cdot \mathbf{E}\big|_{(a/2,0,\pi)} = -16.$ ∎

EXERCISE 3.9 Given $\mathbf{A} = e^{-2y}(\hat{\mathbf{x}}\sin 2x + \hat{\mathbf{y}}\cos 2x)$, find $\nabla \cdot \mathbf{A}$.

Ans. $\nabla \cdot \mathbf{A} = 0.$

EXERCISE 3.10 Given $\mathbf{A} = \hat{\mathbf{r}}r\cos\phi + \hat{\boldsymbol{\phi}}r\sin\phi + \hat{\mathbf{z}}3z$, find $\nabla \cdot \mathbf{A}$ at $(2, 0, 3)$.

Ans. $\nabla \cdot \mathbf{A} = 6.$

EXERCISE 3.11 If $\mathbf{E} = \hat{\mathbf{R}}AR$ in spherical coordinates, calculate the flux of **E** through a spherical surface of radius a, centered at the origin.

Ans. $\displaystyle\oint_S \mathbf{E}\cdot d\mathbf{s} = 4\pi Aa^3.$

EXERCISE 3.12 Verify the divergence theorem by calculating the volume integral of the divergence of the field **E** of Exercise 3.11 over the volume bounded by the surface of radius a.

3-6 Curl of a Vector Field

So far we have defined and discussed two of the three fundamental operators used in vector analysis, the gradient of a scalar and the divergence of a vector.

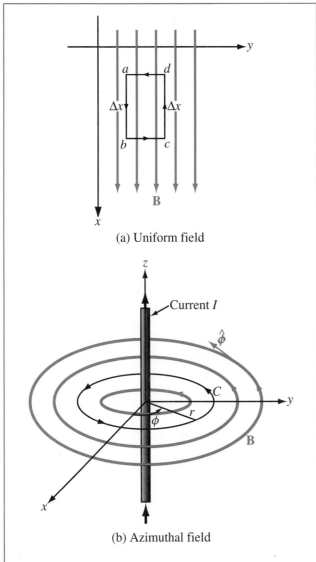

(a) Uniform field

(b) Azimuthal field

Figure 3-22: Circulation is zero for the uniform field in (a), but it is not zero for the azimuthal field in (b).

Now we introduce the *curl operator*. The curl of a vector field **B** describes the rotational property, or the *circulation*, of **B**. For a closed contour C, the circulation of **B** is defined as the line integral of **B** around C. That is,

$$\text{Circulation} = \oint_C \mathbf{B} \cdot d\mathbf{l}. \qquad (3.99)$$

To gain some physical understanding for this definition, we consider two examples. The first example is for a uniform field $\mathbf{B} = \hat{\mathbf{x}} B_0$, whose field lines are as depicted in Fig. 3-22(a). For the rectangular contour *abcd* shown in the figure, we have

$$\text{Circulation} = \int_a^b \hat{\mathbf{x}} B_0 \cdot \hat{\mathbf{x}}\, dx + \int_b^c \hat{\mathbf{x}} B_0 \cdot \hat{\mathbf{y}}\, dy$$

$$+ \int_c^d \hat{\mathbf{x}} B_0 \cdot \hat{\mathbf{x}}\, dx + \int_d^a \hat{\mathbf{x}} B_0 \cdot \hat{\mathbf{y}}\, dy$$

$$= B_0\, \Delta x - B_0\, \Delta x = 0, \qquad (3.100)$$

where $\Delta x = b - a = c - d$ and, because $\hat{\mathbf{x}} \cdot \hat{\mathbf{y}} = 0$, the second and fourth integrals are zero. According to Eq. (3.100), *the circulation of a uniform field is zero.*

Next, we consider the magnetic field **B** induced by an infinite wire carrying a d-c current I. If the current is in free space and it is oriented along the z-direction, then, from Eq. (1.13),

$$\mathbf{B} = \hat{\boldsymbol{\phi}} \frac{\mu_0 I}{2\pi r}, \qquad (3.101)$$

where μ_0 is the permeability of free space and r is the radial distance from the current in the x–y plane. The direction of **B** is along the azimuth direction $\hat{\boldsymbol{\phi}}$. The field lines of **B** are concentric circles around the current source, as shown in Fig. 3-22(b). For a circular contour of radius r, the differential length vector $d\mathbf{l} = \hat{\boldsymbol{\phi}} r\, d\phi$, and the circulation of **B** around C is

$$\text{Circulation} = \oint_C \mathbf{B} \cdot d\mathbf{l}$$

$$= \int_0^{2\pi} \hat{\boldsymbol{\phi}} \frac{\mu_0 I}{2\pi r} \cdot \hat{\boldsymbol{\phi}} r\, d\phi = \mu_0 I. \quad (3.102)$$

In this case, the circulation is not zero, but had the contour C been in any plane perpendicular to the x–y plane, $d\mathbf{l}$ would not have had a $\hat{\boldsymbol{\phi}}$ component, and the integral would have yielded a net circulation of zero. In other words, the magnitude of the circulation of $\mathbf{B}$ depends on the choice of contour. Also, the direction of the contour determines whether the circulation is positive or negative. If we are trying to describe the circulation of a tornado, for example, we would like to choose our contour such that the circulation of the wind field is maximum, and we would like the circulation to have both a magnitude and a direction, with the direction being toward the tornado's vortex. The *curl operator* has been defined to accommodate those properties. The curl of a vector field $\mathbf{B}$, denoted curl $\mathbf{B}$ or $\nabla \times \mathbf{B}$, is defined as

$$\nabla \times \mathbf{B} = \text{curl } \mathbf{B} \triangleq \lim_{\Delta s \to 0} \frac{1}{\Delta s} \left[\hat{\mathbf{n}} \oint_C \mathbf{B} \cdot d\mathbf{l} \right]_{\max} . \quad (3.103)$$

Thus, curl $\mathbf{B}$ is the circulation of $\mathbf{B}$ per unit area, with the area Δs of the contour C being oriented such that the circulation is maximum. The direction of curl $\mathbf{B}$ is $\hat{\mathbf{n}}$, the unit normal of Δs, defined according to the right-hand rule: with the four fingers of the right hand following the contour direction $d\mathbf{l}$, the thumb points along $\hat{\mathbf{n}}$ (Fig. 3-23). The curl of a vector is defined at a point; this follows from the definition given by Eq. (3.103) in which the circulation is normalized to the area Δs, and Δs is made to approach zero. As was mentioned earlier in Section 3-5.2, when we use the notation $\nabla \times \mathbf{B}$ to denote curl $\mathbf{B}$, it should not be interpreted as the cross product of ∇ and $\mathbf{B}$.

For a vector $\mathbf{B}$ given in Cartesian coordinates as

$$\mathbf{B} = \hat{\mathbf{x}} B_x + \hat{\mathbf{y}} B_y + \hat{\mathbf{z}} B_z, \quad (3.104)$$

it can be shown, through a rather lengthy and involved derivation, that the definition given by Eq. (3.103) leads to

$$\nabla \times \mathbf{B} = \hat{\mathbf{x}} \left(\frac{\partial B_z}{\partial y} - \frac{\partial B_y}{\partial z} \right) + \hat{\mathbf{y}} \left(\frac{\partial B_x}{\partial z} - \frac{\partial B_z}{\partial x} \right)$$

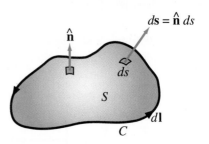

Figure 3-23: The direction of the unit vector $\hat{\mathbf{n}}$ is along the thumb when the other four fingers of the right hand follow $d\mathbf{l}$.

$$+ \hat{\mathbf{z}} \left(\frac{\partial B_y}{\partial x} - \frac{\partial B_x}{\partial y} \right)$$

$$= \begin{vmatrix} \hat{\mathbf{x}} & \hat{\mathbf{y}} & \hat{\mathbf{z}} \\ \dfrac{\partial}{\partial x} & \dfrac{\partial}{\partial y} & \dfrac{\partial}{\partial z} \\ B_x & B_y & B_z \end{vmatrix} . \quad (3.105)$$

Expressions for $\nabla \times \mathbf{B}$ are given on the inside back cover of the book for the three orthogonal coordinate systems considered in this chapter.

3-6.1 Vector Identities Involving the Curl

For any two vectors $\mathbf{A}$ and $\mathbf{B}$,

(1) $\nabla \times (\mathbf{A} + \mathbf{B}) = \nabla \times \mathbf{A} + \nabla \times \mathbf{B}$,   (3.106a)

(2) $\nabla \cdot (\nabla \times \mathbf{A}) = 0$ for any vector $\mathbf{A}$,   (3.106b)

(3) $\nabla \times (\nabla V) = 0$ for any scalar function V.   (3.106c)

Identities (2) and (3) are important properties that we will use in succeeding chapters.

3-6.2 Stokes's Theorem

Stokes's theorem converts the surface integral of the curl of a vector over an open surface S into a line integral of the vector along the contour C bounding the surface S. The geometry is shown in Fig. 3-23. Mathematically, *Stokes's theorem* is given by

$$\int_S (\nabla \times \mathbf{B}) \cdot d\mathbf{s} = \oint_C \mathbf{B} \cdot d\mathbf{l} \quad \text{(Stokes's theorem)},$$

$$(3.107)$$

and its derivation follows from the definition of $\nabla \times \mathbf{B}$ given by Eq. (3.103). The conversion process represented by Eq. (3.107) is used extensively in solutions of electromagnetic problems. If $\nabla \times \mathbf{B} = 0$, the field $\mathbf{B}$ is said to be *conservative* or *irrotational* because its circulation, represented by the right-hand side of Eq. (3.107), is zero.

Example 3-12   Verification of Stokes's Theorem

A vector field is given by $\mathbf{B} = \hat{\mathbf{z}}\cos\phi/r$. Verify Stokes's theorem for a segment of a cylindrical surface defined by $r = 2$, $\pi/3 \le \phi \le \pi/2$, and $0 \le z \le 3$, as depicted in Fig. 3-24.

Solution: Stokes's theorem states that

$$\int_S (\nabla \times \mathbf{B}) \cdot d\mathbf{s} = \oint_C \mathbf{B} \cdot d\mathbf{l}.$$

Left-hand side: With $\mathbf{B}$ having only a component $B_z = \cos\phi/r$, use of the expression for $\nabla \times \mathbf{B}$ in cylindrical coordinates from the inside back cover of the

book gives

$$\nabla \times \mathbf{B} = \hat{\mathbf{r}}\left(\frac{1}{r}\frac{\partial B_z}{\partial\phi} - \frac{\partial B_\phi}{\partial z}\right) + \hat{\boldsymbol{\phi}}\left(\frac{\partial B_r}{\partial z} - \frac{\partial B_z}{\partial r}\right)$$
$$+ \hat{\mathbf{z}}\frac{1}{r}\left(\frac{\partial}{\partial r}(rB_\phi) - \frac{\partial B_r}{\partial\phi}\right)$$
$$= \hat{\mathbf{r}}\frac{1}{r}\frac{\partial}{\partial\phi}\left(\frac{\cos\phi}{r}\right) - \hat{\boldsymbol{\phi}}\frac{\partial}{\partial r}\left(\frac{\cos\phi}{r}\right)$$
$$= -\hat{\mathbf{r}}\frac{\sin\phi}{r^2} + \hat{\boldsymbol{\phi}}\frac{\cos\phi}{r^2}.$$

The integral of $\nabla \times \mathbf{B}$ over the specified surface S is

$$\int_S (\nabla \times \mathbf{B}) \cdot d\mathbf{s}$$
$$= \int_{z=0}^3 \int_{\phi=\pi/3}^{\pi/2} \left(-\hat{\mathbf{r}}\frac{\sin\phi}{r^2} + \hat{\boldsymbol{\phi}}\frac{\cos\phi}{r^2}\right) \cdot \hat{\mathbf{r}} r \, d\phi \, dz$$
$$= \int_0^3 \int_{\pi/3}^{\pi/2} -\frac{\sin\phi}{r} \, d\phi \, dz = -\frac{3}{2r} = -\frac{3}{4},$$

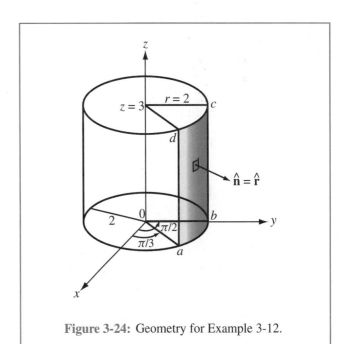

Figure 3-24: Geometry for Example 3-12.

where we used the fact that $r = 2$.

Right-hand side: The surface S is bounded by contour $C = abcd$ in Fig. 3-24. The direction of C is chosen so that it is compatible with the surface normal $\hat{\mathbf{r}}$ by the right-hand rule. Hence,

$$\oint_C \mathbf{B} \cdot d\mathbf{l} = \int_a^b \mathbf{B}_{ab} \cdot d\mathbf{l} + \int_b^c \mathbf{B}_{bc} \cdot d\mathbf{l}$$
$$+ \int_c^d \mathbf{B}_{cd} \cdot d\mathbf{l} + \int_d^a \mathbf{B}_{da} \cdot d\mathbf{l},$$

where $\mathbf{B}_{ab}$, $\mathbf{B}_{bc}$, $\mathbf{B}_{cd}$, and $\mathbf{B}_{da}$ are the expressions for the field $\mathbf{B}$ evaluated for segments ab, bc, cd, and da, respectively. Over segment ab, the dot product of $\mathbf{B}_{ab} = \hat{\mathbf{z}}(\cos\phi)/2$ and $d\mathbf{l} = \hat{\boldsymbol{\phi}} r\, d\phi$ is zero, and the same is true for segment cd. Over segment bc, $\phi = \pi/2$; hence, $\mathbf{B}_{bc} = \hat{\mathbf{z}}(\cos\pi/2)/2 = 0$. For the last segment, $\mathbf{B}_{da} = \hat{\mathbf{z}}(\cos\pi/3)/2 = \hat{\mathbf{z}}/4$ and $d\mathbf{l} = \hat{\mathbf{z}}\, dz$. Hence,

$$\oint_C \mathbf{B} \cdot d\mathbf{l} = \int_d^a \left(\hat{\mathbf{z}}\frac{1}{4}\right) \cdot \hat{\mathbf{z}}\, dz = \int_3^0 \frac{1}{4}\, dz = -\frac{3}{4},$$

which is the same result obtained by evaluating the left-hand side of Stokes's equation. ∎

EXERCISE 3.13 Find $\nabla \times \mathbf{A}$ at $(2, 0, 3)$ in cylindrical coordinates for the vector field

$$\mathbf{A} = \hat{\mathbf{r}}10e^{-2r}\cos\phi + \hat{\mathbf{z}}10\sin\phi.$$

Ans.

$$\nabla \times \mathbf{A} = \left(\hat{\mathbf{r}}\frac{10\cos\phi}{r} + \frac{\hat{\mathbf{z}}\, 10e^{-2r}}{r}\sin\phi\right)\Bigg|_{(2,0,3)} = \hat{\mathbf{r}}5.$$

EXERCISE 3.14 Find $\nabla \times \mathbf{A}$ at $(3, \pi/6, 0)$ in spherical coordinates for the vector field $\mathbf{A} = \hat{\boldsymbol{\theta}}12\sin\theta$.

Ans.

$$\nabla \times \mathbf{A} = \hat{\boldsymbol{\phi}}\frac{12\sin\theta}{R}\Bigg|_{(3,\pi/6,0)} = \hat{\boldsymbol{\phi}}2.$$

3-7 Laplacian Operator

In later chapters, we will sometimes deal with problems involving multiple combinations of operations on scalars and vectors. A frequently encountered combination is the divergence of the gradient of a scalar. For a scalar function V defined in Cartesian coordinates, its gradient is

$$\nabla V = \hat{\mathbf{x}}\frac{\partial V}{\partial x} + \hat{\mathbf{y}}\frac{\partial V}{\partial y} + \hat{\mathbf{z}}\frac{\partial V}{\partial z}$$
$$= \hat{\mathbf{x}}A_x + \hat{\mathbf{y}}A_y + \hat{\mathbf{z}}A_z = \mathbf{A}, \qquad (3.108)$$

where we defined a vector $\mathbf{A}$ with components $A_x = \partial V/\partial x$, $A_y = \partial V/\partial y$, and $A_z = \partial V/\partial z$. The divergence of ∇V is

$$\nabla \cdot (\nabla V) = \nabla \cdot \mathbf{A} = \frac{\partial A_x}{\partial x} + \frac{\partial A_y}{\partial y} + \frac{\partial A_z}{\partial z}$$
$$= \frac{\partial^2 V}{\partial x^2} + \frac{\partial^2 V}{\partial y^2} + \frac{\partial^2 V}{\partial z^2}. \qquad (3.109)$$

For convenience, $\nabla \cdot (\nabla V)$ is called the *Laplacian* of V and is denoted by $\nabla^2 V$ (the symbol ∇^2 is pronounced "del square"). That is,

$$\boxed{\nabla^2 V \triangleq \nabla \cdot (\nabla V) = \frac{\partial^2 V}{\partial x^2} + \frac{\partial^2 V}{\partial y^2} + \frac{\partial^2 V}{\partial z^2}.} \qquad (3.110)$$

As we can see from Eq. (3.110), the Laplacian of a scalar function is a scalar. Expressions for $\nabla^2 V$ in cylindrical and spherical coordinates are given on the inside back cover of the book.

The Laplacian of a scalar can be used to define the Laplacian of a vector. For a vector $\mathbf{E}$ given in Cartesian coordinates by

$$\mathbf{E} = \hat{\mathbf{x}}E_x + \hat{\mathbf{y}}E_y + \hat{\mathbf{z}}E_z, \qquad (3.111)$$

the Laplacian of **E** is defined as

$$\nabla^2 \mathbf{E} = \left(\frac{\partial^2}{\partial x^2} + \frac{\partial^2}{\partial y^2} + \frac{\partial^2}{\partial z^2} \right) \mathbf{E}$$
$$= \hat{\mathbf{x}} \, \nabla^2 E_x + \hat{\mathbf{y}} \, \nabla^2 E_y + \hat{\mathbf{z}} \, \nabla^2 E_z. \quad (3.112)$$

Thus, in Cartesian coordinates the Laplacian of a vector is a vector whose components are equal to the Laplacians of the vector components. Through direct substitution, it can be shown that

$$\boxed{\nabla^2 \mathbf{E} = \nabla(\nabla \cdot \mathbf{E}) - \nabla \times (\nabla \times \mathbf{E}). \quad (3.113)}$$

This identity will prove useful in succeeding chapters.

REVIEW QUESTIONS

Q3.11 What do the magnitude and direction of the gradient of a scalar quantity represent?

Q3.12 Prove the validity of Eq. (3.84c) in Cartesian coordinates.

Q3.13 What is the physical meaning of the divergence of a vector field?

Q3.14 If a vector field is solenoidal at a given point in space, does it necessarily follow that the vector field is zero at that point? Explain.

Q3.15 What is the meaning of the transformation provided by the divergence theorem?

Q3.16 How is the curl of a vector field at a point related to the circulation of the vector field?

Q3.17 What is the meaning of the transformation provided by Stokes's theorem?

Q3.18 When is a vector field "conservative"?

CHAPTER HIGHLIGHTS

- Vector algebra governs the laws of addition, subtraction, and multiplication of vectors, and vector calculus encompasses the laws of differentiation and integration of vectors.

- In a right-handed orthogonal coordinate system, the three base vectors are mutually perpendicular to each other at any point in space, and the cyclic relations governing the cross products of the base vectors obey the right-hand rule.

- The dot product of two vectors produces a scalar, whereas the cross product of two vectors produces another vector.

- A vector expressed in a given coordinate system can be expressed in another coordinate system through the use of transformation relations linking the two coordinate systems.

- The fundamental differential functions in vector calculus are the gradient, the divergence, and the curl.

- The gradient of a scalar function is a vector whose magnitude is equal to the maximum rate of increasing change of the scalar function per unit distance, and its direction is along the direction of maximum increase.

- The divergence of a vector field is a measure of the net outward flux per unit volume through a closed surface surrounding the unit volume.

- The divergence theorem transforms the volume integral of the divergence of a vector field into a surface integral of the field's flux through a closed surface surrounding the volume.

- The curl of a vector field is a measure of the circulation of the vector field per unit area Δs, with the orientation of Δs chosen such that the circulation is maximum.

- Stokes's theorem transforms the surface integral of the curl of a vector field into a line integral of the field over a contour that bounds the surface.

- The Laplacian of a scalar function is defined as the divergence of the gradient of that function.

GLOSSARY OF IMPORTANT TERMS

Provide definitions or explain the meaning of the following terms:

scalar quantity
vector quantity
magnitude
unit vector
base vectors
position vector
distance vector
simple product
scalar (or dot) product
vector (or cross) product
orthogonal coordinate system
Cartesian coordinate system
cylindrical coordinate system
spherical coordinate system
radial distance r
azimuth angle ϕ
zenith angle θ
range R
gradient operator
directional derivative
flux lines
flux density
divergence operator
solenoidal field
divergence theorem
circulation of a vector
curl operator
conservative field
Stokes's theorem
Laplacian operator

PROBLEMS

Section 3-1: Vector Algebra

3.1* Vector **A** starts at point $(1, -1, -2)$ and ends at point $(2, -1, 0)$. Find a unit vector in the direction of **A**.

3.2 Given vectors $\mathbf{A} = \hat{\mathbf{x}}2 - \hat{\mathbf{y}}3 + \hat{\mathbf{z}}$, $\mathbf{B} = \hat{\mathbf{x}}2 - \hat{\mathbf{y}} + \hat{\mathbf{z}}3$, and $\mathbf{C} = \hat{\mathbf{x}}4 + \hat{\mathbf{y}}2 - \hat{\mathbf{z}}2$, show that **C** is perpendicular to both **A** and **B**.

3.3* In Cartesian coordinates, the three corners of a triangle are $P_1(0, 2, 2)$, $P_2(2, -2, 2)$, and $P_3(1, 1, -2)$. Find the area of the triangle.

3.4 Given $\mathbf{A} = \hat{\mathbf{x}}2 - \hat{\mathbf{y}}3 + \hat{\mathbf{z}}1$ and $\mathbf{B} = \hat{\mathbf{x}}B_x + \hat{\mathbf{y}}2 + \hat{\mathbf{z}}B_z$:
(a) Find B_x and B_z if **A** is parallel to **B**.
(b) Find a relation between B_x and B_z if **A** is perpendicular to **B**.

3.5* Given vectors $\mathbf{A} = \hat{\mathbf{x}} + \hat{\mathbf{y}}2 - \hat{\mathbf{z}}3$, $\mathbf{B} = \hat{\mathbf{x}}3 - \hat{\mathbf{y}}4$, and $\mathbf{C} = \hat{\mathbf{y}}3 - \hat{\mathbf{z}}4$, find the following
(a) A and $\hat{\mathbf{a}}$
(b) The component of **B** along **C**
(c) θ_{AC}
(d) $\mathbf{A} \times \mathbf{C}$
(e) $\mathbf{A} \cdot (\mathbf{B} \times \mathbf{C})$
(f) $\mathbf{A} \times (\mathbf{B} \times \mathbf{C})$
(g) $\hat{\mathbf{x}} \times \mathbf{B}$
(h) $(\mathbf{A} \times \hat{\mathbf{y}}) \cdot \hat{\mathbf{z}}$

3.6 Given vectors $\mathbf{A} = \hat{\mathbf{x}}2 - \hat{\mathbf{y}} + \hat{\mathbf{z}}3$ and $\mathbf{B} = \hat{\mathbf{x}}3 - \hat{\mathbf{z}}2$, find a vector **C** whose magnitude is 6 and whose direction is perpendicular to both **A** and **B**.

3.7* Given $\mathbf{A} = \hat{\mathbf{x}}(2x + 3y) - \hat{\mathbf{y}}(2y + 3z) + \hat{\mathbf{z}}(3x - y)$, determine a unit vector parallel to **A** at point $P(1, -1, 2)$.

3.8 By expansion in Cartesian coordinates, prove:

---

*Answer(s) available in Appendix D.
⊛ Solution available in CD-ROM.

(a) The relation for the scalar triple product given by Eq. (3.29).

(b) The relation for the vector triple product given by Eq. (3.33).

3.9* Find an expression for the unit vector directed toward the origin from an arbitrary point on the line described by $x = 1$ and $z = 2$.

3.10 Find an expression for the unit vector directed toward the point P located on the z-axis at a height h above the x–y plane from an arbitrary point $Q(x, y, 2)$ in the plane $z = 2$.

3.11* Find a unit vector parallel to either direction of the line described by

$$2x - z = 4$$

3.12 Two lines in the x–y plane are described by the expressions:

$$\text{Line 1} \quad x + 2y = -6$$
$$\text{Line 2} \quad 3x + 4y = 8$$

Use vector algebra to find the smaller angle between the lines at their intersection point.

3.13* A given line is described by

$$x + 2y = 4$$

Vector **A** starts at the origin and ends at point P on the line such that **A** is orthogonal to the line. Find an expression for **A**.

3.14 Show that, given two vectors **A** and **B**,

(a) The vector **C** defined as the vector component of **B** in the direction of **A** is given by

$$\mathbf{C} = \hat{\mathbf{a}}(\mathbf{B} \cdot \hat{\mathbf{a}}) = \frac{\mathbf{A}(\mathbf{B} \cdot \mathbf{A})}{|\mathbf{A}|^2},$$

where $\hat{\mathbf{a}}$ is the unit vector of **A**.

(b) The vector **D** defined as the vector component of **B** perpendicular to **A** is given by

$$\mathbf{D} = \mathbf{B} - \frac{\mathbf{A}(\mathbf{B} \cdot \mathbf{A})}{|\mathbf{A}|^2}$$

3.15* A certain plane is described by

$$2x + 3y + 4z = 16$$

Find the unit vector normal to the surface in the direction away from the origin.

3.16 Given $\mathbf{B} = \hat{\mathbf{x}}(2z - 3y) + \hat{\mathbf{y}}(2x - 3z) - \hat{\mathbf{z}}(x + y)$, find a unit vector parallel to **B** at point P(1, 0, −1).

3.17 When sketching or demonstrating the spatial variation of a vector field, we often use arrows, as in Fig. 3-25, wherein the length of the arrow is made to be

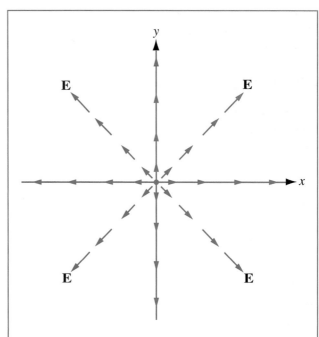

Figure 3-25: Arrow representation for vector field $\mathbf{E} = \hat{\mathbf{r}}r$ (Problem 3.17).

proportional to the strength of the field and the direction of the arrow is the same as that of the field's. The sketch shown in Fig. 3-25, which represents the vector field $\mathbf{E} = \hat{\mathbf{r}}r$, consists of arrows pointing radially away from the origin and their lengths increasing linearly in proportion to their distance away from the origin. Using this arrow representation, sketch each of the following vector fields:

(a) $\mathbf{E}_1 = -\hat{\mathbf{x}}y$

(b) $\mathbf{E}_2 = \hat{\mathbf{y}}x$

(c) $\mathbf{E}_3 = \hat{\mathbf{x}}x + \hat{\mathbf{y}}y$

(d) $\mathbf{E}_4 = \hat{\mathbf{x}}x + \hat{\mathbf{y}}2y$

(e) $\mathbf{E}_5 = \hat{\boldsymbol{\phi}}r$

(f) $\mathbf{E}_6 = \hat{\mathbf{r}}\sin\phi$

3.18 Use arrows to sketch each of the following vector fields:

(a) $\mathbf{E}_1 = \hat{\mathbf{x}}x - \hat{\mathbf{y}}y$

(b) $\mathbf{E}_2 = -\hat{\boldsymbol{\phi}}$

(c) $\mathbf{E}_3 = \hat{\mathbf{y}}(1/x)$

(d) $\mathbf{E}_4 = \hat{\mathbf{r}}\cos\phi$

Sections 3-2 and 3-3: Coordinate Systems

3.19* Convert the coordinates of the following points from Cartesian to cylindrical and spherical coordinates:

(a) $P_1(1, 2, 0)$

(b) $P_2(0, 0, 3)$

(c) $P_3(1, 1, 2)$

(d) $P_4(-3, 3, -3)$

3.20 Convert the coordinates of the following points from cylindrical to Cartesian coordinates:

(a) $P_1(2, \pi/4, -3)$

(b) $P_2(3, 0, 0)$

(c) $P_3(4, \pi, 2)$

3.21* Convert the coordinates of the following points from spherical to cylindrical coordinates:

(a) $P_1(5, 0, 0)$

(b) $P_2(5, 0, \pi)$

(c) $P_3(3, \pi/2, \pi)$

3.22 Use the appropriate expression for the differential surface area $d\mathbf{s}$ to determine the area of each of the following surfaces:

(a) $r = 3; \ 0 \le \phi \le \pi/3; \ -2 \le z \le 2$

(b) $2 \le r \le 5; \ \pi/2 \le \phi \le \pi; \ z = 0$

(c) $2 \le r \le 5; \ \phi = \pi/4; \ -2 \le z \le 2$

(d) $R = 2; \ 0 \le \theta \le \pi/3; \ 0 \le \phi \le \pi$

(e) $0 \le R \le 5; \ \theta = \pi/3; \ 0 \le \phi \le 2\pi$

Also sketch the outline of each surface.

3.23* Find the volumes described by the following:

(a) $2 \le r \le 5; \ \pi/2 \le \phi \le \pi; \ 0 \le z \le 2$

(b) $0 \le R \le 5; \ 0 \le \theta \le \pi/3; \ 0 \le \phi \le 2\pi$

Also sketch the outline of each volume.

3.24 A section of a sphere is described by $0 \le R \le 2$, $0 \le \theta \le 90°$, and $30° \le \phi \le 90°$. Find the following:

(a) The surface area of the spherical section.

(b) The enclosed volume.

Also sketch the outline of the section.

3.25* A vector field is given in cylindrical coordinates by

$$\mathbf{E} = \hat{\mathbf{r}}r\cos\phi + \hat{\boldsymbol{\phi}}r\sin\phi + \hat{\mathbf{z}}z^2$$

Point $P(4, \pi, 2)$ is located on the surface of the cylinder described by $r = 4$. At point P, find:

(a) The vector component of $\mathbf{E}$ perpendicular to the cylinder.

(b) The vector component of $\mathbf{E}$ tangential to the cylinder.

3.26 At a given point in space, vectors **A** and **B** are given in spherical coordinates by

$$\mathbf{A} = \hat{\mathbf{R}}4 + \hat{\boldsymbol{\theta}}2 - \hat{\boldsymbol{\phi}}$$

$$\mathbf{B} = -\hat{\mathbf{R}}2 + \hat{\boldsymbol{\phi}}3$$

Find:

(a) The scalar component, or projection, of **B** in the direction of **A**.

(b) The vector component of **B** in the direction of **A**.

(c) The vector component of **B** perpendicular to **A**.

3.27* Given vectors

$$\mathbf{A} = \hat{\mathbf{r}}(\cos\phi + 3z) - \hat{\boldsymbol{\phi}}(2r + 4\sin\phi) + \hat{\mathbf{z}}(r - 2z)$$

$$\mathbf{B} = -\hat{\mathbf{r}}\sin\phi + \hat{\mathbf{z}}\cos\phi$$

find

(a) θ_{AB} at $(2, \pi/2, 0)$

(b) A unit vector perpendicular to both **A** and **B** at $(2, \pi/3, 1)$

3.28 Find the distance between the following pairs of points:

(a) $P_1(1, 2, 3)$ and $P_2(-2, -3, 2)$ in Cartesian coordinates

(b) $P_3(1, \pi/4, 2)$ and $P_4(3, \pi/4, 4)$ in cylindrical coordinates

(c) $P_5(2, \pi/2, 0)$ and $P_6(3, \pi, 0)$ in spherical coordinates

3.29* Determine the distance between the following pairs of points:

(a) $P_1(1, 1, 2)$ and $P_2(0, 2, 2)$

(b) $P_3(2, \pi/3, 1)$ and $P_4(4, \pi/2, 0)$

(c) $P_5(3, \pi, \pi/2)$ and $P_6(4, \pi/2, \pi)$

3.30 Transform the following vectors into cylindrical coordinates and then evaluate them at the indicated points:

(a) $\mathbf{A} = \hat{\mathbf{x}}(x + y)$ at $P_1(1, 2, 3)$

(b) $\mathbf{B} = \hat{\mathbf{x}}(y - x) + \hat{\mathbf{y}}(x - y)$ at $P_2(1, 0, 2)$

(c) $\mathbf{C} = \hat{\mathbf{x}}y^2/(x^2 + y^2) - \hat{\mathbf{y}}x^2/(x^2 + y^2) + \hat{\mathbf{z}}4$ at $P_3(1, -1, 2)$

(d) $\mathbf{D} = \hat{\mathbf{R}}\sin\theta + \hat{\boldsymbol{\theta}}\cos\theta + \hat{\boldsymbol{\phi}}\cos^2\phi$ at $P_4(2, \pi/2, \pi/4)$

(e) $\mathbf{E} = \hat{\mathbf{R}}\cos\phi + \hat{\boldsymbol{\theta}}\sin\phi + \hat{\boldsymbol{\phi}}\sin^2\theta$ at $P_5(3, \pi/2, \pi)$

3.31* Transform the following vectors into spherical coordinates and then evaluate them at the indicated points:

(a) $\mathbf{A} = \hat{\mathbf{x}}y^2 + \hat{\mathbf{y}}xz + \hat{\mathbf{z}}4$ at $P_1(1, -1, 2)$

(b) $\mathbf{B} = \hat{\mathbf{y}}(x^2 + y^2 + z^2) - \hat{\mathbf{z}}(x^2 + y^2)$ at $P_2(-1, 0, 2)$

(c) $\mathbf{C} = \hat{\mathbf{r}}\cos\phi - \hat{\boldsymbol{\phi}}\sin\phi + \hat{\mathbf{z}}\cos\phi\sin\phi$ at $P_3(2, \pi/4, 2)$

(d) $\mathbf{D} = \hat{\mathbf{x}}y^2/(x^2 + y^2) - \hat{\mathbf{y}}x^2/(x^2 + y^2) + \hat{\mathbf{z}}4$ at $P_4(1, -1, 2)$

Sections 3-4 to 3-7: Gradient, Divergence, and Curl Operators

3.32 Find the gradient of the following scalar functions:

(a) $T = 2/(x^2 + z^2)$

(b) $V = xy^2z^3$

(c) $U = z\cos\phi/(1 + r^2)$

(d) $W = e^{-R}\sin\theta$

(e) $S = x^2e^{-z} + y^2$

(f) $N = r^2\cos\phi$

(g) $M = R\cos\theta\sin\phi$

3.33* The gradient of a scalar function T is given by

$$\nabla T = \hat{\mathbf{z}}e^{-2z}$$

If $T = 10$ at $z = 0$, find $T(z)$.

3.34 Follow a procedure similar to that leading to Eq. (3.82) to derive the expression given by Eq. (3.83) for ∇ in spherical coordinates.

3.35* For the scalar function $V = xy - z^2$, determine its directional derivative along the direction of vector $\mathbf{A} = (\hat{\mathbf{x}} - \hat{\mathbf{y}}z)$ and then evaluate it at $P(1, -1, 2)$.

3.36 For the scalar function $T = e^{-r/5} \cos \phi$, determine its directional derivative along the radial direction $\hat{\mathbf{r}}$ and then evaluate it at $P(2, \pi/4, 3)$.

3.37* For the scalar function $U = \frac{1}{R} \sin^2 \theta$, determine its directional derivative along the range direction $\hat{\mathbf{R}}$ and then evaluate it at $P(4, \pi/4, \pi/2)$.

3.38 Vector field $\mathbf{E}$ is characterized by the following properties: (a) $\mathbf{E}$ points along $\hat{\mathbf{R}}$; (b) the magnitude of $\mathbf{E}$ is a function of only the distance from the origin; (c) $\mathbf{E}$ vanishes at the origin, and (d) $\nabla \cdot \mathbf{E} = 6$, everywhere. Find an expression for $\mathbf{E}$ that satisfies these properties.

3.39* For the vector field $\mathbf{E} = \hat{\mathbf{x}}xz - \hat{\mathbf{y}}yz^2 - \hat{\mathbf{z}}xy$, verify the divergence theorem by computing

(a) The total outward flux flowing through the surface of a cube centered at the origin and with sides equal to 2 units each and parallel to the Cartesian axes.

(b) The integral of $\nabla \cdot \mathbf{E}$ over the cube's volume.

3.40 For the vector field $\mathbf{E} = \hat{\mathbf{r}}10e^{-r} - \hat{\mathbf{z}}3z$, verify the divergence theorem for the cylindrical region enclosed by $r = 2$, $z = 0$, and $z = 4$.

3.41* A vector field $\mathbf{D} = \hat{\mathbf{r}}r^3$ exists in the region between two concentric cylindrical surfaces defined by $r = 1$ and $r = 2$, with both cylinders extending between $z = 0$ and $z = 5$. Verify the divergence theorem by evaluating the following

(a) $\oint_S \mathbf{D} \cdot d\mathbf{s}$

(b) $\int_V \nabla \cdot \mathbf{D} \, dv$

3.42 For the vector field $\mathbf{D} = \hat{\mathbf{R}}3R^2$, evaluate both sides of the divergence theorem for the region enclosed between the spherical shells defined by $R = 1$ and $R = 2$.

3.43* For the vector field $\mathbf{E} = \hat{\mathbf{x}}xy - \hat{\mathbf{y}}(x^2 + 2y^2)$, calculate the following:

(a) $\oint_C \mathbf{E} \cdot d\mathbf{l}$ around the triangular contour shown in Fig. 3-26(a).

(b) $\int_S (\nabla \times \mathbf{E}) \cdot d\mathbf{s}$ over the area of the triangle.

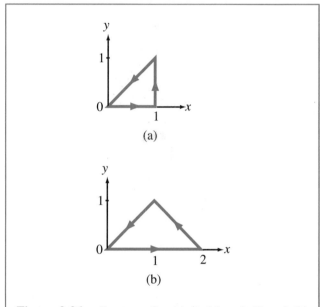

Figure 3-26: Contours for (a) Problem 3.43 and (b) Problem 3.44.

3.44 Repeat Problem 3.43 for the contour shown in Fig. 3-26(b).

3.45* Verify Stokes's theorem for the vector field

$$\mathbf{B} = (\hat{\mathbf{r}}r \cos \phi + \hat{\boldsymbol{\phi}} \sin \phi)$$

by evaluating the following:

(a) $\oint_C \mathbf{B} \cdot d\mathbf{l}$ over the semicircular contour shown in Fig. 3-27(a).

(b) $\int_S (\nabla \times \mathbf{B}) \cdot d\mathbf{s}$ over the surface of the semicircle.

3.46 Repeat Problem 3.45 for the contour shown in Fig. 3-27(b).

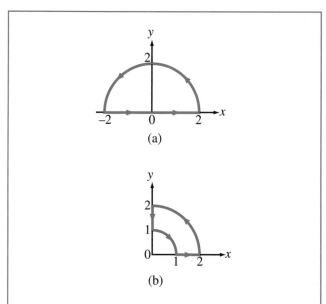

(a)

(b)

Figure 3-27: Contour paths for (a) Problem 3.45 and (b) Problem 3.46.

3.47 Verify Stokes's theorem for the vector field $\mathbf{A} = \hat{\mathbf{R}} \cos\theta + \hat{\boldsymbol{\phi}} \sin\theta$ by evaluating it on the hemisphere of unit radius.

3.48* Determine if each of the following vector fields is solenoidal, conservative, or both:

(a) $\mathbf{A} = \hat{\mathbf{x}}2xy - \hat{\mathbf{y}}y^2$

(b) $\mathbf{B} = \hat{\mathbf{x}}x^2 - \hat{\mathbf{y}}y^2 + \hat{\mathbf{z}}2z$

(c) $\mathbf{C} = \hat{\mathbf{r}}(\sin\phi)/r^2 + \hat{\boldsymbol{\phi}}(\cos\phi)/r^2$

(d) $\mathbf{D} = \hat{\mathbf{R}}/R$

(e) $\mathbf{E} = \hat{\mathbf{r}}\left(3 - \frac{r}{1+r}\right) + \hat{\mathbf{z}}z$

(f) $\mathbf{F} = (\hat{\mathbf{x}}y - \hat{\mathbf{y}}x)/(x^2 + y^2)$

(g) $\mathbf{G} = \hat{\mathbf{x}}(x^2 + z^2) + \hat{\mathbf{y}}(y^2 + x^2) + \hat{\mathbf{z}}(y^2 + z^2)$

(h) $\mathbf{H} = \hat{\mathbf{R}}(Re^{-R})$

3.49 Find the Laplacian of the following scalar functions:

(a) $V = xy^2z^3$

(b) $V = xy + yz + zx$

(c) $V = 1/(x^2 + y^2)$

(d) $V = 5e^{-r}\cos\phi$

(e) $V = 10e^{-R}\sin\theta$

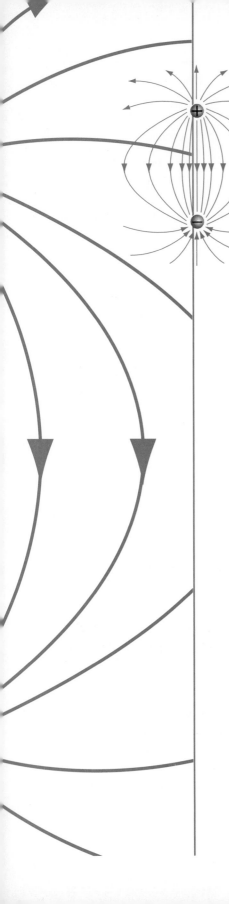

CHAPTER 4

Electrostatics

4-1 Maxwell's Equations

$\mathbf{M}$odern electromagnetism is based on a set of four fundamental relations known as *Maxwell's equations*:

$$\nabla \cdot \mathbf{D} = \rho_v, \qquad (4.1a)$$

$$\nabla \times \mathbf{E} = -\frac{\partial \mathbf{B}}{\partial t}, \qquad (4.1b)$$

$$\nabla \cdot \mathbf{B} = 0, \qquad (4.1c)$$

$$\nabla \times \mathbf{H} = \mathbf{J} + \frac{\partial \mathbf{D}}{\partial t}, \qquad (4.1d)$$

where $\mathbf{E}$ and $\mathbf{D}$ are electric field quantities interrelated by $\mathbf{D} = \varepsilon \mathbf{E}$, with ε being the electrical permittivity of the material; $\mathbf{B}$ and $\mathbf{H}$ are magnetic field quantities interrelated by $\mathbf{B} = \mu \mathbf{H}$, with μ being the magnetic permeability of the material; ρ_v is the electric charge density per unit volume; and $\mathbf{J}$ is the current density per unit area. The field quantities $\mathbf{E}$, $\mathbf{D}$, $\mathbf{B}$, and $\mathbf{H}$ were introduced in Section 1-2, and ρ_v and $\mathbf{J}$ will be defined in Section 4-2. These equations hold in any material, including free space (vacuum), and at any spatial location (x, y, z). In general, all the quantities in Maxwell's equations may be a function of time t. By his formulation of these equations, published in a classic treatise in 1873, James Clerk Maxwell established the first unified theory of electricity and magnetism. His equations, which he deduced from experimental observations reported by Gauss, Ampère, Faraday, and others, not only encapsulate the connection between the electric field and electric charge and between the magnetic field and electric current, but they also define the bilateral coupling between the electric and magnetic field quantities. Together with some auxiliary relations, Maxwell's equations form the fundamental tenets of electromagnetic theory.

In the *static* case, none of the quantities appearing in Maxwell's equations are a function of time (i.e., $\partial/\partial t = 0$). *This happens when all charges are permanently fixed in space, or, if they move, they do so at a steady rate so that ρ_v and $\mathbf{J}$ are constant in time.* Under these circumstances, the time derivatives of $\mathbf{B}$ and $\mathbf{D}$ in Eqs. (4.1b) and (4.1d) are zero, and Maxwell's equations reduce to

Electrostatics

$$\nabla \cdot \mathbf{D} = \rho_v, \qquad (4.2a)$$

$$\nabla \times \mathbf{E} = 0. \qquad (4.2b)$$

Magnetostatics

$$\nabla \cdot \mathbf{B} = 0, \qquad (4.3a)$$

$$\nabla \times \mathbf{H} = \mathbf{J}. \qquad (4.3b)$$

Maxwell's four equations separate into two uncoupled pairs, with the first pair involving only the electric field quantities $\mathbf{E}$ and $\mathbf{D}$ and the second pair involving only the magnetic field quantities $\mathbf{B}$ and $\mathbf{H}$. *The electric and magnetic fields are no longer interconnected in the static case.* This allows us to study electricity and magnetism as two distinct and separate phenomena, as long as the spatial distributions of charge and current flow remain constant in time. We refer to the study of electric and magnetic phenomena under static conditions as *electrostatics* and *magnetostatics*, respectively. Electrostatics is the subject of the present chapter, and in Chapter 5 we learn about magnetostatics. The experience gained through handling the relatively simpler situations in electrostatics and magnetostatics will prove valuable in tackling the more involved material in subsequent chapters, which deals with time-varying fields, charge densities, and currents.

We study electrostatics not only as a prelude to the study of time-varying fields, but also because it is an important field of study in its own right. Many electronic devices and systems are based on the principles of electrostatics. They include x-ray machines, oscilloscopes, ink-jet electrostatic printers, liquid crystal displays, copying machines, capacitance keyboards, and many solid-state control devices. Electrostatics is also used in the design of medical diagnostic sensors, such as the electrocardiogram (for recording the heart's pumping pattern) and the electroencephalogram (for recording brain activity), as well as in numerous industrial applications.

4-2   Charge and Current Distributions

In electromagnetics, we encounter various forms of electric charge distributions, and if the charges are in motion, they constitute current distributions. Charge may be distributed over a volume of space, across a surface, or along a line.

4-2.1   Charge Densities

At the atomic scale, the charge distribution in a material is discrete, meaning that charge exists only where electrons and nuclei are and nowhere else. In electromagnetics, we usually are interested in studying phenomena at a much larger scale, typically three or more orders of magnitudes greater than the spacing between adjacent atoms. At such a macroscopic scale, we can disregard the discontinuous nature of the charge distribution and treat the net charge contained in an elemental volume Δv as if it were uniformly distributed within it. Accordingly, we define the *volume charge density* ρ_v as

$$\rho_v = \lim_{\Delta v \to 0} \frac{\Delta q}{\Delta v} = \frac{dq}{dv} \quad \text{(C/m}^3\text{)}, \qquad (4.4)$$

where Δq is the charge contained in Δv. In general, ρ_v is defined at a given point in space, specified by (x, y, z) in a Cartesian coordinate system, and at a given time t. Thus, $\rho_v = \rho_v(x, y, z, t)$. Physically, ρ_v represents the average charge per unit volume for a volume Δv centered at (x, y, z), with Δv being large enough to contain a large number of atoms and yet small enough to be regarded as a point at the macroscopic scale under consideration. The variation of ρ_v with spatial location is called its *spatial distribution*, or simply its *distribution*. The total charge contained in a given volume v is given by

$$Q = \int_v \rho_v \, dv \quad \text{(C)}. \qquad (4.5)$$

In some cases, particularly when dealing with conductors, electric charge may be distributed across the surface of a material, in which case the relevant quantity of interest is the *surface charge density* ρ_s, defined as

$$\rho_s = \lim_{\Delta s \to 0} \frac{\Delta q}{\Delta s} = \frac{dq}{ds} \quad \text{(C/m}^2\text{)}, \qquad (4.6)$$

where Δq is the charge present across an elemental surface area Δs. Similarly, if the charge is distributed along a line, which need not be straight, we characterize the distribution in terms of the *line charge density* ρ_l, defined as

$$\rho_l = \lim_{\Delta l \to 0} \frac{\Delta q}{\Delta l} = \frac{dq}{dl} \quad \text{(C/m)}. \qquad (4.7)$$

Example 4-1   Line Charge Distribution

Calculate the total charge Q contained in a cylindrical tube of charge oriented along the z-axis as shown in Fig. 4-1(a). The line charge density is $\rho_l = 2z$, where z

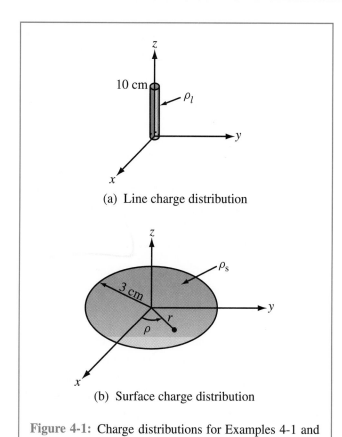

(a) Line charge distribution

(b) Surface charge distribution

Figure 4-1: Charge distributions for Examples 4-1 and 4-2.

Solution: Since ρ_s is symmetrical with respect to the azimuth angle ϕ, its functional form depends only on r and is given by

$$\rho_s = \frac{6r}{3 \times 10^{-2}} = 2 \times 10^2 r \quad (\text{C/m}^2),$$

where r is in meters. In polar coordinates, an element of area is $ds = r\,dr\,d\phi$, and for the disk shown in Fig. 4-1(b), the limits of integration are from 0 to 2π (rad) for ϕ and from 0 to 3×10^{-2} m for r. Hence,

$$Q = \int_S \rho_s \, ds$$

$$= \int_{\phi=0}^{2\pi} \int_{r=0}^{3 \times 10^{-2}} (2 \times 10^2 r) r \, dr \, d\phi$$

$$= 2\pi \times 2 \times 10^2 \left. \frac{r^3}{3} \right|_0^{3 \times 10^{-2}} = 11.31 \quad (\text{mC}). \quad \blacksquare$$

---

is the distance in meters from the bottom end of the tube. The tube length is 10 cm.

Solution: The total charge Q is

$$Q = \int_0^{0.1} \rho_l \, dz = \int_0^{0.1} 2z \, dz = z^2 \big|_0^{0.1} = 10^{-2} \text{ C}. \quad \blacksquare$$

Example 4-2   Surface Charge Distribution

The circular disk of electric charge shown in Fig. 4-1(b) is characterized by an azimuthally symmetric surface charge density that increases linearly with r from zero at the center to 6 C/m^2 at $r = 3$ cm. Find the total charge present on the disk surface.

EXERCISE 4.1 A square plate in the x–y plane is situated in the space defined by -3 m $\leq x \leq 3$ m and -3 m $\leq y \leq 3$ m. Find the total charge on the plate if the surface charge density is given by $\rho_s = 4y^2$ (μC/m^2).

Ans. $Q = 0.432$ (mC).

EXERCISE 4.2 A spherical shell centered at the origin extends between $R = 2$ cm and $R = 3$ cm. If the volume charge density is given by $\rho_v = 3R \times 10^{-4}$ (C/m^3), find the total charge contained in the shell.

Ans. $Q = 0.61$ (nC).

4-2.2 Current Density

Consider a tube of charge with volume charge density ρ_v, as shown in Fig. 4-2(a). The charges are moving with a mean velocity $\mathbf{u}$ along the axis of the tube. Over a period Δt, the charges move a distance $\Delta l = u\,\Delta t$. The amount of charge that crosses the tube's cross-sectional surface $\Delta s'$ in time Δt is therefore

$$\Delta q' = \rho_v\,\Delta \mathcal{V} = \rho_v\,\Delta l\,\Delta s' = \rho_v u\,\Delta s'\,\Delta t. \qquad (4.8)$$

Now consider the more general case where the charges are flowing through a surface Δs whose surface normal $\hat{\mathbf{n}}$ is not necessarily parallel to $\mathbf{u}$, as shown in Fig. 4-2(b). In this case, the amount of charge Δq flowing through Δs is

$$\Delta q = \rho_v \mathbf{u} \cdot \Delta \mathbf{s}\,\Delta t, \qquad (4.9)$$

and the corresponding current is

$$\Delta I = \frac{\Delta q}{\Delta t} = \rho_v \mathbf{u} \cdot \Delta \mathbf{s} = \mathbf{J} \cdot \Delta \mathbf{s}, \qquad (4.10)$$

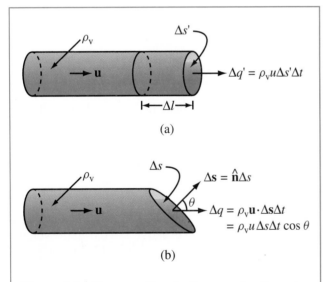

Figure 4-2: Charges with velocity $\mathbf{u}$ moving through a cross section $\Delta s'$ in (a) and Δs in (b).

where

$$\mathbf{J} = \rho_v \mathbf{u} \quad (\text{A/m}^2) \quad (4.11)$$

is defined as the *current density* in ampere per square meter. For an arbitrary surface S, the total current flowing through it is then given by

$$I = \int_S \mathbf{J} \cdot d\mathbf{s} \quad (\text{A}). \quad (4.12)$$

When the current is generated by the actual movement of electrically charged matter, it is called a *convection current*, and $\mathbf{J}$ is called the *convection current density*. A wind-driven charged cloud, for example, gives rise to a convection current. In some cases, the charged matter constituting the convection current consists solely of charged particles, such as the electrons of an electron beam in a cathode ray tube (the picture tube of televisions and computer monitors).

This is distinct from a *conduction current*, where atoms of the conducting medium do not move. In a metal wire, for example, there are equal amounts of positive charges (in atomic nuclei) and negative charges (in the electron shells of the atom). All the positive charges and most of the negative charges cannot move; only those electrons in the outermost electronic shells of the atoms can be easily pushed from one atom to the next if a voltage is applied across the ends of the wire. This movement of electrons from atom to atom gives rise to conduction current. The electrons that emerge from the wire are not necessarily the same electrons that entered the wire at the other end.

Because the two types of current are generated by different physical mechanisms, conduction current obeys Ohm's law, whereas convection current does not. Conduction current is discussed in more detail in Section 4-7.

REVIEW QUESTIONS

Q4.1 What happens to Maxwell's equations under static conditions? Why is that significant?

Q4.2 How is the current density **J** related to the volume charge density ρ_v?

Q4.3 What is the difference between convection current and conduction current?

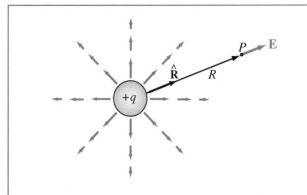

Figure 4-3: Electric-field lines due to a charge q.

4-3 Coulomb's Law

One of the major goals of this chapter is to develop expressions relating the *electric field intensity* **E** and the associated *electric flux density* **D** to any specified distribution of charge. Our discussions, however, will be limited to electrostatic fields induced by static charge distributions.

We begin by reviewing how the electric field was introduced and defined in Section 1-2.2 on the basis of the results of Coulomb's experiments on the electrical force between charged bodies. *Coulomb's law*, which was first introduced for electrical charges in air and later generalized to material media, states that

(**1**) an isolated charge q induces an electric field **E** at every point in space, and at any specific point P, **E** is given by

$$\mathbf{E} = \hat{\mathbf{R}} \frac{q}{4\pi \varepsilon R^2} \quad \text{(V/m)},\qquad(4.13)$$

where $\hat{\mathbf{R}}$ is a unit vector pointing from q to P (Fig. 4-3), R is the distance between them, and ε is the electrical permittivity of the medium containing the observation point P; and

(**2**) in the presence of an electric field **E** at a given point in space, which may be due to a single charge or a distribution of many charges, the force acting on a test charge q', when the charge is placed at that point, is given by

$$\mathbf{F} = q'\mathbf{E} \quad \text{(N)}.\qquad(4.14)$$

With **F** measured in newtons (N) and q' in coulombs (C), the unit of **E** is (N/C), which is shown later in Section 4-5 to be the same as volt per meter (V/m).

For a material with electrical permittivity ε, the electrical field quantities **D** and **E** are related by

$$\mathbf{D} = \varepsilon \mathbf{E}\qquad(4.15)$$

with

$$\varepsilon = \varepsilon_r \varepsilon_0,\qquad(4.16)$$

where

$$\varepsilon_0 = 8.85 \times 10^{-12} \simeq (1/36\pi) \times 10^{-9} \quad \text{(F/m)}$$

is the electrical permittivity of free space, and $\varepsilon_r = \varepsilon/\varepsilon_0$ is called the *relative permittivity* (or *dielectric constant*) of the material. For most materials and under most conditions, ε of the material has a constant value independent of both the magnitude and direction of **E**.

If ε is independent of the magnitude of **E**, *then the material is said to be* **linear** *because* **D** *and* **E** *are related linearly, and if it is independent of the direction of* **E**, *the material is said to be* **isotropic**. Materials do not usually exhibit nonlinear permittivity behavior except when the amplitude of **E** is very high (at levels approaching the dielectric breakdown conditions discussed later in Section 4-8), and anisotropy is peculiar only to certain materials with particular crystalline structures. Hence, except for materials under these very special circumstances, the quantities **D** and **E** are effectively redundant; for a material with known ε, knowledge of either **D** or **E** is sufficient to specify the other in that material.

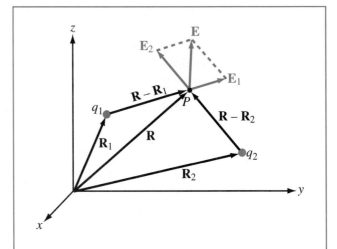

Figure 4-4: The electric field **E** at *P* due to two charges is equal to the vector sum of **E**$_1$ and **E**$_2$.

4-3.1   Electric Field due to Multiple Point Charges

The expression given by Eq. (4.13) for the field **E** due to a single charge can be extended to find the field due to multiple point charges. We begin by considering two point charges, q_1 and q_2, located at position vectors **R**$_1$ and **R**$_2$ from the origin of a given coordinate system, as shown in Fig. 4-4. The electric field **E** is to be evaluated at a point *P* with position vector **R**. At *P*, the electric field **E**$_1$ due to q_1 alone is given by Eq. (4.13) with *R*, the distance between q_1 and *P*, replaced with $|\mathbf{R} - \mathbf{R}_1|$ and the unit vector $\hat{\mathbf{R}}$ replaced with $(\mathbf{R} - \mathbf{R}_1)/|\mathbf{R} - \mathbf{R}_1|$. Thus,

$$\mathbf{E}_1 = \frac{q_1(\mathbf{R} - \mathbf{R}_1)}{4\pi\varepsilon|\mathbf{R} - \mathbf{R}_1|^3} \quad \text{(V/m)}. \qquad (4.17a)$$

Similarly, the electric field due to q_2 alone is

$$\mathbf{E}_2 = \frac{q_2(\mathbf{R} - \mathbf{R}_2)}{4\pi\varepsilon|\mathbf{R} - \mathbf{R}_2|^3} \quad \text{(V/m)}. \qquad (4.17b)$$

The electric field obeys the principle of linear superposition. Consequently, the total electric field **E** at any point

in space is equal to the vector sum of the electric fields induced by all the individual charges. In the present case,

$$\begin{aligned} \mathbf{E} &= \mathbf{E}_1 + \mathbf{E}_2 \\ &= \frac{1}{4\pi\varepsilon}\left[\frac{q_1(\mathbf{R} - \mathbf{R}_1)}{|\mathbf{R} - \mathbf{R}_1|^3} + \frac{q_2(\mathbf{R} - \mathbf{R}_2)}{|\mathbf{R} - \mathbf{R}_2|^3}\right] \end{aligned} \qquad (4.18)$$

Generalizing the preceding result to the case of N point charges, the electric field **E** at position vector **R** caused by charges $q_1, q_2, \ldots, q_N$ located at points with position vectors $\mathbf{R}_1, \mathbf{R}_2, \ldots, \mathbf{R}_N$, is given by

$$\mathbf{E} = \frac{1}{4\pi\varepsilon}\sum_{i=1}^{N}\frac{q_i(\mathbf{R} - \mathbf{R}_i)}{|\mathbf{R} - \mathbf{R}_i|^3} \quad \text{(V/m)}. \qquad (4.19)$$

Example 4-3   Electric Field due to Two Point Charges

Two point charges with $q_1 = 2 \times 10^{-5}$ C and $q_2 = -4 \times 10^{-5}$ C are located in free space at $(1, 3, -1)$ and $(-3, 1, -2)$, respectively, in a Cartesian coordinate system. Find (a) the electric field **E** at $(3, 1, -2)$ and (b) the force on a 8×10^{-5} C charge located at that point. All distances are in meters.

Solution: (a) From Eq. (4.18), the electric field **E** with $\varepsilon = \varepsilon_0$ (free space) is given by

$$\mathbf{E} = \frac{1}{4\pi\varepsilon_0} \left[q_1 \frac{(\mathbf{R} - \mathbf{R}_1)}{|\mathbf{R} - \mathbf{R}_1|^3} + q_2 \frac{(\mathbf{R} - \mathbf{R}_2)}{|\mathbf{R} - \mathbf{R}_2|^3} \right] \quad \text{(V/m)}.$$

The vectors $\mathbf{R}_1$, $\mathbf{R}_2$, and $\mathbf{R}$ are given by

$$\mathbf{R}_1 = \hat{\mathbf{x}} + \hat{\mathbf{y}}3 - \hat{\mathbf{z}},$$
$$\mathbf{R}_2 = -\hat{\mathbf{x}}3 + \hat{\mathbf{y}} - \hat{\mathbf{z}}2,$$
$$\mathbf{R} = \hat{\mathbf{x}}3 + \hat{\mathbf{y}} - \hat{\mathbf{z}}2.$$

Hence,

$$\mathbf{E} = \frac{1}{4\pi\varepsilon_0} \left[\frac{2(\hat{\mathbf{x}}2 - \hat{\mathbf{y}}2 - \hat{\mathbf{z}})}{27} - \frac{4(\hat{\mathbf{x}}6)}{216} \right] \times 10^{-5}$$

$$= \frac{\hat{\mathbf{x}} - \hat{\mathbf{y}}4 - \hat{\mathbf{z}}2}{108\pi\varepsilon_0} \times 10^{-5} \quad \text{(V/m)}.$$

(b)

$$\mathbf{F} = q_3\mathbf{E} = 8 \times 10^{-5} \times \frac{\hat{\mathbf{x}} - \hat{\mathbf{y}}4 - \hat{\mathbf{z}}2}{108\pi\varepsilon_0} \times 10^{-5}$$

$$= \frac{\hat{\mathbf{x}}2 - \hat{\mathbf{y}}8 - \hat{\mathbf{z}}4}{27\pi\varepsilon_0} \times 10^{-10} \quad \text{(N)}. \quad \blacksquare$$

EXERCISE 4.3 Four charges of 10 μC each are located in free space at $(-3, 0, 0)$, $(3, 0, 0)$, $(0, -3, 0)$, and $(0, 3, 0)$ in a Cartesian coordinate system. Find the force on a 20-μC charge located at $(0, 0, 4)$. All distances are in meters.

Ans. $\mathbf{F} = \hat{\mathbf{z}}0.23$ (N).

EXERCISE 4.4 Two identical charges are located on the x-axis at $x = 3$ and $x = 7$. At what point in space is the net electric field zero?

Ans. At point $(5, 0, 0)$.

EXERCISE 4.5 In a hydrogen atom the electron and proton are separated by an average distance of 5.3×10^{-11} m. Find the magnitude of the electrical force F_e between the two particles, and compare it with the gravitational force F_g between them.

Ans. $F_e = 8.2 \times 10^{-8}$ N, and $F_g = 3.6 \times 10^{-47}$ N.

4-3.2 Electric Field due to a Charge Distribution

We now extend the results we obtained for the field caused by discrete point charges to the case of a continuous charge distribution. Consider volume v' shown in Fig. 4-5. It contains a distribution of electric charge characterized by a volume charge density ρ_v, whose magnitude may vary with spatial location within v'. The differential electric field at a point P due to a differential amount of charge $dq = \rho_v\, dv'$ contained in a differential volume dv' is

$$d\mathbf{E} = \hat{\mathbf{R}}' \frac{dq}{4\pi\varepsilon R'^2} = \hat{\mathbf{R}}' \frac{\rho_v\, dv'}{4\pi\varepsilon R'^2}, \quad (4.20)$$

where $\mathbf{R}'$ is the vector from the differential volume dv' to point P. Applying the principle of linear superposition, the total electric field **E** can be obtained by integrating

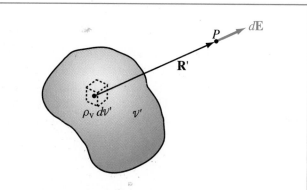

Figure 4-5: Electric field due to a volume charge distribution.

the fields contributed by all the charges making up the charge distribution. Thus,

$$\mathbf{E} = \int_{\nu'} d\mathbf{E} = \frac{1}{4\pi\varepsilon} \int_{\nu'} \hat{\mathbf{R}}' \frac{\rho_v \, dv'}{R'^2}$$

(volume distribution).   (4.21a)

It is important to note that, in general, both R' and $\hat{\mathbf{R}}'$ vary as a function of position over the integration volume ν'.

If the charge is distributed across a surface S' with surface charge density ρ_s, then $dq = \rho_s \, ds'$, and if it is distributed along a line l' with a line charge density ρ_l, then $dq = \rho_l \, dl'$. Accordingly,

$$\mathbf{E} = \frac{1}{4\pi\varepsilon} \int_{S'} \hat{\mathbf{R}}' \frac{\rho_s \, ds'}{R'^2} \quad \text{(surface distribution),} \quad (4.21b)$$

$$\mathbf{E} = \frac{1}{4\pi\varepsilon} \int_{l'} \hat{\mathbf{R}}' \frac{\rho_l \, dl'}{R'^2} \quad \text{(line distribution).} \quad (4.21c)$$

Example 4-4   **Electric Field of a Ring of Charge**

A ring of charge of radius b is characterized by a uniform line charge density of positive polarity ρ_l. With the ring in free space and positioned in the x–y plane as shown in Fig. 4-6, determine the electric field intensity $\mathbf{E}$ at a point $P(0, 0, h)$ along the axis of the ring at a distance h from its center.

Solution: We start by considering the electric field generated by a differential segment of the ring, such as segment 1 located at $(b, \phi, 0)$ in Fig. 4-6(a). The segment has length $dl = b \, d\phi$ and contains charge $dq = \rho_l \, dl = \rho_l b \, d\phi$. The distance vector $\mathbf{R}'_1$ from segment 1 to point $P(0, 0, h)$ is

$$\mathbf{R}'_1 = -\hat{\mathbf{r}}b + \hat{\mathbf{z}}h,$$

from which we have

$$R'_1 = |\mathbf{R}'_1| = \sqrt{b^2 + h^2}, \qquad \hat{\mathbf{R}}'_1 = \frac{\mathbf{R}'_1}{|\mathbf{R}'_1|} = \frac{-\hat{\mathbf{r}}b + \hat{\mathbf{z}}h}{\sqrt{b^2 + h^2}}.$$

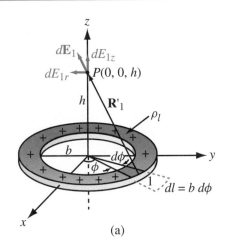

(a)

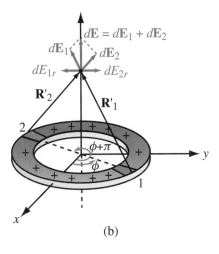

(b)

Figure 4-6: Ring of charge with line density ρ_l. (a) The field $d\mathbf{E}_1$ due to infinitesimal segment 1 and (b) the fields $d\mathbf{E}_1$ and $d\mathbf{E}_2$ due to segments at diametrically opposite locations (Example 4-4).

The electric field at $P(0, 0, h)$ due to the charge of segment 1 is

$$d\mathbf{E}_1 = \frac{1}{4\pi\varepsilon_0} \hat{\mathbf{R}}'_1 \frac{\rho_l \, dl}{R'^2_1} = \frac{\rho_l b}{4\pi\varepsilon_0} \frac{(-\hat{\mathbf{r}}b + \hat{\mathbf{z}}h)}{(b^2 + h^2)^{3/2}} \, d\phi.$$

The field $d\mathbf{E}_1$ has component dE_{1r} along $-\hat{\mathbf{r}}$ and component dE_{1z} along $\hat{\mathbf{z}}$. From symmetry considerations, the field $d\mathbf{E}_2$ generated by segment 2 in Fig. 4-6(b), which is located diametrically opposite the location of segment 1, is identical with $d\mathbf{E}_1$ except that the $\hat{\mathbf{r}}$-component of $d\mathbf{E}_2$ is opposite that of $d\mathbf{E}_1$. Hence, the $\hat{\mathbf{r}}$-components of the sum cancel and the $\hat{\mathbf{z}}$-contributions add. The sum of the two contributions is

$$dE = d\mathbf{E}_1 + d\mathbf{E}_2 = \hat{\mathbf{z}}\,\frac{\rho_l bh}{2\pi\varepsilon_0}\,\frac{d\phi}{(b^2+h^2)^{3/2}}. \qquad (4.22)$$

Since for every ring segment in the semicircle defined over the range $0 \le \phi \le \pi$ (the right-hand half of the circular ring) there is a corresponding segment located diametrically opposite at $(\phi+\pi)$, we can obtain the total field generated by the ring by integrating Eq. (4.22) over a semicircle as follows:

$$\begin{aligned}
\mathbf{E} &= \hat{\mathbf{z}}\,\frac{\rho_l bh}{2\pi\varepsilon_0(b^2+h^2)^{3/2}}\int_0^\pi d\phi \\
&= \hat{\mathbf{z}}\,\frac{\rho_l bh}{2\varepsilon_0(b^2+h^2)^{3/2}} \\
&= \hat{\mathbf{z}}\,\frac{h}{4\pi\varepsilon_0(b^2+h^2)^{3/2}}\,Q, \qquad (4.23)
\end{aligned}$$

where $Q = 2\pi b\rho_l$ is the total charge contained in the ring. ∎

Example 4-5   Electric Field of a Circular Disk of Charge

Find the electric field at a point $P(0,0,h)$ in free space at a height h on the z-axis due to a circular disk of charge in the x–y plane with uniform charge density ρ_s, as shown in Fig. 4-7, and then evaluate $\mathbf{E}$ for the infinite-sheet case by letting $a \to \infty$.

Solution: Building on the expression obtained in Example 4-4 for the on-axis electric field due to a circular ring of charge, we can determine the field due to the circular disk by treating the disk as a set of concentric rings. A ring of radius r and width dr has an area $ds = 2\pi r\,dr$ and contains charge

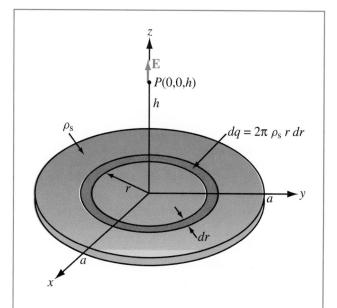

Figure 4-7: Circular disk of charge with surface charge density ρ_s. The electric field at $P(0,0,h)$ points along the z-direction (Example 4-5).

$dq = \rho_s\,ds = 2\pi\rho_s r\,dr$. Upon using this expression in Eq. (4.23) and also replacing b with r, we obtain the following expression for the field due to the ring:

$$d\mathbf{E} = \hat{\mathbf{z}}\,\frac{h}{4\pi\varepsilon_0(r^2+h^2)^{3/2}}\,(2\pi\rho_s r\,dr).$$

The total field at P is obtained by integrating the expression over the limits $r = 0$ to $r = a$:

$$\begin{aligned}
\mathbf{E} &= \hat{\mathbf{z}}\,\frac{\rho_s h}{2\varepsilon_0}\int_0^a \frac{r\,dr}{(r^2+h^2)^{3/2}} \\
&= \pm\hat{\mathbf{z}}\,\frac{\rho_s}{2\varepsilon_0}\left[1 - \frac{|h|}{\sqrt{a^2+h^2}}\right], \qquad (4.24)
\end{aligned}$$

with the plus sign corresponding to when $h > 0$ and the minus sign to when $h < 0$ (below the disk).

For an infinite sheet of charge with $a = \infty$,

$$\mathbf{E} = \hat{\mathbf{z}} \frac{\rho_s}{2\varepsilon_0} \quad \text{(infinite sheet of charge)}. \quad (4.25)$$

We note that for an infinite sheet of charge $\mathbf{E}$ has the same value at any point above the x–y plane. For points located below the x–y plane, the unit vector $\hat{\mathbf{z}}$ in Eq. (4.25) should be replaced with $-\hat{\mathbf{z}}$. ■

REVIEW QUESTIONS

Q4.4 When we say that an electric charge induces an electric field at every point in space, does that mean that the charge "radiates" the electric field? Explain.

Q4.5 If the electric field is zero at a given point in space, does this imply the absence of electric charges?

Q4.6 State the principle of linear superposition as it applies to the electric field due to a distribution of electric charge.

EXERCISE 4.6 An infinite sheet of charge with uniform surface charge density ρ_s is located at $z = 0$ (x–y plane), and another infinite sheet with density $-\rho_s$ is located at $z = 2$ m, both in free space. Determine $\mathbf{E}$ in all regions.

Ans. $\mathbf{E} = 0$ for $z < 0$; $\mathbf{E} = \hat{\mathbf{z}}\rho_s/\varepsilon_0$ for $0 < z < 2$ m; and $\mathbf{E} = 0$ for $z > 2$ m.

4-4 Gauss's Law

We now return to Eq. (4.1a):

$$\nabla \cdot \mathbf{D} = \rho_v \quad \text{(Gauss's law)}, \quad (4.26)$$

which is called the differential form of *Gauss's law*. The adjective "differential" refers to the fact that the divergence operation involves spatial derivatives. As we will see shortly, Eq. (4.26) can be converted and expressed in integral form. When solving electromagnetic problems, we often convert back and forth between the differential and integral forms of equations, depending on which form happens to be the more applicable or convenient to use in each step of a solution. To convert Eq. (4.26) into integral form, we multiply both sides by dv and take the volume integral over an arbitrary volume v. Hence,

$$\int_v \nabla \cdot \mathbf{D} \, dv = \int_v \rho_v \, dv = Q, \quad (4.27)$$

where Q is the total charge enclosed in v. The divergence theorem, given by Eq. (3.98), states that the volume integral of the divergence of any vector over a volume v is equal to the total outward flux of that vector through the surface S enclosing v. Thus, for the vector $\mathbf{D}$,

$$\int_v \nabla \cdot \mathbf{D} \, dv = \oint_S \mathbf{D} \cdot d\mathbf{s}. \quad (4.28)$$

Comparison of Eq. (4.27) with Eq. (4.28) leads to

$$\oint_S \mathbf{D} \cdot d\mathbf{s} = Q \quad \text{(Gauss's law)}. \quad (4.29)$$

The integral form of Gauss's law is illustrated diagrammatically in Fig. 4-8; for each differential surface element $d\mathbf{s}$, $\mathbf{D} \cdot d\mathbf{s}$ is the electric field flux flowing outwardly through $d\mathbf{s}$, and the total flux through the surface S is equal to the enclosed charge Q. The surface S is called a *Gaussian surface*.

When the dimensions of a very small volume Δv containing a total charge q are much smaller than the distance from Δv to the point at which the electric flux density $\mathbf{D}$ is to be evaluated, then q may be regarded as a *point charge*. The integral form of Gauss's law can be applied to determine $\mathbf{D}$ due to a single isolated charge q by constructing a closed, spherical, Gaussian surface S of an

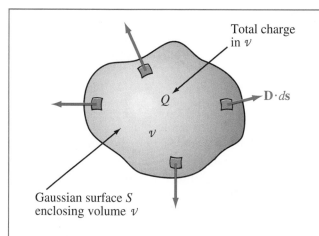

Figure 4-8: Gauss's law states that the outward flux of **D** through a surface is proportional to the enclosed charge Q.

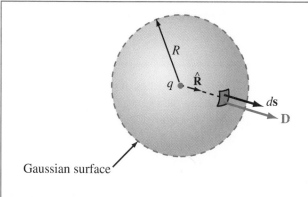

Figure 4-9: Electric field **D** due to point charge q.

arbitrary radius R centered at q, as shown in Fig. 4-9. From symmetry considerations, assuming that q is positive, the direction of **D** must be radially outward along the unit vector $\hat{\mathbf{R}}$, and D_R, the magnitude of **D**, must be the same at all points on the Gaussian surface S. Thus, at any point on the surface, defined by position vector **R**,

$$\mathbf{D(R)} = \hat{\mathbf{R}} D_R, \qquad (4.30)$$

and $d\mathbf{s} = \hat{\mathbf{R}}\, ds$. Applying Gauss's law gives

$$\oint_S \mathbf{D} \cdot d\mathbf{s} = \oint_S \hat{\mathbf{R}} D_R \cdot \hat{\mathbf{R}}\, ds$$

$$= \oint_S D_R\, ds = D_R(4\pi R^2) = q. \qquad (4.31)$$

Solving for D_R and then inserting the result in Eq. (4.30) gives the following expression for the electric field **E** induced by an isolated point charge in a medium with permittivity ε:

$$\mathbf{E(R)} = \frac{\mathbf{D(R)}}{\varepsilon} = \hat{\mathbf{R}} \frac{q}{4\pi \varepsilon R^2} \qquad \text{(V/m)}. \qquad (4.32)$$

This is identical with Eq. (4.13) obtained from Coulomb's law. For this simple case of an isolated point charge, it does not much matter whether Coulomb's law or Gauss's law is used to obtain the expression for **E**. However, it does matter as to which approach we follow when we deal with multiple point charges or continuous charge distributions. Even though Coulomb's law can be used to find **E** for any specified distribution of charge, Gauss's law is easier to apply than Coulomb's law, but its utility is limited to symmetrical charge distributions.

Gauss's law, as given by Eq. (4.29), provides a convenient method for determining the electrostatic flux density **D** when the charge distribution possesses symmetry properties that allow us to make valid assumptions about the variations of the magnitude and direction of **D** as a function of spatial location. Because at every point on the surface the direction of $d\mathbf{s}$ is the outward normal to the surface, only the normal component of **D** at the surface contributes to the integral in Eq. (4.29). To successfully apply Gauss's law, the surface S should be chosen such that, from symmetry considerations, the magnitude of **D** is constant and its direction is normal or tangential at every point of each subsurface of S (the surface of a cube, for example, has six subsurfaces). These aspects are illustrated in Example 4-6.

Example 4-6   Electric Field of an Infinite Line of Charge

Use Gauss's law to obtain an expression for **E** in free space due to an infinitely long line of charge with uniform charge density ρ_l along the z-axis.

Solution: Since the line of charge is infinite in extent and is along the z-axis, symmetry considerations dictate that **D** must be in the radial $\hat{\mathbf{r}}$-direction and must not depend on ϕ or z. Thus, $\mathbf{D} = \hat{\mathbf{r}} D_r$. In Fig. 4-10, we construct a cylindrical Gaussian surface of radius r, concentric around the line of charge. The total charge contained within the cylinder is $Q = \rho_l h$, where h is the height of the cylinder. Since **D** is along $\hat{\mathbf{r}}$, the top and bottom surfaces of the cylinder do not contribute to the surface integral on the left-hand side of Eq. (4.29), and only the side surface contributes to the integral. Hence,

$$\int_{z=0}^{h} \int_{\phi=0}^{2\pi} \hat{\mathbf{r}} D_r \cdot \hat{\mathbf{r}} r \, d\phi \, dz = \rho_l h$$

or

$$2\pi h D_r r = \rho_l h,$$

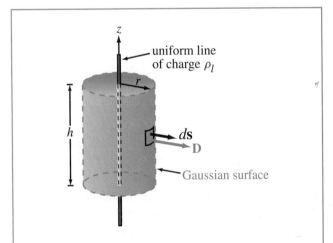

Figure 4-10: Gaussian surface around an infinitely long line of charge (Example 4-6).

which gives the result

$$\boxed{\mathbf{E} = \frac{\mathbf{D}}{\varepsilon_0} = \hat{\mathbf{r}} \frac{D_r}{\varepsilon_0} = \hat{\mathbf{r}} \frac{\rho_l}{2\pi \varepsilon_0 r} \quad \text{(infinite line of charge).}}$$

$$(4.33)$$

Note that Eq. (4.33) is applicable for any infinite line of charge, regardless of its location and direction, as long as $\hat{\mathbf{r}}$ is properly defined as the radial distance vector from the line charge to the observation point (i.e., $\hat{\mathbf{r}}$ is perpendicular to the line of charge). ∎

REVIEW QUESTIONS

Q4.7 Explain Gauss's law. Under what circumstances is it useful?

Q4.8 How should one choose a Gaussian surface?

Q4.9 When is it reasonable to treat a charge distribution as a point charge?

EXERCISE 4.7 Two infinite lines of charge, each carrying a charge density ρ_l, are parallel to the z-axis and located at $x = 1$ and $x = -1$. Determine **E** at an arbitrary point in free space along the y-axis.

Ans. $\mathbf{E} = \hat{\mathbf{y}} \rho_l y / \left[\pi \varepsilon_0 (y^2 + 1) \right].$

EXERCISE 4.8 A thin spherical shell of radius a carries a uniform surface charge density ρ_s. Use Gauss's law to determine **E**.

Ans. $\mathbf{E} = 0$ for $R < a$;
$\mathbf{E} = \hat{\mathbf{R}} \rho_s a^2 / (\varepsilon R^2)$ for $R > a$.

EXERCISE 4.9 A spherical volume of radius a contains a uniform volume charge density ρ_v. Use Gauss's law to determine **D** for (a) $R \le a$ and (b) $R \ge a$.

Ans. (a) $\mathbf{D} = \hat{\mathbf{R}} \rho_v R / 3$,
(b) $\mathbf{D} = \hat{\mathbf{R}} \rho_v a^3 / (3 R^2).$

4-5 Electric Scalar Potential

In electric circuits, we work with voltages and currents. The voltage V between two points in the circuit represents the amount of work, or *potential* energy, required to move a unit charge between the two points. In fact, the term "voltage" is a shortened version of "voltage potential" and is the same as *electric potential*. Even though when we solve a circuit problem we usually do not consider the electric fields present in the circuit, in fact it is the existence of an electric field between two points that gives rise to the voltage difference between them, such as across a resistor or a capacitor. The relationship between the electric field **E** and the electric potential V is the subject of this section.

4-5.1 Electric Potential as a Function of Electric Field

We begin by considering the simple case of a positive charge q in a uniform electric field $\mathbf{E} = -\hat{\mathbf{y}}E$, parallel to the $-y$-direction, as shown in Fig. 4-11. The presence of the field **E** exerts a force $\mathbf{F}_e = q\mathbf{E}$ on the charge in the negative y-direction. If we attempt to move the charge along the positive y-direction (against the force $\mathbf{F}_e$), we will need to provide an external force $\mathbf{F}_{ext}$ to counteract $\mathbf{F}_e$, which requires the expenditure of energy. To move q without any acceleration (at a constant speed),

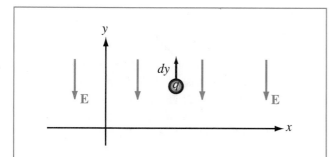

Figure 4-11: Work done in moving a charge q a distance dy against the electric field **E** is $dW = qE\, dy$.

it is necessary that the net force acting on the charge be zero, which means that $\mathbf{F}_{ext} + \mathbf{F}_e = 0$, or

$$\mathbf{F}_{ext} = -\mathbf{F}_e = -q\mathbf{E}. \qquad (4.34)$$

The work done, or energy expended, in moving any object a vector differential distance $d\mathbf{l}$ under the influence of a force $\mathbf{F}_{ext}$ is

$$dW = \mathbf{F}_{ext} \cdot d\mathbf{l} = -q\mathbf{E} \cdot d\mathbf{l} \quad \text{(J)}. \qquad (4.35)$$

Work, or energy, is measured in joules (J). In the present case, if the charge is moved a distance dy along $\hat{\mathbf{y}}$, then

$$dW = -q(-\hat{\mathbf{y}}E) \cdot \hat{\mathbf{y}}\, dy = qE\, dy. \qquad (4.36)$$

The differential electric potential energy dW per unit charge is called the *differential electric potential* (or differential voltage) dV. That is,

$$dV = \frac{dW}{q} = -\mathbf{E} \cdot d\mathbf{l} \quad \text{(J/C or V)}. \qquad (4.37)$$

The unit of V is the volt (V), with $1\text{ V} \triangleq 1\text{ J/C}$, and since V is measured in volts, the electric field is expressed in volts per meter (V/m).

The potential difference between any two points P_2 and P_1 (Fig. 4-12) is obtained by integrating Eq. (4.37) *along any path* between them. That is,

$$\int_{P_1}^{P_2} dV = -\int_{P_1}^{P_2} \mathbf{E} \cdot d\mathbf{l}, \qquad (4.38)$$

or

$$V_{21} = V_2 - V_1 = -\int_{P_1}^{P_2} \mathbf{E} \cdot d\mathbf{l}, \qquad (4.39)$$

where V_1 and V_2 are the electric potentials at points P_1 and P_2, respectively. The result of the line integral on the right-hand side of Eq. (4.39) should be independent of the specific integration path taken between points P_1 and P_2. This requirement is mandated by the law of conservation of energy. To illustrate with an example, let us consider a particle in Earth's gravitational field. If the particle is moved from a height h_1 above Earth's surface to a higher height h_2, the particle gains potential energy in an amount proportional to $(h_2 - h_1)$. If, instead, we were to first raise the particle from height h_1 to a height h_3 greater than h_2, thereby giving it potential energy proportional to $(h_3 - h_1)$, and then we let it drop back to height h_2 by expending an energy amount proportional to $(h_3 - h_2)$, its net gain in potential energy will again be proportional to $(h_2 - h_1)$. The same principle applies to the electric potential energy W and to the potential difference $(V_2 - V_1)$. The voltage difference between two nodes in an electric circuit has the same value regardless of which path in the circuit we follow between the nodes. Moreover, *Kirchhoff's voltage law* states that the net voltage drop around a closed loop is zero. If we go from P_1 to P_2 by path 1 in Fig. 4-12 and then return from P_2 to P_1 by path 2, the right-hand side of Eq. (4.39) becomes a closed contour and the left-hand side becomes zero. In fact, the line integral of the electrostatic field $\mathbf{E}$ around any closed contour C is zero:

$$\oint_C \mathbf{E} \cdot d\mathbf{l} = 0 \quad \text{(Electrostatics).} \quad (4.40)$$

A vector field whose line integral along any closed path is zero is called a *conservative* or an *irrotational* field. Hence, the electrostatic field $\mathbf{E}$ is conservative. As we will see later in Chapter 6, if $\mathbf{E}$ is a time-varying function, it is no longer conservative, and its line integral along a closed path is not necessarily equal to zero.

The conservative property of the electrostatic field can also be deduced from Maxwell's second equation, Eq. (4.1b). If $\partial / \partial t = 0$, then

$$\nabla \times \mathbf{E} = 0. \quad (4.41)$$

If we take the surface integral of $\nabla \times \mathbf{E}$ over an open surface S and then apply Stokes's theorem, given by Eq. (3.107), to convert the surface integral into a line integral, we have

$$\int_S (\nabla \times \mathbf{E}) \cdot d\mathbf{s} = \oint_C \mathbf{E} \cdot d\mathbf{l} = 0, \quad (4.42)$$

where C is a closed contour surrounding S. Thus, Eq. (4.41) is the differential-form equivalent of Eq. (4.40).

We now define what we mean by the electric potential V at a point in space. Before we do so, however, let us revisit our electric-circuit analogue. Absolute voltage at a point in a circuit has no defined meaning, nor does absolute electric potential at a point in space. When we talk of the voltage V of a point in a circuit, we do so in reference to the voltage of some conveniently chosen point to which we have assigned a reference voltage of zero, which we call *ground*. The same principle applies to the electric potential V. Usually, the reference-potential point is chosen to be at infinity. That is, in Eq. (4.39) we

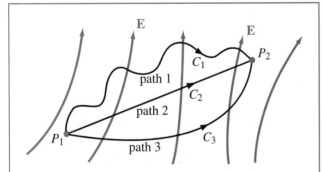

Figure 4-12: In electrostatics, the potential difference between P_2 and P_1 is the same irrespective of the path used for calculating the line integral of the electric field between them.

assume that $V_1 = 0$ when P_1 is at infinity, and therefore the electric potential V at any point P is given by

$$V = -\int_\infty^P \mathbf{E} \cdot d\mathbf{l} \quad \text{(V)}. \quad (4.43)$$

4-5.2 Electric Potential due to Point Charges

For a point charge q located at the origin of a spherical coordinate system, the electric field at a distance R is given by Eq. (4.32) as

$$\mathbf{E} = \hat{\mathbf{R}} \frac{q}{4\pi\varepsilon R^2} \quad \text{(V/m)}. \quad (4.44)$$

As was stated earlier, the choice of integration path between the two end points in Eq. (4.43) is arbitrary. Hence, we will conveniently choose the path to be along the radial direction $\hat{\mathbf{R}}$, in which case $d\mathbf{l} = \hat{\mathbf{R}} \, dR$ and

$$V = -\int_\infty^R \left(\hat{\mathbf{R}} \frac{q}{4\pi\varepsilon R^2} \right) \cdot \hat{\mathbf{R}} \, dR = \frac{q}{4\pi\varepsilon R} \quad \text{(V)}. \quad (4.45)$$

If the charge q is at a location other than the origin, specified by a source position vector $\mathbf{R}_1$, then V at observation position vector $\mathbf{R}$ becomes

$$V(\mathbf{R}) = \frac{q}{4\pi\varepsilon |\mathbf{R} - \mathbf{R}_1|} \quad \text{(V)}, \quad (4.46)$$

where $|\mathbf{R} - \mathbf{R}_1|$ is the distance between the observation point and the location of the charge q. The principle of superposition that we applied previously to the electric field $\mathbf{E}$ also applies to the electric potential V. Hence, for N discrete point charges $q_1, q_2, \ldots, q_N$ having position vectors $\mathbf{R}_1, \mathbf{R}_2, \ldots, \mathbf{R}_N$, the electric potential is

$$V(\mathbf{R}) = \frac{1}{4\pi\varepsilon} \sum_{i=1}^N \frac{q_i}{|\mathbf{R} - \mathbf{R}_i|} \quad \text{(V)}. \quad (4.47)$$

4-5.3 Electric Potential due to Continuous Distributions

For a continuous charge distribution specified over a given volume v', across a surface S', or along a line l', we (1) replace q_i in Eq. (4.47) with, respectively, $\rho_v \, dv'$, $\rho_s \, ds'$, and $\rho_l \, dl'$; (2) convert the summation into an integration; and (3) define $R' = |\mathbf{R} - \mathbf{R}_i|$ as the distance between the integration point and the observation point. These steps lead to the following expressions:

$$V(\mathbf{R}) = \frac{1}{4\pi\varepsilon} \int_{v'} \frac{\rho_v}{R'} \, dv' \quad \text{(volume distribution)}, \quad (4.48a)$$

$$V(\mathbf{R}) = \frac{1}{4\pi\varepsilon} \int_{S'} \frac{\rho_s}{R'} \, ds' \quad \text{(surface distribution)}, \quad (4.48b)$$

$$V(\mathbf{R}) = \frac{1}{4\pi\varepsilon} \int_{l'} \frac{\rho_l}{R'} \, dl' \quad \text{(line distribution)}. \quad (4.48c)$$

4-5.4 Electric Field as a Function of Electric Potential

In Section 4-5.1, we expressed V in terms of a line integral over $\mathbf{E}$. Now we explore the inverse relationship by examining the differential form of V given by Eq. (4.37):

$$dV = -\mathbf{E} \cdot d\mathbf{l}. \quad (4.49)$$

For a scalar function V, Eq. (3.73) gives

$$dV = \nabla V \cdot d\mathbf{l}, \quad (4.50)$$

where ∇V is the gradient of V. Comparison of Eq. (4.49) with Eq. (4.50) leads to

$$\mathbf{E} = -\nabla V. \quad (4.51)$$

This relationship between V and $\mathbf{E}$ in differential form allows us to determine $\mathbf{E}$ for any charge distribution by first calculating V using the expressions given in Sections 4-5.1 to 4-5.3 and then taking the negative gradient of V

to find **E**. The expressions for V, given by Eqs. (4.47) to (4.48c), involve scalar sums and scalar integrals, and as such they are usually much easier to calculate than the vector sums and integrals in the expressions for **E** derived in Section 4-3 on the basis of Coulomb's law. Thus, even though the electric potential approach for finding **E** is a two-step process, it is computationally simpler to apply than the direct method based on Coulomb's law.

Example 4-7   Electric Field of an Electric Dipole

An *electric dipole* consists of two point charges of equal magnitude and opposite polarity, separated by a small distance, as shown in Fig. 4-13(a). Determine V and **E** at any point P in free space, given that P is at a distance $R \gg d$, where d is the spacing between the two charges.

Solution: The electric potential due to a single point charge is given by Eq. (4.45). For the two charges shown in Fig. 4-13(a), application of Eq. (4.47) gives

$$V = \frac{1}{4\pi\varepsilon_0}\left(\frac{q}{R_1} + \frac{-q}{R_2}\right) = \frac{q}{4\pi\varepsilon_0}\left(\frac{R_2 - R_1}{R_1 R_2}\right).$$

Since $d \ll R$, the lines labeled R_1 and R_2 in Fig. 4-13(a) are approximately parallel to each other, in which case the following approximations apply:

$$R_2 - R_1 \simeq d\cos\theta, \qquad R_1 R_2 \simeq R^2.$$

Hence,

$$V = \frac{qd\cos\theta}{4\pi\varepsilon_0 R^2}. \tag{4.52}$$

The numerator of Eq. (4.52) can be written as the dot product of $q\mathbf{d}$, where **d** is the distance vector from charge $-q$ to charge $+q$, and the unit vector $\hat{\mathbf{R}}$ pointing from the center of the dipole toward the observation point P,

$$qd\cos\theta = q\mathbf{d}\cdot\hat{\mathbf{R}} = \mathbf{p}\cdot\hat{\mathbf{R}}, \tag{4.53}$$

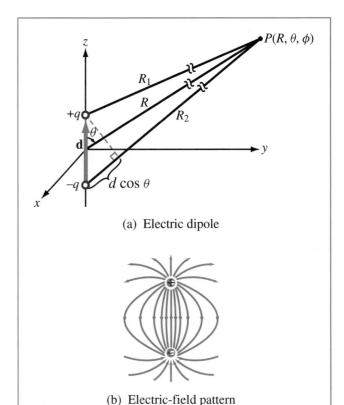

(a) Electric dipole

(b) Electric-field pattern

Figure 4-13: Electric dipole with dipole moment $\mathbf{p} = q\mathbf{d}$ (Example 4-7).

where $\mathbf{p} = q\mathbf{d}$ is called the *dipole moment* of the electric dipole. Using Eq. (4.53) in Eq. (4.52) then gives

$$V = \frac{\mathbf{p}\cdot\hat{\mathbf{R}}}{4\pi\varepsilon_0 R^2} \qquad \text{(electric dipole).} \tag{4.54}$$

In spherical coordinates, Eq. (4.51) is given by

$$\mathbf{E} = -\nabla V$$
$$= -\left(\hat{\mathbf{R}}\frac{\partial V}{\partial R} + \hat{\boldsymbol{\theta}}\frac{1}{R}\frac{\partial V}{\partial\theta} + \hat{\boldsymbol{\phi}}\frac{1}{R\sin\theta}\frac{\partial V}{\partial\phi}\right), \tag{4.55}$$

where we have used the expression for ∇V given on the inside back cover of the book. Upon taking the derivatives of the expression for V given by Eq. (4.52) with respect to R and θ and then substituting the results in Eq. (4.55), we have

$$\mathbf{E} = \frac{qd}{4\pi\varepsilon_0 R^3}\left(\hat{\mathbf{R}}\,2\cos\theta + \hat{\boldsymbol{\theta}}\sin\theta\right) \quad \text{(V/m)}. \quad (4.56)$$

We should note that the expressions for V and $\mathbf{E}$, given by Eqs. (4.54) and (4.56), are applicable only when $R \gg d$. To compute V and $\mathbf{E}$ at points in the vicinity of the two charges making up the dipole, it is necessary to perform the calculations without resorting to the far-distance approximations that led to Eq. (4.52). Such an exact calculation for $\mathbf{E}$ leads to the pattern shown in Fig. 4-13(b).

■

4-5.5 Poisson's Equation

With $\mathbf{D} = \varepsilon\mathbf{E}$, the differential form of Gauss's law given by Eq. (4.26) may be written as

$$\nabla \cdot \mathbf{E} = \frac{\rho_v}{\varepsilon}. \quad (4.57)$$

Inserting Eq. (4.51) in Eq. (4.57) gives

$$\nabla \cdot (\nabla V) = -\frac{\rho_v}{\varepsilon}. \quad (4.58)$$

In view of the definition for the Laplacian of a scalar function V given by Eq. (3.110) as

$$\nabla^2 V = \nabla \cdot (\nabla V) = \frac{\partial^2 V}{\partial x^2} + \frac{\partial^2 V}{\partial y^2} + \frac{\partial^2 V}{\partial z^2}, \quad (4.59)$$

Eq. (4.58) can be cast in the abbreviated form

$$\nabla^2 V = -\frac{\rho_v}{\varepsilon} \quad \text{(Poisson's equation)}. \quad (4.60)$$

This is known as *Poisson's equation*. For a volume v' containing a volume charge density distribution ρ_v, the solution for V derived previously and expressed by Eq. (4.48a) as

$$V = \frac{1}{4\pi\varepsilon}\int_{v'}\frac{\rho_v}{R'}\,dv' \quad (4.61)$$

satisfies Eq. (4.60). If the medium under consideration contains no free charges, Eq. (4.60) reduces to

$$\nabla^2 V = 0 \quad \text{(Laplace's equation)},$$

and it is then referred to as *Laplace's equation*. Poisson's and Laplace's equations are useful for determining the electrostatic potential V in regions at whose boundaries V is known, such as the region between the plates of a capacitor with a specified voltage difference across it.

REVIEW QUESTIONS

Q4.10 What is a conservative field?

Q4.11 Why is the electric potential at a point in space always defined relative to the potential at some reference point?

Q4.12 Explain why Eq. (4.40) is a mathematical statement of Kirchhoff's voltage law.

Q4.13 Why is it usually easier to compute V for a given charge distribution and then find $\mathbf{E}$ from $\mathbf{E} = -\nabla V$ than to compute $\mathbf{E}$ directly by applying Coulomb's law?

Q4.14 What is an electric dipole?

EXERCISE 4.10 Determine the electric potential at the origin in free space due to four charges of $20\,\mu\text{C}$ each located at the corners of a square in the x–y plane and whose center is at the origin. The square has sides of 2 m each.

Ans. $V = \sqrt{2} \times 10^{-5}/(\pi\varepsilon_0)$ (V).

EXERCISE 4.11 A spherical shell of radius R has a uniform surface charge density ρ_s. Determine the electric potential at the center of the shell.

Ans. $V = \rho_s R / \varepsilon$ (V).

4-6 Electrical Properties of Materials

The electromagnetic *constitutive parameters* of a material medium are its electrical permittivity ε, magnetic permeability μ, and conductivity σ. A material is said to be *homogeneous* if its constitutive parameters do not vary from point to point, and it is *isotropic* if its constitutive parameters are independent of direction. Most materials exhibit isotropic properties, but some crystals do not. Throughout this book, all materials are assumed to be homogeneous and isotropic. In this chapter, we are concerned with only ε and σ. Discussion of μ is deferred to Chapter 5.

The conductivity of a material is a measure of how easily electrons can travel through the material under the influence of an external electric field. Materials are classified as *conductors* (metals) or *dielectrics* (insulators) according to the magnitudes of their conductivities. A conductor has a large number of loosely attached electrons in the outermost shells of the atoms. In the absence of an external electric field, these free electrons move in random directions and with varying speeds. Their random motion produces zero average current through the conductor. Upon applying an external electric field, however, the electrons migrate from one atom to the next along a direction opposite that of the external field. Their movement, which is characterized by an average velocity called the *electron drift velocity* $\mathbf{u}_e$, gives rise to a *conduction current*.

In a dielectric, the electrons are tightly held to the atoms, so much so that it is very difficult to detach them even under the influence of an electric field. Consequently, no current flows through the material. A *perfect*

Table 4-1: Conductivity of some common materials at 20° C.

Material	Conductivity, σ (S/m)
Conductors	
Silver	6.2×10^7
Copper	5.8×10^7
Gold	4.1×10^7
Aluminum	3.5×10^7
Iron	10^7
Mercury	10^6
Carbon	3×10^4
Semiconductors	
Pure germanium	2.2
Pure silicon	4.4×10^{-4}
Insulators	
Glass	10^{-12}
Paraffin	10^{-15}
Mica	10^{-15}
Fused quartz	10^{-17}

dielectric is a material with $\sigma = 0$ and, in contrast, a *perfect conductor* is a material with $\sigma = \infty$.

The conductivity σ of most metals is in the range from 10^6 to 10^7 S/m, compared with 10^{-10} to 10^{-17} S/m for good insulators (Table 4-1). Materials whose conductivities fall between those of conductors and insulators are called *semiconductors*. The conductivity of pure germanium, for example, is 2.2 S/m. Tabulated values of σ are given in Appendix B for some common materials at room temperature (20° C), and a subset is given in Table 4-1.

The conductivity of a material depends on several factors, including temperature and the presence of impurities. In general, σ of metals increases with decreasing temperature, and at very low temperatures in the neighborhood of absolute zero, some conductors become *superconductors* because their conductivities become practically infinite.

Q4.15 What are the electromagnetic constitutive parameters of a material medium?

Q4.16 What classifies a material as a conductor, semiconductor, or dielectric? What is a superconductor?

Q4.17 What is the conductivity of a perfect dielectric?

4-7 Conductors

The drift velocity $\mathbf{u}_e$ of electrons in a conducting material is related to the externally applied electric field $\mathbf{E}$ through

$$\mathbf{u}_e = -\mu_e \mathbf{E} \quad \text{(m/s)}, \qquad (4.62\text{a})$$

where μ_e is a material property call the *electron mobility* with units of (m^2/V·s). In a semiconductor, current flow is due to the movement of both electrons and holes, and since holes are positive-charge carriers, the *hole drift velocity* $\mathbf{u}_h$ is in the same direction as $\mathbf{E}$,

$$\mathbf{u}_h = \mu_h \mathbf{E} \quad \text{(m/s)}, \qquad (4.62\text{b})$$

where μ_h is the *hole mobility*. The mobility accounts for the effective mass of a charged particle and the average distance over which the applied electric field can accelerate it before it is stopped by colliding with an atom and must start accelerating over again. From Eq. (4.11), the current density in a medium containing a volume density ρ_v of charges moving with a velocity $\mathbf{u}$ is $\mathbf{J} = \rho_v \mathbf{u}$. In the present case, the current density consists of a component $\mathbf{J}_e$ due to the electrons and a component $\mathbf{J}_h$ due to the holes. Thus, the total *conduction current density* is

$$\mathbf{J} = \mathbf{J}_e + \mathbf{J}_h = \rho_{ve}\mathbf{u}_e + \rho_{vh}\mathbf{u}_h \quad \text{(A/m}^2). \qquad (4.63)$$

Use of Eqs. (4.62a) and (4.62b) gives

$$\mathbf{J} = (-\rho_{ve}\mu_e + \rho_{vh}\mu_h)\mathbf{E}, \qquad (4.64)$$

where $\rho_{ve} = -N_e e$ and $\rho_{vh} = N_h e$, with N_e and N_h being the number of free electrons and the number of free holes per unit volume, and $e = 1.6 \times 10^{-19}$ C is the absolute charge of a single hole or electron. The quantity inside the parentheses in Eq. (4.64) is defined as the *conductivity* of the material, σ. Thus,

$$\sigma = -\rho_{ve}\mu_e + \rho_{vh}\mu_h$$
$$= (N_e\mu_e + N_h\mu_h)e \quad \text{(S/m)} \quad \text{(semiconductor)}, \qquad (4.65)$$

and its unit is siemens per meter (S/m). For a good conductor, $N_h\mu_h \ll N_e\mu_e$, and Eq. (4.65) reduces to

$$\sigma = -\rho_{ve}\mu_e = N_e\mu_e e \quad \text{(S/m)} \quad \text{(conductor)}. \qquad (4.66)$$

In either case, Eq. (4.64) becomes

$$\mathbf{J} = \sigma \mathbf{E} \quad \text{(A/m}^2) \quad \text{(Ohm's law)}, \qquad (4.67)$$

and it is called the *point form of Ohm's law*. Note that, in a perfect dielectric with $\sigma = 0$, $\mathbf{J} = 0$ regardless of $\mathbf{E}$, and in a perfect conductor with $\sigma = \infty$, $\mathbf{E} = \mathbf{J}/\sigma = 0$ regardless of $\mathbf{J}$. That is,

Perfect dielectric:	$\mathbf{J} = 0$
Perfect conductor:	$\mathbf{E} = 0$

Because σ is on the order of 10^6 S/m for most metals, such as silver, copper, gold, and aluminum (Table 4-1), it is common practice to set $\mathbf{E} = 0$ in metal conductors.

A perfect conductor is an *equipotential* medium, meaning that the electric potential is the same at every point in the conductor. This property follows from the fact that V_{21}, the voltage difference between two points in the conductor, is by definition equal to the line integral of $\mathbf{E}$ between the two points, as indicated by

Eq. (4.39), and since $\mathbf{E} = 0$ everywhere in the perfect conductor, the voltage difference $V_{21} = 0$. The fact that the conductor is an equipotential medium, however, does not necessarily imply that the potential difference between the conductor and some other conductor is zero. Each conductor is an equipotential medium, but the presence of different distributions of charges on their surfaces can generate a potential difference between them.

Example 4-8   Conduction Current in a Copper Wire

A 2-mm-diameter copper wire with conductivity of 5.8×10^7 S/m and electron mobility of 0.0032 (m²/V·s) is subjected to an electric field of 20 (mV/m). Find (a) the volume charge density of free electrons, (b) the current density, (c) the current flowing in the wire, (d) the electron drift velocity, and (e) the volume density of free electrons.

Solution

(a)

$$\rho_{ve} = -\frac{\sigma}{\mu_e} = -\frac{5.8 \times 10^7}{0.0032} = -1.81 \times 10^{10} \ (\text{C/m}^3).$$

(b)

$$J = \sigma E = 5.8 \times 10^7 \times 20 \times 10^{-3} = 1.16 \times 10^6 \ (\text{A/m}^2).$$

(c)

$$I = JA$$
$$= J\left(\frac{\pi d^2}{4}\right) = 1.16 \times 10^6 \left(\frac{\pi \times 4 \times 10^{-6}}{4}\right) = 3.64 \ \text{A}.$$

(d)

$$u_e = -\mu_e E = -0.0032 \times 20 \times 10^{-3} = -6.4 \times 10^{-5} \ \text{m/s}.$$

The minus sign indicates that $\mathbf{u}_e$ is in the opposite direction of $\mathbf{E}$.

(e)

$$N_e = -\frac{\rho_{ve}}{e} = \frac{1.81 \times 10^{10}}{1.6 \times 10^{-19}} = 1.13 \times 10^{29} \ \text{electrons/m}^3.$$

∎

EXERCISE 4.12   Determine the density of free electrons in aluminum, given that its conductivity is 3.5×10^7 (S/m) and its electron mobility is 0.0015 (m²/V · s).

Ans. $N_e = 1.46 \times 10^{29}$ electrons/m³.

EXERCISE 4.13 The current flowing through a 100-m-long conducting wire of uniform cross section has a density of 3×10^5 (A/m²). Find the voltage drop across the length of the wire if the wire material has a conductivity of 2×10^7 (S/m).

Ans. $V = 1.5$ (V).

4-7.1   Resistance

By way of demonstrating the use of the point form of Ohm's law, we shall use it to derive an expression for the resistance R of a conductor of length l and uniform cross section A, as shown in Fig. 4-14. The conductor axis is oriented along the x-axis and extends between points x_1 and x_2, with $l = x_2 - x_1$. A voltage V applied across the conductor terminals establishes an electric field $\mathbf{E} = \hat{\mathbf{x}}E_x$; the direction of $\mathbf{E}$ is from the point of higher potential (point 1 in Fig. 4-14) to the point of lower potential (point 2). The relation between V and E_x is obtained by applying Eq. (4.39). Thus,

$$V = V_1 - V_2$$
$$= -\int_{x_2}^{x_1} \mathbf{E} \cdot d\mathbf{l}$$
$$= -\int_{x_2}^{x_1} \hat{\mathbf{x}}E_x \cdot \hat{\mathbf{x}}\, dl = E_x l \quad \text{(V)}. \qquad (4.68)$$

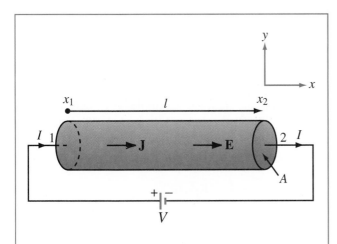

Figure 4-14: Linear resistor of cross section A and length l connected to a d-c voltage source V.

Using Eq. (4.67), the current flowing through the cross section A at x_2 is

$$I = \int_A \mathbf{J} \cdot d\mathbf{s} = \int_A \sigma \mathbf{E} \cdot d\mathbf{s} = \sigma E_x A \quad (A). \quad (4.69)$$

From $R = V/I$, the ratio of Eq. (4.68) to Eq. (4.69) gives

$$R = \frac{l}{\sigma A} \quad (\Omega). \quad (4.70)$$

We now generalize our result for R to any resistor of arbitrary shape by noting that the voltage V across the resistor is equal to the line integral of $\mathbf{E}$ over a path l between two specified points and the current I is equal to the flux of $\mathbf{J}$ through the surface S of the resistor. Thus,

$$R = \frac{V}{I} = \frac{-\int_l \mathbf{E} \cdot d\mathbf{l}}{\int_S \mathbf{J} \cdot d\mathbf{s}} = \frac{-\int_l \mathbf{E} \cdot d\mathbf{l}}{\int_S \sigma \mathbf{E} \cdot d\mathbf{s}}. \quad (4.71)$$

The reciprocal of R is called the *conductance* G, and the unit of G is (Ω^{-1}), or siemens (S). For the linear resistor,

$$G = \frac{1}{R} = \frac{\sigma A}{l} \quad (S). \quad (4.72)$$

Example 4-9 Conductance of Coaxial Cable

The radii of the inner and outer conductors of a coaxial cable of length l are a and b, respectively (Fig. 4-15). The insulation material has conductivity σ. Obtain an expression for G', the conductance per unit length of the insulation layer.

Solution: Let I be the total current flowing from the inner conductor to the outer conductor through the insulation material. At any radial distance r from the axis of the center conductor, the area through which the current flows is $A = 2\pi r l$. Hence,

$$\mathbf{J} = \hat{\mathbf{r}} \frac{I}{A} = \hat{\mathbf{r}} \frac{I}{2\pi r l}, \quad (4.73)$$

and from $\mathbf{J} = \sigma \mathbf{E}$,

$$\mathbf{E} = \hat{\mathbf{r}} \frac{I}{2\pi \sigma r l}. \quad (4.74)$$

In a resistor, the current flows from higher electric potential to lower potential. Hence, if $\mathbf{J}$ is in the $\hat{\mathbf{r}}$-direction, the inner conductor must be at a higher potential than the outer conductor. Accordingly, the voltage difference between the conductors is

$$V_{ab} = -\int_b^a \mathbf{E} \cdot d\mathbf{l} = -\int_b^a \frac{I}{2\pi \sigma l} \frac{\hat{\mathbf{r}} \cdot \hat{\mathbf{r}} \, dr}{r}$$

$$= \frac{I}{2\pi \sigma l} \ln\left(\frac{b}{a}\right). \quad (4.75)$$

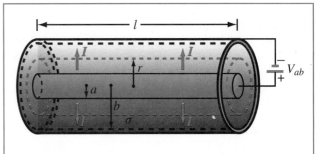

Figure 4-15: Coaxial cable of Example 4-9.

The conductance per unit length is then

$$G' = \frac{G}{l} = \frac{1}{Rl} = \frac{I}{V_{ab}l} = \frac{2\pi\sigma}{\ln(b/a)} \quad \text{(S/m)}. \quad \blacksquare$$

(4.76)

4-7.2 Joule's Law

We now consider the power dissipated in a conducting medium in the presence of an electrostatic field $\mathbf{E}$. The medium contains free electrons and holes with volume charge densities ρ_{ve} and ρ_{vh}, respectively. The electron and hole charge contained in an elemental volume Δv is $q_e = \rho_{ve}\,\Delta v$ and $q_h = \rho_{vh}\,\Delta v$, respectively. The electric forces acting on q_e and q_h are $\mathbf{F}_e = q_e\mathbf{E} = \rho_{ve}\mathbf{E}\,\Delta v$ and $\mathbf{F}_h = q_h\mathbf{E} = \rho_{vh}\mathbf{E}\,\Delta v$. The work (energy) expended by the electric field in moving q_e a differential distance Δl_e and moving q_h a distance Δl_h is

$$\Delta W = \mathbf{F}_e \cdot \Delta\mathbf{l}_e + \mathbf{F}_h \cdot \Delta\mathbf{l}_h. \quad (4.77)$$

Power P, measured in watts (W), is defined as the time rate of change of energy. The change in power corresponding to ΔW is then

$$\begin{aligned}
\Delta P &= \frac{\Delta W}{\Delta t} = \mathbf{F}_e \cdot \frac{\Delta\mathbf{l}_e}{\Delta t} + \mathbf{F}_h \cdot \frac{\Delta\mathbf{l}_h}{\Delta t} \\
&= \mathbf{F}_e \cdot \mathbf{u}_e + \mathbf{F}_h \cdot \mathbf{u}_h \\
&= (\rho_{ve}\mathbf{E}\cdot\mathbf{u}_e + \rho_{vh}\mathbf{E}\cdot\mathbf{u}_h)\,\Delta v \\
&= \mathbf{E}\cdot\mathbf{J}\,\Delta v,
\end{aligned}$$

(4.78)

where $\mathbf{u}_e = \Delta\mathbf{l}_e/\Delta t$ is the electron drift velocity and $\mathbf{u}_h = \Delta\mathbf{l}_h/\Delta t$ is the hole drift velocity. Equation (4.63) was used in the last step of the derivation leading to Eq. (4.78). For a volume v, the total dissipated power is

$$\boxed{P = \int_v \mathbf{E}\cdot\mathbf{J}\,dv \quad \text{(W)} \quad \text{(Joule's law)}, \quad (4.79)}$$

and in view of the relation given by Eq. (4.67),

$$P = \int_v \sigma|\mathbf{E}|^2\,dv \quad \text{(W)}. \quad (4.80)$$

Equation (4.79) is a mathematical statement of *Joule's law*. For the resistor example considered earlier, $|\mathbf{E}| = E_x$ and its volume is $v = lA$. Separating the volume integral in Eq. (4.80) into a product of a surface integral over A and a line integral over l, we have

$$\begin{aligned}
P &= \int_v \sigma|\mathbf{E}|^2\,dv \\
&= \int_A \sigma E_x\,ds \int_l E_x\,dl \\
&= (\sigma E_x A)(E_x l) = IV \quad \text{(W)},
\end{aligned}$$

(4.81)

where use was made of Eq. (4.68) for the voltage V and Eq. (4.69) for the current I. With $V = IR$, we obtain the familiar expression

$$P = I^2 R \quad \text{(W)}. \quad (4.82)$$

REVIEW QUESTIONS

Q4.18 What is the fundamental difference between an insulator, a semiconductor, and a conductor?

Q4.19 Show that the power dissipated in the coaxial cable of Fig. 4-15 is $P = I^2 \ln(b/a)/(2\pi\sigma l)$.

EXERCISE 4.14 A 50-m-long copper wire has a circular cross section with radius $r = 2$ cm. Given that the conductivity of copper is 5.8×10^7 (S/m), determine (a) the resistance R of the wire and (b) the power dissipated in the wire if the voltage across its length is 1.5 (mV).

Ans. (a) $R = 6.9 \times 10^{-4}$ Ω, (b) $P = 3.3$ (mW).

EXERCISE 4.15 Repeat part (b) of Exercise 4.14 by applying Eq. (4.80).

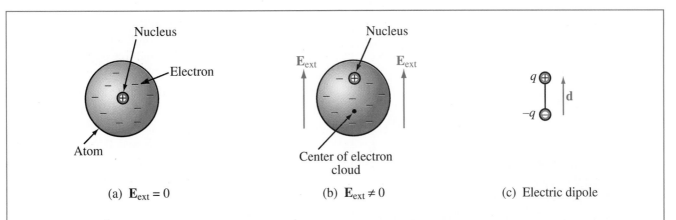

Figure 4-16: In the absence of an external electric field $\mathbf{E}_{ext}$, the center of the electron cloud is co-located with the center of the nucleus, but when a field is applied, the two centers are separated by a distance d.

4-8 Dielectrics

As we discussed previously, the fundamental difference between a conductor and a dielectric is that a conductor has loosely held (free) electrons that can migrate through the crystalline structure of the material, whereas the electrons in the outermost shells of a dielectric are strongly bound to the atom. In the absence of an electric field, the electrons in any material form a symmetrical cloud around the nucleus, with the center of the cloud being at the same location as the center of the nucleus, as shown in Fig. 4-16(a). The electric field generated by the positively charged nucleus attracts and holds the electron cloud around it, and the mutual repulsion of the electron clouds of adjacent atoms gives matter its form. When a conductor is subjected to an externally applied electric field, the most loosely bound electrons in each atom can easily jump from one atom to the next, thereby setting up an electric current. In a dielectric, however, an externally applied electric field $\mathbf{E}_{ext}$ cannot effect mass migration of charges since none are able to move freely, but it can *polarize* the atoms or molecules in the material by distorting the center of the cloud and the location of the nucleus. The polarization process is illustrated in Fig. 4-16(b). The polarized atom or molecule may be represented by

an electric dipole consisting of charge $+q$ at the center of the nucleus and charge $-q$ at the center of the electron cloud [Fig. 4-16(c)]. Each such dipole sets up a small electric field, pointing from the positively charged nucleus to the center of the equally but negatively charged electron cloud. This *induced* electric field, called a *polarization* field, is weaker than and opposite in direction to $\mathbf{E}_{ext}$. Consequently, the net electric field present in the dielectric material is smaller than $\mathbf{E}_{ext}$. At the microscopic level, each dipole exhibits a dipole moment similar to that described in Example 4-7. Within the dielectric material, the dipoles align themselves in a linear arrangement, as shown in Fig. 4-17. Along the upper and lower edges of the material, the dipole arrangement exhibits a positive surface charge density on the upper surface and a negative density on the lower surface.

The relatively simple picture described in Figs. 4-16 and 4-17 pertains to *nonpolar materials* in which the molecules do not have permanent dipole moments. Nonpolar molecules become polarized only when an external electric field is applied, and when the field is terminated, the molecules return to their original unpolarized state. In some materials, such as water, the molecular structure is such that the molecules possess built-in per-

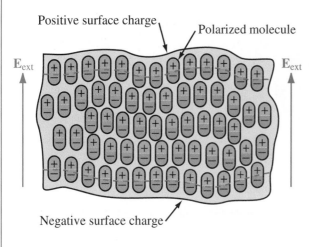

Figure 4-17: A dielectric medium polarized by an external electric field $\mathbf{E}_{ext}$.

A dielectric medium is said to be *linear* if the magnitude of the induced polarization field is directly proportional to the magnitude of $\mathbf{E}$, and it is said to be *isotropic* if the polarization field and $\mathbf{E}$ are in the same direction. In some crystals, the periodic structure of the material allows more polarization to take place along certain directions, such as the crystal axes, than along others. In such *anisotropic* dielectrics, $\mathbf{E}$ and $\mathbf{D}$ may have different directions. A medium is said to be *homogeneous* if its constitutive parameters (ε, μ, and σ) are constant throughout the medium. Our present treatment will be limited to media that are linear, isotropic, and homogeneous. For such media the polarization field is directly proportional to $\mathbf{E}$ and is expressed by the relationship

$$\mathbf{P} = \varepsilon_0 \chi_e \mathbf{E}, \qquad (4.84)$$

where χ_e is called the *electric susceptibility* of the material. Inserting Eq. (4.84) into Eq. (4.83), we have

$$\mathbf{D} = \varepsilon_0 \mathbf{E} + \varepsilon_0 \chi_e \mathbf{E}$$
$$= \varepsilon_0 (1 + \chi_e) \mathbf{E} = \varepsilon \mathbf{E}, \qquad (4.85)$$

which defines the permittivity ε of the material as

$$\varepsilon = \varepsilon_0 (1 + \chi_e). \qquad (4.86)$$

As was mentioned earlier, it is often convenient to characterize the permittivity of a material relative to that of free space, ε_0; this is accommodated by the relative permittivity $\varepsilon_r = \varepsilon/\varepsilon_0$. Values of ε_r are listed in Table 4-2 for a few common materials, and a longer list is given in Appendix B. In free space $\varepsilon_r = 1$, and *for most conductors* $\varepsilon_r \simeq 1$. The dielectric constant of air is approximately 1.0006 at sea level, and it decreases toward unity with increasing altitude. Except in some special circumstances, such as when calculating electromagnetic wave refraction (bending) through the atmosphere over long distances, *air is treated the same as free space.*

The dielectric polarization model presented thus far has placed no restriction on the upper end of the strength of the applied electric field $\mathbf{E}$. In reality, if $\mathbf{E}$ exceeds a

manent dipole moments that are randomly oriented in the absence of an applied electric field. Materials composed of permanent dipoles are called *polar materials*. Owing to their random orientations, the dipoles of polar materials produce no net dipole moment macroscopically (at the macroscopic scale, each point in the material represents a small volume containing thousands of molecules). Under the influence of an applied field, the permanent dipoles tend to align themselves to some extent along the direction of the electric field, in an arrangement somewhat similar to that shown in Fig. 4-17 for nonpolar materials.

Whereas $\mathbf{D}$ and $\mathbf{E}$ are related by ε_0 in free space, the presence of these microscopic dipoles in a dielectric material alters that relationship in that material to

$$\mathbf{D} = \varepsilon_0 \mathbf{E} + \mathbf{P}, \qquad (4.83)$$

where $\mathbf{P}$, called the *electric polarization field*, accounts for the polarization properties of the material. The polarization field is produced by the electric field $\mathbf{E}$ and depends on the material properties.

Table 4-2: Relative permittivity (dielectric constant) and dielectric strength of common materials.

Material	Relative Permittivity, ε_r	Dielectric Strength, E_{ds} (MV/m)
Air (at sea level)	1.0006	3
Petroleum oil	2.1	12
Polystyrene	2.6	20
Glass	4.5–10	25–40
Quartz	3.8–5	30
Bakelite	5	20
Mica	5.4–6	200

$\varepsilon = \varepsilon_r \varepsilon_0$ and $\varepsilon_0 = 8.854 \times 10^{-12}$ F/m.

certain critical value, known as the *dielectric strength* of the material, it will free the electrons completely from the molecules and cause them to accelerate through the material in the form of a conduction current. When this happens, sparking can occur, and the dielectric material can sustain permanent damage due to electron collision with the molecular structure. This abrupt change in behavior is called a *dielectric breakdown*. The dielectric strength E_{ds} is the highest magnitude of **E** that the material can sustain without breakdown. Dielectric breakdown can occur in gas, liquid, and solid dielectrics. The associated field strength depends on the material composition, as well as other factors such as temperature and humidity. The dielectric strength for air is 3 (MV/m); for glass, it is 25 to 40 (MV/m); and for mica, it is 200 (MV/m) [see Table 4-2].

A charged thundercloud with an electric potential V, relative to the ground, induces an electric field $E = V/d$ in the air medium between the ground and the cloud, where d is the height of the cloud base above the ground's surface. If V is sufficiently large so that E exceeds the dielectric strength of air, ionization occurs and discharge (lightning) follows. The *breakdown voltage* V_{br} of a parallel-plate capacitor is discussed in Example 4-11.

REVIEW QUESTIONS

Q4.20 What is a polar material? A nonpolar material?

Q4.21 Do **D** and **E** always point in the same direction? If not, when do they not?

Q4.22 What happens when dielectric breakdown occurs?

4-9 Electric Boundary Conditions

An electric field is said to be spatially continuous if it does not exhibit abrupt changes in either its magnitude or direction as a function of spatial position. Even though the electric field may be continuous in each of two dissimilar media, it may be discontinuous at the boundary between them if surface charge exists along that boundary. Boundary conditions specify how the tangential and normal components of the field in one medium are related to the components of the field across the boundary in another medium. We will derive a general set of boundary conditions, applicable at the interface between *any* two dissimilar media, be they two different dielectrics or a conductor and a dielectric. Also, any of the dielectrics

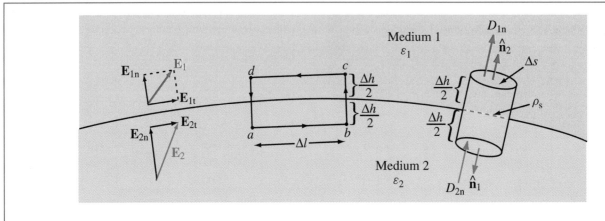

Figure 4-18: Interface between two dielectric media.

may be free space. Even though these boundary conditions will be derived for electrostatic conditions, they will be equally valid for time-varying electric fields. Figure 4-18 shows an interface between medium 1 with permittivity ε_1 and medium 2 with permittivity ε_2. In the general case, the boundary may have a surface charge density ρ_s.

To derive the boundary conditions for the tangential components of **E** and **D**, we begin by constructing the closed rectangular loop *abcda* shown in Fig. 4-18, and then we apply the conservative property of the electric field given by Eq. (4.40), which states that the line integral of the electrostatic field around a closed path is always equal to zero. By letting $\Delta h \rightarrow 0$, the contributions to the line integral by segments *bc* and *da* go to zero. Hence,

$$\oint_C \mathbf{E} \cdot d\mathbf{l} = \int_a^b \mathbf{E}_2 \cdot d\mathbf{l} + \int_c^d \mathbf{E}_1 \cdot d\mathbf{l} = 0, \qquad (4.87)$$

where $\mathbf{E}_1$ and $\mathbf{E}_2$ are the electric fields in media 1 and 2, respectively. In terms of the tangential and normal directions shown in Fig. 4-18,

$$\mathbf{E}_1 = \mathbf{E}_{1t} + \mathbf{E}_{1n}, \qquad (4.88a)$$

$$\mathbf{E}_2 = \mathbf{E}_{2t} + \mathbf{E}_{2n}. \qquad (4.88b)$$

Over segment *ab*, $\mathbf{E}_{2t}$ and $d\mathbf{l}$ have the same direction, but over segment *cd*, $\mathbf{E}_{1t}$ and $d\mathbf{l}$ are in opposite directions. Consequently, Eq. (4.87) gives

$$E_{2t} \, \Delta l - E_{1t} \, \Delta l = 0, \qquad (4.89)$$

or

$$\boxed{E_{1t} = E_{2t} \quad \text{(V/m)}. \quad (4.90)}$$

Accordingly, *the tangential component of the electric field is continuous across the boundary between any two media.* Since $D_{1t} = \varepsilon_1 E_{1t}$ and $D_{2t} = \varepsilon_2 E_{2t}$, the boundary condition on the tangential component of the electric flux density is

$$\boxed{\frac{D_{1t}}{\varepsilon_1} = \frac{D_{2t}}{\varepsilon_2}. \quad (4.91)}$$

Next we apply Gauss's law, as expressed by Eq. (4.29), to determine the boundary conditions on the normal components of **E** and **D**. According to Gauss's law, the total outward flux of **D** through the three surfaces of the small cylinder shown in Fig. 4-18 must equal the total charge enclosed in the cylinder. By letting the cylinder's height

$\Delta h \to 0$, the contribution to the total flux by the side surface goes to zero. Also, even if each of the two media happens to have free or bound volume charge densities, the only charge remaining in the collapsed cylinder is that distributed on the boundary. Thus, $Q = \rho_s \, \Delta s$, and

$$\oint_S \mathbf{D} \cdot d\mathbf{s} = \int_{\text{top}} \mathbf{D}_1 \cdot \hat{\mathbf{n}}_2 \, ds + \int_{\text{bottom}} \mathbf{D}_2 \cdot \hat{\mathbf{n}}_1 \, ds$$

$$= \rho_s \, \Delta s, \qquad (4.92)$$

where $\hat{\mathbf{n}}_1$ and $\hat{\mathbf{n}}_2$ are the outward normal unit vectors of the bottom and top surfaces, respectively. It is important to remember that *the normal unit vector at the surface of any medium is always defined to be in the outward direction away from that medium.* Since $\hat{\mathbf{n}}_1 = -\hat{\mathbf{n}}_2$, Eq. (4.92) simplifies to

$$\boxed{\hat{\mathbf{n}}_2 \cdot (\mathbf{D}_1 - \mathbf{D}_2) = \rho_s \quad (\text{C/m}^2). \quad (4.93)}$$

With D_{1n} and D_{2n} defined as the normal components of $\mathbf{D}_1$ and $\mathbf{D}_2$ along $\hat{\mathbf{n}}_2$, we have

$$\boxed{D_{1n} - D_{2n} = \rho_s \quad (\text{C/m}^2). \quad (4.94)}$$

Thus, *the normal component of **D** changes abruptly at a charged boundary between two different media, and the amount of change is equal to the surface charge density.* The corresponding boundary condition for **E** is

$$\boxed{\varepsilon_1 E_{1n} - \varepsilon_2 E_{2n} = \rho_s. \quad (4.95)}$$

In summary, (1) the conservative property of **E**,

$$\nabla \times \mathbf{E} = 0 \iff \oint_C \mathbf{E} \cdot d\mathbf{l} = 0, \qquad (4.96)$$

led to the result that **E** has a continuous tangential component across a boundary, and (2) the divergence property of **D**,

$$\nabla \cdot \mathbf{D} = \rho_v \iff \oint_S \mathbf{D} \cdot d\mathbf{s} = Q, \qquad (4.97)$$

led to the result that the normal component of **D** changes by ρ_s across the boundary. A summary of the conditions at the boundary between different types of media is given in Table 4-3.

Example 4-10 Application of Boundary Conditions

The x–y plane is a charge-free boundary separating two dielectric media with permittivities ε_1 and ε_2, as shown in Fig. 4-19. If the electric field in medium 1 is $\mathbf{E}_1 = \hat{\mathbf{x}} E_{1x} + \hat{\mathbf{y}} E_{1y} + \hat{\mathbf{z}} E_{1z}$, find (a) the electric field $\mathbf{E}_2$ in medium 2 and (b) the angles θ_1 and θ_2.

Solution: (a) Let $\mathbf{E}_2 = \hat{\mathbf{x}} E_{2x} + \hat{\mathbf{y}} E_{2y} + \hat{\mathbf{z}} E_{2z}$. Our task is to find the components of $\mathbf{E}_2$ in terms of the given components of $\mathbf{E}_1$. The normal to the boundary is $\hat{\mathbf{z}}$. Hence, the x and y components of the fields are tangential to the boundary and the z components are normal to the boundary. At a charge-free interface, the tangential components of **E** and the normal components of **D** are continuous. Consequently,

$$E_{2x} = E_{1x}, \qquad E_{2y} = E_{1y},$$

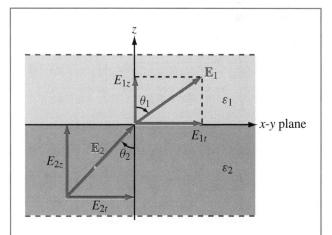

Figure 4-19: Application of boundary conditions at the interface between two dielectric media (Example 4-10).

Table 4-3: Boundary conditions for the electric fields.

Field Component	Any Two Media	Medium 1 Dielectric ε_1	Medium 2 Dielectric ε_2	Medium 1 Dielectric ε_1	Medium 2 Conductor
Tangential **E**	$E_{1t} = E_{2t}$	$E_{1t} = E_{2t}$		$E_{1t} = E_{2t} = 0$	
Tangential **D**	$D_{1t}/\varepsilon_1 = D_{2t}/\varepsilon_2$	$D_{1t}/\varepsilon_1 = D_{2t}/\varepsilon_2$		$D_{1t} = D_{2t} = 0$	
Normal **E**	$\hat{\mathbf{n}} \cdot (\varepsilon_1 \mathbf{E}_1 - \varepsilon_2 \mathbf{E}_2) = \rho_s$	$\varepsilon_1 E_{1n} - \varepsilon_2 E_{2n} = \rho_s$		$E_{1n} = \rho_s/\varepsilon_1$	$E_{2n} = 0$
Normal **D**	$\hat{\mathbf{n}} \cdot (\mathbf{D}_1 - \mathbf{D}_2) = \rho_s$	$D_{1n} - D_{2n} = \rho_s$		$D_{1n} = \rho_s$	$D_{2n} = 0$

Notes: (1) ρ_s is the surface charge density at the boundary; (2) normal components of $\mathbf{E}_1$, $\mathbf{D}_1$, $\mathbf{E}_2$, and $\mathbf{D}_2$ are along $\hat{\mathbf{n}}_2$, the outward normal unit vector of medium 2.

and

$$D_{2z} = D_{1z} \quad \text{or} \quad \varepsilon_2 E_{2z} = \varepsilon_1 E_{1z}.$$

Hence,

$$\mathbf{E}_2 = \hat{\mathbf{x}} E_{1x} + \hat{\mathbf{y}} E_{1y} + \hat{\mathbf{z}} \frac{\varepsilon_1}{\varepsilon_2} E_{1z}. \qquad (4.98)$$

(b) The tangential components of $\mathbf{E}_1$ and $\mathbf{E}_2$ are $E_{1t} = \sqrt{E_{1x}^2 + E_{1y}^2}$ and $E_{2t} = \sqrt{E_{2x}^2 + E_{2y}^2}$. The angles θ_1 and θ_2 are then given by

$$\tan \theta_1 = \frac{E_{1t}}{E_{1z}} = \frac{\sqrt{E_{1x}^2 + E_{1y}^2}}{E_{1z}},$$

$$\tan \theta_2 = \frac{E_{2t}}{E_{2z}} = \frac{\sqrt{E_{2x}^2 + E_{2y}^2}}{E_{2z}} = \frac{\sqrt{E_{1x}^2 + E_{1y}^2}}{(\varepsilon_1/\varepsilon_2) E_{1z}},$$

and the two angles are related by

$$\frac{\tan \theta_2}{\tan \theta_1} = \frac{\varepsilon_2}{\varepsilon_1}. \quad \blacksquare \qquad (4.99)$$

EXERCISE 4.16 With reference to Fig. 4-19, find $\mathbf{E}_1$ if $\mathbf{E}_2 = \hat{\mathbf{x}}2 - \hat{\mathbf{y}}3 + \hat{\mathbf{z}}3$ (V/m), $\varepsilon_1 = 2\varepsilon_0$, and $\varepsilon_2 = 8\varepsilon_0$. Assume the boundary to be charge free.

Ans. $\mathbf{E}_1 = \hat{\mathbf{x}}2 - \hat{\mathbf{y}}3 + \hat{\mathbf{z}}12$ (V/m).

EXERCISE 4.17 Repeat Exercise 4.16 for a boundary with surface charge density $\rho_s = 3.54 \times 10^{-11}$ (C/m²).

Ans. $\mathbf{E}_1 = \hat{\mathbf{x}}2 - \hat{\mathbf{y}}3 + \hat{\mathbf{z}}14$ (V/m).

4-9.1 Dielectric–Conductor Boundary

Consider the case when medium 1 in Fig. 4-18 is a dielectric and medium 2 is a perfect conductor. In a perfect conductor, $\mathbf{E} = \mathbf{D} = 0$ everywhere in the conductor. Hence, $\mathbf{E}_2 = \mathbf{D}_2 = 0$, which requires the tangential and normal components of $\mathbf{E}_2$ and $\mathbf{D}_2$ to be zero. Consequently, from Eq. (4.90) and Eq. (4.94), the fields in the dielectric medium, at the boundary with the conductor, are given by

$$E_{1t} = D_{1t} = 0, \qquad (4.100a)$$

$$D_{1n} = \varepsilon_1 E_{1n} = \rho_s. \qquad (4.100b)$$

These two boundary conditions can be combined into

$$\boxed{\mathbf{D}_1 = \varepsilon_1 \mathbf{E}_1 = \hat{\mathbf{n}} \rho_s \quad \text{(at conductor surface)},} \qquad (4.101)$$

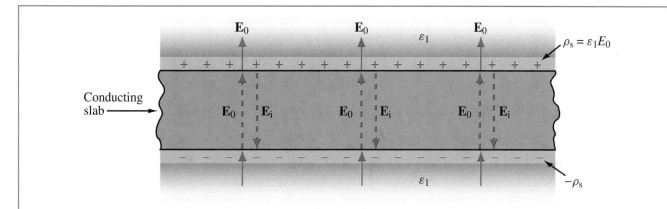

Figure 4-20: When a conducting slab is placed in an external electric field $\mathbf{E}_0$, charges that accumulate on the conductor surfaces induce an internal electric field $\mathbf{E}_i = -\mathbf{E}_0$. Consequently, the total field inside the conductor is zero.

where $\hat{\mathbf{n}}$ is a unit vector directed normally outward from the conducting surface. This means that *the electric field lines point directly away from the conductor surface when ρ_s is positive and directly toward the conductor surface when ρ_s is negative.*

Figure 4-20 shows an infinitely long conducting slab placed in a uniform electric field $\mathbf{E}_0$. The medium above and below the slab has a permittivity ε_1. Because $\mathbf{E}_0$ points away from the upper surface, it induces a positive charge density $\rho_s = \varepsilon_1 |\mathbf{E}_0|$ on the upper surface of the slab. On the bottom surface, $\mathbf{E}_0$ points toward the surface, and therefore the induced charge density is $-\rho_s$. The presence of these surface charges induces an electric field $\mathbf{E}_i$ in the conductor, resulting in a total field $\mathbf{E} = \mathbf{E}_0 + \mathbf{E}_i$. To satisfy the condition that $\mathbf{E}$ must be everywhere zero in the conductor, $\mathbf{E}_i$ must equal $-\mathbf{E}_0$.

If we place a metallic sphere in an electrostatic field, as shown in Fig. 4-21, negative charges will accumulate on the lower hemisphere and positive charges will accumulate on the upper hemisphere. The presence of the sphere causes the field lines to bend to satisfy the condition given

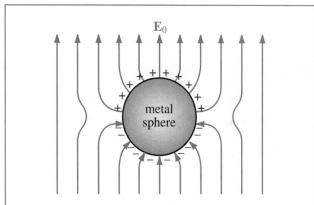

Figure 4-21: Metal sphere placed in an external electric field $\mathbf{E}_0$.

by Eq. (4.101); that is, **E** *is always normal to the surface at the conductor boundary.*

4-9.2 Conductor–Conductor Boundary

We now examine the general case of the boundary between two media neither of which are perfect dielectrics

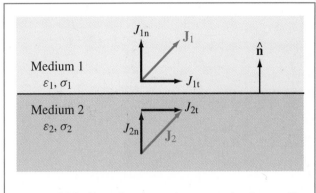

Figure 4-22: Boundary between two conducting media.

or perfect conductors. Depicted in Fig. 4-22, medium 1 has permittivity ε_1 and conductivity σ_1, medium 2 has ε_2 and σ_2, and the interface between them has a surface charge density ρ_s. For the electric fields, Eqs. (4.90) and (4.95) give

$$E_{1t} = E_{2t}, \qquad \varepsilon_1 E_{1n} - \varepsilon_2 E_{2n} = \rho_s. \qquad (4.102)$$

Since we are dealing with conducting media, the electric fields give rise to current densities $\mathbf{J}_1$ and $\mathbf{J}_2$, with $\mathbf{J}_1$ being proportional to $\mathbf{E}_1$ and $\mathbf{J}_2$ being proportional to $\mathbf{E}_2$. From $\mathbf{J} = \sigma \mathbf{E}$, we have

$$\frac{J_{1t}}{\sigma_1} = \frac{J_{2t}}{\sigma_2}, \qquad \varepsilon_1 \frac{J_{1n}}{\sigma_1} - \varepsilon_2 \frac{J_{2n}}{\sigma_2} = \rho_s. \qquad (4.103)$$

The tangential components J_{1t} and J_{2t} represent currents flowing in the two media in a direction parallel to the boundary, and hence no transfer of charge is involved between them. This is not the case for the normal components. If $J_{1n} \neq J_{2n}$, then a different amount of charge arrives at the boundary than leaves it. Hence, ρ_s cannot remain constant with time, which violates the condition of electrostatics requiring all fields and charges to remain constant. Consequently, *the normal component of* $\mathbf{J}$ *has to be continuous across the boundary between two differ-*

ent media under electrostatic conditions. Upon setting $J_{1n} = J_{2n}$ in Eq. (4.103), we have

$$J_{1n} \left(\frac{\varepsilon_1}{\sigma_1} - \frac{\varepsilon_2}{\sigma_2} \right) = \rho_s \quad \text{(electrostatics).} \qquad (4.104)$$

REVIEW QUESTIONS

Q4.23 What are the boundary conditions for the electric field at a conductor–dielectric boundary?

Q4.24 Under electrostatic conditions, we require $J_{1n} = J_{2n}$ at the boundary between two conductors. Why?

4-10 Capacitance

When separated by an insulating (dielectric) medium, any two conducting bodies, regardless of their shapes and sizes, form a *capacitor*. If a d-c voltage source is connected to the conductors, as shown in Fig. 4-23 for two arbitrary conductors, charge of equal and opposite polarity is transferred to the conductors' surfaces. The surface of the conductor connected to the positive side of the source will accumulate charge $+Q$, and charge $-Q$ will accumulate on the surface of the other conductor. From our discussion in Section 4-7, *when a conductor has excess charge, it distributes the charge on its surface in such a manner as to maintain a zero electric field everywhere within the conductor.* This ensures that a conductor is an equipotential body, meaning that the electric potential is the same at every point in the conductor. *Capacitance* of a two-conductor capacitor is defined as

$$C = \frac{Q}{V} \quad \text{(C/V or F),} \qquad (4.105)$$

where V is the potential (voltage) difference between the conductor with charge $+Q$ and the conductor with charge $-Q$. Capacitance is measured in farads (F), which is equivalent to coulombs per volt (C/V).

The presence of free charges on the conductors' surfaces gives rise to an electric field $\mathbf{E}$, as shown in Fig. 4-23; the field lines originate on the positive charges and terminate on the negative charges, and since the tangential component of $\mathbf{E}$ is always equal to zero at a conductor's surface, $\mathbf{E}$ is always perpendicular to the conducting surfaces. The normal component of $\mathbf{E}$ at any point on the surface of either conductor is given by

$$E_n = \hat{\mathbf{n}} \cdot \mathbf{E} = \frac{\rho_s}{\varepsilon} \quad \text{(at conductor surface)}, \quad (4.106)$$

where ρ_s is the surface charge density at that point, $\hat{\mathbf{n}}$ is the outward normal unit vector at the same location, and ε is the permittivity of the dielectric medium separating the conductors. The charge Q is equal to the integral of ρ_s over surface S [Fig. 4-23]:

$$Q = \int_S \rho_s \, ds = \int_S \varepsilon \hat{\mathbf{n}} \cdot \mathbf{E} \, ds = \int_S \varepsilon \mathbf{E} \cdot d\mathbf{s}, \quad (4.107)$$

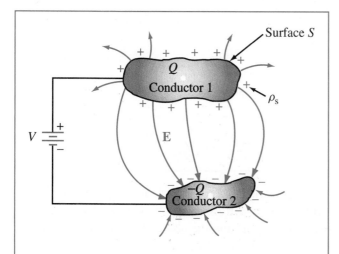

Figure 4-23: A d-c voltage source connected to a capacitor composed of two conducting bodies.

where use was made of Eq. (4.106). The voltage V is related to $\mathbf{E}$ by Eq. (4.39):

$$V = V_{12} = -\int_{P_2}^{P_1} \mathbf{E} \cdot d\mathbf{l}, \quad (4.108)$$

where points P_1 and P_2 are any two points on conductors 1 and 2, respectively. Substituting Eqs. (4.107) and (4.108) into Eq. (4.105) gives

$$C = \frac{\displaystyle\int_S \varepsilon \mathbf{E} \cdot d\mathbf{s}}{\displaystyle -\int_l \mathbf{E} \cdot d\mathbf{l}} \quad \text{(F)}, \quad (4.109)$$

where l is the integration path from conductor 2 to conductor 1. To avoid making sign errors when applying Eq. (4.109), it is important to remember that surface S is the $+Q$ surface and P_1 is on S. Because $\mathbf{E}$ appears in both the numerator and denominator of Eq. (4.109), *the value of C obtained for any specific capacitor configuration is always independent of $\mathbf{E}$.* In fact, C depends only on the capacitor geometry (sizes, shapes and relative positions of the two conductors) and the permittivity of the insulating material.

If the material between the conductors is not a perfect dielectric (i.e., if it has a small conductivity σ), then current can flow through the material between the conductors, and the material will exhibit a resistance R. The general expression for R for a resistor of arbitrary shape is given by Eq. (4.71):

$$R = \frac{\displaystyle -\int_l \mathbf{E} \cdot d\mathbf{l}}{\displaystyle \int_S \sigma \mathbf{E} \cdot d\mathbf{s}} \quad (\Omega). \quad (4.110)$$

For a medium with uniform σ and ε, the product of Eqs. (4.109) and (4.110) gives

$$RC = \frac{\varepsilon}{\sigma}. \quad (4.111)$$

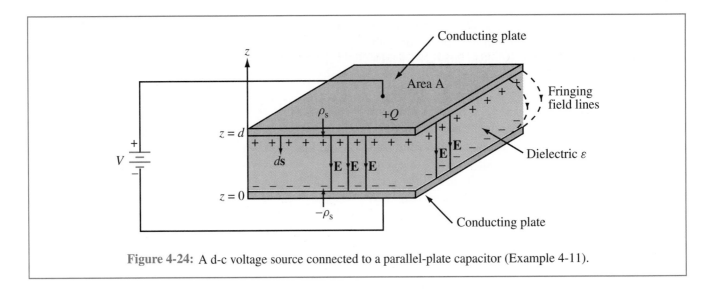

Figure 4-24: A d-c voltage source connected to a parallel-plate capacitor (Example 4-11).

This simple relation allows us to find R if C is known, or vice versa.

Example 4-11   Capacitance and Breakdown Voltage of Parallel-Plate Capacitor

Obtain an expression for the capacitance C of a parallel-plate capacitor comprised of two parallel plates each of surface area A and separated by a distance d. The capacitor is filled with a dielectric material with permittivity ε. Also, determine the breakdown voltage if $d = 1$ cm and the dielectric material is quartz.

Solution: In Fig. 4-24, we place the lower plate of the capacitor in the x–y plane and the upper plate in the plane $z = d$. Because of the applied voltage difference V, charge $+Q$ accumulates uniformly on the top plate and $-Q$ accumulates uniformly on the lower plate. In the dielectric medium between the plates, the charges induce a uniform electric field in the $-\hat{z}$-direction (from positive to negative charges). In addition, some *fringing field lines* will exist near the edges, but their effects may be ignored if the dimensions of the plates are much larger than the separation d between them, because in that case the bulk of the electric field lines will exist in the medium

between the plates. The charge density on the upper plate is $\rho_s = Q/A$. Hence,

$$\mathbf{E} = -\hat{z}E,$$

and from Eq. (4.106), the magnitude of $\mathbf{E}$ at the conductor–dielectric boundary is $E = \rho_s/\varepsilon = Q/\varepsilon A$. From Eq. (4.108), the voltage difference is

$$V = -\int_0^d \mathbf{E} \cdot d\mathbf{l} = -\int_0^d (-\hat{z}E) \cdot \hat{z}\, dz = Ed, \quad (4.112)$$

and the capacitance is

$$C = \frac{Q}{V} = \frac{Q}{Ed} = \frac{\varepsilon A}{d}, \quad (4.113)$$

where use was made of the relation $E = Q/\varepsilon A$.

From $V = Ed$, as given by Eq. (4.112), $V = V_{\text{br}}$ when $E = E_{\text{ds}}$, the dielectric strength of the material. According to Table 4-2, $E_{\text{ds}} = 30$ (MV/m) for quartz. Hence, the breakdown voltage is

$$V_{\text{br}} = E_{\text{ds}}d = 30 \times 10^6 \times 10^{-2} = 3 \times 10^5 \text{ V.} \quad \blacksquare$$

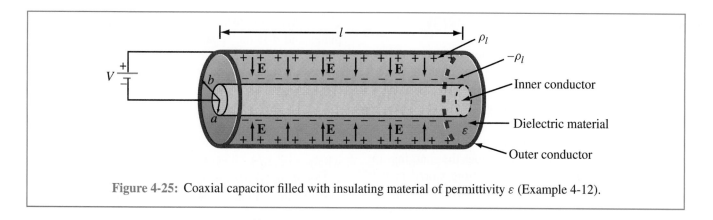

Figure 4-25: Coaxial capacitor filled with insulating material of permittivity ε (Example 4-12).

Example 4-12   Capacitance of Coaxial Line

Obtain an expression for the capacitance of the coaxial line shown in Fig. 4-25.

Solution: For a given voltage V across the capacitor (Fig. 4-25), charge $+Q$ and $-Q$ will accumulate on the surfaces of the outer and inner conductors, respectively. We assume that these charges are uniformly distributed along the length of the conductors with line charge density $\rho_l = Q/l$ on the outer conductor and $-\rho_l$ on the inner conductor. Ignoring fringing fields near the ends of the coaxial line, we can construct a cylindrical Gaussian surface in the dielectric, around the inner conductor, with radius r such that $a < r < b$. The inner conductor is a line charge similar to that of Example 4-6, except that the line charge of the inner conductor is negative. With a minus sign added to the expression for $\mathbf{E}$ given by Eq. (4.33), we have

$$\mathbf{E} = -\hat{\mathbf{r}}\frac{\rho_l}{2\pi\varepsilon r} = -\hat{\mathbf{r}}\frac{Q}{2\pi\varepsilon rl}. \qquad (4.114)$$

The potential difference V between the outer and inner conductors is

$$V = -\int_a^b \mathbf{E}\cdot d\mathbf{l} = -\int_a^b \left(-\hat{\mathbf{r}}\frac{Q}{2\pi\varepsilon rl}\right)\cdot(\hat{\mathbf{r}}\,dr)$$

$$= \frac{Q}{2\pi\varepsilon l}\ln\left(\frac{b}{a}\right). \qquad (4.115)$$

The capacitance C is then given by

$$C = \frac{Q}{V} = \frac{2\pi\varepsilon l}{\ln(b/a)}. \qquad (4.116)$$

The capacitance per unit length of the coaxial line is

$$C' = \frac{C}{l} = \frac{2\pi\varepsilon}{\ln(b/a)} \qquad \text{(F/m)}. \quad \blacksquare \qquad (4.117)$$

REVIEW QUESTIONS

Q4.25 How is the capacitance of a two-conductor structure related to the resistance of the insulating material between the conductors?

Q4.26 What are fringing fields and when may they be ignored?

4-11 Electrostatic Potential Energy

When a source is connected to a capacitor, it expends energy in charging up the capacitor. If the capacitor plates are made of a good conductor with effectively zero resistance and if the dielectric separating the two conductors has negligible conductivity, then no real current can flow through the dielectric, and no ohmic losses occur anywhere in the capacitor. Where then does the charging-up energy go? The energy ends up getting stored in the dielectric medium in the form of *electrostatic potential energy*. The amount of stored energy W_e is related to Q, C, and V.

Under the influence of the electric field in the dielectric medium between the two conductors, charge q accumulates on one of the conductors, and an equal and opposite charge accumulates on the other conductor. In effect, charge q has been transferred from one of the conductors to the other. The voltage v across the capacitor is related to q by

$$v = \frac{q}{C}. \qquad (4.118)$$

From the basic definition of the electric potential V, the amount of work dW_e required to transfer an additional incremental amount of charge dq is

$$dW_e = v \, dq = \frac{q}{C} \, dq. \qquad (4.119)$$

If we start with an uncharged capacitor and charge it up from zero charge until a final charge Q has been reached, then the total amount of work performed is

$$W_e = \int_0^Q \frac{q}{C} \, dq = \frac{1}{2}\frac{Q^2}{C} \qquad \text{(J)}. \qquad (4.120)$$

Using $C = Q/V$, where V is the final voltage, W_e can also be written as

$$W_e = \tfrac{1}{2}CV^2 \qquad \text{(J)}. \qquad (4.121)$$

For the parallel-plate capacitor discussed in Example 4-11, its capacitance is given by Eq. (4.113) as

$C = \varepsilon A/d$, where A is the surface area of each of its plates and d is the separation between them. Also, the voltage V across the capacitor is related to the magnitude of the electric field, E, in the dielectric by $V = Ed$. Using these two expressions in Eq. (4.121) gives

$$W_e = \tfrac{1}{2}\frac{\varepsilon A}{d}(Ed)^2 = \tfrac{1}{2}\varepsilon E^2(Ad) = \tfrac{1}{2}\varepsilon E^2 v, \quad (4.122)$$

where $v = Ad$ is the volume of the capacitor.

The *electrostatic energy density* w_e is defined as the electrostatic potential energy W_e per unit volume:

$$\boxed{w_e = \frac{W_e}{v} = \tfrac{1}{2}\varepsilon E^2 \qquad \text{(J/m}^3). \qquad (4.123)}$$

Even though this expression was derived for a parallel-plate capacitor, it is equally valid for any dielectric medium in an electric field $\mathbf{E}$. Furthermore, for any volume v containing a dielectric ε, the total electrostatic potential energy stored in v is

$$W_e = \frac{1}{2}\int_v \varepsilon E^2 \, dv \qquad \text{(J)}. \qquad (4.124)$$

Returning to the parallel-plate capacitor, the oppositely charged plates are attracted to each other by an electrical force $\mathbf{F}$. The force acting on any system of charges may be obtained from energy considerations. In the material that follows, we will show how $\mathbf{F}$ can be determined from W_e, the electrostatic energy stored in the system by virtue of the presence of the electric charges. The treatment will be general in nature, although we will occasionally use the capacitor example by way of illustration.

If the two plates of the parallel-plate capacitor are allowed to move closer to each other, under the influence of the electric force $\mathbf{F}$, by a differential distance $d\mathbf{l}$, while maintaining the charges on the plates constant, then the mechanical work done by the system (charged capacitor) is

$$dW = \mathbf{F} \cdot d\mathbf{l}. \qquad (4.125)$$

For an isolated system, this mechanical work is provided by *expending* electrostatic energy. Hence, dW is equal to the *loss* of energy stored in the dielectric insulating material of the capacitor, or

$$dW = -dW_e. \qquad (4.126)$$

From Eq. (3.73), dW_e may be written in terms of the gradient of W_e as

$$dW_e = \nabla W_e \cdot d\mathbf{l}. \qquad (4.127)$$

In view of Eq. (4.126), comparison of Eqs. (4.125) and (4.127) leads to

$$\boxed{\mathbf{F} = -\nabla W_e \qquad \text{(N)}. \qquad (4.128)}$$

It is important to note that Eq. (4.128) is predicated on the assumption that the charges in the system are constant.

To apply Eq. (4.128) to the parallel-plate capacitor, we rewrite Eq. (4.120) in the following form:

$$W_e = \frac{1}{2}\frac{Q^2}{C} = \frac{Q^2 z}{2\varepsilon A}, \qquad (4.129)$$

where we have replaced d with the variable z, representing the vertical spacing between the conducting plates. Use of Eq. (4.129) in Eq. (4.128) gives

$$\mathbf{F} = -\nabla W_e = -\hat{\mathbf{z}}\frac{\partial}{\partial z}\left(\frac{Q^2 z}{2\varepsilon A}\right) = -\hat{\mathbf{z}}\left(\frac{Q^2}{2\varepsilon A}\right), \qquad (4.130)$$

and since $Q = \varepsilon A E$, $\mathbf{F}$ can also be written as

$$\mathbf{F} = -\hat{\mathbf{z}}\frac{\varepsilon A E^2}{2} \qquad \text{(parallel-plate capacitor)}. \qquad (4.131)$$

REVIEW QUESTIONS

Q4.27 To bring a charge q from infinity to a given point in space, a certain amount of work W is expended. Where does the energy corresponding to W go?

Q4.28 When a voltage source is connected across a capacitor, what is the direction of the electrical force acting on its two conducting surfaces?

EXERCISE 4.18 The radii of the inner and outer conductors of a coaxial cable are 2 cm and 5 cm, respectively, and the insulating material between them has a relative permittivity of 4. The charge density on the outer conductor is $\rho_l = 10^{-4}$ (C/m). Use the expression for $\mathbf{E}$ derived in Example 4-12 to calculate the total energy stored in a 20-cm length of the cable.

Ans. $W_e = 4.1$ J.

4-12 Image Method

Consider a point charge Q at a distance d above a perfectly conducting plane, as shown in the left-hand section of Fig. 4-26. We want to determine V, $\mathbf{E}$, and $\mathbf{D}$ at any point in the space above the grounded conductor, as well as the distribution of surface charge on the conducting plate. Three different methods have been introduced in this chapter for finding $\mathbf{E}$. The first, based on Coulomb's law, requires knowledge of the magnitudes and locations of all the charges contributing to $\mathbf{E}$ at a given point in space. In the present case, the charge Q will induce an unknown and nonuniform distribution of charge on the surface of the conductor. Hence, we cannot utilize Coulomb's method. The second method is based on the application of Gauss's law, and it is equally difficult to use because it is not clear how one would construct a Gaussian surface such that $\mathbf{E}$ is always totally tangential or totally normal at every point on that surface. In the third method, the electric field is found from $\mathbf{E} = -\nabla V$ after solving Poisson's or Laplace's equation for V, subject to the available boundary conditions; that is, $V = 0$ at any point on the grounded conducting surface and at infinity. Although such an approach is feasible in principle, the solution is quite complicated mathematically.

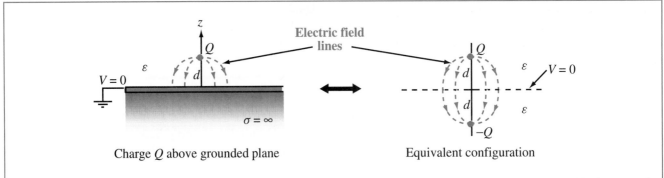

Figure 4-26: By image theory, a charge Q above a grounded perfectly conducting plane is equivalent to Q and its image $-Q$ with the ground plane removed.

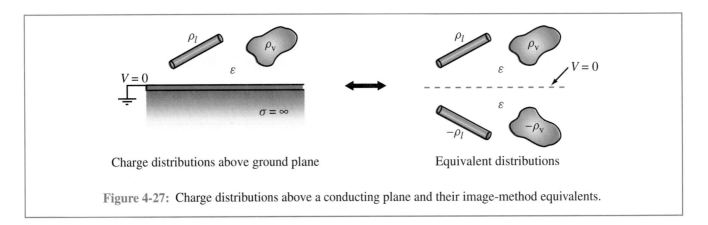

Figure 4-27: Charge distributions above a conducting plane and their image-method equivalents.

Alternatively, the problem at hand can be solved with great ease using *image theory*, which states that *any given charge configuration above an infinite, perfectly conducting plane is electrically equivalent to the combination of the given charge configuration and its image configuration, with the conducting plane removed.* The image-method equivalent of the charge Q above a conducting plane is shown in the right-hand section of Fig. 4-26. It consists of the charge Q itself and an image charge $-Q$ at a distance $2d$ from Q, with nothing else between them. The electric field due to the two isolated charges can now be easily found at any point (x, y, z) by applying Coulomb's method, as demonstrated by Exam-

ple 4-13. The combination of the two charges will always produce a potential $V = 0$ at every point in the plane where the conducting surface had been. If the charge is in the presence of more than one grounded plane, it is necessary to establish images of the charge relative to each of the planes and then to establish images of each of those images against the remaining planes. The process is continued until the condition $V = 0$ is satisfied at all points on all the grounded planes. The image method applies not only to point charges, but also to any distributions of charge, such as the line and volume distributions depicted in Fig. 4-27.

Example 4-13   **Image Method for Charge above Conducting Plane**

Use image theory to determine V and $\mathbf{E}$ at an arbitrary point $P(x, y, z)$ in the region $z > 0$ due to a charge Q in free space at a distance d above a grounded conducting plane.

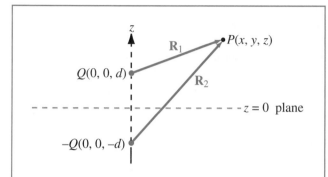

Figure 4-28: Application of the image method for finding $\mathbf{E}$ at point P (Example 4-13).

Solution: In Fig. 4-28, charge Q is at $(0, 0, d)$ and its image $-Q$ is at $(0, 0, -d)$ in Cartesian coordinates. From Eq. (4.19), the electric field at point $P(x, y, z)$ due to the two charges is given by

$$
\begin{aligned}
\mathbf{E} &= \frac{1}{4\pi\varepsilon_0}\left(\frac{Q\mathbf{R}_1}{R_1^3} + \frac{-Q\mathbf{R}_2}{R_2^3}\right) \\
&= \frac{Q}{4\pi\varepsilon_0}\left[\frac{\hat{\mathbf{x}}x + \hat{\mathbf{y}}y + \hat{\mathbf{z}}(z-d)}{[x^2 + y^2 + (z-d)^2]^{3/2}} \right. \\
&\qquad \left. - \frac{\hat{\mathbf{x}}x + \hat{\mathbf{y}}y + \hat{\mathbf{z}}(z+d)}{[x^2 + y^2 + (z+d)^2]^{3/2}}\right]
\end{aligned}
$$

for $z \geq 0$. ∎

EXERCISE 4.19 Use the result of Example 4-13 to find the surface charge density ρ_s on the surface of the conducting plane.

Ans. $\rho_s = -Qd/[2\pi(x^2 + y^2 + d^2)^{3/2}]$.

REVIEW QUESTIONS

Q4.29 What is the fundamental premise of the image method?

Q4.30 Given a certain charge distribution, what are the various approaches described in this chapter for computing the electric field $\mathbf{E}$ at a given point in space?

CHAPTER HIGHLIGHTS

- Maxwell's equations are the fundamental tenets of electromagnetic theory.
- Under static conditions, Maxwell's equations separate into two uncoupled pairs, with one pair pertaining to electrostatics and the other to magnetostatics.
- Coulomb's law provides an explicit expression for the electric field due to a specified charge distribution.
- Gauss's law states that the total electric field flux through a closed surface is equal to the net charge enclosed by the surface.
- The electrostatic field $\mathbf{E}$ at a point is related to the electric potential V at that point by $\mathbf{E} = -\nabla V$, with V being referenced to zero potential at infinity.
- Because most metals have conductivities on the order of 10^6 (S/m), they are treated in practice as perfect conductors. By the same token, insulators with conductivities smaller than 10^{-10} (S/m) are treated as perfect dielectrics.
- Boundary conditions at the interface between two materials specify the relations between the normal and tangential components of $\mathbf{D}$, $\mathbf{E}$, and $\mathbf{J}$ in one of the materials to the corresponding components in the other.

- The capacitance of a two-conductor body and resistance of the medium between them can be computed from knowledge of the electric field in that medium.
- The electrostatic energy density stored in a dielectric medium is $w_e = \frac{1}{2}\varepsilon E^2$ (J/m³).
- When a charge configuration exists above an infinite, perfectly conducting plane, the induced field **E** is the same as that due to the configuration itself and its image with the conducting plane removed.

GLOSSARY OF IMPORTANT TERMS

Provide definitions or explain the meaning of the following terms:

static condition
electrostatics
volume, surface, and line charge densities
current density **J**
conduction current
convection current
electric field intensity **E**
electric flux density **D**
Coulomb's law
Gauss's law
Gaussian surface
conservative field
electric potential
electric dipole
dipole moment **p**
constitutive parameters
conductivity σ
conductor
dielectric material
semiconductor
superconductor
electron drift velocity **u$_e$**
hole drift velocity **u$_h$**
perfect conductor
perfect dielectric

electron mobility μ_e
hole mobility μ_h
Ohm's law
conductance G
Joule's law
boundary conditions
polarization vector **P**
linear material
isotropic material
homogeneous material
electric susceptibility χ_e
permittivity ε
relative permittivity ε_r
dielectric strength
dielectric breakdown
capacitance C
electrostatic potential energy W_e
electrostatic energy density w_e
image method

PROBLEMS

Sections 4-2: Charge and Current Distributions

4.1* A cube 2 m on a side is located in the first octant in a Cartesian coordinate system, with one of its corners at the origin. Find the total charge contained in the cube if the charge density is given by $\rho_v = xy^2e^{-2z}$ (mC/m³).

4.2 Find the total charge contained in a cylindrical volume defined by $r \le 2$ m and $0 \le z \le 3$ m if $\rho_v = 10rz$ (mC/m³).

4.3* Find the total charge contained in a cone defined by $R \le 2$ m and $0 \le \theta \le \pi/4$, given that $\rho_v = 20R^2\cos^2\theta$ (mC/m³).

4.4 If the line charge density is given by $\rho_l = 12y^2$ (mC/m), find the total charge distributed on the y-axis from $y = -5$ to $y = 5$.

---

*Answer(s) available in Appendix D.
⊛ Solution available in CD-ROM.

4.5* Find the total charge on a circular disk defined by $r \leq a$ and $z = 0$ if:

(a) $\rho_s = \rho_{s0} \sin \phi$ (C/m²)

(b) $\rho_s = \rho_{s0} \sin^2 \phi$ (C/m²)

(c) $\rho_s = \rho_{s0} e^{-r}$ (C/m²)

(d) $\rho_s = \rho_{s0} e^{-r} \sin^2 \phi$ (C/m²)

where ρ_{s0} is a constant.

4.6 If $\mathbf{J} = \hat{\mathbf{y}} 2xz$ (A/m²), find the current I flowing through a square with corners at $(0, 0, 0)$, $(2, 0, 0)$, $(2, 0, 2)$, and $(0, 0, 2)$.

4.7* If $\mathbf{J} = \hat{\mathbf{R}} 25/R$ (A/m²), find I through the surface $R = 5$ m.

4.8 An electron beam shaped like a circular cylinder of radius r_0 carries a charge density given by

$$\rho_v = \left(\frac{-\rho_0}{1 + r^2} \right) \quad (\text{C/m}^3)$$

where ρ_0 is a positive constant and the beam's axis is coincident with the z-axis.

(a) Determine the total charge contained in length L of the beam.

(b) If the electrons are moving in the $+z$-direction with uniform speed u, determine the magnitude and direction of the current crossing the z-plane.

Section 4-3: Coulomb's Law

4.9* A square with sides of 2 m has a charge of 20 μC at each of its four corners. Determine the electric field at a point 5 m above the center of the square.

4.10 Three point charges, each with $q = 3$ nC, are located at the corners of a triangle in the x–y plane, with one corner at the origin, another at $(2 \text{ cm}, 0, 0)$, and the third at $(0, 2 \text{ cm}, 0)$. Find the force acting on the charge located at the origin.

4.11* Charge $q_1 = 4 \ \mu$C is located at $(1 \text{ cm}, 1 \text{ cm}, 0)$ and charge q_2 is located at $(0, 0, 4 \text{ cm})$. What should q_2 be so that $\mathbf{E}$ at $(0, 2 \text{ cm}, 0)$ has no y-component?

4.12 A line of charge with uniform density $\rho_l = 4$ (μC/m) exists in air along the z-axis between $z = 0$ and $z = 5$ cm. Find $\mathbf{E}$ at $(0, 10 \text{ cm}, 0)$.

4.13* Electric charge is distributed along an arc located in the x–y plane and defined by $r = 2$ cm and $0 \leq \phi \leq \pi/4$. If $\rho_l = 5$ (μC/m), find $\mathbf{E}$ at $(0, 0, z)$ and then evaluate it at:

(a) The origin.

(b) $z = 5$ cm

(c) $z = -5$ cm

4.14 A line of charge with uniform density ρ_l extends between $z = -L/2$ and $z = L/2$ along the z-axis. Apply Coulomb's law to obtain an expression for the electric field at any point $P(r, \phi, 0)$ on the x–y plane. Show that your result reduces to the expression given by (4.33) as the length L is extended to infinity.

4.15* Repeat Example 4-5 for the circular disk of charge of radius a, but in the present case, assume the surface charge density to vary with r as

$$\rho_s = \rho_{s0} r^2 \quad (\text{C/m}^2)$$

where ρ_{s0} is a constant.

4.16 Multiple charges at different locations are said to be in equilibrium if the force acting on any one of them is identical in magnitude and direction to the force acting on any of the others. Suppose we have two negative charges, one located at the origin and carrying charge $-9e$, and the other located on the positive x-axis at a distance d from the first one and carrying charge $-36e$. Determine the location, polarity and magnitude of a third charge whose placement would bring the entire system into equilibrium.

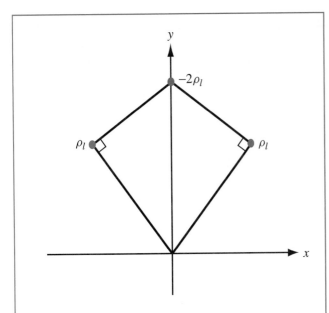

Figure 4-29: Kite-shaped arrangment of line charges for Problem 4.17.

4.17* Three infinite lines of charge, all parallel to the z-axis, are located at the three corners of the kite-shaped arrangement shown in Fig. 4-29. If the two right triangles are symmetrical and of equal corresponding sides, show that the electric field is zero at the origin.

4.18 Three infinite lines of charge, $\rho_{l_1} = 5$ (nC/m), $\rho_{l_2} = -5$ (nC/m), and $\rho_{l_3} = 5$ (nC/m), are all parallel to the z-axis. If they pass through the respective points $(0, -b)$, $(0, 0)$, and $(0, b)$ in the x–y plane, find the electric field at $(a, 0, 0)$. Evaluate your result for $a = 2$ cm and $b = 1$ cm.

4.19 A horizontal strip lying in the x–y plane is of width d in the y-direction and infinitely long in the x-direction. If the strip is in air and has a uniform charge distribution ρ_s, use Coulomb's law to obtain an explicit expression for the electric field at a point P located at a distance h above the centerline of the strip. Extend your result to the special case where d is infinite and compare it with Eq. (4.25).

Section 4-4: Gauss's Law

4.20 Given the electric flux density

$$\mathbf{D} = \hat{\mathbf{x}}2(x + y) + \hat{\mathbf{y}}(3x - 2y) \quad (\text{C/m}^2)$$

determine

(a) ρ_v by applying Eq. (4.26).

(b) The total charge Q enclosed in a cube 2 m on a side, located in the first octant with three of its sides coincident with the x-, y-, and z-axes and one of its corners at the origin.

(c) The total charge Q in the cube, obtained by applying Eq. (4.29).

4.21* Repeat Problem 4.20 for $\mathbf{D} = \hat{\mathbf{x}}xy^2z^3$ (C/m^2).

4.22 Charge Q_1 is uniformly distributed over a thin spherical shell of radius a, and charge Q_2 is uniformly distributed over a second spherical shell of radius b, with $b > a$. Apply Gauss's law to find $\mathbf{E}$ in the regions $R < a$, $a < R < b$, and $R > b$.

4.23* The electric flux density inside a dielectric sphere of radius a centered at the origin is given by

$$\mathbf{D} = \hat{\mathbf{R}}\rho_0 R \quad (\text{C/m}^2)$$

where ρ_0 is a constant. Find the total charge inside the sphere.

4.24 In a certain region of space, the charge density is given in cylindrical coordinates by the function:

$$\rho_v = 20re^{-r} \quad (\text{C/m}^3)$$

Apply Gauss's law to find $\mathbf{D}$.

4.25* An infinitely long cylindrical shell extending between $r = 1$ m and $r = 3$ m contains a uniform charge density ρ_{v0}. Apply Gauss's law to find $\mathbf{D}$ in all regions.

4.26 If the charge density increases linearly with distance from the origin such that $\rho_v = 0$ at the origin and $\rho_v = 10$ C/m^3 at $R = 2$ m, find the corresponding variation of $\mathbf{D}$.

Section 4-5: Electric Potential

4.27 A square in the x–y plane in free space has a point charge of $+Q$ at corner $(a/2, a/2)$, the same at corner $(a/2, -a/2)$, and a point charge of $-Q$ at each of the other two corners.

(a) Find the electric potential at any point P along the x-axis.

(b) Evaluate V at $x = a/2$.

4.28 The circular disk of radius a shown in Fig. 4-7 has uniform charge density ρ_s across its surface.

(a) Obtain an expression for the electric potential V at a point $P(0, 0, z)$ on the z-axis.

(b) Use your result to find $\mathbf{E}$ and then evaluate it for $z = h$. Compare your final expression with (4.24), which was obtained on the basis of Coulomb's law.

4.29* A circular ring of charge of radius a lies in the x–y plane and is centered at the origin. Assume also that the ring is in air and carries a uniform density ρ_l.

(a) Show that the electrical potential at $(0, 0, z)$ is given by $V = \rho_l a/[2\varepsilon_0(a^2 + z^2)^{1/2}]$.

(b) Find the corresponding electric field $\mathbf{E}$.

4.30 Show that the electric potential difference V_{12} between two points in air at radial distances r_1 and r_2 from an infinite line of charge with density ρ_l along the z-axis is $V_{12} = (\rho_l/2\pi\varepsilon_0) \ln(r_2/r_1)$.

4.31* Find the electric potential V at a location a distance b from the origin in the x–y plane due to a line charge with charge density ρ_l and of length l. The line charge is coincident with the z-axis and extends from $z = -l/2$ to $z = l/2$.

4.32 For the electric dipole shown in Fig. 4-13, $d = 1$ cm and $|\mathbf{E}| = 2$ (mV/m) at $R = 1$ m and $\theta = 0°$. Find $\mathbf{E}$ at $R = 2$ m and $\theta = 90°$.

4.33 For each of the distributions of the electric potential V shown in Fig. 4-30, sketch the corresponding distribution of $\mathbf{E}$ (in all cases, the vertical axis is in volts and the horizontal axis is in meters).

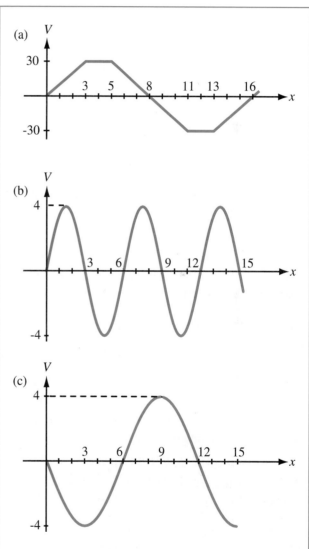

Figure 4-30: Electric potential distributions of Problem 4.33.

4.34   Given the electric field

$$\mathbf{E} = \hat{\mathbf{R}} \frac{12}{R^2} \quad (V/m)$$

find the electric potential of point A with respect to point B where A is at $+2$ m and B at -4 m, both on the z-axis.

4.35*   An infinitely long line of charge with uniform density $\rho_l = 6$ (nC/m) lies in the x–y plane parallel to the y-axis at $x = 2$ m. Find the potential V_{AB} at point $A(3$ m$, 0, 4$ m$)$ in Cartesian coordinates with respect to point $B(0, 0, 0)$ by applying the result of Problem 4.30.

4.36   The x–y plane contains a uniform sheet of charge with $\rho_{s_1} = 0.2$ (nC/m^2). A second sheet with $\rho_{s_2} = -0.2$ (nC/m^2) occupies the plane $z = 6$ m. Find V_{AB}, V_{BC}, and V_{AC} for $A(0, 0, 6$ m$)$, $B(0, 0, 0)$, and $C(0, -2$ m$, 2$ m$)$.

Section 4-7: Conductors

4.37*   A cylindrical bar of silicon has a radius of 2 mm and a length of 5 cm. If a voltage of 5 V is applied between the ends of the bar and $\mu_e = 0.13$ (m^2/V·s), $\mu_h = 0.05$ (m^2/V·s), $N_e = 1.5 \times 10^{16}$ electrons/m^3, and $N_h = N_e$, find the following:

(a) The conductivity of silicon.

(b) The current I flowing in the bar.

(c) The drift velocities $\mathbf{u}_e$ and $\mathbf{u}_h$.

(d) The resistance of the bar.

(e) The power dissipated in the bar.

4.38   Repeat Problem 4.37 for a bar of germanium with $\mu_e = 0.4$ (m^2/V·s), $\mu_h = 0.2$ (m^2/V·s), and $N_e = N_h = 2.4 \times 10^{19}$ electrons or holes/m^3.

4.39   A 100-m-long conductor of uniform cross-section has a voltage drop of 2 V between its ends. If the density of the current flowing through it is 7×10^5 (A/m^2), identify the material of the conductor.

4.40   A coaxial resistor of length l consists of two concentric cylinders. The inner cylinder has radius a and is made of a material with conductivity σ_1, and the outer cylinder, extending between $r = a$ and $r = b$, is made of a material with conductivity σ_2. If the two ends of the resistor are capped with conducting plates, show that the resistance between the two ends is $R = l/[\pi(\sigma_1 a^2 + \sigma_2(b^2 - a^2))]$.

4.41*   Apply the result of Problem 4.40 to find the resistance of a 10-cm-long hollow cylinder (Fig. 4-31) made of carbon with $\sigma = 3 \times 10^4$ (S/m).

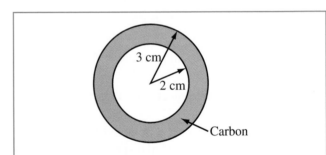

Figure 4-31: Cross-section of hollow cylinder of Problem 4.41.

4.42   A 2×10^{-3}-mm-thick square sheet of aluminum has 10 cm $\times$ 10 cm faces. Find the following:

(a) The resistance between opposite edges on a square face.

(b) The resistance between the two square faces. (See Appendix B for the electrical constants of materials.)

Section 4-9: Boundary Conditions

4.43* With reference to Fig. 4-19, find $\mathbf{E}_1$ if $\mathbf{E}_2 = \hat{\mathbf{x}}3 - \hat{\mathbf{y}}2 + \hat{\mathbf{z}}4$ (V/m), $\varepsilon_1 = 2\varepsilon_0$, $\varepsilon_2 = 18\varepsilon_0$, and the boundary has a surface charge density $\rho_s = 7.08 \times 10^{-11}$ (C/m^2). What angle does $\mathbf{E}_2$ make with the z-axis?

4.44 An infinitely long dielectric cylinder with $\varepsilon_{1r} = 4$ and described by $r \le 10$ cm is surrounded by a material with $\varepsilon_{2r} = 8$. If $\mathbf{E}_1 = \hat{\mathbf{r}}r^2 - \hat{\boldsymbol{\phi}}r\cos\phi + \hat{\mathbf{z}}3$ (V/m) in the cylinder region, find $\mathbf{E}_2$ and $\mathbf{D}_2$ in the surrounding region. Assume that no free charges exist along the cylinder's boundary.

4.45* A 2-cm dielectric sphere with $\varepsilon_{1r} = 3$ is embedded in a medium with $\varepsilon_{2r} = 9$. If $\mathbf{E}_2 = \hat{\mathbf{R}}2\cos\theta - \hat{\boldsymbol{\theta}}3\sin\theta$ (V/m) in the surrounding region, find $\mathbf{E}_1$ and $\mathbf{D}_1$ in the sphere.

4.46 If $\mathbf{E} = \hat{\mathbf{R}}50$ (V/m) at the surface of a 5-cm conducting sphere centered at the origin, what is the total charge Q on the sphere's surface?

4.47* Figure 4-32 shows three planar dielectric slabs of equal thickness but with different dielectric constants. If $\mathbf{E}_0$ in air makes an angle of $45°$ with respect to the z-axis, find the angle of $\mathbf{E}$ in each of the other layers.

Sections 4-10 and 4-11: Capacitance and Electrical Energy

4.48 Determine the force of attraction in a parallel-plate capacitor with $A = 10$ cm^2, $d = 1$ cm, and $\varepsilon_r = 4$ if the voltage across it is 50 V.

4.49* Dielectric breakdown occurs in a material whenever the magnitude of the field $\mathbf{E}$ exceeds the dielectric strength anywhere in that material. In the coaxial capacitor of Example 4-12,

(a) At what value of r is $|E|$ maximum?

(b) What is the breakdown voltage if $a = 1$ cm, $b = 2$ cm, and the dielectric material is mica with $\varepsilon_r = 6$?

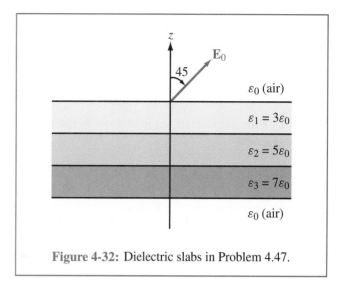

Figure 4-32: Dielectric slabs in Problem 4.47.

4.50 An electron with charge $Q_e = -1.6 \times 10^{-19}$ C and mass $m_e = 9.1 \times 10^{-31}$ kg is injected at a point adjacent to the negatively charged plate in the region between the plates of an air-filled parallel-plate capacitor with separation of 1 cm and rectangular plates each 10 cm^2 in area (Fig. 4-33). If the voltage across the capacitor is 10 V, find the following:

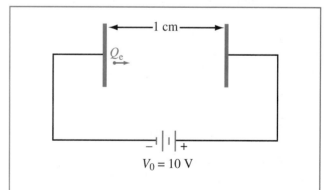

Figure 4-33: Electron between charged plates of Problem 4.50.

(a) The force acting on the electron.

(b) The acceleration of the electron.

(c) The time it takes the electron to reach the positively charged plate, assuming that it starts from rest.

4.51* In a dielectric medium with $\varepsilon_r = 4$, the electric field is given by

$$\mathbf{E} = \hat{\mathbf{x}}(x^2 + 2z) + \hat{\mathbf{y}}x^2 - \hat{\mathbf{z}}(y + z) \quad \text{(V/m)}$$

Calculate the electrostatic energy stored in the region -1 m $\le x \le 1$ m, $0 \le y \le 2$ m, and $0 \le z \le 3$ m.

4.52 Figure 4-34(a) depicts a capacitor consisting of two parallel, conducting plates separated by a distance d.

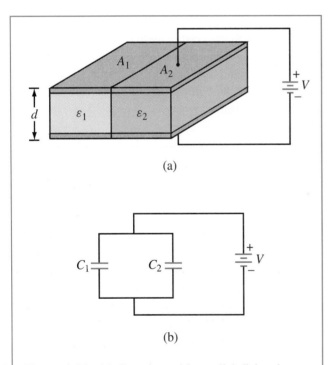

(a)

(b)

Figure 4-34: (a) Capacitor with parallel dielectric section, and (b) equivalent circuit.

The space between the plates contains two adjacent dielectrics, one with permittivity ε_1 and surface area A_1 and another with ε_2 and A_2. The objective of this problem is to show that the capacitance C of the configuration shown in Fig. 4-34(a) is equivalent to two capacitances in parallel, as illustrated in Fig. 4-34(b), with

$$C = C_1 + C_2 \tag{4.132}$$

where

$$C_1 = \frac{\varepsilon_1 A_1}{d} \tag{4.133}$$

$$C_2 = \frac{\varepsilon_2 A_2}{d} \tag{4.134}$$

To this end, proceed as follows:

(a) Find the electric fields $\mathbf{E}_1$ and $\mathbf{E}_2$ in the two dielectric layers.

(b) Calculate the energy stored in each section and use the result to calculate C_1 and C_2.

(c) Use the total energy stored in the capacitor to obtain an expression for C. Show that (4.132) is indeed a valid result.

4.53* Use the result of Problem 4.52 to determine the capacitance for each of the following configurations:

(a) Conducting plates are on top and bottom faces of the rectangular structure in Fig. 4-35(a).

(b) Conducting plates are on front and back faces of the structure in Fig. 4-35(a).

(c) Conducting plates are on top and bottom faces of the cylindrical structure in Fig. 4-35(b).

4.54 The capacitor shown in Fig. 4-36 consists of two parallel dielectric layers. Use energy considerations to show that the equivalent capacitance of the overall capacitor, C, is equal to the series combination of the capacitances of the individual layers, C_1 and C_2, namely

$$C = \frac{C_1 C_2}{C_1 + C_2} \tag{4.135}$$

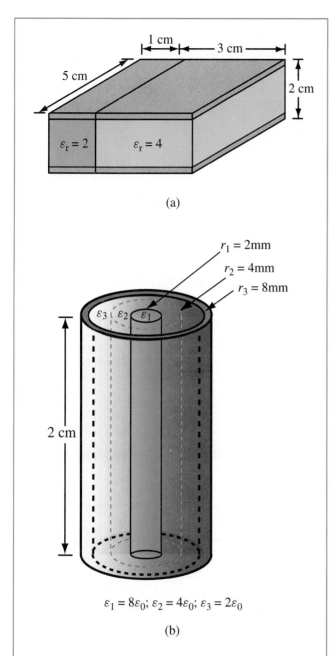

Figure 4-35: Dielectric sections for Problems 4.53 and 4.55.

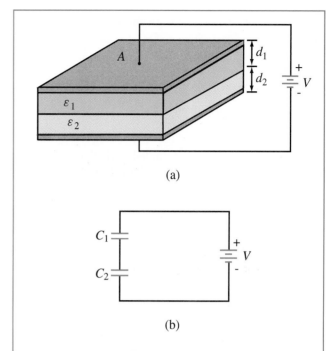

(a)

Figure 4-36: (a) Capacitor with parallel dielectric layers, and (b) equivalent circuit (Problem 4.54).

where

$$C_1 = \varepsilon_1 \frac{A}{d_1}, \qquad C_2 = \varepsilon_2 \frac{A}{d_2}$$

(a) Let V_1 and V_2 be the electric potentials across the upper and lower dielectrics, respectively. What are the corresponding electric fields E_1 and E_2? By applying the appropriate boundary condition at the interface between the two dielectrics, obtain explicit expressions for E_1 and E_2 in terms of ϵ_1, ϵ_2, V, and the indicated dimensions of the capacitor.

(b) Calculate the energy stored in each of the dielectric layers and then use the sum to obtain an expression for C.

(c) Show that C is given by Eq. (4.135).

4.55 Use the expressions given in Problem 4.54 to determine the capacitance for the configurations in Fig. 4-35(a) when the conducting plates are placed on the right and left faces of the structure.

Section 4-12: Image Method

4.56 With reference to Fig. 4-37, charge Q is located at a distance d above a grounded half-plane located in the $x-y$ plane and at a distance d from another grounded half-plane in the $x-z$ plane. Use the image method to

(a) Establish the magnitudes, polarities, and locations of the images of charge Q with respect to each of the two ground planes (as if each is infinite in extent).

(b) Find the electric potential and electric field at an arbitrary point $P(0, y, z)$.

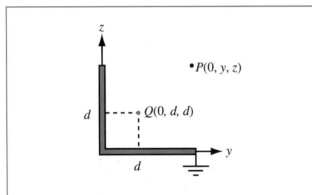

Figure 4-37: Charge Q next to two perpendicular, grounded, conducting half-planes.

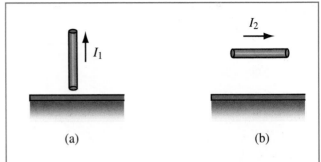

(a)   (b)

Figure 4-38: Currents above a conducting plane (Problem 4.57).

4.58 Use the image method to find the capacitance per unit length of an infinitely long conducting cylinder of radius a situated at a distance d from a parallel conducting plane, as shown in Fig. 4-39.

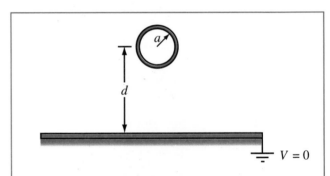

Figure 4-39: Conducting cylinder above a conducting plane (Problem 4.58).

4.57 Conducting wires above a conducting plane carry currents I_1 and I_2 in the directions shown in Fig. 4-38. Keeping in mind that the direction of a current is defined in terms of the movement of positive charges, what are the directions of the image currents corresponding to I_1 and I_2?

CHAPTER 5

Magnetostatics

Overview

OVERVIEW

This chapter on magnetostatics parallels the preceding chapter on electrostatics. Stationary charges produce static electric fields, and steady (nontime varying) currents produce magnetic fields. For $\partial/\partial t = 0$, the magnetic fields in a medium with magnetic permeability μ are governed by the second pair of Maxwell's equations, those given by Eqs. (4.3a and b):

$$\nabla \cdot \mathbf{B} = 0, \qquad (5.1a)$$
$$\nabla \times \mathbf{H} = \mathbf{J}, \qquad (5.1b)$$

where $\mathbf{J}$ is the current density. The magnetic flux density $\mathbf{B}$ and the magnetic field intensity $\mathbf{H}$ are related by

$$\mathbf{B} = \mu \mathbf{H}. \qquad (5.2)$$

When we examined electric fields in a dielectric medium in Chapter 4, we noted that the relation $\mathbf{D} = \varepsilon \mathbf{E}$ is valid only when the medium is linear and isotropic. These properties, which are true for most materials, allow us to treat the permittivity ε as a constant scalar quantity, independent of both the magnitude and direction of $\mathbf{E}$. A similar statement applies to the relation given by Eq. (5.2). With the exception of ferromagnetic materials, for which the relationship between $\mathbf{B}$ and $\mathbf{H}$ is nonlinear, most materials are characterized by constant magnetic permeabilities. *Furthermore, $\mu = \mu_0$ for most dielectrics and metals (excluding ferromagnetic materials).* Our objective in this chapter is to develop an understanding of the relationships between steady currents and the magnetic fields $\mathbf{B}$ and $\mathbf{H}$ for various types of current distributions and in various types of media and to introduce a number of related quantities, such as the magnetic vector potential $\mathbf{A}$, the magnetic energy density w_m, and the inductance of a conducting structure, L. The parallelism between these magnetostatic quantities and their electrostatic counterparts is encapsulated in Table 5-1.

5-1 Magnetic Forces and Torques

The electric field $\mathbf{E}$ at a point in space has been defined as the electric force $\mathbf{F}_e$ per unit charge acting on a test charge when placed at that point. We now define the *magnetic flux density* $\mathbf{B}$ at a point in space in terms of the *magnetic force* $\mathbf{F}_m$ that would be exerted on a charged particle moving with a velocity $\mathbf{u}$ were it to be passing through that point. Based on experiments conducted to determine the motion of charged particles moving in magnetic fields, it was established that the magnetic force $\mathbf{F}_m$ acting on a particle of charge q can be cast in the form

$$\mathbf{F}_m = q\mathbf{u} \times \mathbf{B} \qquad (\text{N}). \qquad (5.3)$$

Accordingly, the strength of $\mathbf{B}$ is measured in newtons/(C·m/s), which also is called the tesla (T) in SI units. For a positively charged particle, the direction of $\mathbf{F}_m$ is in the direction of the cross product $\mathbf{u} \times \mathbf{B}$, which is perpendicular to the plane containing $\mathbf{u}$ and $\mathbf{B}$ and governed by the right-hand rule. If q is negative, the direction of $\mathbf{F}_m$ is reversed, as illustrated in Fig. 5-1. The magnitude of $\mathbf{F}_m$ is given by

$$F_m = quB \sin\theta, \qquad (5.4)$$

where θ is the angle between $\mathbf{u}$ and $\mathbf{B}$. We note that F_m is maximum when $\mathbf{u}$ is perpendicular to $\mathbf{B}$ ($\theta = 90°$), and it is zero when $\mathbf{u}$ is parallel to $\mathbf{B}$ ($\theta = 0$ or $180°$).

Table 5-1: Attributes of electrostatics and magnetostatics.

Attribute	Electrostatics	Magnetostatics
Sources	Stationary charges	Steady currents
Fields	$\mathbf{E}$ and $\mathbf{D}$	$\mathbf{H}$ and $\mathbf{B}$
Constitutive parameter(s)	ε and σ	μ
Governing equations • **Differential form**	$\nabla \cdot \mathbf{D} = \rho_v$ $\nabla \times \mathbf{E} = 0$	$\nabla \cdot \mathbf{B} = 0$ $\nabla \times \mathbf{H} = \mathbf{J}$
• **Integral form**	$\oint_S \mathbf{D} \cdot d\mathbf{s} = Q$ $\oint_C \mathbf{E} \cdot d\mathbf{l} = 0$	$\oint_S \mathbf{B} \cdot d\mathbf{s} = 0$ $\oint_C \mathbf{H} \cdot d\mathbf{l} = I$
Potential	Scalar V, with $\mathbf{E} = -\nabla V$	Vector $\mathbf{A}$, with $\mathbf{B} = \nabla \times \mathbf{A}$
Energy density	$w_e = \frac{1}{2}\varepsilon E^2$	$w_m = \frac{1}{2}\mu H^2$
Force on charge q	$\mathbf{F}_e = q\mathbf{E}$	$\mathbf{F}_m = q\mathbf{u} \times \mathbf{B}$
Circuit element(s)	C and R	L

If a charged particle is in the presence of both an electric field $\mathbf{E}$ and a magnetic field $\mathbf{B}$, then the total *electromagnetic force* acting on it is

$$\mathbf{F} = \mathbf{F}_e + \mathbf{F}_m = q\mathbf{E} + q\mathbf{u} \times \mathbf{B} = q(\mathbf{E} + \mathbf{u} \times \mathbf{B}). \quad (5.5)$$

The force expressed by Eq. (5.5) is known as the *Lorentz force*. Electric and magnetic forces exhibit a number of important differences:

1. Whereas the electric force is always in the direction of the electric field, the magnetic force is always perpendicular to the magnetic field.

2. Whereas the electric force acts on a charged particle whether or not it is moving, the magnetic force acts on it only when it is in motion.

3. Whereas the electric force expends energy in displacing a charged particle, the magnetic force does no work when a particle is displaced.

Our last statement requires further elaboration. Because the magnetic force $\mathbf{F}_m$ is always perpendicular to $\mathbf{u}$, $\mathbf{F}_m \cdot \mathbf{u} = 0$. Hence, the work performed when a particle is displaced by a differential distance $d\mathbf{l} = \mathbf{u}\, dt$ is

$$dW = \mathbf{F}_m \cdot d\mathbf{l} = (\mathbf{F}_m \cdot \mathbf{u})\, dt = 0. \quad (5.6)$$

Since no work is done, a magnetic field cannot change the kinetic energy of a charged particle; *the magnetic field can change the direction of motion of a charged particle, but it cannot change its speed.*

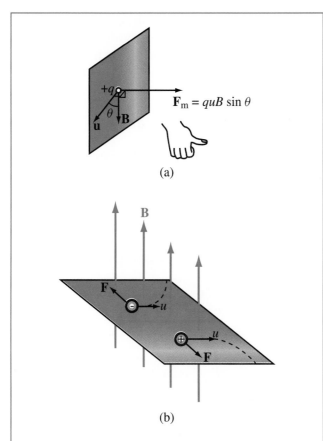

Figure 5-1: The direction of the magnetic force exerted on a charged particle moving in a magnetic field is (a) perpendicular to both **B** and **u** and (b) depends on the charge polarity (positive or negative).

EXERCISE 5.1 An electron moving in the positive x-direction perpendicular to a magnetic field experiences a deflection in the negative z-direction. What is the direction of the magnetic field?

Ans. Positive y-direction.

EXERCISE 5.2 A proton moving with a speed of 2×10^6 m/s through a magnetic field with magnetic flux density of 2.5 T experiences a magnetic force of magnitude 4×10^{-13} N. What is the angle between the magnetic field and the proton's velocity?

Ans. $\theta = 30°$ or $150°$.

EXERCISE 5.3 A charged particle with velocity **u** is moving in a medium containing uniform fields $\mathbf{E} = \hat{\mathbf{x}}E$ and $\mathbf{B} = \hat{\mathbf{y}}B$. What should **u** be so that the particle experiences no net force on it?

Ans. $\mathbf{u} = \hat{\mathbf{z}}E/B$.

5-1.1 Magnetic Force on a Current-Carrying Conductor

A current flowing through a conducting wire consists of charged particles *drifting* through the material of the wire. Consequently, when a current-carrying wire is placed in a magnetic field, it will experience a force equal to the sum of the magnetic forces acting on the charged particles moving within it. Consider, for example, the arrangement shown in Fig. 5-2 in which a vertical wire oriented along the z-direction is placed in a magnetic field **B** (produced by a magnet) oriented along the $-\hat{\mathbf{x}}$-direction (into the page). With no current flowing in the wire, $\mathbf{F}_m = 0$ and the wire maintains its vertical orientation, as shown in Fig. 5-2(a), but when a current is introduced in the wire, the wire deflects to the left ($-\hat{\mathbf{y}}$-direction) if the current's direction is upward ($+\hat{\mathbf{z}}$-direction), and it deflects to the right ($+\hat{\mathbf{y}}$-direction) if the current's direction is downward ($-\hat{\mathbf{z}}$-direction). The directions of these deflections are in accordance with the cross product given by Eq. (5.3).

To quantify the relationship between $\mathbf{F}_m$ and the current I flowing in the wire, let us consider a small segment of the wire of cross-sectional area A and differential length $d\mathbf{l}$, with the direction of $d\mathbf{l}$ denoting the direction of the current. Without loss of generality, we assume that the charge carriers constituting the current I are exclusively electrons, which is always a valid assumption for a good conductor. If the wire contains a free-electron

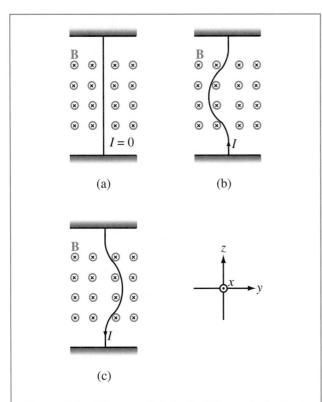

Figure 5-2: When a slightly flexible vertical wire is placed in a magnetic field directed into the page (as denoted by the crosses), it is (a) not deflected when the current through it is zero, (b) deflected to the left when I is upward, and (c) deflected to the right when I is downward.

charge density $\rho_{\text{ve}} = -N_e e$, where N_e is the number of moving electrons per unit volume, then the total amount of moving charge contained in an elemental volume of the wire is

$$dQ = \rho_{\text{ve}} A \, dl = -N_e e A \, dl, \qquad (5.7)$$

and the corresponding magnetic force acting on dQ in the presence of a magnetic field $\mathbf{B}$ is

$$d\mathbf{F}_{\text{m}} = dQ \, \mathbf{u}_e \times \mathbf{B} = -N_e e A \, dl \, \mathbf{u}_e \times \mathbf{B}, \qquad (5.8a)$$

where $\mathbf{u}_e$ is the drift velocity of the electrons. Since the direction of a current is defined as the direction of flow of positive charges, the electron drift velocity $\mathbf{u}_e$ is parallel to $d\mathbf{l}$, but opposite in direction. Thus, $dl \, \mathbf{u}_e = -d\mathbf{l} \, u_e$ and Eq. (5.8a) becomes

$$d\mathbf{F}_{\text{m}} = N_e e A u_e \, d\mathbf{l} \times \mathbf{B}. \qquad (5.8b)$$

From Eqs. (4.11) and (4.12), the current I flowing through a cross-sectional area A due to electrons with density $\rho_{\text{ve}} = -N_e e$, moving with velocity $-u_e$, is $I = \rho_{\text{ve}}(-u_e)A = (-N_e e)(-u_e)A = N_e e A u_e$. Hence, Eq. (5.8b) may be written in the compact form

$$d\mathbf{F}_{\text{m}} = I \, d\mathbf{l} \times \mathbf{B} \qquad (\text{N}). \qquad (5.9)$$

For a closed circuit of contour C carrying a current I, the total magnetic force is

$$\boxed{\mathbf{F}_{\text{m}} = I \oint_C d\mathbf{l} \times \mathbf{B} \qquad (\text{N}). \qquad (5.10)}$$

We will now examine the application of Eq. (5.10) in each of two special situations.

Closed Circuit in a Uniform **B** Field

Consider a closed wire carrying a current I and placed in a uniform external magnetic field $\mathbf{B}$, as shown in Fig. 5-3(a). Since $\mathbf{B}$ is constant, it can be taken outside the integral in Eq. (5.10), in which case we have

$$\mathbf{F}_{\text{m}} = I \left(\oint_C d\mathbf{l} \right) \times \mathbf{B} = 0. \qquad (5.11)$$

This result, which is a consequence of the fact that the vector sum of the displacement vectors $d\mathbf{l}$ over a closed path is equal to zero, states that *the total magnetic force on any closed current loop in a uniform magnetic field is zero.*

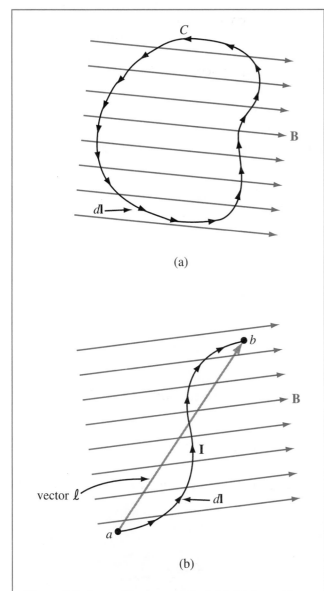

(a)

(b)

Figure 5-3: In a uniform magnetic field, (a) the net force on a closed current loop is zero because the integral of the displacement vector $d\mathbf{l}$ over a closed contour is zero, and (b) the force on a line segment is proportional to the vector between the end point ($\mathbf{F}_m = I\boldsymbol{\ell} \times \mathbf{B}$).

Curved Wire in a Uniform **B** *Field*

If we are interested in the magnetic force exerted on a wire segment, such as that shown in Fig. 5-3(b), when placed in a uniform field **B**, then Eq. (5.10) becomes

$$\mathbf{F}_m = I\left(\int_a^b d\mathbf{l}\right) \times \mathbf{B} = I\boldsymbol{\ell} \times \mathbf{B}, \qquad (5.12)$$

where $\boldsymbol{\ell}$ is the vector directed from a to b, as shown in Fig. 5-3(b). The integral of $d\mathbf{l}$ from a to b has the same value irrespective of the path taken between a and b. For a closed loop, points a and b become the same point, in which case $\boldsymbol{\ell} = 0$ and $\mathbf{F}_m = 0$.

Example 5-1   Force on a Semicircular Conductor

The semicircular conductor shown in Fig. 5-4 lies in the x–y plane and carries a current I. The closed circuit is exposed to a uniform magnetic field $\mathbf{B} = \hat{\mathbf{y}}B_0$. Determine (a) the magnetic force $\mathbf{F}_1$ on the straight section of the wire and (b) the force $\mathbf{F}_2$ on the curved section.

Solution: (a) The straight section of the circuit is of length $2r$, and the current flowing through it is along the

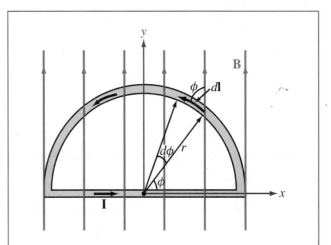

Figure 5-4: Semicircular conductor in a uniform field (Example 5-1).

$+x$-direction. Application of Eq. (5.12) with $\boldsymbol{\ell} = \hat{\mathbf{x}} 2r$ gives

$$\mathbf{F}_1 = \hat{\mathbf{x}}(2Ir) \times \hat{\mathbf{y}} B_0 = \hat{\mathbf{z}} 2Ir B_0 \quad \text{(N)}.$$

(b) Let us consider a segment of differential length dl on the curved part of the circle. The direction of $d\mathbf{l}$ is chosen to coincide with the direction of the current. Since $d\mathbf{l}$ and $\mathbf{B}$ are both in the x–y plane, their cross product $d\mathbf{l} \times \mathbf{B}$ points in the negative z-direction, and the magnitude of $d\mathbf{l} \times \mathbf{B}$ is proportional to $\sin \phi$, where ϕ is the angle between $d\mathbf{l}$ and $\mathbf{B}$. Moreover, the magnitude of $d\mathbf{l}$ is $dl = r\,d\phi$. Hence,

$$\mathbf{F}_2 = I \int_{\phi=0}^{\pi} d\mathbf{l} \times \mathbf{B}$$

$$= -\hat{\mathbf{z}} I \int_{\phi=0}^{\pi} r B_0 \sin \phi \, d\phi = -\hat{\mathbf{z}} 2Ir B_0 \quad \text{(N)}.$$

We note that $\mathbf{F}_2 = -\mathbf{F}_1$, and consequently the net force on the closed loop is zero. ∎

EXERCISE 5.4 A horizontal wire with a mass per unit length of 0.2 kg/m carries a current of 4 A in the $+x$-direction. If the wire is placed in a uniform magnetic flux density $\mathbf{B}$, what should the direction and minimum magnitude of $\mathbf{B}$ be in order to magnetically lift the wire vertically upward? (Hint: The acceleration due to gravity is $\mathbf{g} = -\hat{\mathbf{z}} 9.8 \text{ m/s}^2$.)

Ans. $\mathbf{B} = \hat{\mathbf{y}} 0.49$ T.

REVIEW QUESTIONS

Q5.1 What are the major differences between the behavior of the electric force $\mathbf{F}_e$ and the behavior of the magnetic force $\mathbf{F}_m$?

Q5.2 The ends of a 10-cm-long wire carrying a constant current I are anchored at two points on the x-axis, $x = 0$

and $x = 6$ cm. If the wire lies in the x–y plane and is present in a magnetic field $\mathbf{B} = \hat{\mathbf{y}} B_0$, which of the following arrangements produces a greater magnetic force on the wire: (a) wire is V-shaped with corners at $(0, 0)$, $(3, 4)$, and $(6, 0)$, (b) wire looks like an open rectangle with corners at $(0, 0)$, $(0, 2)$, $(6, 2)$, and $(6, 0)$.

5-1.2 Magnetic Torque on a Current-Carrying Loop

When a force is applied on a rigid body pivoted about a fixed axis, the body will react by rotating about that axis. The strength of the reaction depends on the cross product of the applied force vector $\mathbf{F}$ and the distance vector $\mathbf{d}$, measured *from* a point on the rotation axis (such that $\mathbf{d}$ is perpendicular to the axis) *to* the point of application of $\mathbf{F}$ (Fig. 5-5). The length of $\mathbf{d}$ is called the *moment arm*, and the cross product is called the *torque*:

$$\mathbf{T} = \mathbf{d} \times \mathbf{F} \quad \text{(N·m)}. \quad (5.13)$$

The unit for $\mathbf{T}$ is the same as that for work or energy, but torque does not represent work or energy. The force $\mathbf{F}$

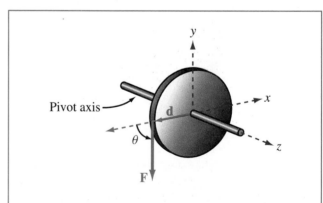

Figure 5-5: The force $\mathbf{F}$ acting on a circular disk pivoted along the z-axis generates a torque $\mathbf{T} = \mathbf{d} \times \mathbf{F}$ that causes the disk to rotate.

applied on the disk shown in Fig. 5-5 lies in the x–y plane and makes an angle θ with $\mathbf{d}$. Hence,

$$\mathbf{T} = \hat{\mathbf{z}} r F \sin\theta, \qquad (5.14)$$

where $|\mathbf{d}| = r$, the radius of the disk, and $F = |\mathbf{F}|$. From Eq. (5.14) we see that a torque along the positive z-direction corresponds to a tendency for the cylinder to rotate in a counterclockwise direction and, conversely, a negative torque corresponds to clockwise rotation. These directions are governed by the following *right-hand rule:* *when the thumb of the right hand is pointed along the direction of the torque, the four fingers indicate the direction that the torque is trying to rotate the body.*

We will now consider the *magnetic torque* exerted on a conducting loop under the influence of magnetic forces. We begin with the simple case where the magnetic field $\mathbf{B}$ is in the plane of the loop, and then we will extend the analysis to the more general case where $\mathbf{B}$ makes an angle θ with the surface normal of the loop.

Magnetic Field in the Plane of the Loop

The rectangular conducting loop shown in Fig. 5-6(a) is made of rigid wire carrying a current I. The loop lies in the x–y plane and is pivoted about the axis shown. Under the influence of an externally generated uniform magnetic field $\mathbf{B} = \hat{\mathbf{x}} B_0$, arms 1 and 3 of the loop are subjected to forces $\mathbf{F}_1$ and $\mathbf{F}_3$, respectively, with

$$\mathbf{F}_1 = I(-\hat{\mathbf{y}} b) \times (\hat{\mathbf{x}} B_0) = \hat{\mathbf{z}} I b B_0, \qquad (5.15a)$$

and

$$\mathbf{F}_3 = I(\hat{\mathbf{y}} b) \times (\hat{\mathbf{x}} B_0) = -\hat{\mathbf{z}} I b B_0. \qquad (5.15b)$$

These results are based on the application of Eq. (5.12). No magnetic force is exerted on either arm 2 or 4 because $\mathbf{B}$ is parallel to the direction of the current flowing in those arms.

The end view of the loop, depicted in Fig. 5-6(b), shows that forces $\mathbf{F}_1$ and $\mathbf{F}_3$ produce a torque about the origin O, causing the loop to rotate in a clockwise direction. The

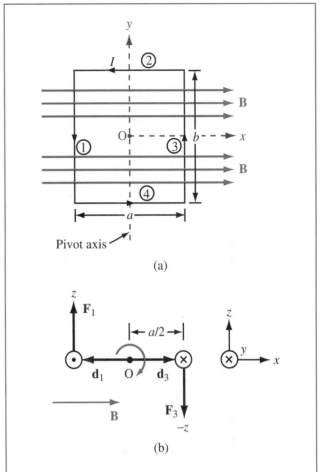

Figure 5-6: Rectangular loop pivoted along the y-axis: (a) front view and (b) bottom view. The combination of forces $\mathbf{F}_1$ and $\mathbf{F}_3$ on the loop generates a torque that tends to rotate the loop in a clockwise direction as shown in (b).

moment arm is $a/2$ for both forces, but $\mathbf{d}_1$ and $\mathbf{d}_3$ are in opposite directions, resulting in a total magnetic torque of

$$\begin{aligned}
\mathbf{T} &= \mathbf{d}_1 \times \mathbf{F}_1 + \mathbf{d}_3 \times \mathbf{F}_3 \\
&= \left(-\hat{\mathbf{x}}\frac{a}{2}\right) \times (\hat{\mathbf{z}} I b B_0) + \left(\hat{\mathbf{x}}\frac{a}{2}\right) \times (-\hat{\mathbf{z}} I b B_0) \\
&= \hat{\mathbf{y}} I a b B_0 = \hat{\mathbf{y}} I A B_0, \qquad (5.16)
\end{aligned}$$

where $A = ab$ is the area of the loop. The right-hand rule tells us that the sense of rotation is clockwise. The result given by Eq. (5.16) is valid only when the magnetic field **B** is parallel to the plane of the loop. As soon as the loop starts to rotate, the torque **T** starts to decrease, and at the end of one quarter of a complete rotation, the torque becomes zero, as discussed next.

Magnetic Field Perpendicular to the Axis of a Rectangular Loop

For the situation represented by Fig. 5-7, where **B** $= \hat{\mathbf{x}}B_0$, the field is still perpendicular to the loop's axis of rotation, but its direction may be at any angle θ with respect to the loop's surface normal $\hat{\mathbf{n}}$, we may now have nonzero forces on all four arms of the rectangular loop. However, forces **F**$_2$ and **F**$_4$ are equal in magnitude and opposite in direction and are along the rotation axis; hence, the net torque contributed by their combination is zero. The directions of the currents in arms 1 and 3 are always perpendicular to **B** regardless of the magnitude of θ. Hence, **F**$_1$ and **F**$_3$ have the same expressions given previously by Eqs. (5.15a and b), and their moment arm is $(a/2) \sin \theta$, as illustrated in Fig. 5-7(b). Consequently, the magnitude of the net torque exerted by the magnetic field about the axis of rotation is the same as that given by Eq. (5.16), but modified by $\sin \theta$:

$$T = IAB_0 \sin \theta. \tag{5.17}$$

According to Eq. (5.17), the torque is maximum when the magnetic field is parallel to the plane of the loop ($\theta = 90°$) and is zero when the field is perpendicular to the plane of the loop ($\theta = 0$). If the loop consists of N turns, each contributing a torque given by Eq. (5.17), then the total torque is

$$T = NIAB_0 \sin \theta. \tag{5.18}$$

The quantity NIA is called the **magnetic moment m** of the loop, and it may be regarded as a vector **m** with direction $\hat{\mathbf{n}}$, where $\hat{\mathbf{n}}$ is the surface normal of the loop and governed by the following **right-hand rule:** *when*

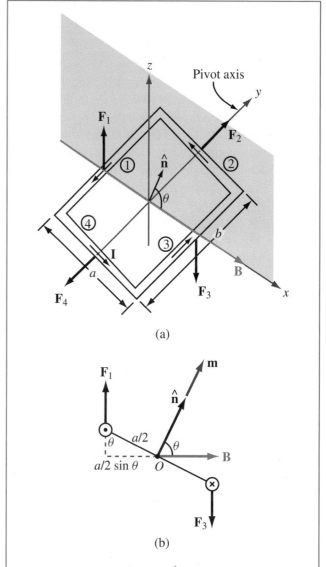

(a)

(b)

Figure 5-7: Rectangular loop in a uniform magnetic field with flux density **B** whose direction is perpendicular to the rotation axis of the loop, but makes an angle θ with the loop's surface normal $\hat{\mathbf{n}}$.

the four fingers of the right hand advance in the direction of the current I, the direction of the thumb specifies the *direction of* n̂. That is,

$$\mathbf{m} \triangleq \hat{\mathbf{n}} NIA \quad (\text{A·m}^2), \quad (5.19)$$

and in terms of **m**, the torque vector **T** can be written as

$$\mathbf{T} = \mathbf{m} \times \mathbf{B} \quad (\text{N·m}). \quad (5.20)$$

Even though the derivation leading to Eq. (5.20) was obtained for **B** being perpendicular to the axis of rotation of a rectangular loop, the expression is valid for any orientation of **B** and for a loop of any shape.

REVIEW QUESTIONS

Q5.3 How is the direction of the magnetic moment of a loop defined?

Q5.4 If one of two wires of equal lengths is formed into a closed square loop and the other is formed into a closed circular loop, and if both wires are carrying equal currents and both loops have their planes parallel to a uniform magnetic field, which loop would experience a greater torque?

EXERCISE 5.5 A square coil of 100 turns and 0.5-m-long sides is in a region with a uniform magnetic flux density of 0.2 T. If the maximum magnetic torque exerted on the coil is 4×10^{-2} (N·m), what is the current flowing in the coil?

Ans. $I = 8$ mA.

5-2 The Biot–Savart Law

In the preceding section, we elected to use the magnetic flux density **B** to denote the presence of a magnetic field in a given region of space. We will now work with the magnetic field intensity **H** instead. We do this in part to remind the reader that **B** and **H** are linearly related for most materials through $\mathbf{B} = \mu\mathbf{H}$, and therefore knowledge of one is synonymous with knowledge of the other (assuming that μ is known).

Through his experiments on the deflection of compass needles by current-carrying wires, Hans Oersted established that currents induce magnetic fields that form closed loops around the wires [see Section 1-2.3]. Building upon Oersted's results, Jean Biot and Felix Savart arrived at an expression that relates the magnetic field **H** at any point in space to the current I that generates **H**. The *Biot–Savart law* states that the differential magnetic field $d\mathbf{H}$ generated by a steady current I flowing through a differential length $d\mathbf{l}$ is given by

$$d\mathbf{H} = \frac{I}{4\pi} \frac{d\mathbf{l} \times \hat{\mathbf{R}}}{R^2} \quad (\text{A/m}), \quad (5.21)$$

where $\mathbf{R} = \hat{\mathbf{R}}R$ is the distance vector between $d\mathbf{l}$ and the observation point P shown in Fig. 5-8. The SI unit for **H** is ampere·m/m² = (A/m). It is important to remember that the direction of the magnetic field is defined such that $d\mathbf{l}$ is along the direction of the current I and the unit vector $\hat{\mathbf{R}}$ points *from* the current element *to* the observation point. According to Eq. (5.21), $d\mathbf{H}$ varies as R^{-2}, which is similar to the distance dependence of the electric field induced by an electric charge. However, unlike the electric field vector **E**, whose direction is *along* the distance vector **R** joining the charge to the observation point, the magnetic field **H** is *orthogonal* to the plane containing the direction of the current element $d\mathbf{l}$ and the distance vector **R**. At point P in Fig. 5-8, the direction of $d\mathbf{H}$ is out of the page, whereas at point P' the direction of $d\mathbf{H}$ is into the page.

To determine the total magnetic field **H** due to a conductor of finite size, we need to sum up the contributions

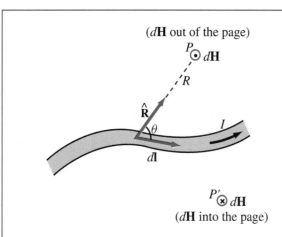

Figure 5-8: Magnetic field $d\mathbf{H}$ generated by a current element $I\,d\mathbf{l}$. The direction of the field induced at point P is opposite that induced at point P'.

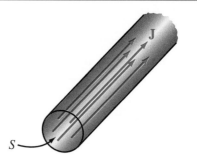

(a) Volume current density $\mathbf{J}$ in (A/m²)

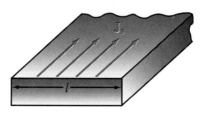

(b) Surface current density $\mathbf{J}_s$ in (A/m)

Figure 5-9: (a) The total current crossing the cross section S of the cylinder is $I = \int_S \mathbf{J} \cdot d\mathbf{s}$. (b) The total current flowing across the surface of the conductor is $I = \int_l J_s\,dl$.

due to all the current elements making up the conductor. Hence, the Biot–Savart law becomes

$$\mathbf{H} = \frac{I}{4\pi} \int_l \frac{d\mathbf{l} \times \hat{\mathbf{R}}}{R^2} \quad \text{(A/m)}, \quad (5.22)$$

where l is the line path along which I exists.

5-2.1 Magnetic Field due to Surface and Volume Current Distributions

The Biot–Savart law may also be expressed in terms of distributed current sources (Fig. 5-9) such as the *volume current density* $\mathbf{J}$, measured in (A/m²), or the *surface current density* $\mathbf{J}_s$, measured in (A/m). The surface current density $\mathbf{J}_s$ applies to currents that flow on the surfaces of conductors in the form of sheets of effectively zero thickness. When the current sources are specified in terms of $\mathbf{J}_s$ over a surface S or in terms of $\mathbf{J}$ over a volume v, we can use the equivalence given by

$$I\,d\mathbf{l} = \mathbf{J}_s\,ds = \mathbf{J}\,dv \quad (5.23)$$

to express the Biot–Savart law as follows:

$$\mathbf{H} = \frac{1}{4\pi} \int_S \frac{\mathbf{J}_s \times \hat{\mathbf{R}}}{R^2}\,ds \quad \text{(for a surface current),}$$
$$(5.24a)$$

$$\mathbf{H} = \frac{1}{4\pi} \int_v \frac{\mathbf{J} \times \hat{\mathbf{R}}}{R^2}\,dv \quad \text{(for a volume current).}$$
$$(5.24b)$$

Example 5-2   Magnetic Field of a Linear Conductor

A linear conductor of length l and carrying a current I is placed along the z-axis as shown in Fig. 5-10.

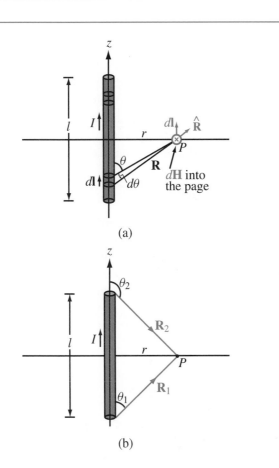

(a)

(b)

Figure 5-10: Linear conductor of length l carrying a current I. (a) The field $d\mathbf{H}$ at point P due to incremental current element $d\mathbf{l}$. (b) Limiting angles θ_1 and θ_2, each measured between vector $I\,d\mathbf{l}$ and the vector connecting the end of the conductor associated with that angle to point P (Example 5-2).

Determine the magnetic flux density $\mathbf{B}$ at a point P located at a distance r in the x–y plane in free space.

Solution: From Fig. 5-10, current element $d\mathbf{l} = \hat{\mathbf{z}}\,dz$ and $d\mathbf{l} \times \hat{\mathbf{R}} = dz\,(\hat{\mathbf{z}} \times \hat{\mathbf{R}}) = \hat{\boldsymbol{\phi}}\sin\theta\,dz$, where $\hat{\boldsymbol{\phi}}$ is the azimuth direction and θ is the angle between $d\mathbf{l}$ and $\hat{\mathbf{R}}$. Hence,

application of Eq. (5.22) gives

$$\mathbf{H} = \frac{I}{4\pi}\int_{z=-l/2}^{z=l/2}\frac{d\mathbf{l}\times\hat{\mathbf{R}}}{R^2} = \hat{\boldsymbol{\phi}}\,\frac{I}{4\pi}\int_{-l/2}^{l/2}\frac{\sin\theta}{R^2}\,dz. \quad (5.25)$$

For convenience, we will convert the integration variable from z to θ by using the transformations

$$R = r\csc\theta, \quad (5.26a)$$

$$z = -r\cot\theta, \quad (5.26b)$$

$$dz = r\csc^2\theta\,d\theta. \quad (5.26c)$$

Upon inserting Eqs. (5.26a) and (5.26c) into Eq. (5.25), we have

$$\mathbf{H} = \hat{\boldsymbol{\phi}}\,\frac{I}{4\pi}\int_{\theta_1}^{\theta_2}\frac{\sin\theta\,r\csc^2\theta\,d\theta}{r^2\csc^2\theta}$$

$$= \hat{\boldsymbol{\phi}}\,\frac{I}{4\pi r}\int_{\theta_1}^{\theta_2}\sin\theta\,d\theta$$

$$= \hat{\boldsymbol{\phi}}\,\frac{I}{4\pi r}(\cos\theta_1 - \cos\theta_2), \quad (5.27)$$

where θ_1 and θ_2 are the limiting angles at $z = -l/2$ and $z = l/2$, respectively. From the right triangle in Fig. 5-10(b),

$$\cos\theta_1 = \frac{l/2}{\sqrt{r^2 + (l/2)^2}}, \quad (5.28a)$$

$$\cos\theta_2 = -\cos\theta_1 = \frac{-l/2}{\sqrt{r^2 + (l/2)^2}}. \quad (5.28b)$$

Hence,

$$\mathbf{B} = \mu_0\mathbf{H} = \hat{\boldsymbol{\phi}}\,\frac{\mu_0 I l}{2\pi r\sqrt{4r^2 + l^2}} \quad \text{(T)}. \quad (5.29)$$

For an infinitely long wire such that $l \gg r$, Eq. (5.29) reduces to

$$\mathbf{B} = \hat{\boldsymbol{\phi}}\,\frac{\mu_0 I}{2\pi r} \quad \text{(infinitely long wire).} \quad \blacksquare \quad (5.30)$$

Example 5-3   Magnetic Field of a Pie-Shaped Loop

Determine the magnetic field at the apex O of the pie-shaped loop shown in Fig. 5-11. Ignore the contributions to the field due to the current in the small arcs near O.

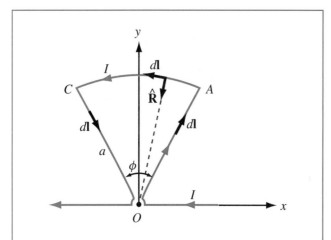

Figure 5-11: Pie-shaped loop of radius a carrying a current I (Example 5-3).

Solution: For the straight segments OA and OC, the magnetic field at O is identically zero. This is because, for all points along these segments, $d\mathbf{l}$ is parallel or anti-parallel to $\hat{\mathbf{R}}$ and hence $d\mathbf{l} \times \hat{\mathbf{R}} = 0$. For segment AC, $d\mathbf{l}$ is perpendicular to $\hat{\mathbf{R}}$ and $d\mathbf{l} \times \hat{\mathbf{R}} = \hat{\mathbf{z}} \, dl = \hat{\mathbf{z}} a \, d\phi$. Consequently, Eq. (5.22) gives

$$\mathbf{H} = \frac{I}{4\pi} \int \frac{\hat{\mathbf{z}} a \, d\phi}{a^2} = \hat{\mathbf{z}} \frac{I}{4\pi a} \phi,$$

where ϕ is in radians.   ■

Example 5-4   Magnetic Field of a Circular Loop

A circular loop of radius a carries a steady current I. Determine the magnetic field $\mathbf{H}$ at a point on the axis of the loop.

Solution: Let us place the loop in the x–y plane, as shown in Fig. 5-12. Our task is to obtain an expression for $\mathbf{H}$ at point $P(0, 0, z)$.

We start by noting that any element $d\mathbf{l}$ on the circular loop is perpendicular to the distance vector $\mathbf{R}$, and that all elements around the loop are at the same distance R from P, with $R = \sqrt{a^2 + z^2}$. From Eq. (5.21), the magnitude of $d\mathbf{H}$ due to element $d\mathbf{l}$ is given by

$$dH = \frac{I}{4\pi R^2} |d\mathbf{l} \times \hat{\mathbf{R}}| = \frac{I \, dl}{4\pi (a^2 + z^2)}, \qquad (5.31)$$

and the direction of $d\mathbf{H}$ is perpendicular to the plane containing $\mathbf{R}$ and $d\mathbf{l}$. As shown in Fig. 5-12, $d\mathbf{H}$ is in the r–z plane, and therefore it has components dH_r and dH_z. If we consider element $d\mathbf{l}'$, located diametrically opposite to $d\mathbf{l}$, we observe that the z-components of the magnetic

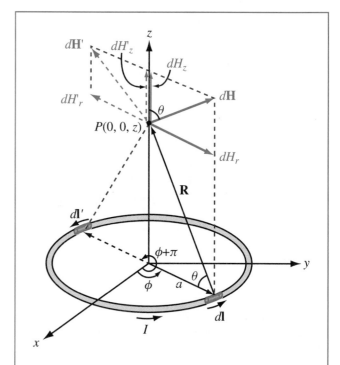

Figure 5-12: Circular loop carrying a current I (Example 5-4).

fields due to $d\mathbf{l}$ and $d\mathbf{l}'$ add because they are in the same direction, but their r-components cancel because they are in opposite directions. Hence, the net magnetic field is along z only. That is,

$$d\mathbf{H} = \hat{\mathbf{z}}\, dH_z = \hat{\mathbf{z}}\, dH \cos\theta = \hat{\mathbf{z}}\frac{I\cos\theta}{4\pi(a^2 + z^2)}\, dl. \quad (5.32)$$

For a fixed point $P(0, 0, z)$ on the axis of the loop, all quantities in Eq. (5.32) are constant, except for dl. Hence, integrating Eq. (5.32) over a circle of radius a gives

$$\mathbf{H} = \hat{\mathbf{z}}\frac{I\cos\theta}{4\pi(a^2 + z^2)}\oint dl = \hat{\mathbf{z}}\frac{I\cos\theta}{4\pi(a^2 + z^2)}(2\pi a).$$
$$(5.33)$$

Upon using the relation $\cos\theta = a/(a^2 + z^2)^{1/2}$, we obtain the final expression:

$$\boxed{\mathbf{H} = \hat{\mathbf{z}}\frac{Ia^2}{2(a^2 + z^2)^{3/2}} \quad \text{(A/m)}. \quad (5.34)}$$

At the center of the loop ($z = 0$), Eq. (5.34) reduces to

$$\mathbf{H} = \hat{\mathbf{z}}\frac{I}{2a} \quad \text{(at } z = 0), \quad (5.35)$$

and at points very far away from the loop such that $z^2 \gg a^2$, Eq. (5.34) can be approximated as

$$\mathbf{H} = \hat{\mathbf{z}}\frac{Ia^2}{2|z|^3} \quad \text{(at } |z| \gg a). \quad \blacksquare \quad (5.36)$$

5-2.2 Magnetic Field of a Magnetic Dipole

In view of the definition given by Eq. (5.19) for the magnetic moment $\mathbf{m}$ of a current loop, a loop with a single turn situated in the x–y plane, such as the one shown in Fig. 5-12, has a magnetic moment $\mathbf{m} = \hat{\mathbf{z}}m$ with $m = I\pi a^2$. Consequently, Eq. (5.36) may be expressed in the form

$$\mathbf{H} = \hat{\mathbf{z}}\frac{m}{2\pi|z|^3} \quad \text{(at } |z| \gg a). \quad (5.37)$$

This expression applies for a point P very far away from the loop, but on the axis of the loop. Had the problem been solved to find $\mathbf{H}$ at any distant point $P'(R', \theta', \phi')$ in a spherical coordinate system, where R' is the distance between the center of the loop and point P', we would have obtained the expression

$$\boxed{\mathbf{H} = \frac{m}{4\pi R'^3}(\hat{\mathbf{R}}\, 2\cos\theta' + \hat{\boldsymbol{\theta}}\sin\theta') \quad (5.38)}$$

for $R' \gg a$. A current loop with dimensions much smaller than the distance between the loop and the observation point is called a *magnetic dipole*. This is because the pattern of its magnetic field is similar to that of a permanent magnet, as well as to the pattern of the electric field of the electric dipole [see Example 4-7]. The similarity is evident from the patterns shown in Fig. 5-13.

REVIEW QUESTIONS

Q5.5 Two infinitely long parallel wires carry currents of equal magnitude. What is the resultant magnetic field due to the two wires at a point midway between the wires, compared with the magnetic field due to one of them alone, if the currents are (a) in the same direction and (b) in opposite directions?

Q5.6 Devise a right-hand rule for the direction of the magnetic field due to a linear current-carrying conductor.

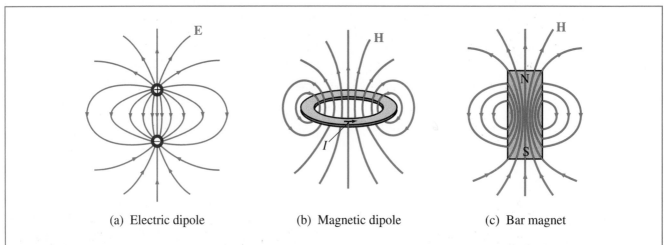

(a) Electric dipole   (b) Magnetic dipole   (c) Bar magnet

Figure 5-13: Patterns of (a) the electric field of an electric dipole, (b) the magnetic field of a magnetic dipole, and (c) the magnetic field of a bar magnet. Far away from the sources, the field patterns are similar in all three cases.

Q5.7 What is a magnetic dipole? Describe its magnetic field distribution.

EXERCISE 5.6 A semiinfinite linear conductor extends between $z = 0$ and $z = \infty$ along the z-axis. If the current I in the conductor flows along the positive z-direction, find $\mathbf{H}$ at a point in the x–y plane at a radial distance r from the conductor.

Ans. $\mathbf{H} = \hat{\boldsymbol{\phi}} \dfrac{I}{4\pi r}$ (A/m).

EXERCISE 5.7 A wire carrying a current of 4 A is formed into a circular loop. If the magnetic field at the center of the loop is 20 A/m, what is the radius of the loop if the loop has (a) only one turn and (b) 10 turns?

Ans.   (a) $a = 10$ cm, (b) $a = 1$ m.

EXERCISE 5.8 A wire is formed into a square loop and placed in the x–y plane with its center at the origin and each of its sides parallel to either the x- or y-axes. Each side is 40 cm in length, and the wire carries a current of 5 A whose direction is clockwise when the loop is viewed

from above. Calculate the magnetic field at the center of the loop.

Ans. $\mathbf{H} = -\hat{\mathbf{z}} \dfrac{4I}{\sqrt{2}\pi l} = -\hat{\mathbf{z}} 11.25$ A/m.

5-3   Magnetic Force between Two Parallel Conductors

In Section 5-1.1 we examined the magnetic force $\mathbf{F}_m$ that acts on a current-carrying conductor when the conductor is placed in an external magnetic field. The current in the conductor, however, also generates its own magnetic field. Hence, if two current-carrying conductors are placed in each other's vicinity, each will exert a magnetic force on the other. Let us consider two very long (or effectively infinitely long), straight, parallel wires in free space, separated by a distance d and carrying currents I_1 and I_2 in the same direction, as shown in Fig. 5-14. Current I_1 is located at $y = -d/2$ and I_2 is located at $y = d/2$, and both point in the z-direction. We denote $\mathbf{B}_1$ as the magnetic field due to current I_1, defined

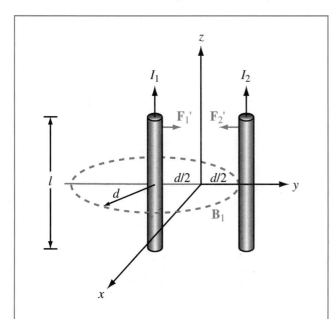

Figure 5-14: Magnetic forces on parallel current-carrying conductors.

at the location of the wire carrying the current I_2 and, conversely, $\mathbf{B}_2$ is the field due to I_2 at the location of the wire carrying current I_1. From Eq. (5.30), with $I = I_1$, $r = d$, and $\hat{\boldsymbol{\phi}} = -\hat{\mathbf{x}}$ at the location of I_2, the field $\mathbf{B}_1$ is

$$\mathbf{B}_1 = -\hat{\mathbf{x}}\frac{\mu_0 I_1}{2\pi d} . \qquad (5.39)$$

The force $\mathbf{F}_2$ exerted on a length l of wire I_2 due to its presence in field $\mathbf{B}_1$ may be obtained by applying Eq. (5.12):

$$\mathbf{F}_2 = I_2 l\hat{\mathbf{z}} \times \mathbf{B}_1 = I_2 l\hat{\mathbf{z}} \times (-\hat{\mathbf{x}})\frac{\mu_0 I_1}{2\pi d}$$

$$= -\hat{\mathbf{y}}\frac{\mu_0 I_1 I_2 l}{2\pi d} , \qquad (5.40)$$

and the corresponding force per unit length is

$$\mathbf{F}_2' = \frac{\mathbf{F}_2}{l} = -\hat{\mathbf{y}}\frac{\mu_0 I_1 I_2}{2\pi d} . \qquad (5.41)$$

A similar analysis performed for the force per unit length exerted on the wire carrying I_1 leads to

$$\mathbf{F}_1' = \hat{\mathbf{y}}\frac{\mu_0 I_1 I_2}{2\pi d} . \qquad (5.42)$$

Thus, $\mathbf{F}_1' = -\mathbf{F}_2'$, which means that the two wires attract each other with equal forces. If the currents are in opposite directions, the wires would repel each other with equal forces.

5-4 Maxwell's Magnetostatic Equations

Thus far, we have defined what we mean by a magnetic flux density $\mathbf{B}$ and the associated magnetic field $\mathbf{H}$, we introduced the formulation provided by the Biot–Savart law for finding the fields $\mathbf{B}$ and $\mathbf{H}$ due to any specified distribution of electric currents, and we examined how magnetic fields can exert magnetic forces on moving charged particles and on current-carrying conductors. We will now examine two additional important properties of magnetostatic fields, those described mathematically by Eqs. (5.1a and b).

5-4.1 Gauss's Law for Magnetism

In Chapter 4 we learned that the net outward flux of the electric flux density $\mathbf{D}$ through a closed surface enclosing a net charge Q is equal to Q. We referred to this property as Gauss's law (for electricity), and we expressed it mathematically in differential and integral forms as

$$\nabla \cdot \mathbf{D} = \rho_v \iff \oint_S \mathbf{D} \cdot d\mathbf{s} = Q. \qquad (5.43)$$

Conversion from differential to integral form was accomplished by applying the divergence theorem to a volume v enclosed by a surface S and containing charge $Q = \int_v \rho_v \, dv$ [Section 4-4]. The magnetic analogue to a point charge is a magnetic pole, but whereas electric charges can exist in isolation, magnetic poles do not. Magnetic poles always occur in pairs; no matter how many times a permanent magnet is subdivided, each new piece will always have a north and a south pole, even if the process were to be continued down to the atomic level. Thus, there is no magnetic equivalence to a charge Q or a charge density ρ_v, and it is therefore not surprising that *Gauss's law for magnetism* is given by

$$\nabla \cdot \mathbf{B} = 0 \iff \oint_S \mathbf{B} \cdot d\mathbf{s} = 0. \quad (5.44)$$

The differential form is one of Maxwell's four equations, and the integral form is obtained with the help of the divergence theorem. Formally, the name "Gauss's law" refers to the electric case, even when no specific reference to electricity is indicated. *The property described by Eq. (5.44) has been called "the law of nonexistence of isolated monopoles," "the law of conservation of magnetic flux," and "Gauss's law for magnetism," among others.* We prefer the last of the three cited names because it reminds us of the parallelism, as well as the differences, between the electric and magnetic properties of nature.

The difference between Gauss's law for electricity and its counterpart for magnetism may be viewed in terms of field lines. Electric field lines originate from positive electric charges and terminate on negative electric charges. Hence, for the electric field lines of the electric dipole shown in Fig. 5-15(a), the electric flux through a closed surface surrounding one of the charges is not zero. In contrast, *magnetic field lines always form continuous closed loops.* As we saw in Section 5-2, the magnetic field lines due to currents do not begin or end at any point; this is true for the linear conductor of Fig. 5-10 and the circular loop of Fig. 5-12, as well as for any distribution of currents. It is also true for a magnet, as illustrated in Fig. 5-15(b) for a bar magnet. Because the magnetic field lines form closed loops, the net magnetic flux through the closed surface surrounding the south pole of the magnet (or through any other closed surface) is always zero, regardless of the shape of that surface.

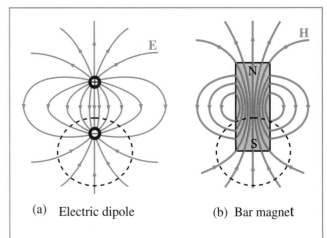

(a)   Electric dipole   (b)   Bar magnet

Figure 5-15: Whereas (a) the net electric flux through a closed surface surrounding a charge is not zero, (b) the net magnetic flux through a closed surface surrounding one of the poles of a magnet is zero.

5-4.2 Ampère's Law

We will now examine the property represented by Eq. (5.1b):

$$\nabla \times \mathbf{H} = \mathbf{J}, \quad (5.45)$$

which is the second of Maxwell's pair of equations characterizing the magnetostatic fields, $\mathbf{B}$ and $\mathbf{H}$. The integral form of Eq. (5.45) is called *Ampère's circuital law* (or simply Ampère's law) under magnetostatic conditions (steady currents). It is obtained by integrating both sides of Eq. (5.45) over an open surface S,

$$\int_S (\nabla \times \mathbf{H}) \cdot d\mathbf{s} = \int_S \mathbf{J} \cdot d\mathbf{s}, \quad (5.46)$$

and then invoking Stokes's theorem given by Eq. (3.107) to obtain the result

$$\oint_C \mathbf{H} \cdot d\mathbf{l} = I \quad \text{(Ampère's law)}, \quad (5.47)$$

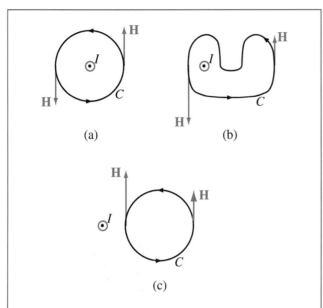

Figure 5-16: Ampère's law states that the line integral of **H** around a closed contour C is equal to the current traversing the surface bounded by the contour. This is true for contours (a) and (b), but the line integral of **H** is zero for the contour in (c) because the current I (denoted by the symbol $\odot$) is not enclosed by the contour C.

where C is the closed contour bounding the surface S and $I = \int \mathbf{J} \cdot d\mathbf{s}$ is the total current flowing through S. *The sign convention for the direction of C is taken so that I and $\mathbf{H}$ satisfy the right-hand rule* defined earlier in connection with the Biot–Savart law. That is, if the direction of I is aligned with the direction of the thumb of the right hand, then the direction of the contour C should be chosen to be along the direction of the other four fingers. In words, *Ampère's circuital law states that the line integral of $\mathbf{H}$ around a closed path is equal to the current traversing the surface bounded by that path.* By way of illustration, for both configurations shown in Figs. 5-16(a) and (b), the line integral of **H** is equal to the current I, even though the paths have very different shapes and the magnitude of **H** is not uniform along the

path of configuration (b). By the same token, because path (c) in Fig. 5-16 does not enclose the current I, its line integral of **H** is identically zero, even though **H** is not zero along the path.

When we examined Gauss's law in Section 4-4, we discovered that in practice its usefulness for calculating the electric flux density **D** is limited to charge distributions that possess a certain degree of symmetry and that the calculation procedure is subject to proper choice of the Gaussian surface enclosing the charges. A similar statement applies to Ampère's law: its usefulness is limited to symmetric current distributions that allow the choice of convenient *Ampèrian contours* around them, as illustrated by Examples 5-5 to 5-7.

Example 5-5 Magnetic Field of a Long Wire

A long (practically infinite) straight wire of radius a carries a steady current I that is uniformly distributed over the cross section of the wire. Determine the magnetic field **H** at a distance r from the axis of the wire both (a) inside the wire ($r \leq a$) and (b) outside the wire ($r \geq a$).

Solution: (a) We choose I to be along the $+z$-direction, as shown in Fig. 5-17(a). To determine $\mathbf{H}_1$ at a distance $r_1 \leq a$, we choose the Ampèrian contour C_1 to be a circular path of radius r_1 as shown in Fig. 5-17(b). In this case, Ampère's law takes the form

$$\oint_{C_1} \mathbf{H}_1 \cdot d\mathbf{l}_1 = I_1,$$

where I_1 is the fraction of the total current I flowing through C_1. From symmetry, $\mathbf{H}_1$ must be constant in magnitude and parallel to the contour at any point along the path. Furthermore, to satisfy the right-hand rule and given that I is along the z-direction, $\mathbf{H}_1$ must be along the $+\phi$-direction in a cylindrical coordinate system. Hence, $\mathbf{H}_1 = \hat{\boldsymbol{\phi}} H_1$. With $d\mathbf{l}_1 = \hat{\boldsymbol{\phi}} r_1 \, d\phi$, the left-hand side of Ampère's law gives

$$\oint_{C_1} \mathbf{H}_1 \cdot d\mathbf{l}_1 = \int_0^{2\pi} H_1(\hat{\boldsymbol{\phi}} \cdot \hat{\boldsymbol{\phi}}) r_1 \, d\phi = 2\pi r_1 H_1.$$

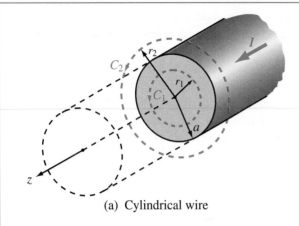

(a) Cylindrical wire

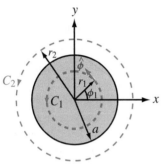

(b) Wire cross section

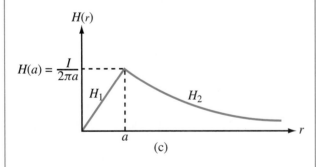

(c)

Figure 5-17: Infinitely long wire of radius a carrying a uniform current I along the $+z$-direction: (a) general configuration showing contours C_1 and C_2; (b) cross-sectional view; and (c) a plot of H versus r (Example 5-5).

The current I_1 flowing through the area enclosed by C_1 is equal to the total current I multiplied by the ratio of the area enclosed by C_1 to the total cross-sectional area of the wire:

$$ I_1 = \left(\frac{\pi r_1^2}{\pi a^2} \right) I = \left(\frac{r_1}{a} \right)^2 I. $$

Equating both sides of Ampère's law and then solving for $\mathbf{H}_1$ leads to

$$ \mathbf{H}_1 = \hat{\boldsymbol{\phi}} H_1 = \hat{\boldsymbol{\phi}} \frac{r_1}{2\pi a^2} I \qquad \text{(for } r_1 \le a\text{).} \qquad (5.48) $$

(b) For $r_2 \ge a$, we choose path C_2, which encloses all the current I. Hence,

$$ \oint_{C_2} \mathbf{H}_2 \cdot d\mathbf{l}_2 = 2\pi r_2 H_2 = I, $$

and

$$ \mathbf{H}_2 = \hat{\boldsymbol{\phi}} H_2 = \hat{\boldsymbol{\phi}} \frac{I}{2\pi r_2} \qquad \text{(for } r_2 \ge a\text{).} \qquad (5.49) $$

If we ignore the subscript 2, we observe that Eq. (5.49) provides the same expression for $\mathbf{B} = \mu_0 \mathbf{H}$ as Eq. (5.30), which we had derived previously on the basis of the Biot–Savart law.

The variation of the magnitude of H as a function of r is plotted in Fig. 5-17(c); H increases linearly between $r = 0$ and $r = a$ (inside the conductor), and then it decreases as $1/r$ outside the conductor. ∎

EXERCISE 5.9 Current I flows in the inner conductor of a long coaxial cable and returns through the outer conductor. What is the magnetic field in the region outside the coaxial cable and why?

Ans. $\mathbf{H} = 0$ outside the coaxial cable because the net current enclosed by the Ampèrian contour is zero.

EXERCISE 5.10 The metal niobium becomes a superconductor with zero electrical resistance when it is cooled to below 9 K, but its superconductive behavior ceases when the magnetic flux density at its surface exceeds 0.12 T. Determine the maximum current that a 0.1-mm-diameter niobium wire can carry and remain superconductive.

Ans. $I = 30$ A.

Example 5-6   **Magnetic Field inside a Toroidal Coil**

A toroidal coil (also called a torus or toroid) is a doughnut-shaped structure (called its core) with closely spaced turns of wire wrapped around it, as shown in Fig. 5-18. For clarity, we show the turns in the figure as spaced apart, but in practice they are wound in a closely spaced arrangement to form approximately circular loops. The toroid is used to magnetically couple multiple circuits and to measure the magnetic properties of materials, as illustrated later in Fig. 5-30. For a toroid with N turns carrying a current I, determine the magnetic field **H** in each of the following three regions: $r < a$, $a < r < b$, and $r > b$, all in the azimuthal plane of the toroid.

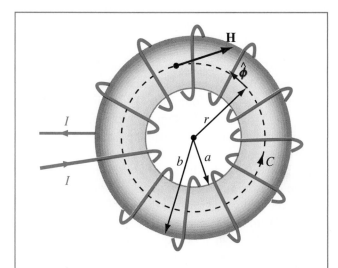

Figure 5-18: Toroidal coil with inner radius a and outer radius b. The wire loops usually are much more closely spaced than shown in the figure (Example 5-6).

Solution: From symmetry, it is clear that **H** has uniform properties in azimuth. If we construct a circular Ampèrian contour with center at the origin and of radius $r < a$, there will be no current flowing through the surface of the contour, and therefore **H** = 0 for $r < a$. Similarly, for an Ampèrian contour with radius $r > b$, the *net* current flowing through its surface is zero because an equal number of current coils cross the surface in both directions. Hence, **H** = 0 in the region exterior to the toroidal coil.

For the region inside the core, we construct a path of radius r, as shown in Fig. 5-18. For each loop, we know from Example 5-4 that the field **H** at the center of the loop points along the axis of the loop, which in this case is the ϕ-direction, and in view of the direction of the current I shown in Fig. 5-18, the right-hand rule tells us that **H** must be in the $-\phi$-direction. Hence, $\mathbf{H} = -\hat{\boldsymbol{\phi}} H$. The total current crossing the surface of the contour with radius r is NI and its direction is into the page. According to the right-hand rule associated with Ampère's law, the current is positive if it crosses the surface of the contour in the direction of the four fingers of the right hand when the thumb is pointing along the direction of the contour C. Hence, in the present situation the current is $-NI$. Application of Ampère's law then gives

$$\oint_C \mathbf{H} \cdot d\mathbf{l} = \int_0^{2\pi} (-\hat{\boldsymbol{\phi}} H) \cdot \hat{\boldsymbol{\phi}} r \, d\phi = -2\pi r H = -NI.$$

Hence, $H = NI/(2\pi r)$ and

$$\mathbf{H} = -\hat{\boldsymbol{\phi}} H = -\hat{\boldsymbol{\phi}} \frac{NI}{2\pi r} \quad \text{(for } a < r < b\text{).} \qquad \blacksquare \quad (5.50)$$

Example 5-7   **Magnetic Field of an Infinite Current Sheet**

The x–y plane of Fig. 5-19 contains an infinite current sheet with surface current density $\mathbf{J}_s = \hat{\mathbf{x}} J_s$. Find the magnetic field **H**.

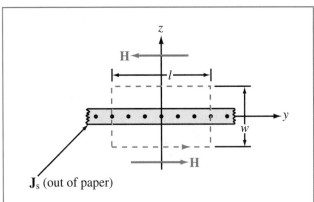

Figure 5-19: A thin current sheet in the $x-y$ plane carrying a surface current density $\mathbf{J}_s = \hat{\mathbf{x}} J_s$ (out of the page) (Example 5-7).

We note that the magnetic field is uniform and everywhere parallel to the current sheet. ∎

Q5.8 What is a fundamental difference between electric field lines and magnetic field lines?

Q5.9 If the line integral of **H** over a closed contour is zero, does it follow that $\mathbf{H} = 0$ at every point on the contour? If not, what then does it imply?

Q5.10 Compare the utility of applying the Biot–Savart law versus applying Ampère's law for computing the magnetic field due to current-carrying conductors.

Q5.11 What is a toroid? What is the magnetic field at points outside the toroid?

Solution: From symmetry considerations and the right-hand rule, **H** must be in the direction shown in the figure. That is,

$$\mathbf{H} = \begin{cases} -\hat{\mathbf{y}} H, & \text{for } z > 0, \\ \hat{\mathbf{y}} H, & \text{for } z < 0. \end{cases}$$

To evaluate the line integral in Ampère's law, we choose a rectangular Ampèrian path around the sheet, with dimensions l and w, as shown in Fig. 5-19. Recalling that J_s represents current per unit length along the y-direction, the total current crossing the surface of the rectangular loop is $I = J_s l$. Hence, applying Ampère's law over the loop, while noting that **H** is zero along the paths of length w, we have

$$\oint_C \mathbf{H} \cdot d\mathbf{l} = 2Hl = J_s l,$$

from which we obtain the result

$$\mathbf{H} = \begin{cases} -\hat{\mathbf{y}} \dfrac{J_s}{2}, & \text{for } z > 0, \\ \hat{\mathbf{y}} \dfrac{J_s}{2}, & \text{for } z < 0. \end{cases} \tag{5.51}$$

5-5 Vector Magnetic Potential

In our treatment of electrostatic fields in Chapter 4, we introduced the electrostatic potential V and defined it in terms of the line integral of the electric field **E**. In differential form, we found that V and **E** are related by $\mathbf{E} = -\nabla V$. This relationship proved useful not only in relating the electric field distribution in circuit elements (such as resistors and capacitors) to the voltage differences across them, but it also proved more convenient in certain cases to first determine the potential V due to a given charge distribution and then apply the relationship $\mathbf{E} = -\nabla V$ to find **E** than to use Coulomb's law directly to find **E**. We will now explore a similar approach in connection with the magnetic flux density **B**.

According to Eq. (5.44), $\nabla \cdot \mathbf{B} = 0$. We wish to define **B** in terms of a magnetic potential with the constraint that such a definition guarantees that the divergence of **B** is always equal to zero. This can be realized by taking advantage of the vector identity given by Eq. (3.106b), which states that, for any vector **A**,

$$\nabla \cdot (\nabla \times \mathbf{A}) = 0. \qquad (5.52)$$

Hence, by defining the *vector magnetic potential* **A** such that

$$\boxed{\mathbf{B} = \nabla \times \mathbf{A} \quad (\text{Wb/m}^2), \quad (5.53)}$$

we are guaranteed that $\nabla \cdot \mathbf{B} = 0$. The SI unit for **B** is the tesla (T). An equivalent unit used in the literature is webers per square meter (Wb/m^2). Consequently, the SI unit for **A** is (Wb/m).

With $\mathbf{B} = \mu \mathbf{H}$, the differential form of Ampère's law given by Eq. (5.45) can be written as

$$\nabla \times \mathbf{B} = \mu \mathbf{J}, \qquad (5.54)$$

where **J** is the current density due to free charges in motion. If we substitute Eq. (5.53) into Eq. (5.54), we have

$$\nabla \times (\nabla \times \mathbf{A}) = \mu \mathbf{J}. \qquad (5.55)$$

For any vector **A**, the Laplacian of **A** obeys the vector identity given by Eq. (3.113), that is,

$$\nabla^2 \mathbf{A} = \nabla(\nabla \cdot \mathbf{A}) - \nabla \times (\nabla \times \mathbf{A}), \qquad (5.56)$$

where, by definition, $\nabla^2 \mathbf{A}$ in Cartesian coordinates is given by

$$\nabla^2 \mathbf{A} = \left(\frac{\partial^2}{\partial x^2} + \frac{\partial^2}{\partial y^2} + \frac{\partial^2}{\partial z^2} \right) \mathbf{A}$$
$$= \hat{\mathbf{x}} \nabla^2 A_x + \hat{\mathbf{y}} \nabla^2 A_y + \hat{\mathbf{z}} \nabla^2 A_z. \qquad (5.57)$$

Combining Eq. (5.55) with Eq. (5.56) gives

$$\nabla(\nabla \cdot \mathbf{A}) - \nabla^2 \mathbf{A} = \mu \mathbf{J}. \qquad (5.58)$$

When we introduced the defining equation for the vector magnetic potential **A**, given by Eq. (5.53), the only constraint we placed on **A** is that its definition satisfy the condition $\nabla \cdot \mathbf{B} = 0$. Equation (5.58) contains a term

involving $\nabla \cdot \mathbf{A}$. It turns out from vector calculus that we have a fair amount of latitude in specifying a value or a functional form for $\nabla \cdot \mathbf{A}$, without conflicting with the requirement represented by Eq. (5.52). The simplest among these allowed specifications is

$$\nabla \cdot \mathbf{A} = 0. \qquad (5.59)$$

Using this choice in Eq. (5.58) leads to the *vector Poisson's equation* given by

$$\boxed{\nabla^2 \mathbf{A} = -\mu \mathbf{J}. \quad (5.60)}$$

Using the definition for $\nabla^2 \mathbf{A}$ given by Eq. (5.57), the vector Poisson's equation can be broken up into three scalar Poisson's equations:

$$\nabla^2 A_x = -\mu J_x, \qquad (5.61a)$$
$$\nabla^2 A_y = -\mu J_y, \qquad (5.61b)$$
$$\nabla^2 A_z = -\mu J_z. \qquad (5.61c)$$

In electrostatics, Poisson's equation for the scalar potential V is given by Eq. (4.60) as

$$\nabla^2 V = -\frac{\rho_v}{\varepsilon}, \qquad (5.62)$$

and its solution for a volume charge distribution ρ_v occupying a volume v' was given by Eq. (4.61) as

$$V = \frac{1}{4\pi \varepsilon} \int_{v'} \frac{\rho_v}{R'} \, dv'. \qquad (5.63)$$

Poisson's equations for A_x, A_y, and A_z are mathematically identical in form to Eq. (5.62). Hence, for a current density **J** with x-component J_x distributed over a volume v', the solution for Eq. (5.61a) is

$$A_x = \frac{\mu}{4\pi} \int_{v'} \frac{J_x}{R'} \, dv' \quad (\text{Wb/m}), \qquad (5.64)$$

and similar solutions can be written for A_y in terms of J_y and A_z in terms of J_z. The three solutions can be combined into a vector equation of the form

$$\mathbf{A} = \frac{\mu}{4\pi} \int_{v'} \frac{\mathbf{J}}{R'} \, dv' \quad \text{(Wb/m)}. \quad (5.65)$$

In view of Eq. (5.23), if the current distribution is given in the form of a surface current density $\mathbf{J}_s$ over a surface S', then $\mathbf{J} \, dv'$ should be replaced with $\mathbf{J}_s \, ds'$ and v' should be replaced with S'; and, similarly, for a line distribution, $\mathbf{J} \, dv'$ should be replaced with $I \, d\mathbf{l}'$ and the integration should be performed over the associated line path l'.

In addition to the Biot–Savart law and Ampère's law, the vector magnetic potential provides a third approach for computing the magnetic field due to current-carrying conductors. For a specified current distribution, Eq. (5.65) can be used to find $\mathbf{A}$, and then Eq. (5.53) can be used to find $\mathbf{B}$. Except for simple current distributions with symmetrical geometries that lend themselves to the application of Ampère's law, the choice usually is between the approaches provided by the Biot–Savart law and the vector magnetic potential, and among those two the latter often is easier to apply because it is easier to perform the integration given by Eq. (5.65) than to perform the integration given by Eq. (5.22).

The *magnetic flux* Φ linking a surface S is defined as the total magnetic flux density passing through S, or

$$\Phi = \int_S \mathbf{B} \cdot d\mathbf{s} \quad \text{(Wb)}. \quad (5.66)$$

If we insert Eq. (5.53) into Eq. (5.66) and then invoke Stokes's theorem, we have

$$\Phi = \int_S (\nabla \times \mathbf{A}) \cdot d\mathbf{s} = \oint_C \mathbf{A} \cdot d\mathbf{l} \quad \text{(Wb)}, \quad (5.67)$$

where C is the contour bounding the surface S. Thus, Φ can be determined by either Eq. (5.66) or Eq. (5.67), whichever is easier to integrate for the specific problem under consideration.

5-6 Magnetic Properties of Materials

According to our preceding discussions, because the pattern of the magnetic field generated by a current loop is similar to that exhibited by a permanent magnet, the loop is regarded as a magnetic dipole with a north pole and a south pole [see Section 5-2.2 and Fig. 5-13]. The magnetic moment $\mathbf{m}$ of a loop of area A has a magnitude $m = IA$, and the direction of $\mathbf{m}$ is normal to the plane of the loop in accordance with the right-hand rule. Magnetization in a material substance is associated with atomic current loops generated by two principal mechanisms: (1) orbital motions of the electrons around the nucleus and similar motions of the protons around each other in the nucleus and (2) electron spin. The magnetic moment of an electron is due to the combination of its orbital motion and its spinning motion about its own axis. As we will see later, the magnetic moment of the nucleus is much smaller than that of an electron, and therefore the total magnetic moment of an atom is dominated by the sum of the magnetic moments of its electrons. The magnetic behavior of a material is governed by the interaction of the magnetic dipole moments of its atoms with an external magnetic field. This behavior, which depends on the crystalline structure of the material, is used as a basis for classifying materials as *diamagnetic*, *paramagnetic*, or *ferromagnetic*. The atoms of a diamagnetic material have no permanent magnetic dipole moments. In contrast, both paramagnetic and ferromagnetic materials have atoms with permanent magnetic dipole moments, but, as will be explained later, the atoms of materials belonging to these two classes have very different organizational structures.

5-6.1 Orbital and Spin Magnetic Moments

To keep the following presentation simple, we will begin our discussion with a classical model of the atom, in which the electrons are assumed to move in circular motion about the nucleus, and we will then extend the results by incorporating the predictions provided by the

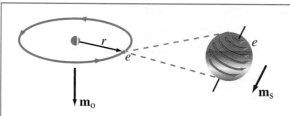

(a) Orbiting electron   (b) Spinning electron

Figure 5-20: An electron generates (a) an orbital magnetic moment $\mathbf{m}_o$ as it rotates around the nucleus and (b) a spin magnetic moment $\mathbf{m}_s$, as it spins about its own axis.

more correct quantum-mechanical model of matter. An electron with charge of $-e$ moving with a constant velocity u in a circular orbit of radius r [Fig. 5-20(a)] completes one revolution in time $T = 2\pi r/u$. This circular motion of the electron constitutes a tiny current loop with current I given by

$$I = -\frac{e}{T} = -\frac{eu}{2\pi r}. \qquad (5.68)$$

The magnitude of the associated *orbital magnetic moment* $\mathbf{m}_o$ is

$$m_o = IA = \left(-\frac{eu}{2\pi r}\right)(\pi r^2)$$
$$= -\frac{eur}{2} = -\left(\frac{e}{2m_e}\right)L_e, \qquad (5.69)$$

where $L_e = m_e u r$ is the angular momentum of the electron and m_e is its mass. According to quantum physics, the orbital angular momentum is quantized; specifically, L_e is always some integer multiple of $\hbar = h/2\pi$, where h is Planck's constant. That is, $L_e = 0, \hbar, 2\hbar, \ldots$. Consequently, the smallest nonzero magnitude of the orbital magnetic moment of an electron is

$$m_o = -\frac{e\hbar}{2m_e}. \qquad (5.70)$$

Despite the fact that all substances contain electrons and the electrons exhibit magnetic dipole moments, most substances are effectively nonmagnetic. This is because, in the absence of an external magnetic field, the atoms of most materials are oriented *randomly*, as a result of which the net magnetic moment generated by their electrons is either zero or very small.

In addition to the magnetic moment produced by its orbital motion, an electron generates a *spin magnetic moment* $\mathbf{m}_s$ due to its spinning motion about its own axis [Fig. 5-20(b)]. The magnitude of $\mathbf{m}_s$ predicted by quantum theory is

$$m_s = -\frac{e\hbar}{2m_e}, \qquad (5.71)$$

which is equal to the minimum orbital magnetic moment m_o. The electrons of an atom with an even number of electrons usually exist in pairs, with the members of a pair having opposite spin directions, thereby canceling each others' spin magnetic moments. If the number of electrons is odd, the atom will have a nonzero spin magnetic moment due to its unpaired electron.

According to Eq. (5.71), the spin magnetic moment of an electron is inversely proportional to the electron mass m_e. The nucleus of an atom also exhibits a spinning motion, but because its mass is much greater than that of an electron, its magnetic moment is on the order of 10^{-3} of that of an electron.

5-6.2 Magnetic Permeability

The *magnetization vector* $\mathbf{M}$ of a material is defined as the vector sum of the magnetic dipole moments of the atoms contained in a unit volume of the material. The magnetic flux density corresponding to $\mathbf{M}$ is $\mathbf{B}_m = \mu_0 \mathbf{M}$. In the presence of an externally applied magnetic field $\mathbf{H}$, the total magnetic flux density in the material is

$$\mathbf{B} = \mu_0 \mathbf{H} + \mu_0 \mathbf{M} = \mu_0 (\mathbf{H} + \mathbf{M}), \qquad (5.72)$$

where the first term represents the contribution of the external field and the second term represents the contribution of the magnetization of the material. In general, a material becomes magnetized in response to the external field **H**. Hence, **M** can be expressed as

$$\mathbf{M} = \chi_m \mathbf{H},\qquad(5.73)$$

where χ_m is a dimensionless quantity called the ***magnetic susceptibility*** of the material. For diamagnetic and paramagnetic materials, χ_m is a constant at a given temperature, resulting in a linear relationship between **M** and **H**. This is not the case for ferromagnetic substances; the relationship between **M** and **H** not only is nonlinear, but also depends on the previous "history" of the material, as explained in the next section.

While keeping this fact in mind, let us combine Eqs. (5.72) and (5.73) to get

$$\mathbf{B} = \mu_0(\mathbf{H} + \chi_m \mathbf{H}) = \mu_0(1 + \chi_m)\mathbf{H},\qquad(5.74)$$

or

$$\mathbf{B} = \mu \mathbf{H},\qquad(5.75)$$

where μ, the ***magnetic permeability*** of the material, is given in terms of χ_m by

$$\boxed{\mu = \mu_0(1 + \chi_m)\quad\text{(H/m)}.\qquad(5.76)}$$

Often it is convenient to define the magnetic properties of a material in terms of the ***relative permeability*** μ_r:

$$\boxed{\mu_r = \frac{\mu}{\mu_0} = 1 + \chi_m,\qquad(5.77)}$$

where μ_0 is the permeability of free space. A material usually is classified as diamagnetic, paramagnetic, or ferromagnetic on the basis of the value of its χ_m, as shown in Table 5-2. Diamagnetic materials have negative susceptibilities and paramagnetic materials have positive susceptibilities. However, the absolute magnitude of χ_m

is on the order of 10^{-5} for both classes of materials, which allows us to ignore χ_m relative to 1 in Eq. (5.77). *This gives $\mu_r \simeq 1$ or $\mu \simeq \mu_0$ for diamagnetic and paramagnetic substances, which include dielectric materials and most metals. In contrast, $|\mu_r| \gg 1$ for ferromagnetic materials; $|\mu_r|$ of purified iron, for example, is on the order of 2×10^5.* Ferromagnetic materials are discussed next.

EXERCISE 5.11 The magnetic vector **M** is the vector sum of the magnetic moments of all the atoms contained in a unit volume ($1\,\text{m}^3$). If a certain type of iron with 8.5×10^{28} atoms/m^3 contributes one electron per atom to align its spin magnetic moment along the direction of the applied field, find (a) the spin magnetic moment of a single electron, given that $m_e = 9.1 \times 10^{-31}$ (kg) and $\hbar = 1.06 \times 10^{-34}$ (J·s), and (b) the magnitude of **M**.

Ans. (a) $m_s = 9.27 \times 10^{-24}$ (A·m^2), (b) $M = 7.88 \times 10^5$ (A/m).

5-6.3 Magnetic Hysteresis of Ferromagnetic Materials

Ferromagnetic materials, which include iron, nickel, and cobalt, exhibit strong magnetic properties due to the fact that their magnetic moments tend to align readily along the direction of an external magnetic field. Moreover, such materials remain partially magnetized even after the removal of the external field. Because of this property, ferromagnetic materials are used in the fabrication of permanent magnets.

Common to all ferromagnetic materials is a characteristic feature described by ***magnetized domains***. A magnetized domain of a material is a microscopic region (on the order of 10^{-10} m^3) within which the magnetic moments of all its atoms (typically on the order of 10^{19} atoms) are aligned parallel to each other. This permanent alignment is attributed to strong coupling forces between

Table 5-2: Properties of magnetic materials.

	Diamagnetism	Paramagnetism	Ferromagnetism
Permanent magnetic dipole moment	No	Yes, but weak	Yes, and strong
Primary magnetization mechanism	Electron orbital magnetic moment	Electron spin magnetic moment	Magnetized domains
Direction of induced magnetic field (relative to external field)	Opposite	Same	Hysteresis [see Fig. 5-22]
Common substances	Bismuth, copper, diamond, gold, lead, mercury, silver, silicon	Aluminum, calcium, chromium, magnesium, niobium, platinum, tungsten	Iron, nickel, cobalt
Typical value of χ_m **Typical value of** μ_r	$\approx -10^{-5}$ ≈ 1	$\approx 10^{-5}$ ≈ 1	$\lvert\chi_m\rvert \gg 1$ and hysteretic $\lvert\mu_r\rvert \gg 1$ and hysteretic

the magnetic dipole moments constituting an individual domain. In the absence of an external magnetic field, the domains take on random orientations relative to each other, as shown in Fig. 5-21(a), resulting in a net magnetization of zero. The *domain walls* forming the boundaries between adjacent domains consist of thin transition regions. When an unmagnetized sample of a ferromagnetic material is placed in an external magnetic field, the domains will align partially with the external field, as illustrated in Fig. 5-21(b). A quantitative understanding of how these domains form and how they behave under the influence of an external magnetic field requires a heavy dose of quantum mechanics, which is outside the scope of the present treatment. Hence, we will confine our discussion to a qualitative description of the magnetization process and its implications.

The magnetization behavior of a ferromagnetic material is described in terms of its B–H *magnetization curve*, where H is the amplitude of the externally applied magnetic field, $\mathbf{H}$, and B is the amplitude of the total magnetic flux density $\mathbf{B}$ present within the material. According to

Eq. (5.72), the flux $\mathbf{B}$ consists of a contribution $\mu_0\mathbf{H}$ due to the external field and a contribution $\mu_0\mathbf{M}$ due to the magnetization field induced in the material. Suppose that we start with an unmagnetized sample of iron, and let us assume that we have an experimental arrangement capable of measuring $\mathbf{B}$ and $\mathbf{H}$. The unmagnetized state is denoted by point O in Fig. 5-22. As we start to increase $\mathbf{H}$ continuously, $\mathbf{B}$ increases also, and the response follows the curve from point O to point A_1, at which nearly all the domains have become aligned with $\mathbf{H}$. Point A_1 represents a saturation condition. If we now start to decrease $\mathbf{H}$ from its value at point A_1 back to zero, the magnetization curve follows the path from A_1 to A_2. At point A_2, the external field $\mathbf{H}$ is zero, but the flux density $\mathbf{B}$ in the material is not zero. This value of $\mathbf{B}$ is called the *residual flux density* B_r. The iron material is now magnetized and can serve as a permanent magnet owing to the fact that a large fraction of its magnetization domains have remained aligned. Reversing the direction of $\mathbf{H}$ and increasing its intensity causes $\mathbf{B}$ to decrease from B_r at point A_2 to zero at point A_3, and if the

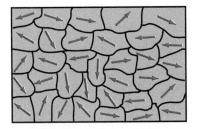

(a) Unmagnetized domains

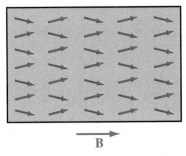

B

(b) Magnetized domains

Figure 5-21: Comparison of (a) unmagnetized and (b) magnetized domains in a ferromagnetic material.

intensity of **H** is increased further (while maintaining its negative direction), the magnetization curve moves to the saturation condition at point A_4. Finally, as **H** is made to return to zero and is then increased again in the positive direction, the curve follows the path from A_4 to A_1. This process is called *magnetic hysteresis*. The term hysteresis means "to lag behind." The *hysteresis loop* shows that the magnetization process in ferromagnetic materials depends not only on the external magnetic field **H**, but on the magnetic history of the material as well. The specific shape and extent of the hysteresis loop depend on the properties of the ferromagnetic material and on the peak-to-peak range over which **H** is made to vary. Materials characterized by wide hysteresis loops are called *hard ferromagnetic materials* [Fig. 5-23(a)]. These materials cannot be easily demagnetized by an external magnetic field because they have a large residual magnetization B_r. Hard ferromagnetic materials are used in the fabrication of permanent magnets for motors and generators. *Soft ferromagnetic materials* have narrow hysteresis loops [Fig. 5-23(b)], and hence they can be more easily magnetized and demagnetized. To demagnetize any ferromagnetic material, the material

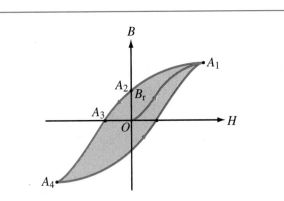

Figure 5-22: Typical hysteresis curve for a ferromagnetic material.

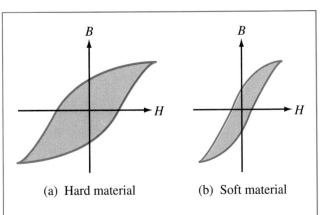

(a) Hard material   (b) Soft material

Figure 5-23: Comparison of hysteresis curves for (a) a hard ferromagnetic material and (b) a soft ferromagnetic material.

is subjected to several hysteresis cycles while gradually decreasing the peak-to-peak range of the applied field.

REVIEW QUESTIONS

Q5.12   What are the three types of magnetic materials and what are typical values of their relative magnetic permeabilities?

Q5.13   What causes magnetic hysteresis in ferromagnetic materials?

Q5.14   What does the magnetization curve describe? What is the difference between the magnetization curves of hard and soft ferromagnetic materials?

5-7 Magnetic Boundary Conditions

In Section 4-9, we derived a set of boundary conditions that describes how, at the boundary between two dissimilar contiguous media, the electric field quantities **D** and **E** in the first medium are related to those in the second medium. We will now derive a similar set of boundary conditions for the magnetic field quantities **B** and **H**. By applying Gauss's law to a pill box that straddles the boundary, we determined that the difference between the normal components of the electric flux densities in the two media is equal to the surface charge density ρ_s. That is,

$$\oint_S \mathbf{D} \cdot d\mathbf{s} = Q \implies \boxed{D_{1n} - D_{2n} = \rho_s.} \quad (5.78)$$

By analogy, application of Gauss's law for magnetism, as expressed by Eq. (5.44), would lead to the conclusion that

$$\oint_S \mathbf{B} \cdot d\mathbf{s} = 0 \implies \boxed{B_{1n} = B_{2n}.} \quad (5.79)$$

This result states that *the normal component of* **B** *is continuous across the boundary between two adjacent media.* In view of the relations $\mathbf{B}_1 = \mu_1 \mathbf{H}_1$ and $\mathbf{B}_2 = \mu_2 \mathbf{H}_2$ for linear, isotropic media, the boundary condition for **H** corresponding to Eq. (5.79) is

$$\boxed{\mu_1 H_{1n} = \mu_2 H_{2n}.} \quad (5.80)$$

Comparison of Eqs. (5.78) and (5.79) tells us that, *whereas the normal component of* **B** *is continuous across the boundary, the normal component of* **D** *may not be (unless $\rho_s = 0$).* A similar reversal applies to the tangential components of the electric and magnetic fields **E** and **H**: *whereas the tangential component of* **E** *is continuous across the boundary, the tangential component of* **H** *may not be (unless the surface current density* $\mathbf{J}_s = 0$*).* To obtain the boundary condition for the tangential component of **H**, we follow the same basic procedure that we used previously in Section 4-9 to establish the boundary condition for the tangential component of **E**. With reference to Fig. 5-24, if we apply Ampère's law [Eq. (5.47)] to a closed rectangular path with sides Δl and Δh, and then we let $\Delta h \to 0$, we end up with the result

$$\oint_C \mathbf{H} \cdot d\mathbf{l} = \int_a^b \mathbf{H}_2 \cdot d\mathbf{l} + \int_c^d \mathbf{H}_1 \cdot d\mathbf{l} = I, \quad (5.81)$$

where $\mathbf{H}_1$ and $\mathbf{H}_2$ are the magnetic fields in media 1 and 2, respectively. According to Ampère's law, I is the net current crossing the surface of the loop in the direction specified by the right-hand rule (I is in the direction of the thumb when the fingers of the right hand extend in the direction of the loop C). For the directions of $\mathbf{H}_1$ and $\mathbf{H}_2$ and the direction of the integration contour C indicated in Fig. 5-24, the component of $\mathbf{H}_2$ tangential to the boundary, $\mathbf{H}_{2t}$, is parallel to and in the same direction as $d\mathbf{l}$ over segment ab, but the tangential component of $\mathbf{H}_1$ is antiparallel to $d\mathbf{l}$ over segment cd. Furthermore, as we let Δh of the loop approach zero, the surface of the loop approaches a thin line of length Δl. Hence, the

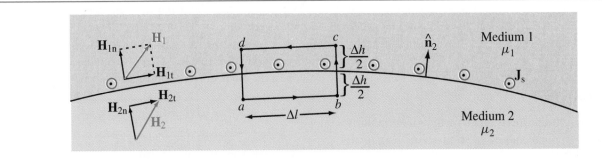

Figure 5-24: Boundary between medium 1 with μ_1 and medium 2 with μ_2.

total current flowing through this thin line is $I = J_s \, \Delta l$, where J_s is the magnitude of the normal component of the surface current density traversing the loop. In view of these considerations, Eq. (5.81) becomes

$$H_{2t} \, \Delta l - H_{1t} \, \Delta l = J_s \, \Delta l \qquad (5.82)$$

or

$$H_{2t} - H_{1t} = J_s \quad \text{(A/m)}. \qquad (5.83)$$

This result can be generalized to a vector form that incorporates the directional relationship defined by the right-hand rule,

$$\boxed{\hat{\mathbf{n}}_2 \times (\mathbf{H}_1 - \mathbf{H}_2) = \mathbf{J}_s,} \qquad (5.84)$$

where $\hat{\mathbf{n}}_2$ is the normal unit vector pointing away from medium 2 (Fig. 5-24). Surface currents can exist only on the surfaces of perfect conductors and superconductors. Hence, *at the interface between media with finite conductivities,* $\mathbf{J}_s = 0$ and

$$\boxed{H_{1t} = H_{2t}.} \qquad (5.85)$$

EXERCISE 5.12 With reference to Fig. 5-24, determine the angle between $\mathbf{H}_1$ and $\hat{\mathbf{n}}_2 = \hat{\mathbf{z}}$ if $\mathbf{H}_2 = (\hat{\mathbf{x}}3 + \hat{\mathbf{z}}2)$ (A/m), $\mu_{r_1} = 2$, and $\mu_{r_2} = 8$, and $\mathbf{J}_s = 0$.

Ans. $\theta = 20.6°$.

5-8   Inductance

An inductor is the magnetic analogue of an electrical capacitor. Just as a capacitor can store electric energy in the electric field present in the medium between its conducting surfaces, an inductor can store magnetic energy in the volume comprising the inductors. A typical example of an inductor is a coil consisting of multiple turns of wire wound in a helical geometry around a cylindrical core, as shown in Fig. 5-25(a). Such a structure is called a *solenoid*. The core may be air filled or may contain a magnetic material with magnetic permeability μ. If the wire carries a current I and the turns are closely spaced, the solenoid can produce a relatively uniform magnetic field within its interior region, and its magnetic field pattern resembles that of a permanent magnet, as illustrated by the field lines in Fig. 5-25(b).

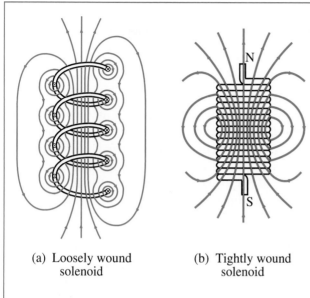

(a) Loosely wound
solenoid

(b) Tightly wound
solenoid

Figure 5-25: Magnetic field lines of (a) a loosely wound solenoid and (b) a tightly wound solenoid.

loop of $n\,dz$ turns and carrying a current $I' = In\,dz$, then the induced field at point P is

$$d\mathbf{B} = \mu\,d\mathbf{H} = \hat{\mathbf{z}}\frac{\mu nIa^2}{2(a^2 + z^2)^{3/2}}\,dz. \qquad (5.87)$$

The total field $\mathbf{B}$ at P is obtained by integrating the contributions from the entire length of the solenoid. This is facilitated by expressing the variable z in terms of the angle θ. That is,

$$z = a\tan\theta, \qquad (5.88a)$$
$$a^2 + z^2 = a^2 + a^2\tan^2\theta = a^2\sec^2\theta, \qquad (5.88b)$$
$$dz = a\sec^2\theta\,d\theta. \qquad (5.88c)$$

Upon substituting the last two expressions in Eq. (5.87) and integrating from θ_1 to θ_2, we have

$$\mathbf{B} = \hat{\mathbf{z}}\frac{\mu nIa^2}{2}\int_{\theta_1}^{\theta_2}\frac{a\sec^2\theta\,d\theta}{a^3\sec^3\theta}$$

$$= \hat{\mathbf{z}}\frac{\mu nI}{2}(\sin\theta_2 - \sin\theta_1). \qquad (5.89)$$

5-8.1 Magnetic Field in a Solenoid

We precede our discussion of inductance by deriving an expression for the magnetic flux density $\mathbf{B}$ in the interior region of a tightly wound solenoid with n turns per unit length. Even though the turns are slightly helical in shape, we will treat them as circular loops, as shown in Fig. 5-26. The solenoid is of length l and radius a and carries a current I. Let us start by considering the magnetic flux density $\mathbf{B}$ at point P, located on the axis of the solenoid. In Example 5-4, we derived the following expression for the magnetic field $\mathbf{H}$ at a distance z along the axis of a circular loop of radius a:

$$\mathbf{H} = \hat{\mathbf{z}}\frac{I'a^2}{2(a^2 + z^2)^{3/2}}, \qquad (5.86)$$

where I' is the current carried by the loop. If we treat an incremental length dz of the solenoid as an equivalent

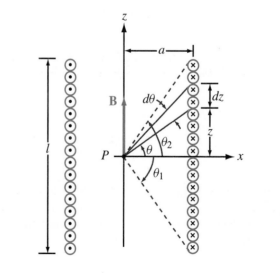

Figure 5-26: Solenoid cross section showing geometry for calculating $\mathbf{H}$ at a point P on the solenoid axis.

If the solenoid length l is much larger than its radius a, then $\theta_1 \simeq -90°$ and $\theta_2 \simeq 90°$, in which case Eq. (5.89) reduces to

$$\mathbf{B} \simeq \hat{\mathbf{z}}\mu n I = \frac{\hat{\mathbf{z}}\mu N I}{l} \quad \text{(long solenoid with } l/a \gg 1\text{),}$$
$$(5.90)$$

where $N = nl$ is the total number of turns over the length l. Even though the expression given by Eq. (5.90) was derived for the field $\mathbf{B}$ at the midpoint of the solenoid, it is approximately valid at all points in the interior of the solenoid, except near the ends.

We now return to a discussion of inductance, which includes *self-inductance*, representing the magnetic flux linkage of a coil or circuit with itself, and *mutual inductance*, which involves the magnetic flux linkage in a circuit due to the magnetic field generated by a current in another circuit. Usually, when the term *inductance* is used, the intended reference is to self-inductance.

EXERCISE 5.13 Use Eq. (5.89) to obtain an expression for $\mathbf{B}$ at a point on the axis of a very long solenoid but situated at its end points. How does $\mathbf{B}$ at the end points compare to $\mathbf{B}$ at the midpoint of the solenoid?

Ans. $\mathbf{B} = \hat{\mathbf{z}}(\mu N I / 2l)$ at the end points, which is half as large as $\mathbf{B}$ at the midpoint.

5-8.2 Self-inductance

From Eq. (5.66), the magnetic flux Φ linking a surface S is given by

$$\Phi = \int_S \mathbf{B} \cdot d\mathbf{s} \quad \text{(Wb).} \quad (5.91)$$

In a solenoid with an approximately uniform magnetic field given by Eq. (5.90), the flux linking a single loop is

$$\Phi = \int_S \hat{\mathbf{z}}\left(\mu \frac{N}{l} I\right) \cdot \hat{\mathbf{z}} \, ds = \mu \frac{N}{l} I S, \quad (5.92)$$

where S is the cross-sectional area of the loop. *Magnetic flux linkage* Λ is defined as the total magnetic flux linking a given circuit or conducting structure. If the structure consists of a single conductor with multiple loops, as in the case of the solenoid, Λ is equal to the flux linking all loops of the structure. For a solenoid with N turns,

$$\Lambda = N\Phi = \mu \frac{N^2}{l} I S \quad \text{(Wb).} \quad (5.93)$$

If, on the other hand, the structure consists of two *separate* conductors, as in the case of the parallel-wire and coaxial transmission lines shown in Fig. 5-27, the flux linkage Λ associated with a length l of either line refers to the flux Φ through the closed surface between the two conductors, as highlighted by the shaded areas in Fig. 5-27. In reality, there is also some magnetic flux that passes through the conductors themselves, but it may be ignored by assuming that the currents flow only on the surfaces of the conductors, in which case the magnetic fields inside the conductors are zero. This assumption is justified by the fact that our interest in calculating Λ is for the purpose of determining the inductance of a given structure, and inductance is of interest primarily in the a-c case (i.e., time-varying currents, voltages, and fields). As we will see later in Section 7-5, the current flowing in a conductor under a-c conditions is concentrated within a very thin layer on the skin of the conductor. For the parallel-wire transmission line, the currents flow on the outer surfaces of the wires, and for the coaxial line, the current flows on the outer surface of the inner conductor and on the inner surface of the outer conductor (the current-carrying surfaces are those adjacent to the electric and magnetic fields present in the region between the conductors).

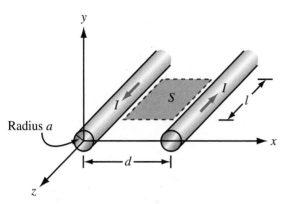

(a) Parallel-wire transmission line

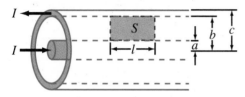

(b) Coaxial transmission line

Figure 5-27: To compute the inductance per unit length of a two-conductor transmission line, we need to determine the magnetic flux through the area S between the conductors.

The *self-inductance* of any conducting structure is defined as the ratio of the magnetic flux linkage Λ to the current I flowing through the structure:

$$L = \frac{\Lambda}{I} \quad \text{(H)}. \quad (5.94)$$

The SI unit for inductance is the henry (H), which is equivalent to webers per ampere (Wb/A).

For a solenoid, use of Eq. (5.93) gives

$$L = \mu \frac{N^2}{l} S \quad \text{(solenoid)}, \quad (5.95)$$

and for two-conductor configurations similar to those of Fig. 5-27,

$$L = \frac{\Lambda}{I} = \frac{\Phi}{I} = \frac{1}{I} \int_S \mathbf{B} \cdot d\mathbf{s}. \quad (5.96)$$

Example 5-8   Inductance of a Coaxial Transmission Line

Develop an expression for the inductance per unit length of a coaxial transmission line. The conductors have radii a and b, as shown in Fig. 5-28, and the insulating material has a linear permeability μ.

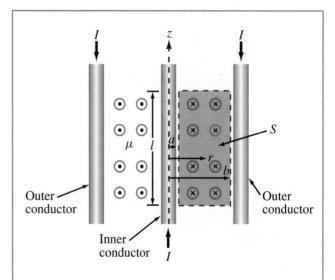

Figure 5-28: Cross-sectional view of coaxial transmission line (Example 5-8).

Solution: Due to the current I in the inner conductor, the magnetic field generated in the region with permeability μ between the two conductors is given by Eq. (5.30) as

$$\mathbf{B} = \hat{\boldsymbol{\phi}} \frac{\mu I}{2\pi r} , \qquad (5.97)$$

where r is the radial distance from the axis of the coaxial line. Let us choose a transmission-line segment of length l as shown in Fig. 5-28. Over the planar surface S between the conductors, $\mathbf{B}$ is everywhere perpendicular to the surface. Hence, the flux through S is

$$\Phi = l \int_a^b B \, dr = l \int_a^b \frac{\mu I}{2\pi r} \, dr = \frac{\mu I l}{2\pi} \ln\left(\frac{b}{a}\right). \qquad (5.98)$$

Using Eq. (5.96), the inductance per unit length of the coaxial transmission line is given by

$$\boxed{L' = \frac{L}{l} = \frac{\Phi}{lI} = \frac{\mu}{2\pi} \ln\left(\frac{b}{a}\right). \quad \blacksquare \quad (5.99)}$$

5-8.3   Mutual Inductance

Magnetic coupling between two different conducting structures is described in terms of the mutual inductance between them. For simplicity, let us assume that we have two closed loops with surfaces S_1 and S_2 and a current I_1 flowing through the first loop, as shown in Fig. 5-29. The magnetic field $\mathbf{B}_1$ generated by I_1 results in a flux Φ_{12} through loop 2, given by

$$\Phi_{12} = \int_{S_2} \mathbf{B}_1 \cdot d\mathbf{s}, \qquad (5.100)$$

and if loop 2 consists of N_2 turns all coupled by $\mathbf{B}_1$ in exactly the same way, then the total magnetic flux linkage through loop 2 due to $\mathbf{B}_1$ is

$$\Lambda_{12} = N_2 \Phi_{12} = N_2 \int_{S_2} \mathbf{B}_1 \cdot d\mathbf{s}. \qquad (5.101)$$

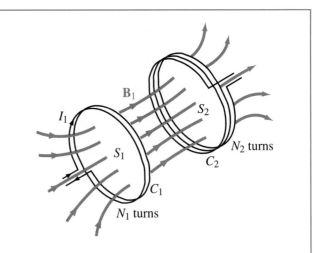

Figure 5-29: Magnetic field lines generated by current I_1 in loop 1 linking surface S_2 of loop 2.

The mutual inductance associated with this magnetic coupling is given by

$$\boxed{L_{12} = \frac{\Lambda_{12}}{I_1} = \frac{N_2}{I_1} \int_{S_2} \mathbf{B}_1 \cdot d\mathbf{s} \quad \text{(H)}. \qquad (5.102)}$$

Mutual inductance is important in transformers wherein the windings of two or more circuits share a common magnetic core, as illustrated by the toroidal arrangement shown in Fig. 5-30.

REVIEW QUESTIONS

Q5.15 What is the magnetic field like in the interior of a long solenoid?

Q5.16 What is the difference between self-inductance and mutual inductance?

Q5.17 How is the inductance of a solenoid related to its number of turns N?

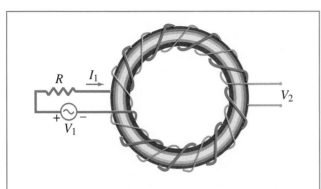

Figure 5-30: Toroidal coil with two windings used as a transformer.

5-9 Magnetic Energy

When we introduced electrostatic energy in Section 4-11, we did so by examining what happens to the energy expended in charging up a capacitor from zero voltage to some final voltage V. We will now use a similar analogy by considering an inductor with inductance L connected to a current source. Suppose that we were to increase the current i flowing through the inductor from zero to a final value I. From circuit theory, we know that the voltage v across the inductor is given by $v = L\,di/dt$. Power p is equal to the product of v and i, and the time integral of power is work, or energy. Hence, the total energy in joules (J) expended in building up the current in the inductor is

$$W_m = \int p\,dt = \int iv\,dt = L \int_0^I i\,di$$
$$= \tfrac{1}{2}LI^2 \quad \text{(J)}. \quad (5.103)$$

For reasons that will become apparent shortly, we call this the *magnetic energy* stored in the inductor.

Let us consider the solenoid inductor. Its inductance is given by Eq. (5.95) as $L = \mu N^2 S/l$, and the magnitude of the magnetic flux density in its interior is given from Eq. (5.90) by $B = \mu NI/l$. Equivalently, $I = Bl/(\mu N)$.

If we use these expressions for L and I in Eq. (5.103), we get

$$W_m = \frac{1}{2}LI^2 = \frac{1}{2}\left(\mu \frac{N^2}{l}S\right)\left(\frac{Bl}{\mu N}\right)^2$$
$$= \frac{1}{2}\frac{B^2}{\mu}(lS) = \frac{1}{2}\mu H^2 \nu, \quad (5.104)$$

where $\nu = lS$ is the volume of the interior of the solenoid and $H = B/\mu$. The *magnetic energy density* w_m is defined as the magnetic energy W_m per unit volume,

$$\boxed{w_m = \frac{W_m}{\nu} = \frac{1}{2}\mu H^2 \quad \text{(J/m}^3\text{)}. \quad (5.105)}$$

Even though this expression was derived for a solenoid inductor, it is equally valid for any medium with magnetic field $\mathbf{H}$. Furthermore, for any volume ν containing a material with permeability μ (including free space with permeability μ_0), the total magnetic energy stored in the medium due to the presence of a magnetic field $\mathbf{H}$ is

$$W_m = \frac{1}{2}\int_\nu \mu H^2\,d\nu \quad \text{(J)}. \quad (5.106)$$

Example 5-9 Magnetic Energy in a Coaxial Cable

Derive an expression for the magnetic energy stored in a coaxial cable of length l and inner and outer radii a and b. The insulation material has permeability μ.

Solution: From Eq. (5.97), the magnitude of the magnetic field in the insulating material is given by

$$H = \frac{B}{\mu} = \frac{I}{2\pi r},$$

where r is the radial distance from the center of the inner conductor, as shown in Fig. 5-28. The magnetic energy stored in the coaxial cable is then given by

$$W_m = \frac{1}{2}\int_\nu \mu H^2\,d\nu = \frac{\mu I^2}{8\pi^2}\int_\nu \frac{1}{r^2}\,d\nu.$$

Since H is a function of r only, we choose dv to be a cylindrical shell of length l, radius r, and thickness dr along the radial direction. Thus, $dv = 2\pi r l \, dr$ and

$$W_{\mathrm{m}} = \frac{\mu I^2}{8\pi^2} \int_a^b \frac{1}{r^2} \cdot 2\pi r l \, dr$$

$$= \frac{\mu I^2 l}{4\pi} \ln\left(\frac{b}{a}\right) \qquad \text{(J).} \quad \blacksquare$$

CHAPTER HIGHLIGHTS

- Magnetic force acting on a charged particle q moving with a velocity $\mathbf{u}$ in a region containing a magnetic flux density $\mathbf{B}$ is $\mathbf{F}_{\mathrm{m}} = q\mathbf{u} \times \mathbf{B}$.
- The total electromagnetic force, known as the Lorentz force, acting on a moving charge in the presence of both electric and magnetic fields is $\mathbf{F} = q(\mathbf{E} + \mathbf{u} \times \mathbf{B})$.
- Magnetic forces acting on current loops can generate magnetic torques.
- The magnetic field intensity induced by a current element is defined by the Biot–Savart law.
- Gauss's law for magnetism states that the net magnetic flux flowing out of any closed surface is zero.
- Ampère's law states that the line integral of $\mathbf{H}$ over a closed contour is equal to the net current crossing the surface bounded by the contour.
- Vector magnetic potential $\mathbf{A}$ is related to $\mathbf{B}$ by $\mathbf{B} = \nabla \times \mathbf{A}$.
- Materials are classified as diamagnetic, paramagnetic, or ferromagnetic, depending on their crystalline structure and behavior under the influence of an external magnetic field.
- Diamagnetic and paramagnetic materials exhibit a linear behavior between $\mathbf{B}$ and $\mathbf{H}$, with $\mu \simeq \mu_0$ for both.
- Ferromagnetic materials exhibit a nonlinear hysteretic behavior between $\mathbf{B}$ and $\mathbf{H}$ and, for some, their μ may be as large as $10^5 \mu_0$.

- At the boundary between two different media, the normal component of $\mathbf{B}$ is continuous, and the tangential components of $\mathbf{H}$ are related by $H_{2\mathrm{t}} - H_{1\mathrm{t}} = J_{\mathrm{s}}$, where J_{s} is the surface current density flowing in a direction orthogonal to $H_{1\mathrm{t}}$ and $H_{2\mathrm{t}}$.
- Inductance of a circuit is defined as the ratio of magnetic flux linking the circuit to the current flowing through it.
- Magnetic energy density is given by $w_{\mathrm{m}} = \frac{1}{2}\mu H^2$.

GLOSSARY OF IMPORTANT TERMS

Provide definitions or explain the meaning of the following terms:

magnetic flux density $\mathbf{B}$
magnetic force $\mathbf{F}_{\mathrm{m}}$
Lorentz force $\mathbf{F}$
moment arm $\mathbf{d}$
torque $\mathbf{T}$
magnetic moment $\mathbf{m}$
Biot–Savart law
current density (volume) $\mathbf{J}$
surface current density $\mathbf{J}_{\mathrm{s}}$
magnetic dipole
Gauss's law for magnetism
Ampère's law
toroidal coil
vector Poisson's equation
magnetic flux Φ
orbital and spin magnetic moments
magnetization vector $\mathbf{M}$
magnetic susceptibility χ_{m}
diamagnetic, paramagnetic, ferromagnetic
magnetized domains
magnetization curve
magnetic hysteresis
hard and soft ferromagnetic materials
inductance (self- and mutual)

solenoid
magnetic energy W_m
magnetic energy density w_m

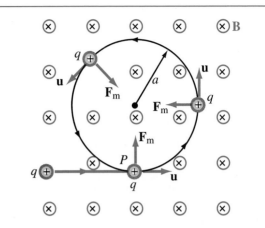

Figure 5-31: Particle of charge q projected with velocity **u** into a medium with a uniform field **B** perpendicular to **u** (Problem 5.2).

PROBLEMS

Section 5-1: Forces and Torques

5.1* An electron with a speed of 4×10^6 m/s is projected along the positive x-direction into a medium containing a uniform magnetic flux density $\mathbf{B} = (\hat{\mathbf{x}}2 - \hat{\mathbf{z}}3)$ T. Given that $e = 1.6 \times 10^{-19}$ C and the mass of an electron is $m_e = 9.1 \times 10^{-31}$ kg, determine the initial acceleration vector of the electron (at the moment it is projected into the medium).

5.2 When a particle with charge q and mass m is introduced into a medium with a uniform field **B** such that the initial velocity of the particle **u** is perpendicular to **B**, as shown in Fig. 5-31, the magnetic force exerted on the particle causes it to move in a circle of radius a. By equating $\mathbf{F}_m$ to the centripetal force on the particle, determine a in terms of q, m, u, and **B**.

5.3* The circuit shown in Fig. 5-32 uses two identical springs to support a 10-cm-long horizontal wire with a mass of 5 g. In the absence of a magnetic field, the weight of the wire causes the springs to stretch a distance of 0.2 cm each. When a uniform magnetic field is turned on in the region containing the horizontal wire, the springs are observed to stretch an additional 0.5 cm. What is the intensity of the magnetic flux density **B**?

5.4 The rectangular loop shown in Fig. 5-33 consists of 20 closely wrapped turns and is hinged along the z-axis. The plane of the loop makes an angle of 30° with the y-axis, and the current in the windings is 0.5 A. What is the magnitude of the torque exerted on the loop in

---

*Answer(s) available in Appendix D.

⊕ Solution available in CD-ROM.

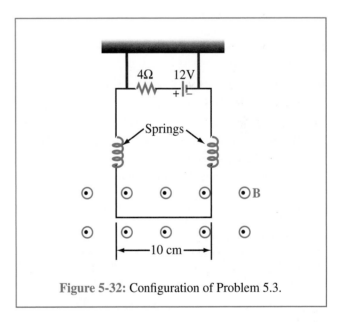

Figure 5-32: Configuration of Problem 5.3.

the presence of a uniform field $\mathbf{B} = \hat{\mathbf{y}} 1.2$ T? When viewed from above, is the expected direction of rotation clockwise or counterclockwise?

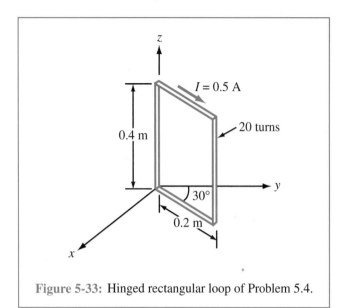

Figure 5-33: Hinged rectangular loop of Problem 5.4.

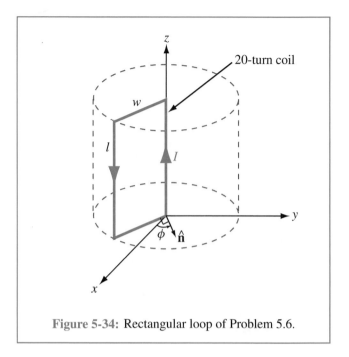

Figure 5-34: Rectangular loop of Problem 5.6.

5.5* In a cylindrical coordinate system, a 2-m-long straight wire carrying a current of 5 A in the positive z-direction is located at $r = 4$ cm, $\phi = \pi/2$, and $-1\text{ m} \le z \le 1\text{ m}$.

(a) If $\mathbf{B} = \hat{\mathbf{r}}\,0.2\cos\phi$ (T), what is the magnetic force acting on the wire?

(b) How much work is required to rotate the wire once about the z-axis in the negative ϕ-direction (while maintaining $r = 4$ cm)?

(c) At what angle ϕ is the force a maximum?

5.6 A 20-turn rectangular coil with sides $l = 15$ cm and $w = 5$ cm is placed in the y–z plane as shown in Fig. 5-34.

(a) If the coil, which carries a current $I = 10$ A, is in the presence of a magnetic flux density

$$\mathbf{B} = 2 \times 10^{-2}(\hat{\mathbf{x}} + \hat{\mathbf{y}}2) \quad (T)$$

determine the torque acting on the coil.

(b) At what angle ϕ is the torque zero?

(c) At what angle ϕ is the torque maximum? Determine its value.

Section 5-2: Biot–Savart Law

5.7* An 8 cm $\times$ 12 cm rectangular loop of wire is situated in the x–y plane with the center of the loop at the origin and its long sides parallel to the x-axis. The loop has a current of 25 A flowing clockwise (when viewed from above). Determine the magnetic field at the center of the loop.

5.8 Use the approach outlined in Example 5-2 to develop an expression for the magnetic field $\mathbf{H}$ at an arbitrary point P due to the linear conductor defined by the geometry shown in Fig. 5-35. If the conductor extends between $z_1 = 3$ m and $z_2 = 7$ m and carries a current $I = 5$ A, find $\mathbf{H}$ at $P(2, \phi, 0)$.

5.9* The loop shown in Fig. 5-36 consists of radial lines and segments of circles whose centers are at point P. Determine the magnetic field $\mathbf{H}$ at P in terms of a, b, θ, and I.

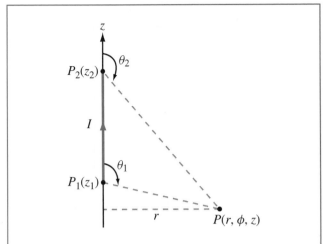

Figure 5-35: Current-carrying linear conductor of Problem 5.8.

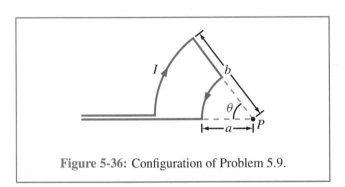

Figure 5-36: Configuration of Problem 5.9.

5.10 An infinitely long, thin conducting sheet defined over the space $0 \leq x \leq w$ and $-\infty \leq y \leq \infty$ is carrying a current with a uniform surface current density $\mathbf{J}_s = \hat{\mathbf{y}}5$ (A/m). Obtain an expression for the magnetic field at point $P(0, 0, z)$ in Cartesian coordinates.

5.11* An infinitely long wire carrying a 50-A current in the positive x-direction is placed along the x-axis in the vicinity of a 10-turn circular loop located in the x–y plane as shown in Fig. 5-37. If the magnetic field at the center of the loop is zero, what is the direction and magnitude of the current flowing in the loop?

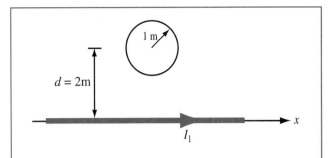

Figure 5-37: Circular loop next to a linear current (Problem 5.11).

5.12 Two infinitely long, parallel wires are carrying 6-A currents in opposite directions. Determine the magnetic flux density at point P in Fig. 5-38.

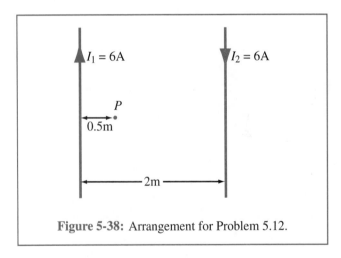

Figure 5-38: Arrangement for Problem 5.12.

5.13* A long, East-West–oriented power cable carrying an unknown current I is at a height of 8 m above the Earth's surface. If the magnetic flux density recorded by a magnetic-field meter placed at the surface is 12 μT when the current is flowing through the cable and 20 μT when the current is zero, what is the magnitude of I?

5.14 Two parallel, circular loops carrying a current of 20 A each are arranged as shown in Fig. 5-39. The first loop is situated in the x–y plane with its center at the origin, and the second loop's center is at $z = 2$ m. If the two loops have the same radius $a = 3$ m, determine the magnetic field at:

(a) $z = 0$

(b) $z = 1$ m

(c) $z = 2$ m

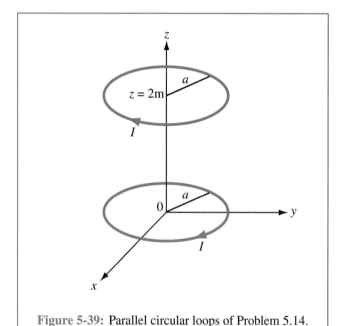

Figure 5-39: Parallel circular loops of Problem 5.14.

Section 5-3: Forces between Currents

5.15* The long, straight conductor shown in Fig. 5-40 lies in the plane of the rectangular loop at a distance $d = 0.1$ m. The loop has dimensions $a = 0.2$ m and $b = 0.5$ m, and the currents are $I_1 = 10$ A and $I_2 = 15$ A. Determine the net magnetic force acting on the loop.

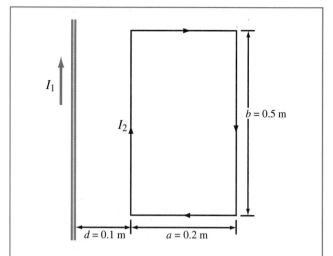

Figure 5-40: Current loop next to a conducting wire (Problem 5.15).

5.16 In the arrangement shown in Fig. 5-41, each of the two long, parallel conductors carries a current I, is supported by 8-cm-long strings, and has a mass per unit length of 0.3 g/cm. Due to the repulsive force acting on the conductors, the angle θ between the supporting strings is 10°. Determine the magnitude of I and the relative directions of the currents in the two conductors.

5.17* An infinitely long, thin conducting sheet of width w along the x-direction lies in the x–y plane and carries a current I in the $-y$-direction. Determine the following:

(a) The magnetic field at a point P midway between the edges of the sheet and at a height h above it. (Fig. 5-42)

(b) The force per unit length exerted on an infinitely long wire passing through point P and parallel to the sheet if the current through the wire is equal in magnitude but opposite in direction to that carried by the sheet.

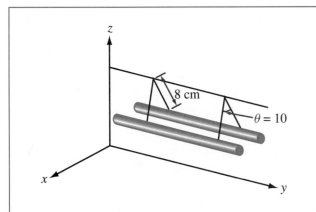

Figure 5-41: Parallel conductors supported by strings (Problem 5.16).

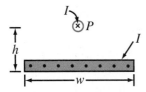

Figure 5-42: A linear current source above a current sheet (Problem 5.17).

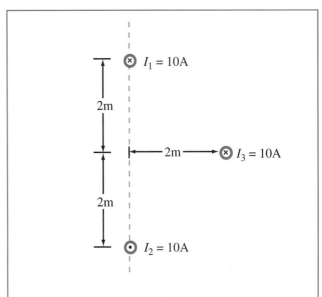

Figure 5-43: Three parallel wires of Problem 5.18.

radius a, and the inner and outer radii of the outer conductor are b and c, respectively.

(a) Determine the magnetic field in each of the following

5.18 Three long, parallel wires are arranged as shown in Fig. 5-43. Determine the force per unit length acting on the wire carrying I_3.

5.19* A square loop placed as shown in Fig. 5-44 has 2-m sides and carries a current $I_1 = 5$ A. If a straight, long conductor carrying a current $I_2 = 10$ A is introduced and placed just above the midpoints of two of the loop's sides, determine the net force acting on the loop.

Section 5-4: Gauss's Law for Magnetism and Ampère's Law

5.20 Current I flows along the positive z-direction in the inner conductor of a long coaxial cable and returns through the outer conductor. The inner conductor has

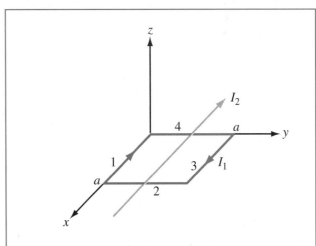

Figure 5-44: Long wire carrying current I_2, just above a square loop carrying I_1 (Problem 5.19).

regions: $0 \leq r \leq a$, $a \leq r \leq b$, $b \leq r \leq c$, and $r \geq c$.

(b) Plot the magnitude of **H** as a function of r over the range from $r = 0$ to $r = 10$ cm, given that $I = 10$ A, $a = 2$ cm, $b = 4$ cm, and $c = 5$ cm.

5.21* A long cylindrical conductor whose axis is coincident with the z-axis has a radius a and carries a current characterized by a current density $\mathbf{J} = \hat{\mathbf{z}} J_0 / r$, where J_0 is a constant and r is the radial distance from the cylinder's axis. Obtain an expression for the magnetic field **H** for

(a) $0 \leq r \leq a$

(b) $r > a$

5.22 Repeat Problem 5.21 for a current density $\mathbf{J} = \hat{\mathbf{z}} J_0 e^{-r}$.

5.23* In a certain conducting region, the magnetic field is given in cylindrical coordinates by

$$\mathbf{H} = \hat{\boldsymbol{\phi}} \frac{4}{r}[1 - (1 + 2r)e^{-2r}]$$

Find the current density **J**.

Section 5-5: Magnetic Potential

5.24 With reference to Fig. 5-10:

(a) Derive an expression for the vector magnetic potential **A** at a point P located at a distance r from the wire in the x–y plane.

(b) Derive **B** from **A**. Show that your result is identical with the expression given by Eq. (5.29), which was derived by applying the Biot–Savart law.

5.25* In a given region of space, the vector magnetic potential is given by $\mathbf{A} = \hat{\mathbf{x}} 5 \cos \pi y + \hat{\mathbf{z}}(2 + \sin \pi x)$ (Wb/m).

(a) Determine **B**.

(b) Use Eq. (5.66) to calculate the magnetic flux passing through a square loop with 0.25-m-long edges if the loop is in the x–y plane, its center is at the origin, and its edges are parallel to the x- and y-axes.

(c) Calculate Φ again using Eq. (5.67).

5.26 A uniform current density given by

$$\mathbf{J} = \hat{\mathbf{z}} J_0 \quad (\text{A/m}^2)$$

gives rise to a vector magnetic potential

$$\mathbf{A} = -\hat{\mathbf{z}} \frac{\mu_0 J_0}{4}(x^2 + y^2) \quad (\text{Wb/m})$$

(a) Apply the vector Poisson's equation to confirm the above statement.

(b) Use the expression for **A** to find **H**.

(c) Use the expression for **J** in conjunction with Ampère's law to find **H**. Compare your result with that obtained in part (b).

5.27* A thin current element extending between $z = -L/2$ and $z = L/2$ carries a current I along $+\hat{\mathbf{z}}$ through a circular cross-section of radius a.

(a) Find **A** at a point P located very far from the origin (assume R is so much larger than L that point P may be considered to be at approximately the same distance from every point along the current element).

(b) Determine the corresponding **H**.

Section 5-6: Magnetic Properties of Materials

5.28 In the model of the hydrogen atom proposed by Bohr in 1913, the electron moves around the nucleus at a speed of 2×10^6 m/s in a circular orbit of radius 5×10^{-11} m. What is the magnitude of the magnetic moment generated by the electron's motion?

5.29* Iron contains 8.5×10^{28} atoms/m^3. At saturation, the alignment of the electrons' spin magnetic moments in iron can contribute 1.5 T to the total magnetic flux density **B**. If the spin magnetic moment of a single electron is 9.27×10^{-24} (A·m^2), how many electrons per atom contribute to the saturated field?

Section 5-7: Magnetic Boundary Conditions

5.30 The x–y plane separates two magnetic media with magnetic permeabilities μ_1 and μ_2, as shown in Fig. 5-45. If there is no surface current at the interface and the magnetic field in medium 1 is

$$\mathbf{H}_1 = \hat{\mathbf{x}} H_{1x} + \hat{\mathbf{y}} H_{1y} + \hat{\mathbf{z}} H_{1z}$$

find:

(a) $\mathbf{H}_2$

(b) θ_1 and θ_2

(c) Evaluate $\mathbf{H}_2, \theta_1$, and θ_2 for $H_{1x} = 3$ (A/m), $H_{1y} = 0$, $H_{1z} = 4$ (A/m), $\mu_1 = \mu_0$, and $\mu_2 = 4\mu_0$

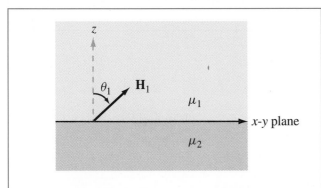

Figure 5-45: Adjacent magnetic media (Problem 5.30).

5.31* Given that a current sheet with surface current density $\mathbf{J}_s = \hat{\mathbf{x}} 4$ (A/m) exists at $y = 0$, the interface between two magnetic media, and $\mathbf{H}_1 = \hat{\mathbf{z}} 8$ (A/m) in medium 1 ($y > 0$), determine $\mathbf{H}_2$ in medium 2 ($y < 0$).

5.32 In Fig. 5-46, the plane defined by $x - y = 1$ separates medium 1 of permeability μ_1 from medium 2 of permeability μ_2. If no surface current exists on the boundary and

$$\mathbf{B}_1 = \hat{\mathbf{x}} 2 + \hat{\mathbf{y}} 3 \quad \text{(T)}$$

find $\mathbf{B}_2$ and then evaluate your result for $\mu_1 = 5\mu_2$.

Hint: Start by deriving the equation for the unit vector normal to the given plane.

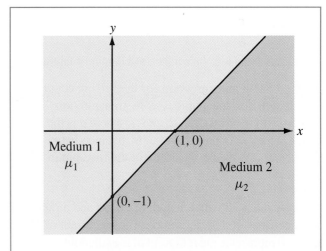

Figure 5-46: Magnetic media separated by the plane $x - y = 1$ (Problem 5.32).

5.33 The plane boundary defined by $z = 0$ separates air from a block of iron. If $\mathbf{B}_1 = \hat{\mathbf{x}} 4 - \hat{\mathbf{y}} 6 + \hat{\mathbf{z}} 8$ in air ($z \geq 0$), find $\mathbf{B}_2$ in iron ($z \leq 0$), given that $\mu = 5000\mu_0$ for iron.

5.34 Show that if no surface current densities exist at the parallel interfaces shown in Fig. 5-47, the relationship between θ_4 and θ_1 is independent of μ_2.

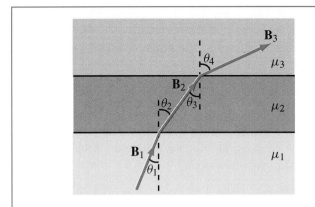

Figure 5-47: Three magnetic media with parallel interfaces (Problem 5.34).

Sections 5-8 and 5-9: Inductance and Magnetic Energy

5.35* Obtain an expression for the self-inductance per unit length for the parallel wire transmission line of Fig. 5-27(a) in terms of a, d, and μ, where a is the radius of the wires, d is the axis-to-axis distance between the wires, and μ is the permeability of the medium in which they reside.

5.36 A solenoid with a length of 20 cm and a radius of 5 cm consists of 400 turns and carries a current of 12 A. If $z = 0$ represents the midpoint of the solenoid, generate a plot for $|\mathbf{H}(z)|$ as a function of z along the axis of the solenoid for the range -20 cm $\le z \le 20$ cm in 1-cm steps.

5.37* In terms of the d-c current I, how much magnetic energy is stored in the insulating medium of a 2-m-long, air-filled section of a coaxial transmission line, given that the radius of the inner conductor is 5 cm and the inner radius of the outer conductor is 10 cm?

5.38 The rectangular loop shown in Fig. 5-48 is coplanar with the long, straight wire carrying the current $I = 20$ A. Determine the magnetic flux through the loop.

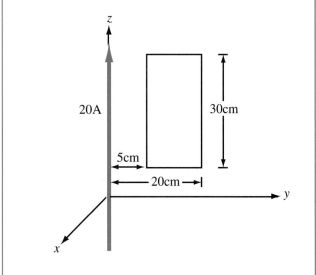

Figure 5-48: Loop and wire arrangement for Problem 5.38.

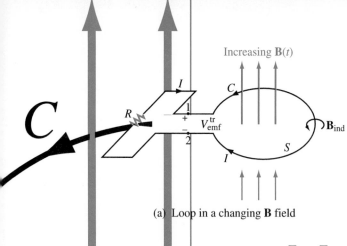

(a) Loop in a changing **B** field

CHAPTER 6

Maxwell's Equations for Time-Varying Fields

Dynamics Fields

DYNAMIC FIELDS

Electric charges induce electric fields and electric currents induce magnetic fields. Those are the lessons we learned in the preceding two chapters. As long as the charge and current distributions remain constant in time, so will the fields that they induce. However, if the charge and current sources were to vary with time t, not only will the fields also vary with time, but much more happens. The electric and magnetic fields become interconnected, and the coupling between them produces electromagnetic waves capable of traveling through free space and in material media. Electromagnetic waves, which include light waves, x-rays, infrared waves, gamma rays, and radio waves [see Fig. 1-15], are an important part of our physical world, and their uses are manifested in many fields of science and technology.

To study time-varying electromagnetic phenomena, we need to use Maxwell's equations as an integrated unit. These equations, which were first introduced in the opening section of Chapter 4, are given in both differential and integral form in Table 6-1. Whereas in the static case ($\partial/\partial t = 0$) we were able to use the first pair of Maxwell's equations to study electrical phenomena in Chapter 4 and the second pair to study magnetic phenomena in Chapter 5, in the dynamic case we have to deal with the coupling that exists between the electric and magnetic fields, as expressed by the second and fourth equations in Table 6-1. The first equation represents Gauss's law, and it is equally valid for static and dynamic fields. The same is true for the third equation, $\nabla \cdot \mathbf{B} = 0$, which basically states that there are no such things as magnetic charges. The second and fourth equations, however, exhibit different meanings for static and dynamic fields. In the dynamic case, a time-varying magnetic field gives rise to an electric field (Faraday's law) and, conversely, a time-varying electric field gives rise to a magnetic field (Ampère's law).

Some of the results we will obtain in this and succeeding chapters might contradict statements made and conclusions reached in Chapter 4 and 5. This is because the earlier material pertains to the special case of steady currents and static charges. When $\partial/\partial t$ is set equal to zero, the results and expressions for the fields under dynamic conditions will reduce to those applicable under static conditions.

We begin this chapter with examinations of Faraday's and Ampère's laws and some of their practical applications. We will then combine Maxwell's equations to obtain relations among the charge and current sources, ρ_v and $\mathbf{J}$, the scalar and vector potentials, V and $\mathbf{A}$, and the electromagnetic fields, $\mathbf{E}$, $\mathbf{D}$, $\mathbf{H}$, and $\mathbf{B}$, for the time-varying case in general and for sinusoidal-time variations in particular.

6-1 Faraday's Law

The close connection between electricity and magnetism was established by Oersted, who demonstrated that a wire carrying an electric current exerts a force on a compass needle and that the needle always turns so as to point in the $\hat{\boldsymbol{\phi}}$-direction when the current is along the $\hat{\mathbf{z}}$-direction. The force acting on the compass needle is due to the magnetic field produced by the current in the wire. Following this discovery, Michael Faraday developed the following hypothesis: if a current can produce a magnetic field, then the converse should also be true: a magnetic field should produce a current in a wire. To prove his hypothesis, he conducted numerous experiments in his

Table 6-1: Maxwell's equations.

Reference	Differential Form	Integral Form	
Gauss's law	$\nabla \cdot \mathbf{D} = \rho_v$	$\oint_S \mathbf{D} \cdot d\mathbf{s} = Q$	(6.1)
Faraday's law	$\nabla \times \mathbf{E} = -\dfrac{\partial \mathbf{B}}{\partial t}$	$\oint_C \mathbf{E} \cdot d\mathbf{l} = -\displaystyle\int_S \dfrac{\partial \mathbf{B}}{\partial t} \cdot d\mathbf{s}$	(6.2)*
No magnetic charges	$\nabla \cdot \mathbf{B} = 0$	$\oint_S \mathbf{B} \cdot d\mathbf{s} = 0$	(6.3)
(Gauss's law for magnetism)			
Ampère's law	$\nabla \times \mathbf{H} = \mathbf{J} + \dfrac{\partial \mathbf{D}}{\partial t}$	$\oint_C \mathbf{H} \cdot d\mathbf{l} = \displaystyle\int_S \left(\mathbf{J} + \dfrac{\partial \mathbf{D}}{\partial t} \right) \cdot d\mathbf{s}$	(6.4)

*For a stationary surface S.

laboratory in London over a period of about 10 years, all aimed at making magnetic fields induce currents in wires. Similar work was being conducted by Joseph Henry in Albany, New York. Wires were placed next to permanent magnets of all different sizes, but no currents were detected in the wires. Current was passed through a wire while another wire was placed parallel to it, with the expectation that the magnetic field of the current-carrying wire would induce a current in the other wire, but again the result was negative. Eventually, these types of experiments led to the true answer, which both Faraday and Henry discovered independently at about the same time (1831). They discovered that indeed *magnetic fields can produce an electric current in a closed loop, but only if the magnetic flux linking the surface area of the loop changes with time.* The key to the induction process is *change*. To explain how the induction process works, let us consider the arrangement shown in Fig. 6-1. A square conducting loop connected to a galvanometer, which is a sensitive instrument used in the 1800s to detect the flow of current in a circuit, is placed next to a conducting coil

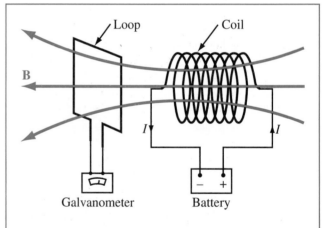

Figure 6-1: The galvanometer shows a deflection whenever the magnetic flux passing through the square loop changes with time.

connected to a battery. The current in the coil produces a magnetic field **B** whose lines pass through the loop as shown in Fig. 6-1. In Section 5-5, we defined the magnetic flux Φ passing through a loop as the integral of the normal component of the magnetic flux density over the surface area of the loop, S, or

$$\Phi = \int_S \mathbf{B} \cdot d\mathbf{s} \quad \text{(Wb)}. \quad (6.5)$$

Under stationary conditions, the d-c current in the coil produces a constant magnetic field $\mathbf{B}$, which in turn produces a constant flux through the loop. When the flux is constant, no current is detected by the galvanometer. However, when the battery is disconnected, thereby interrupting the flow of current in the coil, the magnetic field drops to zero, and the consequent change in magnetic flux causes a momentary deflection of the galvanometer needle. When the battery is reconnected, the galvanometer again exhibits a momentary deflection, but in the opposite direction. Thus, current is induced in the loop when the magnetic flux changes, and the direction of the current depends on whether the flux is increasing (as when the battery is being connected) or decreasing (as when the battery is being disconnected). It was further discovered that current can also flow in the loop, while the battery is connected to the coil, if the loop is turned around suddenly or while moving it closer to or away from the coil. The physical movement of the loop changes the amount of flux linking its surface S, even though the field $\mathbf{B}$ due to the coil has not changed.

A galvanometer is the predecessor of the voltmeter and ammeter. When a galvanometer detects the flow of current through the coil, it means that a voltage has been induced across the galvanometer terminals. This voltage is called the *electromotive force* (emf), V_{emf}, and the process is called *electromagnetic induction*. The emf induced in a closed conducting loop of N turns is given by

$$V_{emf} = -N\frac{d\Phi}{dt} = -N\frac{d}{dt}\int_S \mathbf{B} \cdot d\mathbf{s} \quad \text{(V)}. \quad (6.6)$$

Even though the results leading to Eq. (6.6) were also discovered independently by Henry, Eq. (6.6) is attributed to Faraday and is known as *Faraday's law*. The significance

of the negative sign in Eq. (6.6) will be explained in the next section.

We note that the derivative in Eq. (6.6) is a total time derivative that operates on the magnetic field $\mathbf{B}$, as well as the differential surface area $d\mathbf{s}$. Accordingly, an emf can be generated in a closed conducting loop under any of the following three conditions:

1. *A time-varying magnetic field linking a stationary loop;* the induced emf is then called the **transformer emf**, V_{emf}^{tr}.

2. *A moving loop with a time-varying area (relative to the normal component of $\mathbf{B}$) in a static field $\mathbf{B}$;* the induced emf is then called the **motional emf**, V_{emf}^{m}.

3. *A moving loop in a time-varying field $\mathbf{B}$.*

The total emf is given by

$$V_{emf} = V_{emf}^{tr} + V_{emf}^{m}, \quad (6.7)$$

with $V_{emf}^{m} = 0$ if the loop is stationary (case (1)) and $V_{emf}^{tr} = 0$ if $\mathbf{B}$ is static (case (2)). For case (3), neither term is zero. Each of the three cases will be examined separately in the following sections.

6-2 Stationary Loop in a Time-Varying Magnetic Field

The single-turn, conducting, circular loop with contour C and surface area S shown in Fig. 6-2(a) is in a time-varying magnetic field $\mathbf{B}(t)$. As was stated earlier, the emf induced when S is stationary and the field is time varying is called the **transformer emf** and is denoted V_{emf}^{tr}. Since the loop is stationary, d/dt in Eq. (6.6) now operates on $\mathbf{B}(t)$ only. Hence,

$$V_{emf}^{tr} = -N\int_S \frac{\partial \mathbf{B}}{\partial t} \cdot d\mathbf{s}, \quad (6.8)$$

where the full derivative d/dt has been moved inside the integral and changed into a partial derivative $\partial/\partial t$ to

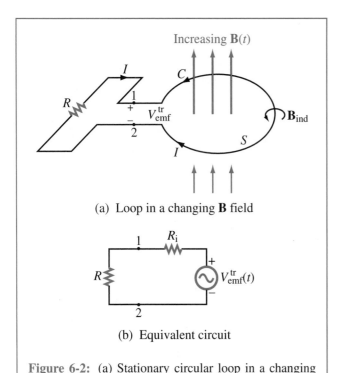

(a) Loop in a changing **B** field

(b) Equivalent circuit

Figure 6-2: (a) Stationary circular loop in a changing magnetic field **B**(t), and (b) its equivalent circuit.

If the loop has an internal resistance R_i, the circuit in Fig. 6-2(a) can be represented by the equivalent circuit shown in Fig. 6-2(b), in which case the current I flowing through the circuit is given by

$$I = \frac{V_{emf}^{tr}}{R + R_i}.$$
(6.9)

For good conductors, R_i usually is very small, and it may be ignored in comparison with practical values of R, except when $R = 0$ (loop is shorted across its ends).

The polarity of V_{emf}^{tr} and hence the direction of I is governed by **Lenz's law**, which states that *the current in the loop is always in such a direction as to oppose the change of magnetic flux* $\Phi(t)$ *that produced it.* The current I induces a magnetic field of its own, **B**$_{ind}$, with a corresponding flux Φ_{ind}. The direction of **B**$_{ind}$ is governed by the right-hand rule; if I is in a clockwise direction, then **B**$_{ind}$ points downward through S and, conversely, if I is in a counterclockwise direction, then **B**$_{ind}$ points upward through S. If the original field **B**(t) is increasing, which means that $d\Phi/dt > 0$, then according to Lenz's law, I has to be in the direction shown in Fig. 6-2(a) in order for **B**$_{ind}$ to be in opposition to **B**(t). Consequently, terminal 2 would be at a higher potential than terminal 1, and V_{emf}^{tr} would have a negative value. However, if **B**(t) were to remain in the same direction but to decrease in magnitude, then $d\Phi/dt$ would become negative, the current would have to reverse direction, and its induced field **B**$_{ind}$ would be in the same direction as **B**(t) so as to oppose the change (decrease) of **B**(t). In that case, V_{emf}^{tr} would be positive. It is important to remember that **B**$_{ind}$ serves to oppose the *change* in **B**(t), and not necessarily **B**(t) itself.

Despite the presence of the small opening between terminals 1 and 2 of the loop in Fig. 6-2(a), we shall treat the loop as a closed path with contour C. We do this in order to establish the link between **B** and the electric field **E** associated with the induced emf, V_{emf}^{tr}. Also, at any point along the loop, the field E is related to the

signify that it operates on **B** only. The transformer emf is the voltage difference that would appear across the small opening between terminals 1 and 2, even in the absence of the resistor R. That is, $V_{emf}^{tr} = V_{12}$, where V_{12} is the open-circuit voltage across the open ends of the loop. Under d-c conditions, $V_{emf}^{tr} = 0$. For the loop shown in Fig. 6-2(a) and the associated definition for V_{emf}^{tr} given by Eq. (6.8), the direction of $d\mathbf{s}$, the loop's differential surface normal, can be chosen to be either upward or downward. The two choices are associated with opposite designations of the polarities of terminals 1 and 2 in Fig. 6-2(a). The connection between the direction of $d\mathbf{s}$ and the polarity of V_{emf}^{tr} is governed by the following right-hand rule: if $d\mathbf{s}$ points along the thumb of the right hand, then the direction of the contour C indicated by the four fingers is such that it always passes across the opening from the positive terminal of V_{emf}^{tr} to the negative terminal.

current I flowing through the loop. For contour C, $V_{\text{emf}}^{\text{tr}}$ is related to $\mathbf{E}$ by

$$V_{\text{emf}}^{\text{tr}} = \oint_C \mathbf{E} \cdot d\mathbf{l}. \qquad (6.10)$$

For $N = 1$ (a loop with one turn), equating Eqs. (6.8) and (6.10) gives

$$\oint_C \mathbf{E} \cdot d\mathbf{l} = -\int_S \frac{\partial \mathbf{B}}{\partial t} \cdot d\mathbf{s}, \qquad (6.11)$$

which is the integral form of Faraday's law given in Table 6-1. We should keep in mind that the direction of the contour C and the direction of $d\mathbf{s}$ are related by the right-hand rule.

By applying Stokes's theorem to the left-hand side of Eq. (6.11), we have

$$\int_S (\nabla \times \mathbf{E}) \cdot d\mathbf{s} = -\int_S \frac{\partial \mathbf{B}}{\partial t} \cdot d\mathbf{s}, \qquad (6.12)$$

and in order for the two integrals to be equal, their integrands have to be equal, which gives

$$\boxed{\nabla \times \mathbf{E} = -\frac{\partial \mathbf{B}}{\partial t} \qquad \text{(Faraday's law).} \qquad (6.13)}$$

This differential form of Faraday's law states that a time-varying magnetic field induces an electric field $\mathbf{E}$ whose curl is equal to the negative of the time derivative of $\mathbf{B}$. Even though the derivation leading to Faraday's law started out by considering the field associated with a physical circuit, Eq. (6.13) applies at any point in space, whether or not a physical circuit exists at that point.

Example 6-1 Inductor in a Changing Magnetic Field

An inductor is formed by winding N turns of a thin conducting wire into a circular loop of radius a. The inductor loop is in the x–y plane with its center at the origin, and it is connected to a resistor R, as shown in Fig. 6-3. In the presence of a magnetic field given by $\mathbf{B} = B_0(\hat{\mathbf{y}}2 + \hat{\mathbf{z}}3) \sin \omega t$, where ω is the angular frequency, find

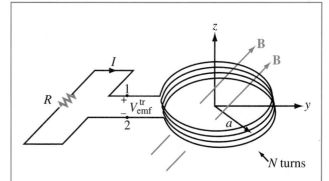

Figure 6-3: Circular loop with N turns in the x–y plane. The magnetic field is $\mathbf{B} = B_0(\hat{\mathbf{y}}2 + \hat{\mathbf{z}}3) \sin \omega t$ (Example 6-1).

(a) the magnetic flux linking a single turn of the inductor,
(b) the transformer emf, given that $N = 10$, $B_0 = 0.2$ T, $a = 10$ cm, and $\omega = 10^3$ rad/s,
(c) the polarity of $V_{\text{emf}}^{\text{tr}}$ at $t = 0$, and
(d) the induced current in the circuit for $R = 1$ kΩ (assume the wire resistance to be negligibly small).

Solution: (a) The magnetic flux linking each turn of the inductor is

$$\Phi = \int_S \mathbf{B} \cdot d\mathbf{s}$$

$$= \int_S [B_0(\hat{\mathbf{y}}2 + \hat{\mathbf{z}}3) \sin \omega t] \cdot \hat{\mathbf{z}} \, ds = 3\pi a^2 B_0 \sin \omega t.$$

(b) To find $V_{\text{emf}}^{\text{tr}}$, we can apply Eq. (6.8) or we can apply the general expression given by Eq. (6.6) directly. The latter approach gives

$$V_{\text{emf}}^{\text{tr}} = -N \frac{d\Phi}{dt}$$

$$= -\frac{d}{dt}(3\pi N a^2 B_0 \sin \omega t) = -3\pi N \omega a^2 B_0 \cos \omega t.$$

For $N = 10$, $a = 0.1$ m, $\omega = 10^3$ rad/s, and $B_0 = 0.2$ T,

$$V_{\text{emf}}^{\text{tr}} = -188.5 \cos 10^3 t \qquad \text{(V).}$$

(c) At $t = 0$, $d\Phi/dt > 0$ and $V_{emf}^{tr} = -188.5$ V. Since the flux is increasing, the current I must be in the direction shown in Fig. 6-3 in order to satisfy Lenz's law. Consequently, point 2 is at a higher potential than point 1 and

$$V_{emf}^{tr} = V_1 - V_2 = -188.5 \quad (V).$$

(d) The current I is given by

$$I = \frac{V_2 - V_1}{R} = \frac{188.5}{10^3}\cos 10^3 t = 0.19\cos 10^3 t \quad (A).$$

∎

EXERCISE 6.1 For the loop shown in Fig. 6-3, what is V_{emf}^{tr} if $\mathbf{B} = \hat{\mathbf{y}}B_0\cos\omega t$? Explain.

Ans. $V_{emf}^{tr} = 0$ because $\mathbf{B}$ is orthogonal to the loop's surface normal $d\mathbf{s}$.

EXERCISE 6.2 Suppose that the loop of Example 6-1 is replaced with a 10-turn square loop centered at the origin and having 20-cm sides oriented parallel to the x- and y-axes. If $\mathbf{B} = \hat{\mathbf{z}}B_0 x^2\cos 10^3 t$ and $B_0 = 100$ T, find the current in the circuit.

Ans. $I = -133\sin 10^3 t$ (mA).

Example 6-2   Lenz's Law

Determine the voltages V_1 and V_2 across the 2-Ω and 4-Ω resistors shown in Fig. 6-4. The loop is located in the x–y plane, its area is 4 m², the magnetic flux density is $\mathbf{B} = -\hat{\mathbf{z}}0.3t$ (T), and the internal resistance of the wire may be ignored.

Solution: The flux flowing through the loop is

$$\Phi = \int_S \mathbf{B}\cdot d\mathbf{s} = \int_S (-\hat{\mathbf{z}}0.3t)\cdot\hat{\mathbf{z}}\,ds$$
$$= -0.3t \times 4 = -1.2t \quad (Wb),$$

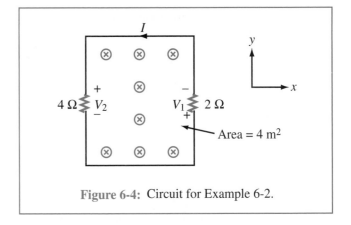

Figure 6-4: Circuit for Example 6-2.

and the corresponding transformer emf is

$$V_{emf}^{tr} = -\frac{d\Phi}{dt} = 1.2 \quad (V).$$

Since the magnetic flux through the loop is along the $-z$-direction (into the page) and it is increasing in magnitude with time t, Lenz's law states that the induced current I should be in a direction such that the magnetic flux density $\mathbf{B}_{ind}$ induced by I counteracts the direction of change of Φ. Hence, I has to be in the direction shown in the circuit because the corresponding $\mathbf{B}_{ind}$ is along the $+z$-direction in the region inside the loop area. This, in turn, means that V_1 and V_2 are positive voltages.

The total voltage of 1.2 V is distributed across two resistors in series. Consequently,

$$I = \frac{V_{emf}^{tr}}{R_1 + R_2} = \frac{1.2}{2 + 4} = 0.2 \text{ A},$$

and

$$V_1 = IR_1 = 0.2 \times 2 = 0.4 \text{ V},$$
$$V_2 = IR_2 = 0.2 \times 4 = 0.8 \text{ V}. \quad∎$$

REVIEW QUESTIONS

Q6.1 Explain Faraday's law and the function of Lenz's law.

Q6.2 Under what circumstances is the net voltage around a closed loop equal to zero?

Q6.3 Suppose the magnetic flux density linking the loop of Fig. 6-4 (Example 6-2) is given by $\mathbf{B} = -\hat{\mathbf{z}}\,0.3e^{-t}$ (T). What would the direction of the current be, relative to that shown in Fig. 6-4, for $t \geq 0$? Explain.

6-3 The Ideal Transformer

The transformer shown in Fig. 6-5(a) consists of two coils wound around a common magnetic core. The coil of the primary circuit has N_1 turns and that of the secondary circuit has N_2 turns. The primary coil is connected to an a-c voltage source $V_1(t)$ and the secondary coil is connected to a load resistor R_L. In an ideal transformer the core has infinite permeability ($\mu = \infty$), and the magnetic flux is confined within the core. The directions of the currents flowing in the two coils, I_1 and I_2, are defined such that, when I_1 and I_2 are both positive, the flux generated by I_2 is opposite that generated by I_1. *The transformer gets its name from the fact that it is used to transform currents, voltages, and impedances between its primary and secondary circuits.*

On the primary side of the transformer, the voltage source V_1 generates a current I_1 in the primary coil, which establishes a flux Φ in the magnetic core. The flux Φ and the voltage V_1 are related by Faraday's law:

$$V_1 = -N_1 \frac{d\Phi}{dt}, \tag{6.14}$$

and, similarly, on the secondary side,

$$V_2 = -N_2 \frac{d\Phi}{dt}. \tag{6.15}$$

The combination of Eqs. (6.14) and (6.15) gives

$$\boxed{\frac{V_1}{V_2} = \frac{N_1}{N_2}.} \tag{6.16}$$

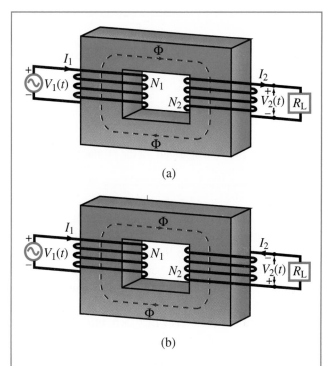

(a)

(b)

Figure 6-5: In a transformer, the directions of I_1 and I_2 are such that the flux Φ generated by one of them is opposite that generated by the other. The direction of the secondary winding in (b) is opposite that in (a), and so are the direction of I_2 and the polarity of V_2.

In an ideal lossless transformer, all the instantaneous power supplied by the source connected to the primary coil is delivered to the load on the secondary side. Thus, no power is lost in the core, and

$$P_1 = P_2. \tag{6.17}$$

Since $P_1 = I_1 V_1$ and $P_2 = I_2 V_2$, and in view of Eq. (6.16), we have

$$\boxed{\frac{I_1}{I_2} = \frac{N_2}{N_1}.} \tag{6.18}$$

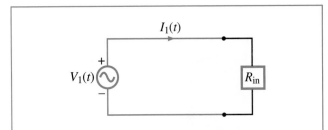

Figure 6-6: Equivalent circuit for the primary side of the transformer.

Thus, whereas the ratio of the voltages given by Eq. (6.16) is proportional to the corresponding turns ratio, the ratio of the currents is equal to the inverse of the turns ratio. If $N_1/N_2 = 0.1$, V_2 of the secondary circuit would be 10 times V_1 of the primary circuit, but I_2 would be only $I_1/10$.

The transformer shown in Fig. 6-5(b) is identical to that in Fig. 6-5(a) except for the direction of the windings of the secondary coil. Because of this change, the direction of I_2 and the polarity of V_2 in Fig. 6-5(b) are the reverse of those in Fig. 6-5(a).

The voltage and current in the secondary circuit in Fig. 6-5(a) are related by $V_2 = I_2 R_L$. To the input circuit, the transformer may be represented by an equivalent input resistance R_{in}, as shown in Fig. 6-6, defined as

$$R_{in} = \frac{V_1}{I_1}. \tag{6.19}$$

Use of Eqs. (6.16) and (6.18) gives

$$R_{in} = \frac{V_2}{I_2}\left(\frac{N_1}{N_2}\right)^2 = \left(\frac{N_1}{N_2}\right)^2 R_L. \tag{6.20}$$

When the load is an impedance Z_L and V_1 is a sinusoidal source, the input resistance representation can be extended to an equivalent input impedance Z_{in} given by

$$Z_{in} = \left(\frac{N_1}{N_2}\right)^2 Z_L. \tag{6.21}$$

6-4 Moving Conductor in a Static Magnetic Field

Consider a wire of length l moving across a static magnetic field $\mathbf{B} = \hat{\mathbf{z}}B_0$ at a constant velocity $\mathbf{u}$, as shown in Fig. 6-7. The conducting wire contains free electrons. From Eq. (5.3), the magnetic force $\mathbf{F}_m$ acting on any charged particle q moving with a velocity $\mathbf{u}$ in a magnetic field $\mathbf{B}$ is given by

$$\mathbf{F}_m = q(\mathbf{u} \times \mathbf{B}). \tag{6.22}$$

This magnetic force is equivalent to the electrical force that would be exerted on the particle by an electric field $\mathbf{E}_m$ given by

$$\mathbf{E}_m = \frac{\mathbf{F}_m}{q} = \mathbf{u} \times \mathbf{B}. \tag{6.23}$$

The field $\mathbf{E}_m$ generated by the motion of the charged particle is called a *motional electric field*, and it is in a direction perpendicular to the plane containing $\mathbf{u}$ and $\mathbf{B}$. For the wire shown in Fig. 6-7, $\mathbf{E}_m$ is along $-\hat{\mathbf{y}}$. The magnetic force acting on the electrons in the wire causes them to move in the direction of $-\mathbf{E}_m$; that is, toward the end labeled 1 in Fig. 6-7. This in turn induces a

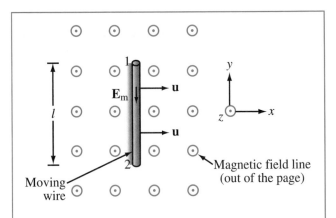

Figure 6-7: Conducting wire moving in a static magnetic field.

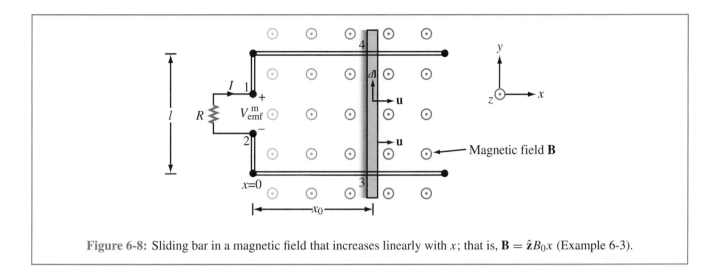

Figure 6-8: Sliding bar in a magnetic field that increases linearly with x; that is, $\mathbf{B} = \hat{\mathbf{z}}B_0 x$ (Example 6-3).

voltage difference between ends 1 and 2, with end 2 being at the higher potential. The induced voltage is called a *motional emf*, $V_{\text{emf}}^{\text{m}}$, and is defined as the line integral of $\mathbf{E}_{\text{m}}$ between ends 2 and 1 of the wire,

$$V_{\text{emf}}^{\text{m}} = V_{12} = \int_2^1 \mathbf{E}_{\text{m}} \cdot d\mathbf{l} = \int_2^1 (\mathbf{u} \times \mathbf{B}) \cdot d\mathbf{l}. \quad (6.24)$$

For the conducting wire, $\mathbf{u} \times \mathbf{B} = \hat{\mathbf{x}}u \times \hat{\mathbf{z}}B_0 = -\hat{\mathbf{y}}uB_0$ and $d\mathbf{l} = \hat{\mathbf{y}}\,dl$. Hence,

$$V_{\text{emf}}^{\text{m}} = V_{12} = -uB_0 l. \quad (6.25)$$

In general, if any segment of a closed circuit with contour C moves with a velocity $\mathbf{u}$ across a static magnetic field $\mathbf{B}$, then the induced motional emf is given by

$$V_{\text{emf}}^{\text{m}} = \oint_C (\mathbf{u} \times \mathbf{B}) \cdot d\mathbf{l}. \quad (6.26)$$

Only those segments of the circuit that cross magnetic field lines contribute to $V_{\text{emf}}^{\text{m}}$.

Example 6-3 Sliding Bar

The rectangular loop shown in Fig. 6-8 has a constant width l, but its length x_0 increases with time as a conducting

bar slides at a uniform velocity $\mathbf{u}$ in a static magnetic field $\mathbf{B} = \hat{\mathbf{z}}B_0 x$. Note that $\mathbf{B}$ increases linearly with x. The bar starts from $x = 0$ at $t = 0$. Find the motional emf between terminals 1 and 2 and the current I flowing through the resistor R. Assume that the loop resistance $R_{\text{i}} \ll R$.

Solution: This problem can be solved by using the motional emf expression given by Eq. (6.26) or by applying the general formula of Faraday's law. We will show that the two approaches yield the same result.

The sliding bar, being the only part of the circuit that crosses the lines of the field $\mathbf{B}$, is the only part of contour 2341 that contributes to $V_{\text{emf}}^{\text{m}}$. Hence, at $x = x_0$,

$$V_{\text{emf}}^{\text{m}} = V_{12} = V_{43} = \int_3^4 (\mathbf{u} \times \mathbf{B}) \cdot d\mathbf{l}$$

$$= \int_3^4 (\hat{\mathbf{x}}u \times \hat{\mathbf{z}}B_0 x_0) \cdot \hat{\mathbf{y}}\,dl = -uB_0 x_0 l.$$

The length of the loop is related to u by $x_0 = ut$. Hence,

$$V_{\text{emf}}^{\text{m}} = -B_0 u^2 lt \quad \text{(V)}. \quad (6.27)$$

Since $\mathbf{B}$ is static, $V_{\text{emf}}^{\text{tr}} = 0$ and $V_{\text{emf}} = V_{\text{emf}}^{\text{m}}$ only. To verify that the same result can be obtained by the general form of

Faraday's law, we start by finding the flux Φ through the surface of the loop. Thus,

$$\Phi = \int_S \mathbf{B} \cdot d\mathbf{s}$$

$$= \int_S (\hat{\mathbf{z}} B_0 x) \cdot \hat{\mathbf{z}} \, dx \, dy$$

$$= B_0 l \int_0^{x_0} x \, dx = \frac{B_0 l x_0^2}{2} . \qquad (6.28)$$

Substituting $x_0 = ut$ in Eq. (6.28) and then taking the negative derivative with respect to time gives

$$V_{\text{emf}} = -\frac{d\Phi}{dt} = -\frac{d}{dt}\left(\frac{B_0 l u^2 t^2}{2}\right)$$

$$= -B_0 u^2 lt \quad (\text{V}), \qquad (6.29)$$

which is identical with Eq. (6.27). Since V_{12} is negative, the current $I = B_0 u^2 lt / R$ flows in the direction shown in Fig. 6-8. ■

Example 6-4   Moving Loop

The rectangular loop shown in Fig. 6-9 is situated in the x–y plane and moves away from the origin at a velocity $\mathbf{u} = \hat{\mathbf{y}} 5$ (m/s) in a magnetic field given by

$$\mathbf{B}(y) = \hat{\mathbf{z}} 0.2 e^{-0.1y} \quad (\text{T}).$$

If $R = 5\,\Omega$, find the current I at the instant that the loop sides are at $y_1 = 2$ m and $y_2 = 2.5$ m. The loop resistance may be ignored.

Solution: Since $\mathbf{u} \times \mathbf{B}$ is along $\hat{\mathbf{x}}$, voltages are induced across only the sides oriented along $\hat{\mathbf{x}}$, namely the side between points 1 and 2 and the side between points 3 and 4. Had $\mathbf{B}$ been uniform, the induced voltages would have been the same and the net voltage across the resistor would have been zero. In the present case, however, $\mathbf{B}$ decreases exponentially with y, thereby assuming a different value over side 1-2 than over side 3-4. Side 1-2 is at $y_1 = 2$ m, and the corresponding magnetic field is

$$\mathbf{B}(y_1) = \hat{\mathbf{z}} 0.2 e^{-0.1 y_1} = \hat{\mathbf{z}} 0.2 e^{-0.2} \quad (\text{T}).$$

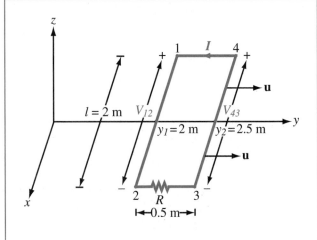

Figure 6-9: Moving loop of Example 6-4.

The induced voltage V_{12} is then given by

$$V_{12} = \int_2^1 [\mathbf{u} \times \mathbf{B}(y_1)] \cdot d\mathbf{l}$$

$$= \int_{l/2}^{-l/2} (\hat{\mathbf{y}} 5 \times \hat{\mathbf{z}} 0.2 e^{-0.2}) \cdot \hat{\mathbf{x}} \, dx$$

$$= -e^{-0.2} l = -2e^{-0.2} = -1.637 \quad (\text{V}).$$

Similarly,

$$V_{43} = -u\, B(y_2)\, l = -5 \times 0.2 e^{-0.25} \times 2$$

$$= -1.558 \quad (\text{V}).$$

Consequently, the current is in the direction shown in the figure and its magnitude is

$$I = \frac{V_{43} - V_{12}}{R} = \frac{0.079}{5} = 15.8\,(\text{mA}). \quad \blacksquare$$

Example 6-5   Moving Rod Next to a Wire

The wire shown in Fig. 6-10 is in free space and carries a current $I = 10$ A. A 30-cm-long metal rod moves at a constant velocity $\mathbf{u} = \hat{\mathbf{z}} 5$ m/s. Find V_{12}.

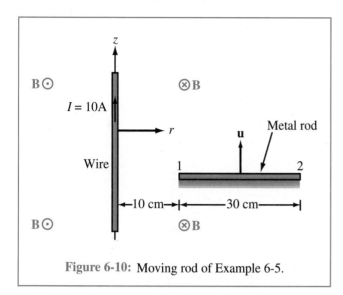

Figure 6-10: Moving rod of Example 6-5.

Solution: The current I induces a magnetic field

$$\mathbf{B} = \hat{\boldsymbol{\phi}} \frac{\mu_0 I}{2\pi r} ,$$

where r is the radial distance from the wire and the direction of $\hat{\boldsymbol{\phi}}$ is into the page at the metal rod side of the wire. The movement of the rod in the presence of the field $\mathbf{B}$ induces a motional emf given by

$$V_{12} = \int_{40\,\text{cm}}^{10\,\text{cm}} (\mathbf{u} \times \mathbf{B}) \cdot d\mathbf{l}$$

$$= \int_{40\,\text{cm}}^{10\,\text{cm}} \left(\hat{\mathbf{z}}5 \times \hat{\boldsymbol{\phi}} \frac{\mu_0 I}{2\pi r} \right) \cdot \hat{\mathbf{r}} \, dr$$

$$= -\frac{5\mu_0 I}{2\pi} \int_{40\,\text{cm}}^{10\,\text{cm}} \frac{dr}{r}$$

$$= -\frac{5 \times 4\pi \times 10^{-7} \times 10}{2\pi} \times \ln\left(\frac{10}{40}\right) = 13.9 \, (\mu\text{V}). \blacksquare$$

EXERCISE 6.3 For the moving loop of Fig. 6-9, find I at the instant that the loop sides are at $y_1 = 4$ m and

$y_2 = 4.5$ m. Also, reverse the direction of motion such that $\mathbf{u} = -\hat{\mathbf{y}}5$ (m/s).

Ans. $I = -13$ (mA).

EXERCISE 6.4 Suppose that we turn the loop of Fig. 6-9 so that its surface is parallel to the x–z plane. What would I be in that case?

Ans. $I = 0$.

REVIEW QUESTIONS

Q6.4 Suppose that no friction is involved in sliding the conducting bar of Fig. 6-8 and that the horizontal arms of the circuit are very long. Hence, if the bar is given an initial push, it should continue moving at a constant velocity, and its movement generates electrical energy in the form of an induced emf, indefinitely. Is this a valid argument? If not, why not? How can we generate electrical energy without having to supply an equal amount of energy by other means?

Q6.5 Is the current flowing in the rod of Fig. 6-10 a steady current? Examine the force on a charge q at ends 1 and 2 and compare.

6-5 The Electromagnetic Generator

The electromagnetic generator is the converse of the electromagnetic motor. The principles of operation of both instruments may be explained with the help of Fig. 6-11. A permanent magnet is used to produce a static magnetic field $\mathbf{B}$ in the slot between the two poles of the magnet. When a current is passed through the conducting loop, as depicted in Fig. 6-11(a), the current flows in opposite directions in segments 1–2 and 3–4 of the loop. The induced magnetic forces on the two segments are also opposite, resulting in a torque that causes the loop to rotate about its axis. Thus, in a motor, electrical energy supplied by a voltage

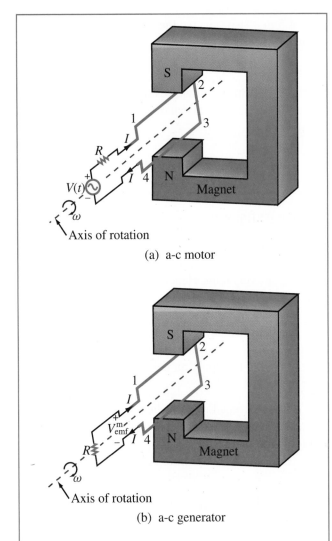

(a) a-c motor

(b) a-c generator

Figure 6-11: Principles of the a-c motor and the a-c generator. In (a) the magnetic torque on the wires causes the loop to rotate, and in (b) the rotating loop generates an emf.

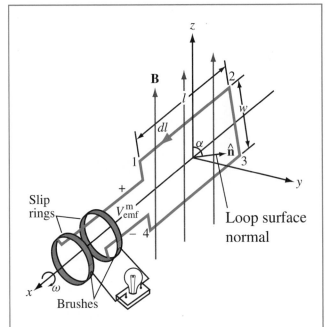

Figure 6-12: A loop rotating in a magnetic field induces an emf.

If, instead of passing a current through the loop to make it turn, the loop is made to rotate by an external force, the movement of the loop in the magnetic field will produce a motional emf, $V_{\text{emf}}^{\text{m}}$, as shown in Fig. 6-11(b). Hence, the motor has become a generator, and mechanical energy is being converted into electrical energy.

Let us examine the operation of the electromagnetic generator in more detail using the coordinate system shown in Fig. 6-12. The magnetic field is given by

$$\mathbf{B} = \hat{\mathbf{z}}B_0, \tag{6.30}$$

and the axis of rotation of the conducting loop is along the x-axis. Segments 1–2 and 3–4 of the loop are of length l each, and both cross the magnetic flux lines as the loop rotates. The other two segments are each of width w, and neither crosses the **B** lines when the loop rotates. Hence, only segments 1–2 and 3–4 contribute to the generation of the motional emf, $V_{\text{emf}}^{\text{m}}$.

source is converted into mechanical energy in the form of a rotating loop, which can be coupled to pulleys, gears, or other movable objects.

As the loop rotates with an angular velocity ω about its own axis, segment 1–2 moves with a velocity $\mathbf{u}$ given by

$$\mathbf{u} = \hat{\mathbf{n}}\omega \frac{w}{2}, \qquad (6.31)$$

where $\hat{\mathbf{n}}$, the surface normal to the loop, makes an angle α with the z-axis. Hence,

$$\hat{\mathbf{n}} \times \hat{\mathbf{z}} = \hat{\mathbf{x}} \sin \alpha. \qquad (6.32)$$

Segment 3–4 moves with a velocity $-\mathbf{u}$. Application of Eq. (6.26), consistent with our choice of $\hat{\mathbf{n}}$, gives

$$V_{\text{emf}}^{\text{m}} = V_{14} = \int_2^1 (\mathbf{u} \times \mathbf{B}) \cdot d\mathbf{l} + \int_4^3 (\mathbf{u} \times \mathbf{B}) \cdot d\mathbf{l}$$

$$= \int_{-l/2}^{l/2} \left[\left(\hat{\mathbf{n}}\omega \frac{w}{2} \right) \times \hat{\mathbf{z}} B_0 \right] \cdot \hat{\mathbf{x}} \, dx$$

$$+ \int_{l/2}^{-l/2} \left[\left(-\hat{\mathbf{n}}\omega \frac{w}{2} \right) \times \hat{\mathbf{z}} B_0 \right] \cdot \hat{\mathbf{x}} \, dx. \quad (6.33)$$

Using Eq. (6.32) in Eq. (6.33), we obtain the result

$$V_{\text{emf}}^{\text{m}} = wl\omega B_0 \sin \alpha = A\omega B_0 \sin \alpha, \qquad (6.34)$$

where $A = wl$ is the surface area of the loop. The angle α is related to ω by

$$\alpha = \omega t + C_0, \qquad (6.35)$$

where C_0 is a constant determined by initial conditions. For example, if $\alpha = 0$ at $t = 0$, then $C_0 = 0$. In general,

$$V_{\text{emf}}^{\text{m}} = A\omega B_0 \sin(\omega t + C_0) \quad \text{(V)}. \qquad (6.36)$$

This result can also be obtained by applying the general form of Faraday's law given by Eq. (6.6). The flux linking the surface of the loop is

$$\Phi = \int_S \mathbf{B} \cdot d\mathbf{s} = \int_S \hat{\mathbf{z}} B_0 \cdot \hat{\mathbf{n}} \, ds$$

$$= B_0 A \cos \alpha$$

$$= B_0 A \cos(\omega t + C_0), \qquad (6.37)$$

and

$$V_{\text{emf}} = -\frac{d\Phi}{dt} = -\frac{d}{dt} [B_0 A \cos(\omega t + C_0)]$$

$$= A\omega B_0 \sin(\omega t + C_0), \qquad (6.38)$$

which is identical with the result given by Eq. (6.36).

REVIEW QUESTIONS

Q6.6 Contrast the operation of an a-c motor with that of an a-c generator.

Q6.7 The rotating loop of Fig. 6-12 had a single turn. What would be the emf generated by a loop with 10 turns?

Q6.8 The magnetic flux linking the loop shown in Fig. 6-12 is a maximum when $\alpha = 0$ (loop in x–y plane), and yet according to Eq. (6.34), the induced emf is zero when $\alpha = 0$. Conversely, when $\alpha = 90°$, the flux linking the loop is zero, but $V_{\text{emf}}^{\text{m}}$ is at a maximum. Is this consistent with your expectations? Why?

6-6 Moving Conductor in a Time-Varying Magnetic Field

For the general case of a single-turn conducting loop moving in a time-varying magnetic field, the induced emf is the sum of a transformer component and a motional component. Thus, the sum of Eqs. (6.8) and (6.26) gives

$$V_{\text{emf}} = V_{\text{emf}}^{\text{tr}} + V_{\text{emf}}^{\text{m}}$$

$$= \oint_C \mathbf{E} \cdot d\mathbf{l}$$

$$= -\int_S \frac{\partial \mathbf{B}}{\partial t} \cdot d\mathbf{s} + \oint_C (\mathbf{u} \times \mathbf{B}) \cdot d\mathbf{l}. \qquad (6.39)$$

V_{emf} is also given by the general expression of Faraday's law:

$$V_{emf} = -\frac{d\Phi}{dt} = -\frac{d}{dt}\int_S \mathbf{B}\cdot d\mathbf{s}. \quad (6.40)$$

In fact, it can be shown mathematically that the right-hand side of Eq. (6.39) is equivalent to the right-hand side of Eq. (6.40). For a particular problem, the choice between using Eq. (6.39) or Eq. (6.40) is usually made on the basis of which is the easier to apply. If the loop consists of N turns, the terms on the right-hand sides of Eqs. (6.39) and (6.40) should be multiplied by N.

Example 6-6 **Electromagnetic Generator**

Find the induced voltage when the rotating loop of the electromagnetic generator of Section 6-5 is in a magnetic field $\mathbf{B} = \hat{\mathbf{z}}B_0 \cos \omega t$. Assume that $\alpha = 0$ at $t = 0$.

Solution: In this case the flux Φ is given by Eq. (6.37) with B_0 replaced with $B_0 \cos \omega t$. Thus,

$$\Phi = B_0 A \cos^2 \omega t,$$

and

$$V_{emf} = -\frac{\partial \Phi}{\partial t}$$

$$= -\frac{\partial}{\partial t}(B_0 A \cos^2 \omega t)$$

$$= 2B_0 A \omega \cos \omega t \sin \omega t = B_0 A \omega \sin 2\omega t. \quad \blacksquare$$

6-7 Displacement Current

From Table 6-1, Ampère's law in differential form is given by

$$\nabla \times \mathbf{H} = \mathbf{J} + \frac{\partial \mathbf{D}}{\partial t} \quad \text{(Ampère's law).} \quad (6.41)$$

If we take the surface integral of both sides of Eq. (6.41) over an arbitrary open surface S with contour C, we have

$$\int_S (\nabla \times \mathbf{H})\cdot d\mathbf{s} = \int_S \mathbf{J}\cdot d\mathbf{s} + \int_S \frac{\partial \mathbf{D}}{\partial t}\cdot d\mathbf{s}. \quad (6.42)$$

The surface integral of $\mathbf{J}$ is equal to the conduction current I_c flowing through S, and the surface integral of $\nabla \times \mathbf{H}$ can be converted into a line integral of $\mathbf{H}$ over the contour C by invoking Stokes's theorem. Hence,

$$\oint_C \mathbf{H}\cdot d\mathbf{l} = I_c + \int_S \frac{\partial \mathbf{D}}{\partial t}\cdot d\mathbf{s} \quad \text{(Ampère's law)}. (6.43)$$

The second term on the right-hand side of Eq. (6.43) has to have the same unit (amperes) as the current I_c, and because it is proportional to the time derivative of the electric flux density $\mathbf{D}$ (which is also called the electric displacement), it is called the *displacement current* I_d. That is,

$$I_d \triangleq \int_S \mathbf{J}_d\cdot d\mathbf{s} = \int_S \frac{\partial \mathbf{D}}{\partial t}\cdot d\mathbf{s}, \quad (6.44)$$

where $\mathbf{J}_d = \partial \mathbf{D}/\partial t$ represents a *displacement current density*. In view of Eq. (6.44),

$$\oint_C \mathbf{H}\cdot d\mathbf{l} = I_c + I_d = I, \quad (6.45)$$

where I is the total current. In electrostatics, $\partial \mathbf{D}/\partial t = 0$ and therefore $I_d = 0$ and $I = I_c$. The concept of displacement current was first introduced by James Clerk Maxwell in 1873 in his successful attempt to establish a

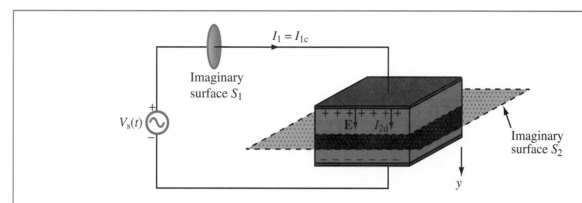

Figure 6-13: The displacement current I_{2d} in the insulating material of the capacitor is equal to the conducting current I_{1c} in the wire.

unified connection between electric and magnetic fields under time-varying conditions.

The parallel-plate capacitor is commonly used as a convenient example to illustrate the physical meaning of the displacement current I_d. The simple circuit shown in Fig. 6-13 consists of a capacitor and an a-c source with voltage $V_s(t)$ given by

$$V_s(t) = V_0 \cos \omega t \quad \text{(V)}. \quad (6.46)$$

According to Eq. (6.45), the total current flowing through any surface consists, in general, of a conduction current I_c and a displacement current I_d. Let us find I_c and I_d through each of the following two imaginary surfaces: (1) the cross section of the conducting wire, S_1, and (2) the cross section of the capacitor (surface S_2 in Fig. 6-13). We shall denote the conduction and displacement currents in the wire I_{1c} and I_{1d} and those through the capacitor I_{2c} and I_{2d}.

In a perfect conductor, $\mathbf{D} = \mathbf{E} = 0$; hence, Eq. (6.44) gives $I_{1d} = 0$ in the wire. As to I_{1c}, we know from circuit theory that it is related to the voltage across the capacitor V_c by

$$I_{1c} = C \frac{dV_c}{dt} = C \frac{d}{dt}(V_0 \cos \omega t) = -CV_0\omega \sin \omega t, \quad (6.47)$$

where we used the fact that $V_c = V_s(t)$. With $I_{1d} = 0$, the total current in the wire is simply $I_1 = I_{1c} = -CV_0\omega \sin \omega t$.

We now consider the currents flowing through surface S_2 in Fig. 6-13, which is an imaginary, open surface parallel to the capacitor plates and situated somewhere between them. The space between the two plates, each of which is of area A, is filled with a perfect dielectric material with permittivity ε. Since electrical charges cannot physically move through a dielectric medium, conduction cannot take place between the conducting plates of the capacitor, and therefore $I_{2c} = 0$. To determine I_{2d}, we need to apply Eq. (6.44). From Example 4-11, the electric field $\mathbf{E}$ in the dielectric spacing is related to the voltage V_c across the capacitor by

$$\mathbf{E} = \hat{\mathbf{y}}\frac{V_c}{d} = \hat{\mathbf{y}}\frac{V_0}{d} \cos \omega t, \quad (6.48)$$

where d is the spacing between the plates and $\hat{\mathbf{y}}$ is the direction from the higher-potential plate toward the lower-potential plate. The displacement current I_{2d} in the direction shown in Fig. 6-13 is obtained by applying

Eq. (6.44) with $d\mathbf{s} = \hat{\mathbf{y}} \, ds$:

$$
\begin{aligned}
I_{2d} &= \int_S \frac{\partial \mathbf{D}}{\partial t} \cdot d\mathbf{s} \\
&= \int_A \left[\frac{\partial}{\partial t} \left(\hat{\mathbf{y}} \frac{\varepsilon V_0}{d} \cos \omega t \right) \right] \cdot (\hat{\mathbf{y}} \, ds) \\
&= -\frac{\varepsilon A}{d} V_0 \omega \sin \omega t = -C V_0 \omega \sin \omega t, \quad (6.49)
\end{aligned}
$$

where we used the relation $C = \varepsilon A / d$ for the capacitance of the parallel-plate capacitor. The expression for I_{2d} in the dielectric region between the conducting plates is identical with that given by Eq. (6.47) for the conduction current I_{1c} in the wire. The fact that these two currents are equal ensures the continuity of current flow through the circuit. *Even though the displacement current does not carry real charge, it nonetheless behaves like a real current.*

In the capacitor example, we treated the wire as a perfect conductor, and we assumed that the spacing between the capacitor plates is a perfect dielectric. If the wire has a finite conductivity σ_w, then $\mathbf{D}$ in the wire would not be zero, and therefore the current I_1 would consist of a conduction current I_{1c} as well as a displacement current I_{1d}; that is, $I_1 = I_{1c} + I_{1d}$. By the same token, if the dielectric spacing material has a nonzero conductivity σ_d, then charges would be able to flow between the two plates and I_{2c} would not be zero. In that case, the total current flowing through the capacitor would be $I_2 = I_{2c} + I_{2d}$, and it would be equal to the total current in the wire. That is, $I_1 = I_2$.

Example 6-7   Displacement Current Density

The conduction current flowing through a wire with conductivity $\sigma = 2 \times 10^7$ S/m and relative permittivity $\varepsilon_r = 1$ is given by $I_c = 2 \sin \omega t$ (mA). If $\omega = 10^9$ rad/s, find the displacement current.

Solution: The conduction current $I_c = JA = \sigma EA$, where A is the cross section of the wire. Hence,

$$
\begin{aligned}
E &= \frac{I_c}{\sigma A} = \frac{2 \times 10^{-3} \sin \omega t}{2 \times 10^7 A} \\
&= \frac{1 \times 10^{-10}}{A} \sin \omega t \quad \text{(V/m)}.
\end{aligned}
$$

Application of Eq. (6.44), with $D = \varepsilon E$, leads to

$$
\begin{aligned}
I_d &= J_d A \\
&= \varepsilon A \, \partial E / \partial t \\
&= \varepsilon A \frac{\partial}{\partial t} \left(\frac{1 \times 10^{-10}}{A} \sin \omega t \right) \\
&= \varepsilon \omega \times 10^{-10} \cos \omega t = 0.885 \times 10^{-12} \cos \omega t \quad \text{(A)},
\end{aligned}
$$

where we used $\omega = 10^9$ rad/s and $\varepsilon = \varepsilon_0 = 8.85 \times 10^{-12}$ F/m. Note that I_c and I_d are in phase quadrature (90° phase shift between them). Also, I_d is about nine orders of magnitude smaller than I_c, which is why the displacement current usually is ignored in good conductors. ∎

EXERCISE 6.5 A poor conductor is characterized by a conductivity $\sigma = 100$ (S/m) and permittivity $\varepsilon = 4\varepsilon_0$. At what angular frequency ω is the amplitude of the conduction current density $\mathbf{J}$ equal to the amplitude of the displacement current density $\mathbf{J}_d$?

Ans. $\omega = 2.82 \times 10^{12}$ (rad/s).

6-8 Boundary Conditions for Electromagnetics

In Chapters 4 and 5 we applied the integral form of Maxwell's equations under static conditions to obtain boundary conditions that the tangential and normal components of $\mathbf{E}, \mathbf{D}, \mathbf{B},$ and $\mathbf{H}$ must satisfy at the interface between two contiguous media. These conditions are given in

Table 6-2: Boundary conditions for the electric and magnetic fields.

Field Components	General Form	Medium 1 Dielectric	Medium 2 Dielectric	Medium 1 Dielectric	Medium 2 Conductor
Tangential E	$\hat{n}_2 \times (\mathbf{E}_1 - \mathbf{E}_2) = 0$	$E_{1t} = E_{2t}$		$E_{1t} = E_{2t} = 0$	
Normal D	$\hat{n}_2 \cdot (\mathbf{D}_1 - \mathbf{D}_2) = \rho_s$	$D_{1n} - D_{2n} = \rho_s$		$D_{1n} = \rho_s$	$D_{2n} = 0$
Tangential H	$\hat{n}_2 \times (\mathbf{H}_1 - \mathbf{H}_2) = \mathbf{J}_s$	$H_{1t} = H_{2t}$		$H_{1t} = J_s$	$H_{2t} = 0$
Normal B	$\hat{n}_2 \cdot (\mathbf{B}_1 - \mathbf{B}_2) = 0$	$B_{1n} = B_{2n}$		$B_{1n} = B_{2n} = 0$	

Notes: (1) ρ_s is the surface charge density at the boundary; (2) $\mathbf{J}_s$ is the surface current density at the boundary; (3) normal components of all fields are along $\hat{n}_2$, the outward unit vector of medium 2; (4) $E_{1t} = E_{2t}$ implies that the tangential components are equal in magnitude and parallel in direction; (5) direction of $\mathbf{J}_s$ is orthogonal to $(\mathbf{H}_1 - \mathbf{H}_2)$.

Section 4-9 for **E** and **D** and in Section 5-7 for **B** and **H**. In the dynamic case, Maxwell's equations [Table 6-1] include two new terms, $\partial \mathbf{B}/\partial t$ in Faraday's law and $\partial \mathbf{D}/\partial t$ in Ampère's law. Nevertheless, *the boundary conditions derived previously for electrostatics and magnetostatics remain valid for time-varying fields as well.* This is because, if we were to apply the same procedures outlined in the above-referenced sections, we would find that the combination of the aforementioned terms vanish as the areas of the rectangular loops in Figs. 4-18 and 5-24 are made to approach zero.

For easy access, the combined set of electromagnetic boundary conditions is given in Table 6-2.

REVIEW QUESTIONS

Q6.9 When conduction current flows through a material, a certain number of charges enter the material on one end and an equal number leave on the other end. What's the situation like for the displacement current through a perfect dielectric?

Q6.10 Verify that the integral form of Ampère's law given by Eq. (6.43) leads to the boundary condition that the tangential component of **H** is continuous across the boundary between two dielectric media.

6-9 Charge—Current Continuity Relation

Under static conditions, the charge density ρ_v and the current density **J** at a given point in a material are totally independent of one another. This is not true in the time-varying case. To show the connection between ρ_v and **J**, we start by considering an arbitrary volume v bounded by a closed surface S, as shown in Fig. 6-14. The net positive charge contained in v is Q. Since, according to the law of conservation of electric charge [Section 1-2.2], charge can neither be created nor destroyed, the only way Q can increase is as a result of a net inward flux of positive charge into the volume v and, by the same token, for Q to decrease there has to be a net outward flux of charge from v. The inward and outward fluxes of charge constitute currents flowing across the surface S into and out of v, respectively. We define I as the *net current flowing across S out of v.* Accordingly, I is equal to the *negative* rate of change of Q:

$$I = -\frac{dQ}{dt} = -\frac{d}{dt} \int_v \rho_v \, dv, \qquad (6.50)$$

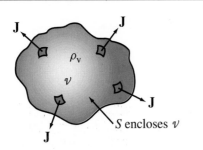

Figure 6-14: Total current flowing out of a volume $\mathcal{V}$ is equal to the flux of current density **J** through the surface S, which in turn is equal to the rate of decrease of the charge enclosed in $\mathcal{V}$.

where ρ_v is the volume charge density in $\mathcal{V}$. According to Eq. (4.12), the current I is also defined as the outward flux of the current density vector **J** through the surface S. Hence,

$$\oint_S \mathbf{J} \cdot d\mathbf{s} = -\frac{d}{dt} \int_{\mathcal{V}} \rho_v \, d\mathcal{V}. \qquad (6.51)$$

By applying the divergence theorem given by Eq. (3.98), we can convert the surface integral of **J** into a volume integral of its divergence $\nabla \cdot \mathbf{J}$, which then gives

$$\oint_S \mathbf{J} \cdot d\mathbf{s} = \int_{\mathcal{V}} \nabla \cdot \mathbf{J} \, d\mathcal{V} = -\frac{d}{dt} \int_{\mathcal{V}} \rho_v \, d\mathcal{V}. \qquad (6.52)$$

For a stationary volume $\mathcal{V}$, the time derivative operates on ρ_v only. Hence, we can move it inside the integral and express it as a partial derivative of ρ_v:

$$\int_{\mathcal{V}} \nabla \cdot \mathbf{J} \, d\mathcal{V} = -\int_{\mathcal{V}} \frac{\partial \rho_v}{\partial t} \, d\mathcal{V}. \qquad (6.53)$$

In order for the volume integrals on the two sides of Eq. (6.53) to be equal for any volume $\mathcal{V}$, their integrands have to be equal at every point within $\mathcal{V}$. Hence,

$$\boxed{\nabla \cdot \mathbf{J} = -\frac{\partial \rho_v}{\partial t}, \qquad (6.54)}$$

which is known as the *charge–current continuity relation* or simply as the *charge continuity equation*.

If the volume charge density within an elemental volume $\Delta \mathcal{V}$ (such as a small cylinder) is not a function of time (i.e., $\partial \rho_v / \partial t = 0$), it means that the net current flowing out of $\Delta \mathcal{V}$ is zero or, equivalently, that the current flowing into $\Delta \mathcal{V}$ is equal to the current flowing out of it. In this case, Eq. (6.54) becomes

$$\nabla \cdot \mathbf{J} = 0, \qquad (6.55)$$

and its integral-form equivalent [from Eq. (6.51)] is

$$\boxed{\oint_S \mathbf{J} \cdot d\mathbf{s} = 0 \quad \text{(Kirchhoff's current law)}. \qquad (6.56)}$$

Let us examine the meaning of Eq. (6.56) by considering a junction (or node) connecting two or more branches in an electric circuit. No matter how small, the junction has a volume $\mathcal{V}$ enclosed by a surface S. The junction shown in Fig. 6-15 has been drawn as a cube, and its dimensions have been artificially enlarged to facilitate the present discussion. The junction has six faces (surfaces), which collectively constitute the surface S associated with the closed-surface integration given by Eq. (6.56). For

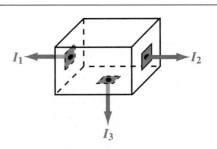

Figure 6-15: Kirchhoff's current law states that the algebraic sum of all the currents flowing out of a junction is zero.

each face, the integration represents the current flowing out through that face. Thus, Eq. (6.56) can be rewritten as

$$\sum_i I_i = 0 \quad \text{(Kirchhoff's current law)}, \quad (6.57)$$

where I_i is the current flowing outward through the ith face. For the junction of Fig. 6-15, Eq. (6.57) translates into $(I_1 + I_2 + I_3) = 0$. In its general form, Eq. (6.57) is an expression of *Kirchhoff's current law*, which states that in an electric circuit *the algebraic sum of all the currents flowing out of a junction is zero.*

6-10 Free-Charge Dissipation in a Conductor

We stated earlier that current flow in a conductor is realized by the movement of loosely attached electrons under the influence of an externally applied electric field. These electrons, however, *are not excess charges*; they are balanced by an equal amount of positive charge in the atoms' nuclei. In other words, the conductor material is electrically neutral, and the net charge density in the conductor is zero ($\rho_v = 0$). What happens then if an excess free charge q is introduced at some interior point in a conductor? The excess charge will give rise to an electric field, which will force the charges of the host material nearest to the excess charge to rearrange their locations, which in turn cause other charges to move, and the process will continue until neutrality is reestablished in the conductor material and a charge equal to q is supplied to the conductor's surface.

How fast does the excess charge dissipate? To answer this question, let us introduce a volume charge density ρ_{vo} at the interior of a conductor and then find out the rate at which it decays down to zero. From Eq. (6.54), the continuity equation is given by

$$\nabla \cdot \mathbf{J} = -\frac{\partial \rho_v}{\partial t} . \quad (6.58)$$

In a conductor, the point form of Ohm's law, given by Eq. (4.67), states that $\mathbf{J} = \sigma \mathbf{E}$. Hence,

$$\sigma \nabla \cdot \mathbf{E} = -\frac{\partial \rho_v}{\partial t} . \quad (6.59)$$

Next, we use Eq. (6.1), $\nabla \cdot \mathbf{E} = \rho_v / \varepsilon$, to obtain the partial differential equation

$$\frac{\partial \rho_v}{\partial t} + \frac{\sigma}{\varepsilon} \rho_v = 0. \quad (6.60)$$

Given that $\rho_v = \rho_{vo}$ at $t = 0$, the solution of Eq. (6.60) is

$$\rho_v(t) = \rho_{vo} e^{-(\sigma/\varepsilon)t} = \rho_{vo} e^{-t/\tau_r} \quad (\text{C/m}^3), \quad (6.61)$$

where $\tau_r = \varepsilon/\sigma$ is called the *relaxation time constant*. We see from Eq. (6.61) that the initial excess charge ρ_{vo} decays exponentially at a rate τ_r. At $t = \tau_r$, the initial charge ρ_{vo} will have decayed to $1/e \approx 37\%$ of its initial value, and at $t = 3\tau_r$, it will have decayed to $e^{-3} \approx 5\%$ of its initial value at $t = 0$. For copper, with $\varepsilon \simeq \varepsilon_0 = 8.854 \times 10^{-12}$ F/m and $\sigma = 5.8 \times 10^7$ S/m, $\tau_r = 1.53 \times 10^{-19}$ s. Thus, the charge dissipation process in a conductor is extremely fast. In contrast, the decay rate is very slow in a good insulator. For a material like mica with $\varepsilon = 6\varepsilon_0$ and $\sigma = 10^{-15}$ S/m, $\tau_r = 5.31 \times 10^4$ s, or approximately 14.8 hours.

REVIEW QUESTIONS

Q6.11 Explain how the charge continuity equation leads to Kirchhoff's current law.

Q6.12 How long is the relaxation time constant for charge dissipation in a perfect conductor? In a perfect dielectric?

EXERCISE 6.6 Determine (a) the relaxation time constant and (b) the time it takes for a charge density to decay to 1% of its initial value in quartz, given that $\varepsilon_r = 5$ and $\sigma = 10^{-17}$ S/m.

Ans. (a) $\tau_r = 51.2$ days, (b) 236 days.

6-11 Electromagnetic Potentials

Through our discussion of Faraday's and Ampère's laws, we examined two aspects of the interconnection that exists between the electric and magnetic fields when the fields are time varying. We will now examine the implications of this interconnection with regard to the electric scalar potential V and the vector magnetic potential $\mathbf{A}$.

For $\partial/\partial t = 0$, Faraday's law reduces to

$$\nabla \times \mathbf{E} = 0 \quad \text{(electrostatics)}, \qquad (6.62)$$

which states that the electrostatic field $\mathbf{E}$ is conservative. According to the rules of vector calculus, if a vector field $\mathbf{E}$ is conservative, it can be expressed as the gradient of a scalar. Hence, in Chapter 4 we defined $\mathbf{E}$ as

$$\mathbf{E} = -\nabla V \quad \text{(electrostatics)}. \qquad (6.63)$$

In the dynamic case, Faraday's law is given by

$$\nabla \times \mathbf{E} = -\frac{\partial \mathbf{B}}{\partial t}, \qquad (6.64)$$

and in view of the relation $\mathbf{B} = \nabla \times \mathbf{A}$, Eq. (6.64) becomes

$$\nabla \times \mathbf{E} = -\frac{\partial}{\partial t}(\nabla \times \mathbf{A}), \qquad (6.65)$$

which can be rewritten as

$$\nabla \times \left(\mathbf{E} + \frac{\partial \mathbf{A}}{\partial t}\right) = 0 \quad \text{(dynamic case)}. \qquad (6.66)$$

Let us for the moment define

$$\mathbf{E}' = \mathbf{E} + \frac{\partial \mathbf{A}}{\partial t}, \qquad (6.67)$$

in which case Eq. (6.66) becomes

$$\nabla \times \mathbf{E}' = 0. \qquad (6.68)$$

Following the same logic that led to Eq. (6.63) from Eq. (6.62), we define

$$\mathbf{E}' = -\nabla V. \qquad (6.69)$$

Upon substituting Eq. (6.67) for $\mathbf{E}'$ in Eq. (6.69) and then solving for $\mathbf{E}$, we have

$$\boxed{\mathbf{E} = -\nabla V - \frac{\partial \mathbf{A}}{\partial t} \quad \text{(dynamic case).} \qquad (6.70)}$$

Equation (6.70) reduces to Eq. (6.63) in the static case.

When the scalar potential V and the vector potential $\mathbf{A}$ are known, $\mathbf{E}$ can be obtained from Eq. (6.70), and $\mathbf{B}$ can be obtained from

$$\boxed{\mathbf{B} = \nabla \times \mathbf{A}. \qquad (6.71)}$$

Next we examine the relations between the potentials, V and $\mathbf{A}$, and their sources, the charge and current distributions ρ_v and $\mathbf{J}$, in the time-varying case.

6-11.1 Retarded Potentials

Consider the situation depicted in Fig. 6-16. A charge distribution ρ_v exists over a volume v' centered at the origin of a coordinate system. The surrounding medium is a perfect dielectric with permittivity ε. From Eq. (4.48a), the electric potential $V(\mathbf{R})$ at an observation point in space specified by the position vector $\mathbf{R}$ is given by

$$V(\mathbf{R}) = \frac{1}{4\pi\varepsilon} \int_{v'} \frac{\rho_v(\mathbf{R_i})}{R'} \, dv', \qquad (6.72)$$

where $\mathbf{R_i}$ denotes the position vector of an elemental volume $\Delta v'$ containing charge density ρ_v, and $R' = |\mathbf{R} - \mathbf{R_i}|$ is the distance between $\Delta v'$ and the observation point. If

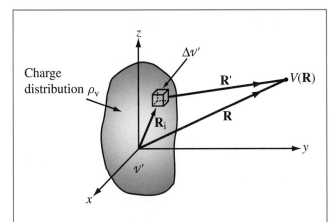

Figure 6-16: Electric potential $V(\mathbf{R})$ due to a charge distribution ρ_v over a volume v'.

the charge distribution is a time-varying function, we may be tempted to rewrite Eq. (6.72) for the dynamic case as

$$V(\mathbf{R}, t) = \frac{1}{4\pi\varepsilon} \int_{v'} \frac{\rho_v(\mathbf{R}_i, t)}{R'} \, dv', \qquad (6.73)$$

but such a form does not account for "reaction time." If V_1 is the potential due to a certain distribution ρ_{v1}, and if ρ_{v1} were to suddenly change to ρ_{v2}, it will take a finite amount of time before V_1 at distance R' changes to V_2. In other words, $V(\mathbf{R}, t)$ cannot change instantaneously. The delay time is equal to $t' = R'/u_p$, where u_p is the velocity of propagation in the medium between the charge distribution and the observation point. Thus, $V(\mathbf{R}, t)$ at time t corresponds to ρ_v at an earlier time, that is, $(t - t')$. Hence, Eq. (6.73) should be rewritten as

$$V(\mathbf{R}, t) = \frac{1}{4\pi\varepsilon} \int_{v'} \frac{\rho_v(\mathbf{R}_i, \ t - R'/u_p)}{R'} \, dv' \qquad \text{(V)}, \qquad (6.74)$$

and $V(\mathbf{R}, t)$ is appropriately called the *retarded scalar potential*. If the propagation medium is vacuum, u_p is equal to the velocity of light c.

Similarly, the *retarded vector potential* $\mathbf{A}(\mathbf{R}, t)$ is related to the distribution of current density $\mathbf{J}$ by

$$\mathbf{A}(\mathbf{R}, t) = \frac{\mu}{4\pi} \int_{v'} \frac{\mathbf{J}(\mathbf{R}_i, \ t - R'/u_p)}{R'} \, dv' \qquad \text{(Wb/m)}. \qquad (6.75)$$

This expression is obtained by extending the expression for the magnetostatic vector potential $\mathbf{A}(\mathbf{R})$ given by Eq. (5.65) to the time-varying case.

6-11.2 Time-Harmonic Potentials

The expressions given by Eqs. (6.74) and (6.75) for the retarded scalar and vector potentials are valid under both static and dynamic conditions and for any type of time dependence that the source functions ρ_v and $\mathbf{J}$ may exhibit. In the dynamic case, ρ_v and $\mathbf{J}$ are linked to one another by the charge continuity relation given by Eq. (6.54). Hence, in general, both quantities will have the same functional dependence on time t and, by extension, so will the potentials V and $\mathbf{A}$ and the fields $\mathbf{E}$, $\mathbf{D}$, $\mathbf{B}$, and $\mathbf{H}$. Furthermore, because V and $\mathbf{A}$ are linearly dependent on ρ_v and $\mathbf{J}$, respectively, and also $\mathbf{E}$ and $\mathbf{B}$ are linearly dependent on V and $\mathbf{A}$, the relationships interconnecting all these quantities obey the rules of linear systems. When a system is linear, we can take advantage of sinusoidal-time functions to determine the response of the system due to a source with any type of time dependence. As was noted in Section 1-6, if the time dependence is described by a (nonsinusoidal) periodic time function, it can always be expanded into a Fourier series of sinusoidal components, and if the time function is nonperiodic, it can be represented by a Fourier integral. In either case, if the response of the linear system is known for the steady-state sinusoidal case, the principle of superposition can be used to determine the response for the specified time function. Thus, the sinusoidal response of the system constitutes a fundamental building block determining the response due to a source described by any function of time. The term *time harmonic* is often used in this context as a synonym for "steady-state sinusoidal time dependence."

In this subsection, we will examine the time-harmonic responses of the retarded scalar and vector potentials. Suppose that $\rho_v(\mathbf{R}_i, t)$ is a sinusoidal-time function with angular frequency ω, given by

$$\rho_v(\mathbf{R}_i, t) = \rho_v(\mathbf{R}_i) \cos \omega t. \qquad (6.76)$$

Phasor analysis, which was first introduced in Section 1-6 and then used extensively in Chapter 2 to study wave propagation on transmission lines, is a useful tool for solving time-harmonic problems. In phasor notation, $\rho_v(\mathbf{R}_i, t)$ is written in the form

$$\rho_v(\mathbf{R}_i, t) = \mathfrak{Re} \left[\tilde{\rho}_v(\mathbf{R}_i) e^{j\omega t} \right], \qquad (6.77)$$

where $\tilde{\rho}_v(\mathbf{R}_i)$ is defined as the phasor function corresponding to the time function $\rho_v(\mathbf{R}_i, t)$. The phasor function is denoted with a tilde ($\sim$) over the letter for easy recognition. Comparison of Eqs. (6.76) and (6.77) shows that in the present case $\tilde{\rho}_v(\mathbf{R}_i) = \rho_v(\mathbf{R}_i)$.

Next, we express the retarded charge density $\rho_v(\mathbf{R}_i, t - R'/u_p)$ in phasor form by replacing t with $(t - R'/u_p)$ in Eq. (6.77):

$$\rho_v(\mathbf{R}_i, t - R'/u_p) = \mathfrak{Re} \left[\tilde{\rho}_v(\mathbf{R}_i) e^{j\omega(t - R'/u_p)} \right]$$

$$= \mathfrak{Re} \left[\tilde{\rho}_v(\mathbf{R}_i) e^{-j\omega R'/u_p} e^{j\omega t} \right]$$

$$= \mathfrak{Re} \left[\tilde{\rho}_v(\mathbf{R}_i) e^{-jkR'} e^{j\omega t} \right] \quad (6.78)$$

where

$$k = \frac{\omega}{u_p} \qquad (6.79)$$

is called the *wavenumber* or phase constant of the propagation medium. (In general, the phase constant is denoted by the symbol "β", but for lossless dielectric media, it is commonly denoted by the symbol "k" and called the

wavenumber.) Similarly, we define the phasor $\tilde{V}(\mathbf{R})$ of the time function $V(\mathbf{R}, t)$ according to

$$V(\mathbf{R}, t) = \mathfrak{Re} \left[\tilde{V}(\mathbf{R}) e^{j\omega t} \right]. \qquad (6.80)$$

Using Eqs. (6.78) and (6.80) in Eq. (6.74) gives

$$\mathfrak{Re} \left[\tilde{V}(\mathbf{R}) e^{j\omega t} \right] =$$

$$\mathfrak{Re} \left[\frac{1}{4\pi\varepsilon} \int_{v'} \frac{\tilde{\rho}_v(\mathbf{R}_i) e^{-jkR'}}{R'} e^{j\omega t} \, dv' \right]. \quad (6.81)$$

By equating the quantities inside the square brackets on the two sides of Eq. (6.81) and then deleting $e^{j\omega t}$ from both sides, we obtain the phasor-domain expression

$$\boxed{\tilde{V}(\mathbf{R}) = \frac{1}{4\pi\varepsilon} \int_{v'} \frac{\tilde{\rho}_v(\mathbf{R}_i) e^{-jkR'}}{R'} \, dv' \quad \text{(V)}. \quad (6.82)}$$

For any given charge distribution, Eq. (6.82) can be used to compute $\tilde{V}(\mathbf{R})$, and then the resultant expression can be used in Eq. (6.80) to find $V(\mathbf{R}, t)$. In like manner, the expression for $\mathbf{A}(\mathbf{R}, t)$ given by Eq. (6.75) can be transformed into

$$\mathbf{A}(\mathbf{R}, t) = \mathfrak{Re} \left[\tilde{\mathbf{A}}(\mathbf{R}) e^{j\omega t} \right] \qquad (6.83)$$

with

$$\boxed{\tilde{\mathbf{A}}(\mathbf{R}) = \frac{\mu}{4\pi} \int_{v'} \frac{\tilde{\mathbf{J}}(\mathbf{R}_i) e^{-jkR'}}{R'} \, dv', \quad (6.84)}$$

where $\tilde{\mathbf{J}}(\mathbf{R}_i)$ is the phasor function corresponding to $\mathbf{J}(\mathbf{R}_i, t)$.

The magnetic field phasor $\tilde{\mathbf{H}}$ corresponding to $\tilde{\mathbf{A}}$ is given by

$$\tilde{\mathbf{H}} = \frac{1}{\mu} \nabla \times \tilde{\mathbf{A}}. \qquad (6.85)$$

Recalling that differentiation in the time domain is equivalent to multiplication by $j\omega$ in the phasor domain, in a nonconducting medium ($\mathbf{J} = 0$), Ampère's law given by Eq. (6.4) becomes

$$\nabla \times \widetilde{\mathbf{H}} = j\omega\varepsilon\widetilde{\mathbf{E}} \quad \text{or} \quad \widetilde{\mathbf{E}} = \frac{1}{j\omega\varepsilon}\nabla \times \widetilde{\mathbf{H}}. \quad (6.86)$$

Hence, given a time-harmonic current-density distribution with phasor $\widetilde{\mathbf{J}}$, Eqs. (6.84) to (6.86) can be used successively to determine both $\widetilde{\mathbf{E}}$ and $\widetilde{\mathbf{H}}$. The phasor vectors $\widetilde{\mathbf{E}}$ and $\widetilde{\mathbf{H}}$ also are related by the phasor form of Faraday's law:

$$\nabla \times \widetilde{\mathbf{E}} = -j\omega\mu\widetilde{\mathbf{H}} \quad \text{or} \quad \widetilde{\mathbf{H}} = -\frac{1}{j\omega\mu}\nabla \times \widetilde{\mathbf{E}}. \quad (6.87)$$

Example 6-8

In a nonconducting medium with $\varepsilon = 16\varepsilon_0$ and $\mu = \mu_0$, the electric field intensity of an electromagnetic wave is given by

$$\mathbf{E}(z, t) = \hat{\mathbf{x}}\, 10 \sin(10^{10}t - kz) \quad \text{(V/m)}. \quad (6.88)$$

Determine the associated magnetic field intensity $\mathbf{H}$ and find the value of k.

Solution: We begin by finding the phasor $\widetilde{\mathbf{E}}(z)$ of $\mathbf{E}(z, t)$. Since $\mathbf{E}(z, t)$ is given as a sine function and phasors are defined in this book with reference to the cosine function, we rewrite Eq. (6.88) as

$$\mathbf{E}(z, t) = \hat{\mathbf{x}}\, 10 \cos(10^{10}t - kz - \pi/2) \quad \text{(V/m)}$$
$$= \mathfrak{Re}\left[\widetilde{\mathbf{E}}(z)\, e^{j\omega t}\right], \quad (6.89)$$

with $\omega = 10^{10}$ (rad/s) and

$$\widetilde{\mathbf{E}}(z) = \hat{\mathbf{x}}\, 10 e^{-jkz} e^{-j\pi/2} = -\hat{\mathbf{x}}\, j 10 e^{-jkz}. \quad (6.90)$$

To find both $\widetilde{\mathbf{H}}(z)$ and k, we will perform a "circle": we will use the given expression for $\widetilde{\mathbf{E}}(z)$ in Faraday's law to find $\widetilde{\mathbf{H}}(z)$; then we will use $\widetilde{\mathbf{H}}(z)$ in Ampère's law to find $\widetilde{\mathbf{E}}(z)$,

which we will then compare with the original expression for $\widetilde{\mathbf{E}}(z)$; and the comparison will yield the value of k. Application of Eq. (6.87) gives

$$\widetilde{\mathbf{H}}(z) = -\frac{1}{j\omega\mu}\nabla \times \widetilde{\mathbf{E}}$$

$$= -\frac{1}{j\omega\mu}\begin{vmatrix} \hat{\mathbf{x}} & \hat{\mathbf{y}} & \hat{\mathbf{z}} \\ \partial/\partial x & \partial/\partial y & \partial/\partial z \\ -j10e^{-jkz} & 0 & 0 \end{vmatrix}$$

$$= -\frac{1}{j\omega\mu}\left[\hat{\mathbf{y}}\frac{\partial}{\partial z}(-j10e^{-jkz})\right]$$

$$= -\hat{\mathbf{y}}j\frac{10k}{\omega\mu}e^{-jkz}. \quad (6.91)$$

So far, we have used Eq. (6.90) for $\widetilde{\mathbf{E}}(z)$ to find $\widetilde{\mathbf{H}}(z)$, but k remains unknown. To find k, we use $\widetilde{\mathbf{H}}(z)$ in Eq. (6.86) to find $\widetilde{\mathbf{E}}(z)$:

$$\widetilde{\mathbf{E}}(z) = \frac{1}{j\omega\varepsilon}\nabla \times \widetilde{\mathbf{H}}$$

$$= \frac{1}{j\omega\varepsilon}\left[-\hat{\mathbf{x}}\frac{\partial}{\partial z}\left(-j\frac{10k}{\omega\mu}e^{-jkz}\right)\right]$$

$$= -\hat{\mathbf{x}}j\frac{10k^2}{\omega^2\mu\varepsilon}e^{-jkz}. \quad (6.92)$$

Equating Eqs. (6.90) and (6.92) leads to

$$k^2 = \omega^2\mu\varepsilon,$$

or

$$k = \omega\sqrt{\mu\varepsilon}$$
$$= 4\omega\sqrt{\mu_0\varepsilon_0}$$
$$= \frac{4\omega}{c} = \frac{4 \times 10^{10}}{3 \times 10^8} = 133 \quad \text{(rad/m)}. \quad (6.93)$$

With k known, the instantaneous magnetic field intensity is then given by

$$\mathbf{H}(z, t) = \mathfrak{Re}\left[\widetilde{\mathbf{H}}(z)\, e^{j\omega t}\right]$$

$$= \mathfrak{Re}\left[-\hat{\mathbf{y}}\, j\, \frac{10k}{\omega\mu}\, e^{-jkz} e^{j\omega t}\right]$$

$$= \hat{\mathbf{y}}\, 0.11 \sin(10^{10}t - 133z) \quad \text{(A/m)}. \quad (6.94)$$

We note that k has the same expression as the phase constant of a lossless transmission line [Eq. (2.39)]. ■

EXERCISE 6.7 The magnetic field intensity of an electromagnetic wave propagating in a lossless medium with $\varepsilon = 9\varepsilon_0$ and $\mu = \mu_0$ is given by

$$\mathbf{H}(z, t) = \hat{\mathbf{x}}\, 0.3 \cos(10^8 t - kz + \pi/4) \quad \text{(A/m)}.$$

Find $\mathbf{E}(z, t)$ and k.

Ans. $\mathbf{E}(z, t) = -\hat{\mathbf{y}}\, 37.7 \cos(10^8 t - z + \pi/4)$ (V/m); $k = 1$ (rad/m).

CHAPTER HIGHLIGHTS

- Faraday's law states that a voltage is induced across the terminals of a loop if the magnetic flux linking its surface changes with time.

- In an ideal transformer, the ratios of the primary to secondary voltages, currents, and impedances are governed by the turns ratio.

- Displacement current accounts for the "apparent" flow of charges through a dielectric. In reality, charges of opposite polarity accumulate along the two ends of a dielectric, giving the appearance of current flow through it.

- Boundary conditions for the electromagnetic fields at the interface between two different media are the same for both static and dynamic conditions.

- The charge continuity equation is a mathematical statement of the law of conservation of electric charge.

- Excess charges in the interior of a good conductor dissipate very quickly; through a rearrangement process, the excess charge is transferred to the surface of the conductor.

- In the dynamic case, the electric field $\mathbf{E}$ is related to both the scalar electric potential V and the magnetic vector potential $\mathbf{A}$.

- The retarded scalar and vector potentials at a given observation point take into account the finite time required for propagation between their sources, the charge and current distributions, and the location of the observation point.

GLOSSARY OF IMPORTANT TERMS

Provide definitions or explain the meaning of the following terms:

Faraday's law
electromotive force V_{emf}
electromagnetic induction
transformer emf $V_{\text{emf}}^{\text{tr}}$
motional emf $V_{\text{emf}}^{\text{m}}$
Lenz's law
displacement current I_{d}
charge continuity equation
Kirchhoff's current law
charge dissipation
relaxation time constant
wavenumber k
retarded potential

PROBLEMS

Sections 6-1 to 6-6: Faraday's Law and its Applications

6.1* The switch in the bottom loop of Fig. 6-17 is closed at $t = 0$ and then opened at a later time t_1. What is the direction of the current I in the top loop (clockwise or counterclockwise) at each of these two times?

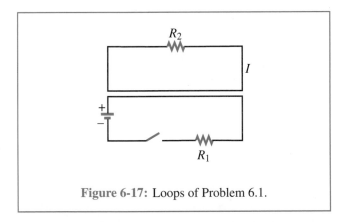

Figure 6-17: Loops of Problem 6.1.

6.2 The loop in Fig. 6-18 is in the x–y plane and $\mathbf{B} = \hat{\mathbf{z}} B_0 \sin \omega t$ with B_0 positive. What is the direction of I ($\hat{\boldsymbol{\phi}}$ or $-\hat{\boldsymbol{\phi}}$) at:

(a) $t = 0$

(b) $\omega t = \pi/4$

(c) $\omega t = \pi/2$

6.3* A coil consists of 100 turns of wire wrapped around a square frame of sides 0.25 m. The coil is centered at the origin with each of its sides parallel to the x- or y-axis. Find the induced emf across the open-circuited ends of the coil if the magnetic field is given by

---

*Answer(s) available in Appendix D.

🌐 Solution available in CD-ROM.

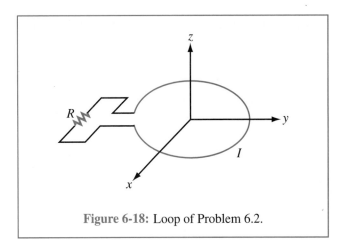

Figure 6-18: Loop of Problem 6.2.

(a) $\mathbf{B} = \hat{\mathbf{z}} 10 e^{-2t}$ (T)

(b) $\mathbf{B} = \hat{\mathbf{z}} 10 \cos x \, \cos 10^3 t$ (T)

(c) $\mathbf{B} = \hat{\mathbf{z}} 10 \cos x \, \sin 2y \, \cos 10^3 t$ (T)

6.4 A stationary conducting loop with an internal resistance of 0.5 Ω is placed in a time-varying magnetic field. When the loop is closed, a current of 2.5 A flows through it. What will the current be if the loop is opened to create a small gap and a 2-Ω resistor is connected across its open ends?

6.5* A circular-loop TV antenna with 0.01-m^2 area is in the presence of a uniform-amplitude 300-MHz signal. When oriented for maximum response, the loop develops an emf with a peak value of 20 (mV). What is the peak magnitude of $\mathbf{B}$ of the incident wave?

6.6 The square loop shown in Fig. 6-19 is coplanar with a long, straight wire carrying a current

$$i(t) = 2.5 \cos 2\pi \times 10^4 t \quad \text{(A)}$$

(a) Determine the emf induced across a small gap created in the loop.

(b) Determine the direction and magnitude of the current that would flow through a 4-Ω resistor connected across the gap. The loop has an internal resistance of 1 Ω.

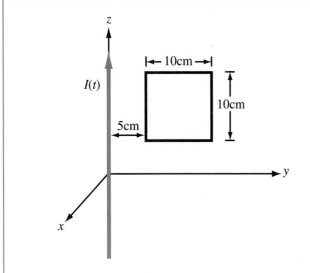

Figure 6-19: Loop coplanar with long wire (Problem 6.6).

6.7* The rectangular conducting loop shown in Fig. 6-20 rotates at 6,000 revolutions per minute in a uniform magnetic flux density given by

$$\mathbf{B} = \hat{\mathbf{y}}\, 50 \quad \text{(mT)}$$

Determine the current induced in the loop if its internal resistance is 0.5 Ω.

6.8 A rectangular conducting loop 5 cm ×10 cm with a small air gap in one of its sides is spinning at 7200 revolutions per minute. If the field **B** is normal to the loop

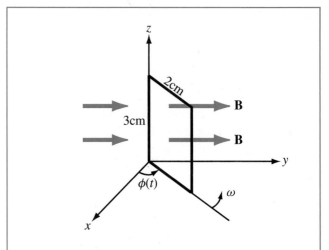

Figure 6-20: Rotating loop in a magnetic field (Problem 6.7).

axis and its magnitude is 5×10^{-6} T, what is the peak voltage induced across the air gap?

6.9* A 50-cm-long metal rod rotates about the z-axis at 180 revolutions per minute, with end 1 fixed at the origin as shown in Fig. 6-21. Determine the induced emf V_{12} if $\mathbf{B} = \hat{\mathbf{z}}\, 3 \times 10^{-4}$ T.

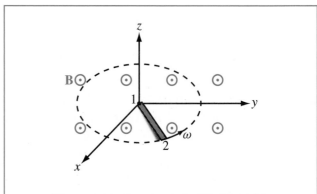

Figure 6-21: Rotating rod of Problem 6.9.

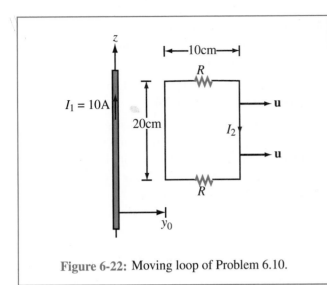

Figure 6-22: Moving loop of Problem 6.10.

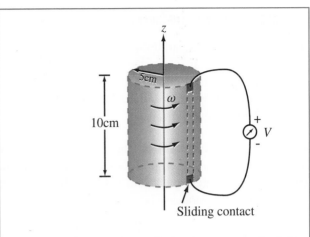

Figure 6-23: Rotating cylinder in a magnetic field (Problem 6.11).

6.10 The loop shown in Fig. 6-22 moves away from a wire carrying a current $I_1 = 10$ A at a constant velocity $\mathbf{u} = \hat{\mathbf{y}}5$ (m/s). If $R = 10\ \Omega$ and the direction of I_2 is as defined in the figure, find I_2 as a function of y_0, the distance between the wire and the loop. Ignore the internal resistance of the loop.

6.11* The conducting cylinder shown in Fig. 6-23 rotates about its axis at 1,200 revolutions per minute in a radial field given by

$$\mathbf{B} = \hat{\mathbf{r}}6 \quad \text{(T)}$$

The cylinder, whose radius is 5 cm and height is 10 cm, has sliding contacts at its top and bottom connected to a voltmeter. Determine the induced voltage.

6.12 The electromagnetic generator shown in Fig. 6-12 is connected to an electric bulb with a resistance of 100 Ω. If the loop area is 0.1 m^2 and it rotates at 3,600 revolutions per minute in a uniform magnetic flux density $B_0 = 0.2$ T, determine the amplitude of the current generated in the light bulb.

6.13* The circular disk shown in Fig. 6-24 lies in the x–y plane and rotates with uniform angular velocity ω about the z-axis. The disk is of radius a and is present in a uniform magnetic flux density $\mathbf{B} = \hat{\mathbf{z}}B_0$. Obtain an expression for the emf induced at the rim relative to the center of the disk.

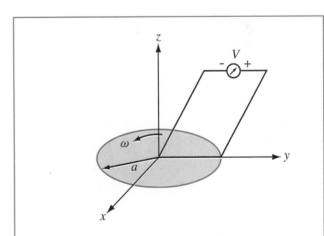

Figure 6-24: Rotating circular disk in a magnetic field (Problem 6.13).

Section 6-7: Displacement Current

6.14 The plates of a parallel-plate capacitor have areas of 10 cm^2 each and are separated by 1 cm. The capacitor is filled with a dielectric material with $\varepsilon = 4\varepsilon_0$, and the voltage across it is given by $V(t) = 20 \cos 2\pi \times 10^6 t$ (V). Find the displacement current.

6.15* A coaxial capacitor of length $l = 6$ cm uses an insulating dielectric material with $\epsilon_r = 9$. The radii of the cylindrical conductors are 0.5 cm and 1 cm. If the voltage applied across the capacitor is

$$V(t) = 100 \sin(120\pi t) \quad \text{(V)}$$

what is the displacement current?

6.16 The parallel-plate capacitor shown in Fig. 6-25 is filled with a lossy dielectric material of relative permittivity ϵ_r and conductivity σ. The separation between the plates is d and each plate is of area A. The capacitor is connected to a time-varying voltage source $V(t)$.

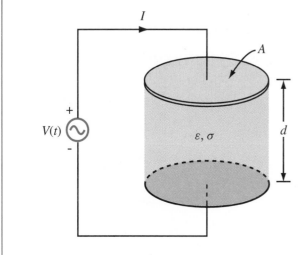

Figure 6-25: Parallel-plate capacitor containing a lossy dielectric material (Problem 6.16).

(a) Obtain an expression for I_c, the conduction current flowing between the plates inside the capacitor, in terms of the given quantities.

(b) Obtain an expression for I_d, the displacement current flowing inside the capacitor.

(c) Based on your expressions for parts (a) and (b), give an equivalent-circuit representation for the capacitor.

(d) Evaluate the values of the circuit elements for $A = 2$ cm^2, $d = 0.5$ cm, $\epsilon_r = 4$, $\sigma = 2.5$ (S/m), and $V(t) = 10 \cos(3\pi \times 10^3 t)$ (V).

6.17 An electromagnetic wave propagating in seawater has an electric field with a time variation given by $\mathbf{E} = \hat{\mathbf{z}}E_0 \cos \omega t$. If the permittivity of water is $81\varepsilon_0$ and its conductivity is 4 (S/m), find the ratio of the magnitudes of the conduction current density to displacement current density at each of the following frequencies:

(a) 1 kHz

(b) 1 MHz

(c) 1 GHz

(d) 100 GHz

Sections 6-9 and 6-10: Continuity Equation and Charge Dissipation

6.18 At $t = 0$, charge density ρ_{v0} was introduced into the interior of a material with a relative permittivity $\varepsilon_r = 4$. If at $t = 1$ μs the charge density has dissipated down to $10^{-3}\rho_{v0}$, what is the conductivity of the material?

6.19* If the current density in a conducting medium is given by

$$\mathbf{J}(x, y, z; t) = (\hat{\mathbf{x}}z - \hat{\mathbf{y}}3y^2 + \hat{\mathbf{z}}2x) \cos \omega t$$

determine the corresponding charge distribution $\rho_v(x, y, z; t)$.

6.20 In a certain medium, the direction of current density **J** points in the radial direction in cylindrical coordinates and its magnitude is independent of both ϕ and z. Determine **J**, given that the charge density in the medium is

$$\rho_v = \rho_0 r \cos \omega t \quad (\text{C/m}^3)$$

6.21 If we were to characterize how good a material is as an insulator by its resistance to dissipating charge, which of the following two materials is the better insulator?

Dry Soil: $\epsilon_r = 2.5$, $\sigma = 10^{-4}$ (S/m)
Fresh Water: $\epsilon_r = 80$, $\sigma = 10^{-3}$ (S/m)

Sections 6-11: Electromagnetic Potentials

6.22 The electric field of an electromagnetic wave propagating in air is given by

$$\mathbf{E}(z, t) = \hat{\mathbf{x}} 4 \cos(6 \times 10^8 t - 2z)$$
$$+ \hat{\mathbf{y}} 3 \sin(6 \times 10^8 t - 2z) \quad (\text{V/m})$$

Find the associated magnetic field $\mathbf{H}(z, t)$.

6.23* The magnetic field in a dielectric material with $\varepsilon = 4\varepsilon_0$, $\mu = \mu_0$, and $\sigma = 0$ is given by

$$\mathbf{H}(y, t) = \hat{\mathbf{x}} 5 \cos(2\pi \times 10^7 t + ky) \quad (\text{A/m})$$

Find k and the associated electric field $\mathbf{E}$.

6.24 Given an electric field

$$\mathbf{E} = \hat{\mathbf{x}} E_0 \sin ay \cos(\omega t - kz)$$

where E_0, a, ω, and k are constants, find **H**.

6.25* The electric field radiated by a short dipole antenna is given in spherical coordinates by

$$\mathbf{E}(R, \theta; t) =$$
$$\hat{\theta} \frac{2 \times 10^{-2}}{R} \sin \theta \; \cos(6\pi \times 10^8 t - 2\pi R) \quad (\text{V/m})$$

Find $\mathbf{H}(R, \theta; t)$.

6.26 A Hertzian dipole is a short conducting wire carrying an approximately constant current over its length l. If such a dipole is placed along the z-axis with its midpoint at the origin, and if the current flowing through it is $i(t) = I_0 \cos \omega t$, find the following:

(a) The retarded vector potential $\tilde{\mathbf{A}}(R, \theta, \phi)$ at an observation point $Q(R, \theta, \phi)$ in a spherical coordinate system.
(b) The magnetic field phasor $\tilde{\mathbf{H}}(R, \theta, \phi)$.

Assume l to be sufficiently small so that the observation point is approximately equidistant to all points on the dipole; that is, assume $R' \simeq R$.

6.27 The magnetic field in a given dielectric medium is given by

$$\mathbf{H} = \hat{\mathbf{y}} 6 \cos 2z \sin(2 \times 10^7 t - 0.1x) \quad (\text{A/m})$$

where x and z are in meters. Determine:

(a) **E**.
(b) The displacement current density $\mathbf{J}_d$.
(c) The charge density ρ_v.

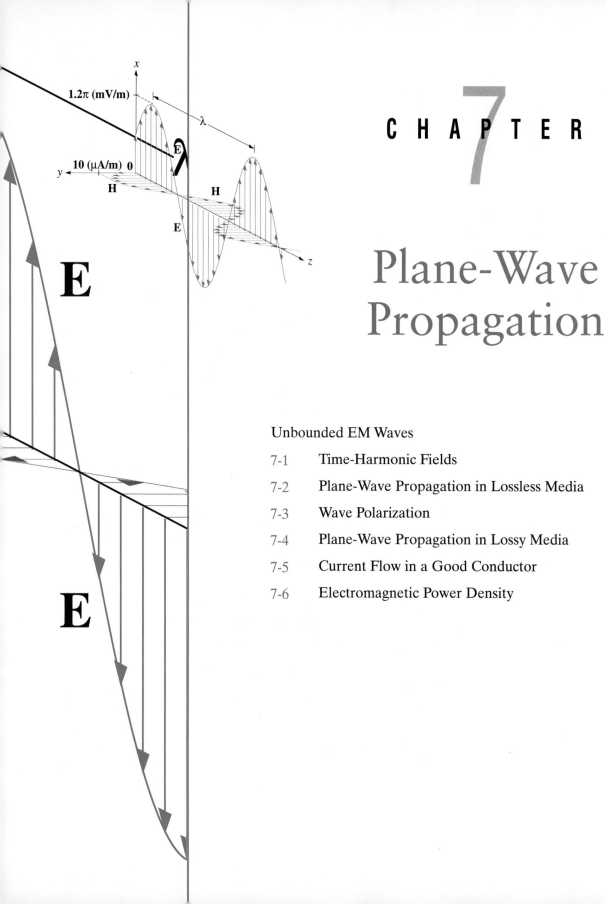

UNBOUNDED EM WAVES

We established in Chapter 6 that a time-varying electric field $\mathbf{E}(t)$ produces a time-varying magnetic field $\mathbf{H}(t)$ and, conversely, a time-varying magnetic field produces an electric field. This cyclic pattern generates electromagnetic (EM) waves capable of propagating through free space and in material media. When its propagation is guided by a material structure, such as a transmission line, the EM wave is said to be traveling in a *guided medium*. Earth's surface and ionosphere constitute parallel boundaries of a natural guiding structure for the propagation of short-wave radio transmissions in the HF band* (3 to 30 MHz); the ionosphere is a good reflector at these frequencies, thereby allowing the waves to zigzag between the two boundaries [Fig. 7-1]. EM waves also can travel in ***unbounded media***; light waves emitted by the sun and radio transmissions by antennas are typical examples.

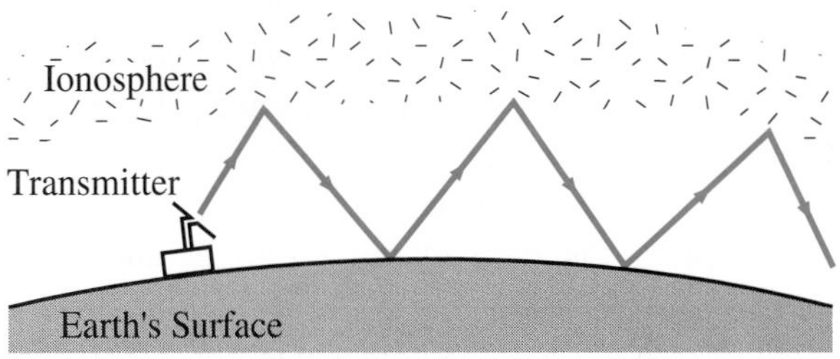

When we discussed wave propagation on a transmission line in Chapter 2, we dealt with voltages and currents. For a transmission-line circuit such as that shown in Fig. 7-2, the a-c voltage source excites an incident wave that travels down the coaxial line toward the load, and unless the load is properly matched to the line, part (or all) of the incident wave is reflected back toward the generator. At any point z on the line, the instantaneous total voltage $v(z, t)$ is the sum of the incident and reflected waves, both of which vary sinusoidally with time. Associated with the voltage difference between the inner and outer conductors of the coaxial line is a radial electric field $\mathbf{E}$ that exists in the dielectric material between the conductors, and since $v(z, t)$ varies sinusoidally with time, so does $\mathbf{E}(z, t)$. Furthermore, the

---

*See Fig. 1-16.

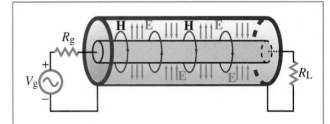

Figure 7-1: The atmospheric layer bounded by the ionosphere at the top and Earth's surface at the bottom forms a guiding structure for the propagation of radio waves in the HF band.

Figure 7-2: A guided electromagnetic wave traveling in a coaxial transmission line consists of time-varying electric and magnetic fields in the dielectric medium between the inner and outer conductors.

current flowing through the inner conductor induces an azimuthal magnetic field $\mathbf{H}(z, t)$ in the dielectric material surrounding it. These coupled fields, $\mathbf{E}(z, t)$ and $\mathbf{H}(z, t)$, constitute electromagnetic waves. Thus, we can model wave propagation on such a transmission line either in terms of the voltages across the line and the currents through its conductors or in terms of the electric and magnetic fields in the dielectric medium between the conductors.

In this chapter we focus our attention on wave propagation in unbounded media. We will consider both lossless and lossy media. Wave propagation in a *lossless medium* (perfect dielectric such as air) is similar to that on a lossless transmission line. In a *lossy medium* characterized by a nonzero conductivity, such as water, part of the power carried by the EM wave gets converted into heat, just like what happens to a wave propagating on a lossy transmission line.

When energy is emitted by a source, such as an antenna, it expands outwardly from the source in the form of *spherical waves*, as depicted in Fig. 7-3(a). Even though the antenna may radiate more energy along some directions than along others, the spherical wave travels at the same speed in all directions and therefore expands at the same rate. To an observer very far away from the source, the *wavefront* of the spherical wave appears approximately *planar*, as if it were part of a *uniform plane wave* with uniform properties at all points in the plane tangent to the wavefront [Fig. 7-3(b)]. Plane-wave propagation can be accommodated by Cartesian coordinates, which are easier to work with mathematically than the spherical coordinates needed for describing the propagation of a spherical wave. Hence, even though strictly speaking a uniform plane wave cannot exist, we will nonetheless use it in this chapter to develop a physical understanding for wave propagation in lossless and lossy media, and then in Chapter 8 we will examine how waves, both planar and spherical, are reflected by and transmitted through boundaries between dissimilar media. The processes of radiation and reception of waves by antennas are treated in Chapter 9.

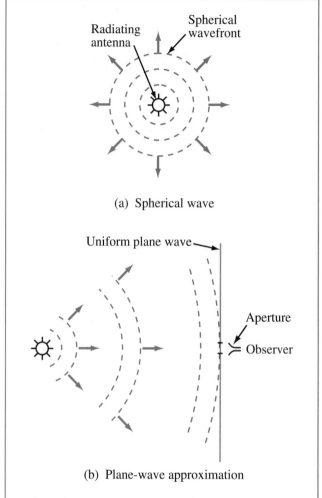

(a) Spherical wave

(b) Plane-wave approximation

Figure 7-3: Waves radiated by an EM source, such as a light bulb or an antenna, have spherical wavefronts, as in (a); to a distant observer, however, the wavefront across the observer's aperture appears approximately planar, as in (b).

7-1 Time-Harmonic Fields

In the time-varying case, the electric and magnetic fields, $\mathbf{E}$, $\mathbf{D}$, $\mathbf{B}$, and $\mathbf{H}$, and their sources, the charge density ρ_v and current density $\mathbf{J}$, are each, in general, a function of the spatial coordinates (x, y, z) and the time variable t. If their time variation is a sinusoidal function with an angular frequency ω, each of these quantities can be represented by a time-independent phasor that depends on (x, y, z) only. Thus, the vector phasor $\widetilde{\mathbf{E}}(x, y, z)$ corresponding to the instantaneous field $\mathbf{E}(x, y, z; t)$ is defined according to

$$\mathbf{E}(x, y, z; t) = \mathfrak{Re}\left[\widetilde{\mathbf{E}}(x, y, z)\, e^{j\omega t}\right], \qquad (7.1)$$

and similar definitions apply to the other fields and to ρ_v and $\mathbf{J}$. For a linear, isotropic, and homogeneous medium characterized by electrical permittivity ε, magnetic permeability μ, and conductivity σ, and recalling that differentiation in the time domain corresponds to multiplication by $j\omega$ in the phasor domain, Maxwell's equations given by Eqs. (6.1) to (6.4) assume the following form in the phasor domain:

$$\boxed{\begin{aligned} \nabla \cdot \widetilde{\mathbf{E}} &= \tilde{\rho}_v/\varepsilon, & (7.2a) \\ \nabla \times \widetilde{\mathbf{E}} &= -j\omega\mu\widetilde{\mathbf{H}}, & (7.2b) \\ \nabla \cdot \widetilde{\mathbf{H}} &= 0, & (7.2c) \\ \nabla \times \widetilde{\mathbf{H}} &= \widetilde{\mathbf{J}} + j\omega\varepsilon\widetilde{\mathbf{E}}, & (7.2d) \end{aligned}}$$

where we have used the relations $\mathbf{D} = \varepsilon\mathbf{E}$ and $\mathbf{B} = \mu\mathbf{H}$. This set of equations defines the starting point for the subject matter treated in this chapter.

7-1.1 Complex Permittivity

In a medium with conductivity σ, the current density $\mathbf{J}$ is related to $\mathbf{E}$ by $\mathbf{J} = \sigma\mathbf{E}$. Consequently, Eq. (7.2d) may be written as

$$\begin{aligned} \nabla \times \widetilde{\mathbf{H}} &= \widetilde{\mathbf{J}} + j\omega\varepsilon\widetilde{\mathbf{E}} \\ &= (\sigma + j\omega\varepsilon)\widetilde{\mathbf{E}} = j\omega\left(\varepsilon - j\frac{\sigma}{\omega}\right)\widetilde{\mathbf{E}}. \quad (7.3) \end{aligned}$$

By introducing the *complex permittivity* ε_c defined as

$$\boxed{\varepsilon_c \triangleq \varepsilon - j\frac{\sigma}{\omega}, \qquad (7.4)}$$

Eq. (7.3) can be rewritten as

$$\nabla \times \widetilde{\mathbf{H}} = j\omega\varepsilon_c\widetilde{\mathbf{E}}. \qquad (7.5)$$

The complex permittivity ε_c given by Eq. (7.4) is often written in terms of a real part ε' and an imaginary part ε''. Thus,

$$\varepsilon_c = \varepsilon - j\frac{\sigma}{\omega} \triangleq \varepsilon' - j\varepsilon'', \qquad (7.6)$$

with

$$\varepsilon' = \varepsilon, \qquad (7.7a)$$

$$\varepsilon'' = \frac{\sigma}{\omega}. \qquad (7.7b)$$

For a lossless medium with $\sigma = 0$, it follows that $\varepsilon'' = 0$ and $\varepsilon_c = \varepsilon' = \varepsilon$.

7-1.2 Wave Equations for a Charge-Free Medium

A medium is said to be charge free if it contains no excess charges, that is, if $\rho_v = 0$. Upon replacing Eq. (7.2d) with Eq. (7.5) and setting $\rho_v = 0$ in Eq. (7.2a), Maxwell's equations for a charge-free medium become

$$\boxed{\begin{aligned} \nabla \cdot \widetilde{\mathbf{E}} &= 0, & (7.8a) \\ \nabla \times \widetilde{\mathbf{E}} &= -j\omega\mu\widetilde{\mathbf{H}}, & (7.8b) \\ \nabla \cdot \widetilde{\mathbf{H}} &= 0, & (7.8c) \\ \nabla \times \widetilde{\mathbf{H}} &= j\omega\varepsilon_c\widetilde{\mathbf{E}}. & (7.8d) \end{aligned}}$$

To describe the propagation of an EM wave in a charge-free medium, we need to derive wave equations for $\widetilde{\mathbf{E}}$ and $\widetilde{\mathbf{H}}$ and then solve them to obtain explicit expressions for $\widetilde{\mathbf{E}}$ and $\widetilde{\mathbf{H}}$ as a function of the spatial variables (x, y, z). To this end, we start by taking the curl of both sides of Eq. (7.8b) to get

$$\nabla \times (\nabla \times \widetilde{\mathbf{E}}) = -j\omega\mu(\nabla \times \widetilde{\mathbf{H}}). \qquad (7.9)$$

Upon substituting Eq. (7.8d) into Eq. (7.9) we have

$$\nabla \times (\nabla \times \widetilde{\mathbf{E}}) = -j\omega\mu(j\omega\varepsilon_c\widetilde{\mathbf{E}}) = \omega^2\mu\varepsilon_c\widetilde{\mathbf{E}}. \quad (7.10)$$

From Eq. (3.113), the curl of the curl of $\widetilde{\mathbf{E}}$ is given by

$$\nabla \times (\nabla \times \widetilde{\mathbf{E}}) = \nabla(\nabla \cdot \widetilde{\mathbf{E}}) - \nabla^2\widetilde{\mathbf{E}}, \qquad (7.11)$$

where $\nabla^2\widetilde{\mathbf{E}}$ is the Laplacian of $\widetilde{\mathbf{E}}$ and is given in Cartesian coordinates by

$$\nabla^2\widetilde{\mathbf{E}} = \left(\frac{\partial^2}{\partial x^2} + \frac{\partial^2}{\partial y^2} + \frac{\partial^2}{\partial z^2}\right)\widetilde{\mathbf{E}}. \qquad (7.12)$$

In view of Eq. (7.8a), which states that $\nabla \cdot \widetilde{\mathbf{E}} = 0$, the use of Eq. (7.11) in Eq. (7.10) gives

$$\nabla^2\widetilde{\mathbf{E}} + \omega^2\mu\varepsilon_c\widetilde{\mathbf{E}} = 0, \qquad (7.13)$$

which is called the *homogeneous wave equation for* $\widetilde{\mathbf{E}}$. Upon introducing the *propagation constant* γ defined such that

$$\gamma^2 \triangleq -\omega^2\mu\varepsilon_c, \qquad (7.14)$$

Eq. (7.13) can be rewritten as

$$\boxed{\nabla^2\widetilde{\mathbf{E}} - \gamma^2\widetilde{\mathbf{E}} = 0. \qquad (7.15)}$$

In deriving Eq. (7.15), we started by taking the curl of both sides of Eq. (7.8b) and then we used Eq. (7.8d) to obtain an equation in $\widetilde{\mathbf{E}}$ only. If we reverse the process, that is, if we start by taking the curl of both sides of Eq. (7.8d) and then use Eq. (7.8b), we obtain the wave equation for $\widetilde{\mathbf{H}}$:

$$\boxed{\nabla^2\widetilde{\mathbf{H}} - \gamma^2\widetilde{\mathbf{H}} = 0. \qquad (7.16)}$$

Since the wave equations for $\widetilde{\mathbf{E}}$ and $\widetilde{\mathbf{H}}$ are of the same form, their solutions will have the same form also.

7-2 Plane-Wave Propagation in Lossless Media

The propagation properties of an electromagnetic wave, such as its phase velocity u_p and wavelength λ, are governed by the angular frequency ω and the three constitutive parameters of the medium: ε, μ, and σ. If the medium is *nonconducting* ($\sigma = 0$), the wave does not suffer any attenuation as it travels through the medium, and hence the medium is said to be *lossless*. From Eq. (7.4), $\varepsilon_c = \varepsilon$ in a lossless medium, in which case Eq. (7.14) becomes

$$\gamma^2 = -\omega^2\mu\varepsilon. \qquad (7.17)$$

When the medium is lossless, it is customary to introduce the *wavenumber* k defined by

$$\boxed{k = \omega\sqrt{\mu\varepsilon}. \qquad (7.18)}$$

In view of Eq. (7.17), $\gamma^2 = -k^2$ and Eq. (7.15) becomes

$$\nabla^2\widetilde{\mathbf{E}} + k^2\widetilde{\mathbf{E}} = 0. \qquad (7.19)$$

7-2.1 Uniform Plane Waves

For an electric field phasor given in Cartesian coordinates by

$$\widetilde{\mathbf{E}} = \hat{\mathbf{x}}\widetilde{E}_x + \hat{\mathbf{y}}\widetilde{E}_y + \hat{\mathbf{z}}\widetilde{E}_z, \qquad (7.20)$$

substitution of Eq. (7.12) into Eq. (7.19) gives

$$\left(\frac{\partial^2}{\partial x^2} + \frac{\partial^2}{\partial y^2} + \frac{\partial^2}{\partial z^2}\right)(\hat{\mathbf{x}}\widetilde{E}_x + \hat{\mathbf{y}}\widetilde{E}_y + \hat{\mathbf{z}}\widetilde{E}_z)$$
$$+ k^2(\hat{\mathbf{x}}\widetilde{E}_x + \hat{\mathbf{y}}\widetilde{E}_y + \hat{\mathbf{z}}\widetilde{E}_z) = 0. \quad (7.21)$$

To satisfy Eq. (7.21), each vector component on the left-hand side of the equation has to equal zero. Hence,

$$\left(\frac{\partial^2}{\partial x^2} + \frac{\partial^2}{\partial y^2} + \frac{\partial^2}{\partial z^2} + k^2\right)\widetilde{E}_x = 0, \qquad (7.22)$$

and similar expressions apply to $\widetilde{E}_y$ and $\widetilde{E}_z$.

A *uniform plane wave* is characterized by electric and magnetic fields that have uniform properties at all points across an infinite plane [see Fig. 7-3(b)]. If this is the x–y plane, then $\mathbf{E}$ and $\mathbf{H}$ do not vary with x and y. Hence, $\partial \widetilde{E}_x / \partial x = 0$ and $\partial \widetilde{E}_x / \partial y = 0$, in which case Eq. (7.22) reduces to

$$\frac{d^2 \widetilde{E}_x}{dz^2} + k^2 \widetilde{E}_x = 0, \tag{7.23}$$

and similar expressions apply to $\widetilde{E}_y$, $\widetilde{H}_x$, and $\widetilde{H}_y$. The remaining components of $\widetilde{\mathbf{E}}$ and $\widetilde{\mathbf{H}}$ are zero; that is, $\widetilde{E}_z = \widetilde{H}_z = 0$. To show that $\widetilde{E}_z = 0$, let us consider the z-component of Eq. (7.8d),

$$\hat{\mathbf{z}} \left(\frac{\partial \widetilde{H}_y}{\partial x} - \frac{\partial \widetilde{H}_x}{\partial y} \right) = \hat{\mathbf{z}} j \omega \varepsilon \widetilde{E}_z. \tag{7.24}$$

Since $\partial \widetilde{H}_y / \partial x = \partial \widetilde{H}_x / \partial y = 0$, it follows that $\widetilde{E}_z = 0$. A similar examination involving Eq. (7.8b) reveals that $\widetilde{H}_z = 0$. This means that *a plane wave has no electric- or magnetic-field components along its direction of propagation.*

For the phasor quantity $\widetilde{E}_x$, the general solution of the ordinary differential equation given by Eq. (7.23) is

$$\widetilde{E}_x(z) = \widetilde{E}_x^+(z) + \widetilde{E}_x^-(z) = E_{x0}^+ e^{-jkz} + E_{x0}^- e^{jkz}, \tag{7.25}$$

where E_{x0}^+ and E_{x0}^- are constants to be determined from boundary conditions. The solution given by Eq. (7.25) is similar in form to the solution for the phasor voltage $\widetilde{V}(z)$ given by Eq. (2.44a) for the lossless transmission line. The first term in Eq. (7.25), containing the negative exponential e^{-jkz}, represents a wave with amplitude E_{x0}^+ traveling in the $+z$-direction, and the second term (with e^{jkz}) represents a wave with amplitude E_{x0}^- traveling in the $-z$-direction. Let us assume for the time being that $\widetilde{\mathbf{E}}$ has only a component along x (i.e., $\widetilde{E}_y = 0$) and that $\widetilde{E}_x$ consists of a wave traveling in the $+z$-direction only (i.e., $E_{x0}^- = 0$). Hence,

$$\widetilde{\mathbf{E}}(z) = \hat{\mathbf{x}} \widetilde{E}_x^+(z) = \hat{\mathbf{x}} E_{x0}^+ e^{-jkz}. \tag{7.26}$$

With $\widetilde{E}_y = \widetilde{E}_z = 0$, we apply Eq. (7.8b) to find the magnetic field $\widetilde{\mathbf{H}}(z)$:

$$\nabla \times \widetilde{\mathbf{E}} = \begin{vmatrix} \hat{\mathbf{x}} & \hat{\mathbf{y}} & \hat{\mathbf{z}} \\ \dfrac{\partial}{\partial x} & \dfrac{\partial}{\partial y} & \dfrac{\partial}{\partial z} \\ \widetilde{E}_x^+(z) & 0 & 0 \end{vmatrix}$$

$$= -j\omega\mu (\hat{\mathbf{x}} \widetilde{H}_x + \hat{\mathbf{y}} \widetilde{H}_y + \hat{\mathbf{z}} \widetilde{H}_z). \tag{7.27}$$

For a plane wave traveling in the z-direction,

$$\partial E_x^+(z) / \partial x = \partial E_x^+(z) / \partial y = 0.$$

Hence, Eq. (7.27) gives

$$\widetilde{H}_x = 0, \tag{7.28a}$$

$$\widetilde{H}_y = \frac{1}{-j\omega\mu} \frac{\partial \widetilde{E}_x^+(z)}{\partial z}, \tag{7.28b}$$

$$\widetilde{H}_z = \frac{1}{-j\omega\mu} \frac{\partial E_x^+(z)}{\partial y} = 0. \tag{7.28c}$$

Use of Eq. (7.26) in Eq. (7.28b) gives

$$\widetilde{H}_y(z) = \frac{k}{\omega\mu} E_{x0}^+ e^{-jkz} = H_{y0}^+ e^{-jkz}, \tag{7.29}$$

where H_{y0}^+ is the amplitude of $\widetilde{H}_y(z)$ and is given by

$$H_{y0}^+ = \frac{k}{\omega\mu} E_{x0}^+. \tag{7.30}$$

For a wave traveling from the source toward the load on a transmission line, the amplitudes of its voltage and current phasors, V_0^+ and I_0^+, are related by the characteristic impedance of the line, Z_0. A similar connection exists between the electric and magnetic fields of an electromagnetic wave. The *intrinsic impedance* of a lossless medium is defined as

$$\eta \triangleq \frac{\omega\mu}{k} = \frac{\omega\mu}{\omega\sqrt{\mu\varepsilon}} = \sqrt{\frac{\mu}{\varepsilon}} \quad (\Omega), \tag{7.31}$$

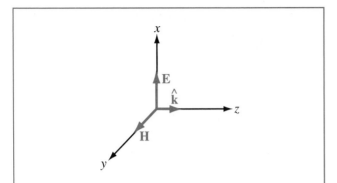

Figure 7-4: A transverse electromagnetic (TEM) wave propagating in the direction $\hat{\mathbf{k}} = \hat{\mathbf{z}}$. For all TEM waves, $\hat{\mathbf{k}}$ is parallel to $\mathbf{E} \times \mathbf{H}$.

where we used the expression for k given by Eq. (7.18). In view of Eq. (7.31), we can summarize our results as

$$\widetilde{\mathbf{E}}(z) = \hat{\mathbf{x}}\widetilde{E}_x^+(z) = \hat{\mathbf{x}}E_{x0}^+ e^{-jkz}, \qquad (7.32a)$$

$$\widetilde{\mathbf{H}}(z) = \hat{\mathbf{y}}\frac{\widetilde{E}_x^+(z)}{\eta} = \hat{\mathbf{y}}\frac{E_{x0}^+}{\eta}e^{-jkz}. \qquad (7.32b)$$

The electric and magnetic fields are perpendicular to each other, and both are perpendicular to the direction of wave travel (Fig. 7-4). These directional properties characterize a *transverse electromagnetic (TEM)* wave. Other examples of TEM waves include cylindrical waves traveling along coaxial transmission lines ($\mathbf{E}$ is along $\hat{\mathbf{r}}$, $\mathbf{H}$ is along $\hat{\boldsymbol{\phi}}$, and direction of travel is along $\hat{\mathbf{z}}$) and spherical waves radiated by antennas.

In the general case, E_{x0}^+ may be a complex quantity composed of a magnitude $|E_{x0}^+|$ and a phase angle ϕ^+. That is,

$$E_{x0}^+ = |E_{x0}^+|e^{j\phi^+}. \qquad (7.33)$$

The instantaneous electric and magnetic fields are then given by

$$\mathbf{E}(z, t) = \mathfrak{Re}\left[\widetilde{\mathbf{E}}(z)\,e^{j\omega t}\right]$$

$$= \hat{\mathbf{x}}|E_{x0}^+|\cos(\omega t - kz + \phi^+) \quad \text{(V/m)}, \qquad (7.34a)$$

$$\mathbf{H}(z, t) = \mathfrak{Re}\left[\widetilde{\mathbf{H}}(z)\,e^{j\omega t}\right]$$

$$= \hat{\mathbf{y}}\frac{|E_{x0}^+|}{\eta}\cos(\omega t - kz + \phi^+) \quad \text{(A/m)}. \qquad (7.34b)$$

Because $\mathbf{E}(z, t)$ and $\mathbf{H}(z, t)$ exhibit the same functional dependence on z and t, they are said to be *in phase*; when the amplitude of one of them is a maximum, the amplitude of the other is a maximum also. This in-phase property is characteristic of waves propagating in lossless media. Their time variation is defined by the oscillation frequency $f = \omega/2\pi$, and their spatial variation is characterized by the wavelength λ. From the material on wave motion given in Section 1-3, we deduce that the *phase velocity* of the wave is given by

$$u_{\mathrm{p}} = \frac{\omega}{k} = \frac{\omega}{\omega\sqrt{\mu\varepsilon}} = \frac{1}{\sqrt{\mu\varepsilon}} \quad \text{(m/s)}, \qquad (7.35)$$

and the wavelength is

$$\lambda = \frac{2\pi}{k} = \frac{u_{\mathrm{p}}}{f} \quad \text{(m)}. \qquad (7.36)$$

If the medium is vacuum, $\varepsilon = \varepsilon_0$ and $\mu = \mu_0$, in which case the phase velocity u_{p} and the intrinsic impedance η given by Eq. (7.31) become

$$u_{\mathrm{p}} = c = \frac{1}{\sqrt{\mu_0\varepsilon_0}} = 3 \times 10^8 \quad \text{(m/s)}, \qquad (7.37)$$

$$\eta = \eta_0 \triangleq \sqrt{\frac{\mu_0}{\varepsilon_0}} = 377 \ (\Omega) \approx 120\pi \quad (\Omega), \qquad (7.38)$$

where c is the velocity of light and η_0 is called the *intrinsic impedance of free space*.

Example 7-1   EM Plane Wave in Air

This example is analogous to the "Sound Wave in Water" problem described by Example 1-1.

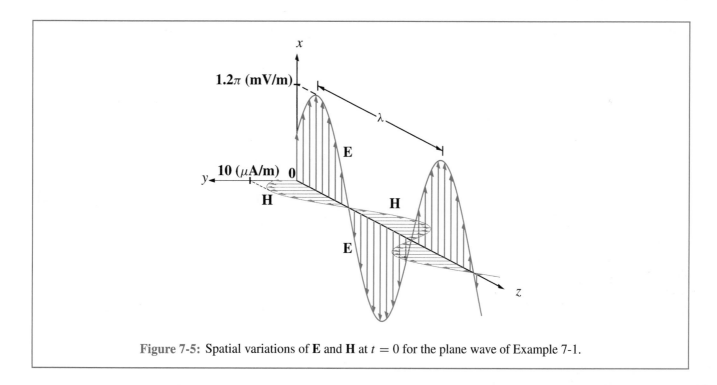

Figure 7-5: Spatial variations of **E** and **H** at $t = 0$ for the plane wave of Example 7-1.

The electric field of a 1-MHz plane wave traveling in the $+z$-direction in air points along the x-direction. If the peak value of E is 1.2π (mV/m) and E is maximum at $t = 0$ and $z = 50$ m, obtain expressions for $\mathbf{E}(z, t)$ and $\mathbf{H}(z, t)$, and then plot these variations as a function of z at $t = 0$.

Solution: At $f = 1$ MHz, the wavelength in air is given by

$$\lambda = \frac{c}{f} = \frac{3 \times 10^8}{1 \times 10^6} = 300 \text{ m},$$

and the corresponding wavenumber is $k = (2\pi/300)$ (rad/m). The general expression for an x-directed electric field traveling in the $+z$-direction is given by Eq. (7.34a)

as

$$\mathbf{E}(z, t) = \hat{\mathbf{x}} |E_{x0}^+| \cos(\omega t - kz + \phi^+)$$
$$= \hat{\mathbf{x}} \, 1.2\pi \cos\left(2\pi \times 10^6 t - \frac{2\pi z}{300} + \phi^+\right) \text{ (mV/m)}.$$

The field $\mathbf{E}(z, t)$ is maximum when the argument of the cosine function equals zero or multiples of 2π. At $t = 0$ and $z = 50$ m, this condition is

$$-\frac{2\pi \times 50}{300} + \phi^+ = 0 \quad \text{or} \quad \phi^+ = \frac{\pi}{3} \,.$$

Hence,

$$\mathbf{E}(z, t) = \hat{\mathbf{x}} \, 1.2\pi \cos\left(2\pi \times 10^6 t - \frac{2\pi z}{300} + \frac{\pi}{3}\right) \text{ (mV/m)},$$

and from Eq. (7.34b) we have

$$\mathbf{H}(z,t) = \hat{\mathbf{y}} \frac{E(z,t)}{\eta_0}$$

$$= \hat{\mathbf{y}} \, 10 \cos\left(2\pi \times 10^6 t - \frac{2\pi z}{300} + \frac{\pi}{3}\right) \quad (\mu\text{A/m}),$$

where we have used the approximation $\eta_0 \simeq 120\pi$ (Ω).
At $t = 0$,

$$\mathbf{E}(z,0) = \hat{\mathbf{x}} \, 1.2\pi \cos\left(\frac{2\pi z}{300} - \frac{\pi}{3}\right) \quad (\text{mV/m}),$$

$$\mathbf{H}(z,0) = \hat{\mathbf{y}} \, 10 \cos\left(\frac{2\pi z}{300} - \frac{\pi}{3}\right) \quad (\mu\text{A/m}).$$

Plots of $\mathbf{E}(z,0)$ and $\mathbf{H}(z,0)$ are shown in Fig. 7-5 as a function of z. ∎

7-2.2 General Relation between E and H

It can be shown that, for any uniform plane wave traveling in an arbitrary direction denoted by the unit vector $\hat{\mathbf{k}}$, the magnetic field phasor $\widetilde{\mathbf{H}}$ is interrelated to the electric field phasor $\widetilde{\mathbf{E}}$ by

$$\boxed{\begin{aligned} \widetilde{\mathbf{H}} &= \frac{1}{\eta} \hat{\mathbf{k}} \times \widetilde{\mathbf{E}}, \qquad &(7.39a) \\[2mm] \widetilde{\mathbf{E}} &= -\eta \hat{\mathbf{k}} \times \widetilde{\mathbf{H}}. \qquad &(7.39b) \end{aligned}}$$

The cross product can be phrased in terms of the following right-hand rule: *when we rotate the four fingers of the right hand from the direction of* **E** *toward the direction of* **H**, *the thumb will point in the direction of wave travel,* $\hat{\mathbf{k}}$. The relations given by Eqs. (7.39a and b) are valid not only for lossless media, but for lossy media as well. As we will see later in Section 7-4, the expression for η of a lossy medium is different from that given by Eq. (7.31). As long as the expression used for η is appropriate for the medium in which the wave is traveling, the relations given by Eqs. (7.39a and b) always hold.

Let us apply Eq. (7.39a) to the wave given by Eq. (7.32a). The direction of propagation $\hat{\mathbf{k}} = \hat{\mathbf{z}}$ and $\widetilde{\mathbf{E}} = \hat{\mathbf{x}} \widetilde{E}_x^+(z)$. Hence,

$$\widetilde{\mathbf{H}} = \frac{1}{\eta} \hat{\mathbf{k}} \times \widetilde{\mathbf{E}} = \frac{1}{\eta} (\hat{\mathbf{z}} \times \hat{\mathbf{x}}) \, \widetilde{E}_x^+(z) = \hat{\mathbf{y}} \frac{\widetilde{E}_x^+(z)}{\eta}, \quad (7.40)$$

which is the same as the result given by Eq. (7.32b). For a wave traveling in the $-z$-direction with an electric field given by

$$\widetilde{\mathbf{E}} = \hat{\mathbf{x}} \widetilde{E}_x^-(z) = \hat{\mathbf{x}} E_{x0}^- e^{jkz}, \qquad (7.41)$$

application of Eq. (7.39a) gives

$$\widetilde{\mathbf{H}} = \frac{1}{\eta}(-\hat{\mathbf{z}} \times \hat{\mathbf{x}}) \, \widetilde{E}_x^-(z) = -\hat{\mathbf{y}} \frac{\widetilde{E}_x^-(z)}{\eta} = -\hat{\mathbf{y}} \frac{E_{x0}^-}{\eta} e^{jkz}.$$
$$(7.42)$$

Hence, in this case, $\widetilde{\mathbf{H}}$ points in the negative y-direction.

In general, a uniform plane wave traveling in the $+z$-direction may have both x- and y-components, in which case $\widetilde{\mathbf{E}}$ is given by

$$\widetilde{\mathbf{E}} = \hat{\mathbf{x}} \widetilde{E}_x^+(z) + \hat{\mathbf{y}} \widetilde{E}_y^+(z), \qquad (7.43a)$$

and the associated magnetic field is

$$\widetilde{\mathbf{H}} = \hat{\mathbf{x}} \widetilde{H}_x^+(z) + \hat{\mathbf{y}} \widetilde{H}_y^+(z). \qquad (7.43b)$$

Application of Eq. (7.39a) gives

$$\widetilde{\mathbf{H}} = \frac{1}{\eta} \hat{\mathbf{z}} \times \widetilde{\mathbf{E}} = -\hat{\mathbf{x}} \frac{\widetilde{E}_y^+(z)}{\eta} + \hat{\mathbf{y}} \frac{\widetilde{E}_x^+(z)}{\eta}. \quad (7.44)$$

By equating Eq. (7.43b) to Eq. (7.44), we have

$$\widetilde{H}_x^+(z) = -\frac{\widetilde{E}_y^+(z)}{\eta}, \qquad \widetilde{H}_y^+ = \frac{\widetilde{E}_x^+(z)}{\eta}. \quad (7.45)$$

These results are illustrated in Fig. 7-6. The wave may be considered the sum of two waves, one with (E_x^+, H_y^+) components and another with (E_y^+, H_x^+) components. In general, a TEM wave may have an electric field in any direction in the plane orthogonal to the direction of wave travel, and the associated magnetic field is also in the same plane and its direction is dictated by Eq. (7.39a).

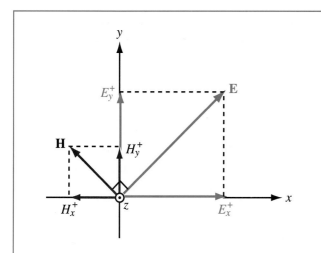

Figure 7-6: The wave $(\mathbf{E}, \mathbf{H})$ is equivalent to the sum of two waves, one with fields (E_x^+, H_y^+) and another with (E_y^+, H_x^+), with both traveling in the $+z$-direction.

REVIEW QUESTIONS

Q7.1 What is a uniform plane wave? Describe its properties, both physically and in mathematical terms. Under what conditions is it appropriate to treat a spherical wave as a plane wave?

Q7.2 Since $\widetilde{\mathbf{E}}$ and $\widetilde{\mathbf{H}}$ are governed by wave equations of the same form [Eqs. (7.15) and (7.16)], does it follow that $\widetilde{\mathbf{E}} = \widetilde{\mathbf{H}}$? Explain.

Q7.3 If a TEM wave is traveling in the $\hat{\mathbf{y}}$-direction, can its electric field have components along $\hat{\mathbf{x}}$, $\hat{\mathbf{y}}$, and $\hat{\mathbf{z}}$? Explain.

EXERCISE 7.1 A 10-MHz uniform plane wave is traveling in a nonmagnetic medium with $\mu = \mu_0$ and $\varepsilon_{\mathrm{r}} = 9$. Find (a) the phase velocity, (b) the wavenumber, (c) the wavelength in the medium, and (d) the intrinsic impedance of the medium.

Ans. (a) $u_{\mathrm{p}} = 1 \times 10^8$ m/s, (b) $k = 0.2\pi$ rad/m, (c) $\lambda = 10$ m, (d) $\eta = 125.67 \ \Omega$.

EXERCISE 7.2 The electric field phasor of a uniform plane wave traveling in a lossless medium with an intrinsic impedance of 188.5 Ω is given by $\widetilde{\mathbf{E}} = \hat{\mathbf{z}} 10 e^{-j4\pi y}$ (mV/m). Determine (a) the associated magnetic field phasor and (b) the instantaneous expression for $\mathbf{E}(y, t)$ if the medium is nonmagnetic ($\mu = \mu_0$).

Ans. (a) $\widetilde{\mathbf{H}} = \hat{\mathbf{x}} 53 e^{-j4\pi y}$ (μA/m), (b) $\mathbf{E}(y, t) = \hat{\mathbf{z}} 10 \cos(6\pi \times 10^8 t - 4\pi y)$ (mV/m).

EXERCISE 7.3 If the magnetic field phasor of a plane wave traveling in a medium with intrinsic impedance $\eta = 100 \ \Omega$ is given by $\widetilde{\mathbf{H}} = (\hat{\mathbf{y}} 10 + \hat{\mathbf{z}} 20) e^{-j4x}$ (mA/m), find the associated electric field phasor.

Ans. $\widetilde{\mathbf{E}} = (-\hat{\mathbf{z}} + \hat{\mathbf{y}} 2) e^{-j4x}$ (V/m).

EXERCISE 7.4 Repeat Exercise 7.3 for a magnetic field given by $\widetilde{\mathbf{H}} = \hat{\mathbf{y}}(10 e^{-j3x} - 20 e^{j3x})$ (mA/m).

Ans. $\widetilde{\mathbf{E}} = -\hat{\mathbf{z}}(e^{-j3x} + 2 e^{j3x})$ (V/m).

7-3 Wave Polarization

The *polarization* of a uniform plane wave *describes the shape and locus of the tip of the* $\mathbf{E}$ *vector (in the plane orthogonal to the direction of propagation) at a given point in space as a function of time.* In the most general case, the locus of $\mathbf{E}$ is an ellipse, and the wave is called *elliptically polarized.* Under certain conditions, the ellipse may degenerate into a circle or a segment of a straight line, in which case the *polarization state* is then called *circular* or *linear*, respectively.

As was shown in Section 7-2, the z-components of the electric and magnetic fields of a z-propagating plane wave are both zero. Hence, the electric field phasor $\widetilde{\mathbf{E}}(z)$ may consist of an x-component, $\widetilde{E}_x(z)$, and a y-component, $\widetilde{E}_y(z)$:

$$\widetilde{\mathbf{E}}(z) = \hat{\mathbf{x}} \widetilde{E}_x(z) + \hat{\mathbf{y}} \widetilde{E}_y(z), \qquad (7.46)$$

with

$$\widetilde{E}_x(z) = E_{x0}e^{-jkz}, \qquad (7.47a)$$

$$\widetilde{E}_y(z) = E_{y0}e^{-jkz}, \qquad (7.47b)$$

where E_{x0} and E_{y0} are the complex amplitudes of $\widetilde{E}_x(z)$ and $\widetilde{E}_y(z)$, respectively. For the sake of simplicity, the plus sign superscript has been suppressed throughout; the negative sign in e^{-jkz} is sufficient to remind us that the wave is traveling in the positive z-direction.

The two amplitudes E_{x0} and E_{y0} are, in general, complex quantities, with each characterized by a magnitude and a phase angle. The phase of a wave is defined relative to a reference condition, such as $z = 0$ and $t = 0$ or any other combination of z and t. *Wave polarization depends on the phase of E_{y0} relative to that of E_{x0}, but not on the absolute phases of E_{x0} and E_{y0}.* Hence, for convenience, we will choose the phase of E_{x0} as our reference (thereby assigning E_{x0} a phase angle of zero), and we will denote the phase of E_{y0}, relative to that of E_{x0}, as δ. Thus, δ is the *phase difference* between the y-component of $\widetilde{\mathbf{E}}$ and its x-component. Accordingly, we define E_{x0} and E_{y0} as

$$E_{x0} = a_x, \qquad (7.48a)$$

$$E_{y0} = a_y e^{j\delta}, \qquad (7.48b)$$

where $a_x = |E_{x0}|$ and $a_y = |E_{y0}|$ are the magnitudes of E_{x0} and E_{y0}, respectively. Thus, by definition, a_x and a_y may not assume negative values. Using Eqs. (7.48a) and (7.48b) in Eqs. (7.47a) and (7.47b), the total electric field phasor is then given by

$$\widetilde{\mathbf{E}}(z) = (\hat{\mathbf{x}}a_x + \hat{\mathbf{y}}a_y e^{j\delta})e^{-jkz}, \qquad (7.49)$$

and the corresponding instantaneous field is

$$\mathbf{E}(z, t) = \mathfrak{Re}\left[\widetilde{\mathbf{E}}(z)\, e^{j\omega t}\right]$$

$$= \hat{\mathbf{x}}a_x \cos(\omega t - kz)$$

$$+ \hat{\mathbf{y}}a_y \cos(\omega t - kz + \delta). \qquad (7.50)$$

When characterizing the behavior of an EM wave, two properties of particular interest are the intensity and

direction of its electric field. The intensity of $\mathbf{E}(z, t)$ is given by its modulus $|\mathbf{E}(z, t)|$, namely

$$|\mathbf{E}(z, t)| = [E_x^2(z, t) + E_y^2(z, t)]^{1/2}$$

$$= [a_x^2 \cos^2(\omega t - kz)$$

$$+ a_y^2 \cos^2(\omega t - kz + \delta)]^{1/2}. \quad (7.51)$$

The electric field $\mathbf{E}(z, t)$ has components along the x- and y-directions. At a specific position z, the direction of $\mathbf{E}(z, t)$ is defined in the x–y plane (at that value of z) by the inclination angle ψ, defined with respect to the zero-phase reference component of $\mathbf{E}(z, t)$, which is the x-component in the present case. Thus,

$$\psi(z, t) \triangleq \tan^{-1}\left(\frac{E_y(z, t)}{E_x(z, t)}\right). \qquad (7.52)$$

In the general case, both the intensity of $\mathbf{E}(z, t)$ and its direction are functions of z and t. Next, we examine some special cases.

7-3.1   Linear Polarization

The polarization state of a wave traveling in the z-direction is determined by tracing the tip of $\mathbf{E}(z, t)$ as a function of time in a plane orthogonal to the direction of wave travel. For convenience and without loss of generality, we usually choose the $z = 0$ plane. *A wave is said to be linearly polarized if $E_x(z, t)$ and $E_y(z, t)$ are in phase (i.e., $\delta = 0$) or out of phase ($\delta = \pi$).* This is because, at a specified value of z, say $z = 0$, the tip of $\mathbf{E}(0, t)$ traces a straight line in the x–y plane. At $z = 0$ and for $\delta = 0$ or π, Eq. (7.50) simplifies to

$$\mathbf{E}(0, t) = (\hat{\mathbf{x}}a_x + \hat{\mathbf{y}}a_y) \cos \omega t \quad \text{(in-phase)}, \qquad (7.53a)$$

$$\mathbf{E}(0, t) = (\hat{\mathbf{x}}a_x - \hat{\mathbf{y}}a_y) \cos \omega t \quad \text{(out-of-phase)}. \qquad (7.53b)$$

Let us examine the out-of-phase case. At $\omega t = 0$, $\mathbf{E}(0, 0) = \hat{\mathbf{x}}a_x - \hat{\mathbf{y}}a_y$, which means that $\mathbf{E}$ extends from the origin in Fig. 7-7 to the point $(a_x, -a_y)$ in the fourth quadrant in the x–y plane. As t increases, the modulus

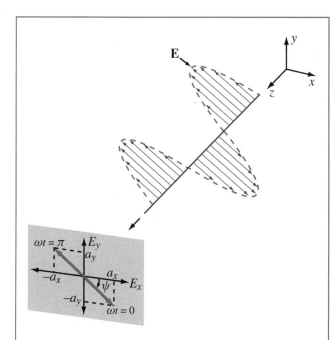

Figure 7-7: Linearly polarized wave traveling in the $+z$-direction (out of the page).

of $\mathbf{E}(0, t)$ varies as $\cos \omega t$:

$$|\mathbf{E}(0, t)| = [a_x^2 + a_y^2]^{1/2} \cos \omega t, \qquad (7.54a)$$

and the inclination angle defined by Eq. (7.52) is given by

$$\psi = \tan^{-1}\left(\frac{-a_y}{a_x}\right) \quad \text{(out-of-phase)}, \qquad (7.54b)$$

which is independent of both z and t. Thus, the length of the vector representing $\mathbf{E}(0, t)$ decreases to zero at $\omega t = \pi/2$, then reverses direction and increases in magnitude to $[a_x^2 + a_y^2]^{1/2}$ in the second quadrant of the x–y plane at $\omega t = \pi$. Since ψ is independent of both z and t, $\mathbf{E}(z, t)$ maintains a direction along the line making an angle ψ with the x-axis.

If $a_y = 0$, $\psi = 0°$ or $180°$, and the wave becomes x-polarized, and if $a_x = 0$, $\psi = 90°$ or $-90°$, and the wave becomes y-polarized.

7-3.2 Circular Polarization

We now consider the special case when the magnitudes of the x- and y-components of $\widetilde{\mathbf{E}}(z)$ are equal, and the phase difference $\delta = \pm\pi/2$. For reasons that will become evident shortly, the wave polarization is called *left-hand circular* when $\delta = \pi/2$, and it is called *right-hand circular* when $\delta = -\pi/2$.

Left-Hand Circular (LHC) Polarization

For $a_x = a_y = a$ and $\delta = \pi/2$, Eqs. (7.49) and (7.50) become

$$\widetilde{\mathbf{E}}(z) = (\hat{\mathbf{x}}a + \hat{\mathbf{y}}ae^{j\pi/2})e^{-jkz}$$
$$= a(\hat{\mathbf{x}} + j\hat{\mathbf{y}})e^{-jkz}, \qquad (7.55a)$$
$$\mathbf{E}(z, t) = \Re\mathfrak{e}\left[\widetilde{\mathbf{E}}(z)\, e^{j\omega t}\right]$$
$$= \hat{\mathbf{x}}a \cos(\omega t - kz) + \hat{\mathbf{y}}a \cos(\omega t - kz + \pi/2)$$
$$= \hat{\mathbf{x}}a \cos(\omega t - kz) - \hat{\mathbf{y}}a \sin(\omega t - kz). \quad (7.55b)$$

The corresponding modulus and inclination angle are given by

$$|\mathbf{E}(z, t)| = \left[E_x^2(z, t) + E_y^2(z, t)\right]^{1/2}$$
$$= [a^2 \cos^2(\omega t - kz) + a^2 \sin^2(\omega t - kz)]^{1/2}$$
$$= a, \qquad (7.56a)$$
$$\psi(z, t) = \tan^{-1}\left[\frac{E_y(z, t)}{E_x(z, t)}\right]$$
$$= \tan^{-1}\left[\frac{-a \sin(\omega t - kz)}{a \cos(\omega t - kz)}\right]$$
$$= -(\omega t - kz). \qquad (7.56b)$$

We observe that in this case the modulus of $\mathbf{E}$ is independent of both z and t, whereas ψ depends on both variables. These functional dependencies are the

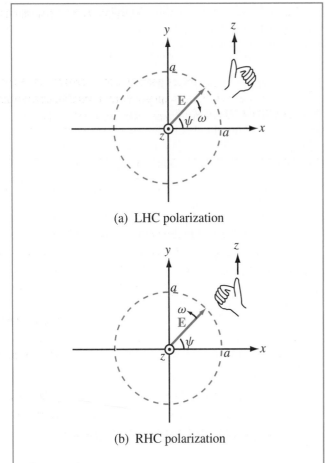

(a) LHC polarization

(b) RHC polarization

Figure 7-8: Circularly polarized plane waves propagating in the $+z$-direction (out of the page).

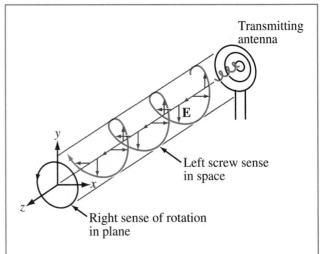

Figure 7-9: Right-hand circularly polarized wave radiated by a helical antenna.

converse of those applicable to the linear-polarization case.

At a fixed position z, say $z = 0$, Eq. (7.56b) gives $\psi = -\omega t$; the negative sign means that the inclination angle decreases with increasing time. As illustrated in Fig. 7-8(a), the tip of $\mathbf{E}(t)$ traces a circle in the x–y plane, and it rotates in a clockwise direction as a function of time (when viewing the wave approaching). Such a wave is called *left-hand circularly polarized* because, when the thumb of the left hand points along the direction of propagation of the

wave (the z-direction in this case), the other four fingers point in the direction of rotation of $\mathbf{E}$.

Right-Hand Circular (RHC) Polarization

For $a_x = a_y = a$ and $\delta = -\pi/2$, we have

$$|\mathbf{E}(z, t)| = a, \qquad \psi = (\omega t - kz). \qquad (7.57)$$

The trace of $\mathbf{E}$ as a function of t is shown in Fig. 7-8(b) for $z = 0$. For RHC polarization, the fingers of the right hand point in the direction of rotation of $\mathbf{E}$ when the thumb is along the propagation direction. Figure 7-9 depicts a right-hand circularly polarized wave radiated by a helical antenna. Note that *polarization handedness is defined in terms of the rotation of $\mathbf{E}$ as a function of time in a fixed plane orthogonal to the direction of propagation,* which is opposite of the direction of rotation of $\mathbf{E}$ as a function of distance at a fixed point in time.

Example 7-2   RHC Polarized Wave

An RHC polarized plane wave with electric field modulus of 3 (mV/m) is traveling in the $+y$-direction in

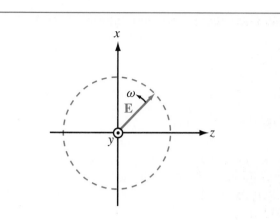

Figure 7-10: Right-hand circularly polarized wave of Example 7-2.

a dielectric medium with $\varepsilon = 4\varepsilon_0$, $\mu = \mu_0$, and $\sigma = 0$. If the wave frequency is 100 MHz, obtain expressions for $\mathbf{E}(y,t)$ and $\mathbf{H}(y,t)$.

Solution: Since the wave is traveling in the $+y$-direction, its field components must be along the x- and z-directions. The rotation of $\mathbf{E}(y,t)$ is depicted in Fig. 7-10, where $\hat{\mathbf{y}}$ is out of the page. By comparison with the RHC polarized wave shown in Fig. 7-8(b), we assign the z-component of $\widetilde{\mathbf{E}}(y)$ a phase angle of zero and the x-component a phase shift $\delta = -\pi/2$. Both components have a magnitude $a = 3$ (mV/m). Hence,

$$\widetilde{\mathbf{E}}(y) = \hat{\mathbf{x}}\widetilde{E}_x + \hat{\mathbf{z}}\widetilde{E}_z$$
$$= \hat{\mathbf{x}}ae^{-j\pi/2}e^{-jky} + \hat{\mathbf{z}}ae^{-jky}$$
$$= (-\hat{\mathbf{x}}j + \hat{\mathbf{z}})3e^{-jky} \quad \text{(mV/m)},$$

and application of (7.39a) gives

$$\widetilde{\mathbf{H}}(y) = \frac{1}{\eta}\hat{\mathbf{y}} \times \widetilde{\mathbf{E}}(y)$$
$$= \frac{1}{\eta}\hat{\mathbf{y}} \times (-\hat{\mathbf{x}}j + \hat{\mathbf{z}})3e^{-jky}$$
$$= \frac{3}{\eta}(\hat{\mathbf{z}}j + \hat{\mathbf{x}})e^{-jky} \quad \text{(mA/m)}.$$

With $\omega = 2\pi f = 2\pi \times 10^8$ (rad/s), the wavenumber k is

$$k = \frac{\omega\sqrt{\varepsilon_r}}{c} = \frac{2\pi \times 10^8 \sqrt{4}}{3 \times 10^8} = \frac{4}{3}\pi \quad \text{(rad/m)},$$

and the intrinsic impedance η is given by

$$\eta = \frac{\eta_0}{\sqrt{\varepsilon_r}} \simeq \frac{120\pi}{\sqrt{4}} = 60\pi \quad (\Omega).$$

The instantaneous functions $\mathbf{E}(y,t)$ and $\mathbf{H}(y,t)$ are then given by

$$\mathbf{E}(y,t) = \mathfrak{Re}\left[\widetilde{\mathbf{E}}(y)e^{j\omega t}\right]$$
$$= \mathfrak{Re}\left[(-\hat{\mathbf{x}}j + \hat{\mathbf{z}})3e^{-jky}e^{j\omega t}\right]$$
$$= 3[\hat{\mathbf{x}}\sin(\omega t - ky) + \hat{\mathbf{z}}\cos(\omega t - ky)] \quad \text{(mV/m)},$$
$$\mathbf{H}(y,t) = \mathfrak{Re}\left[\widetilde{\mathbf{H}}(y)e^{j\omega t}\right]$$
$$= \mathfrak{Re}\left[\frac{3}{\eta}(\hat{\mathbf{z}}j + \hat{\mathbf{x}})e^{-jky}e^{j\omega t}\right]$$
$$= \frac{1}{20\pi}[\hat{\mathbf{x}}\cos(\omega t - ky) - \hat{\mathbf{z}}\sin(\omega t - ky)] \quad \text{(mA/m)},$$

with $\omega = 2\pi \times 10^8$ (rad/s) and $k = 4\pi/3$ (rad/m). ∎

7-3.3 Elliptical Polarization

In the most general case, where $a_x \neq 0$, $a_y \neq 0$, and $\delta \neq 0$, the tip of $\mathbf{E}$ traces an ellipse in the x–y plane, and the wave is said to be elliptically polarized. The shape of the ellipse and its handedness (left-hand or right-hand rotation) are determined by the values of the ratio (a_y/a_x) and the polarization phase difference δ.

The polarization ellipse shown in Fig. 7-11 has a major axis a_ξ along the ξ-direction and a minor axis a_η along the η-direction. The *rotation angle* γ is defined as the angle between the major axis of the ellipse and a reference direction, chosen here to be the x-axis, with γ being bounded within the range $-\pi/2 \leq \gamma \leq \pi/2$. The shape

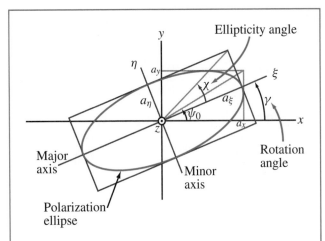

Figure 7-11: Polarization ellipse in the x–y plane, with the wave traveling in the z-direction (out of the page).

of the ellipse and its handedness are characterized by the *ellipticity angle* χ, defined as follows:

$$\tan \chi = \pm \frac{a_\eta}{a_\xi} = \pm \frac{1}{R}, \qquad (7.58)$$

with the plus sign corresponding to left-handed rotation and the minus sign corresponding to right-handed rotation. The limits for χ are $-\pi/4 \leq \chi \leq \pi/4$. The quantity $R = a_\xi/a_\eta$ is called the **axial ratio** of the polarization ellipse, and it varies between 1 for circular polarization and ∞ for linear polarization. The polarization angles γ and χ are related to the wave parameters a_x, a_y, and δ by*

$$\tan 2\gamma = (\tan 2\psi_0) \cos \delta \quad (-\pi/2 \leq \gamma \leq \pi/2), (7.59a)$$

$$\sin 2\chi = (\sin 2\psi_0) \sin \delta \quad (-\pi/4 \leq \chi \leq \pi/4), (7.59b)$$

where ψ_0 is an auxiliary angle defined by

$$\tan \psi_0 = \frac{a_y}{a_x} \quad \left(0 \leq \psi_0 \leq \frac{\pi}{2}\right). \qquad (7.60)$$

---

*From M. Born and E. Wolf, *Principles of Optics*, New York: Macmillan, 1965, p. 27.

Sketches of the polarization ellipse are shown in Fig. 7-12 for various combinations of the angles (γ, χ). The ellipse reduces to a circle for $\chi = \pm 45°$ and to a line for $\chi = 0$. *Positive values of χ, corresponding to $\sin \delta > 0$, are associated with left-handed rotation, and negative values of χ, corresponding to $\sin \delta < 0$, are associated with right-handed rotation.*

Since the magnitudes a_x and a_y are, by definition, nonnegative numbers, the ratio a_y/a_x may vary between zero for an x-polarized linear polarization and ∞ for a y-polarized linear polarization. Consequently, the angle ψ_0 is limited to the range $0 \leq \psi_0 \leq 90°$. Application of Eq. (7.59a) leads to two possible solutions for the value of γ, both of which fall within the defined range from $-\pi/2$ to $\pi/2$. The correct choice is governed by the following rule:

$$\gamma > 0 \text{ if } \cos \delta > 0,$$
$$\gamma < 0 \text{ if } \cos \delta < 0.$$

In summary, the sign of the rotation angle γ is the same as the sign of $\cos \delta$ and the sign of the ellipticity angle χ is the same as the sign of $\sin \delta$.

Example 7-3   Polarization State

Determine the polarization state of a plane wave with electric field

$$\mathbf{E}(z, t) = \hat{\mathbf{x}} 3 \cos(\omega t - kz + 30°)$$
$$- \hat{\mathbf{y}} 4 \sin(\omega t - kz + 45°) \quad \text{(mV/m)}.$$

Solution: We begin by converting the second term into a cosine reference,

$$\mathbf{E} = \hat{\mathbf{x}} 3 \cos(\omega t - kz + 30°)$$
$$- \hat{\mathbf{y}} 4 \cos(\omega t - kz + 45° - 90°)$$
$$= \hat{\mathbf{x}} 3 \cos(\omega t - kz + 30°) - \hat{\mathbf{y}} 4 \cos(\omega t - kz - 45°).$$

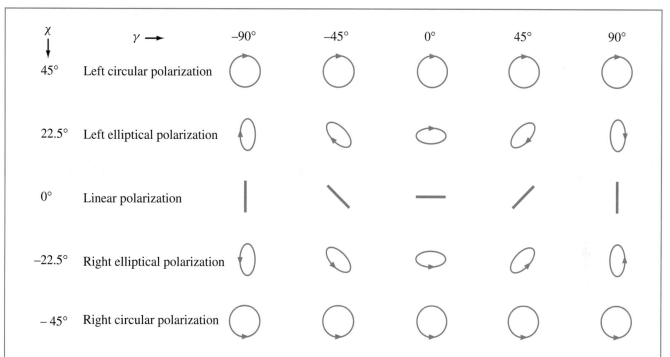

Figure 7-12: Polarization states for various combinations of the polarization angles (γ, χ) for a wave traveling out of the page.

The corresponding field phasor $\widetilde{\mathbf{E}}(z)$ is

$$\widetilde{\mathbf{E}}(z) = \hat{\mathbf{x}}\, 3e^{-jkz}e^{j30°} - \hat{\mathbf{y}}\, 4e^{-jkz}e^{-j45°}$$

$$= \hat{\mathbf{x}}\, 3e^{-jkz}e^{j30°} + \hat{\mathbf{y}}\, 4e^{-jkz}e^{-j45°}e^{j180°}$$

$$= \hat{\mathbf{x}}\, 3e^{-jkz}e^{j30°} + \hat{\mathbf{y}}\, 4e^{-jkz}e^{j135°},$$

where we have replaced the negative sign of the second term with $e^{j180°}$ in order to have positive amplitudes for both terms, thereby allowing us to use the definitions given in Section 7-3.3. According to the expression for $\widetilde{\mathbf{E}}(z)$, the phase angles of the x- and y-components are $\delta_x = 30°$ and $\delta_y = 135°$, giving a phase difference $\delta = \delta_y - \delta_x = 135° - 30° = 105°$. The auxiliary angle ψ_0 is obtained from

$$\psi_0 = \tan^{-1}\left(\frac{a_y}{a_x}\right) = \tan^{-1}\left(\frac{4}{3}\right) = 53.1°.$$

From Eq. (7.59a),

$$\tan 2\gamma = (\tan 2\psi_0)\cos\delta$$

$$= \tan 106.2° \cos 105°$$

$$= 0.89,$$

which gives two solutions for γ, namely $\gamma = 20.8°$ and $\gamma = -69.2°$. Since $\cos\delta < 0$, the correct value of γ is $-69.2°$. From Eq. (7.59b),

$$\sin 2\chi = (\sin 2\psi_0)\sin\delta$$

$$= \sin 106.2° \sin 105°$$

$$= 0.93 \quad \text{or} \quad \chi = 34.0°.$$

The magnitude of χ indicates that the wave is elliptically polarized and its positive polarity specifies its rotation as left handed. ∎

REVIEW QUESTIONS

Q7.4   An elliptically polarized wave is characterized by amplitudes a_x and a_y and by the phase difference δ. If a_x and a_y are both nonzero, what should δ be in order for the polarization state to reduce to linear polarization?

Q7.5   Which of the following two descriptions defines an RHC polarized wave: A wave incident upon an observer is RHC polarized if its electric field appears to the observer to rotate in a counterclockwise direction (a) as a function of time in a fixed plane perpendicular to the direction of wave travel or (b) as a function of travel distance at a fixed time t?

EXERCISE 7.5   The electric field of a plane wave is given by

$$\mathbf{E}(z,t) = \hat{\mathbf{x}}\, 3\cos(\omega t - kz) + \hat{\mathbf{y}}\, 4\cos(\omega t - kz) \quad \text{(V/m)}.$$

Determine (a) the polarization state, (b) the modulus of $\mathbf{E}$, and (c) the inclination angle.

Ans.   (a) Linear, (b) $|\mathbf{E}| = 5\cos(\omega t - kz)$ (V/m), (c) $\psi_0 = 53.1°$.

EXERCISE 7.6   If the electric field phasor of a TEM wave is given by $\widetilde{\mathbf{E}} = (\hat{\mathbf{y}} - \hat{\mathbf{z}}j)e^{-jkx}$, determine the polarization state.

Ans.   RHC polarization.

7-4 Plane-Wave Propagation in Lossy Media

To examine wave propagation in a conducting medium, we return to the wave equation given by Eq. (7.15),

$$\nabla^2 \widetilde{\mathbf{E}} - \gamma^2 \widetilde{\mathbf{E}} = 0, \tag{7.61}$$

with

$$\gamma^2 = -\omega^2 \mu \varepsilon_c = -\omega^2 \mu (\varepsilon' - j\varepsilon''), \tag{7.62}$$

where $\varepsilon' = \varepsilon$ and $\varepsilon'' = \sigma/\omega$. Since γ is complex, we express it as

$$\gamma = \alpha + j\beta, \tag{7.63}$$

where α is the *attenuation constant* of the medium and β is its *phase constant*. By replacing γ with $(\alpha + j\beta)$ in Eq. (7.62), we have

$$(\alpha + j\beta)^2 = (\alpha^2 - \beta^2) + j2\alpha\beta$$
$$= -\omega^2 \mu \varepsilon' + j\omega^2 \mu \varepsilon''. \tag{7.64}$$

The rules of complex algebra require the real and imaginary parts on one side of an equation to be respectively equal to the real and imaginary parts on the other side. Hence,

$$\alpha^2 - \beta^2 = -\omega^2 \mu \varepsilon', \tag{7.65a}$$

$$2\alpha\beta = \omega^2 \mu \varepsilon''. \tag{7.65b}$$

Solving these two equations for α and β gives

$$\alpha = \omega \left\{ \frac{\mu \varepsilon'}{2} \left[\sqrt{1 + \left(\frac{\varepsilon''}{\varepsilon'} \right)^2} - 1 \right] \right\}^{1/2} \quad \text{(Np/m)}, \tag{7.66a}$$

$$\beta = \omega \left\{ \frac{\mu \varepsilon'}{2} \left[\sqrt{1 + \left(\frac{\varepsilon''}{\varepsilon'} \right)^2} + 1 \right] \right\}^{1/2} \quad \text{(rad/m)}. \tag{7.66b}$$

For a uniform plane wave with an electric field $\widetilde{\mathbf{E}} = \hat{\mathbf{x}}\, \widetilde{E}_x(z)$ traveling in the $+z$-direction, the wave equation given by Eq. (7.61) reduces to

$$\frac{d^2 \widetilde{E}_x(z)}{dz^2} - \gamma^2 \widetilde{E}_x(z) = 0. \tag{7.67}$$

The solution of this wave equation leads to

$$\widetilde{\mathbf{E}}(z) = \hat{\mathbf{x}}\widetilde{E}_x(z) = \hat{\mathbf{x}}E_{x0}e^{-\gamma z} = \hat{\mathbf{x}}E_{x0}e^{-\alpha z}e^{-j\beta z}. \quad (7.68)$$

The associated magnetic field $\widetilde{\mathbf{H}}$ can be determined either (1) by applying Eq. (7.2b): $\nabla \times \widetilde{\mathbf{E}} = -j\omega\mu\widetilde{\mathbf{H}}$, (2) by applying Eq. (7.39a): $\widetilde{\mathbf{H}} = (\hat{\mathbf{k}} \times \widetilde{\mathbf{E}})/\eta_c$, where η_c is the *intrinsic impedance of the lossy medium*, or (3) by analogy with the lossless case. Any one of these approaches gives

$$\widetilde{\mathbf{H}}(z) = \hat{\mathbf{y}}\,\widetilde{H}_y(z) = \hat{\mathbf{y}}\,\frac{\widetilde{E}_x(z)}{\eta_c} = \hat{\mathbf{y}}\,\frac{E_{x0}}{\eta_c}e^{-\alpha z}e^{-j\beta z}, \quad (7.69)$$

where

$$\eta_c = \sqrt{\frac{\mu}{\varepsilon_c}} = \sqrt{\frac{\mu}{\varepsilon'}}\left(1 - j\frac{\varepsilon''}{\varepsilon'}\right)^{-1/2} \quad (\Omega). \quad (7.70)$$

We noted earlier that in a nonconducting medium, $\mathbf{E}(z, t)$ is in phase with $\mathbf{H}(z, t)$, but because η_c is a complex quantity in a conducting medium, the fields no longer have equal phase (as will be illustrated in Example 7-4). From Eq. (7.68), the magnitude of $\widetilde{E}_x(z)$ is given by

$$|\widetilde{E}_x(z)| = |E_{x0}e^{-\alpha z}e^{-j\beta z}| = |E_{x0}|e^{-\alpha z}, \quad (7.71)$$

which decreases exponentially with z at a rate specified by the attenuation constant α. Since $\widetilde{H}_y = \widetilde{E}_x/\eta_c$, the magnitude of $\widetilde{H}_y$ also attenuates as $e^{-\alpha z}$. The attenuation process converts part of the energy carried by the electromagnetic wave into heat as a result of conduction in the medium. Through a distance $z = \delta_s$ such that

$$\delta_s = \frac{1}{\alpha} \quad (\text{m}), \quad (7.72)$$

the wave magnitude decreases by a factor of $e^{-1} \approx 0.37$ compared with its value at $z = 0$, as shown in Fig. 7-13. This distance δ_s, called the *skin depth* of the medium, characterizes how well an electromagnetic wave can

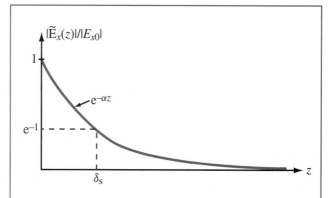

Figure 7-13: Attenuation of the magnitude of $\widetilde{E}_x(z)$ with distance z. The skin depth δ_s is the value of z at which $|\widetilde{E}_x(z)|/|E_{x0}| = e^{-1}$, or $z = \delta_s = 1/\alpha$.

penetrate into a conducting medium. In a perfect dielectric, $\sigma = 0$; hence, $\alpha = 0$ and therefore $\delta_s = \infty$. Thus, in free space, a plane wave can propagate with no loss in magnitude indefinitely. On the other extreme, if the medium is a perfect conductor with $\sigma = \infty$, use of $\varepsilon'' = \sigma/\omega$ in Eq. (7.66a) leads to $\alpha = \infty$ and hence $\delta_s = 0$. In a coaxial cable, if the outer conductor is designed to be several skin depths thick, it serves to prevent the energy inside the cable from leaking outward, as well as to shield against the penetration of outside electromagnetic energy into the cable.

The expressions given by Eqs. (7.66a), (7.66b), and (7.70) for α, β, and η_c are valid for any linear, isotropic, and homogeneous medium. If the medium is a perfect dielectric ($\sigma = 0$), these expressions reduce to the lossless case [Section 7-2], wherein $\alpha = 0$, $\beta = k = \omega\sqrt{\mu\varepsilon}$, and $\eta_c = \eta$. For a lossy medium, the ratio $\varepsilon''/\varepsilon' = \sigma/\omega\varepsilon$ appears in all these expressions and plays an important role in determining how lossy a medium is. When $\varepsilon''/\varepsilon' \ll 1$, the medium is called a *low-loss dielectric*, and when $\varepsilon''/\varepsilon' \gg 1$, the medium is characterized as a *good conductor*. In practice, the medium may be regarded as a low-loss dielectric if $\varepsilon''/\varepsilon' < 10^{-2}$, as a good conductor if $\varepsilon''/\varepsilon' > 10^2$, and as a *quasi-conductor* if $10^{-2} \leq \varepsilon''/\varepsilon' \leq 10^2$.

7-4.1 Low-Loss Dielectric

From Eq. (7.62), the general expression for γ is given by

$$\gamma = j\omega\sqrt{\mu\varepsilon'}\left(1 - j\frac{\varepsilon''}{\varepsilon'}\right)^{1/2}. \qquad (7.73)$$

For any quantity $|x| \ll 1$, the function $(1 - x)^{1/2}$ can be approximated by the first two terms of its binomial series; that is, $(1 - x)^{1/2} \simeq 1 - x/2$. By applying such an expansion to Eq. (7.73) for a low-loss dielectric with $x = j\varepsilon''/\varepsilon'$ and $\varepsilon''/\varepsilon' \ll 1$, we have

$$\gamma \cong j\omega\sqrt{\mu\varepsilon'}\left(1 - j\frac{\varepsilon''}{2\varepsilon'}\right). \qquad (7.74)$$

The real and imaginary parts of Eq. (7.74) give

$$\alpha \cong \frac{\omega\varepsilon''}{2}\sqrt{\frac{\mu}{\varepsilon'}} = \frac{\sigma}{2}\sqrt{\frac{\mu}{\varepsilon}} \quad \text{(Np/m)}, \qquad (7.75a)$$

$$\beta \cong \omega\sqrt{\mu\varepsilon'} = \omega\sqrt{\mu\varepsilon} \quad \text{(rad/m)}. \qquad (7.75b)$$

We note that the expression for β is the same as that for the wavenumber k of a lossless medium. Applying the binomial approximation $(1 - x)^{-1/2} \simeq (1 + x/2)$ to Eq. (7.70) leads to

$$\eta_c \cong \sqrt{\frac{\mu}{\varepsilon'}}\left(1 + j\frac{\varepsilon''}{2\varepsilon'}\right) = \sqrt{\frac{\mu}{\varepsilon}}\left(1 + j\frac{\sigma}{2\omega\varepsilon}\right). \qquad (7.76a)$$

In practice, these approximate expressions for α, β, and η_c are used whenever $\varepsilon''/\varepsilon' = \sigma/\omega\varepsilon < 1/100$, in which case the second term in Eq. (7.76a) may be ignored. Thus,

$$\eta_c \cong \sqrt{\frac{\mu}{\varepsilon}}, \qquad (7.76b)$$

which is the same as the expression given by Eq. (7.31) for the lossless case.

7-4.2 Good Conductor

We now examine the case of a good conductor characterized by $\varepsilon''/\varepsilon' > 100$. Under this condition, Eqs. (7.66a), (7.66b), and (7.70) can be approximated as

$$\alpha \cong \omega\sqrt{\frac{\mu\varepsilon''}{2}} = \omega\sqrt{\frac{\mu\sigma}{2\omega}} = \sqrt{\pi f\mu\sigma} \quad \text{(Np/m)}, \qquad (7.77a)$$

$$\beta = \alpha \cong \sqrt{\pi f\mu\sigma} \quad \text{(rad/m)}, \qquad (7.77b)$$

$$\eta_c \cong \sqrt{j\frac{\mu}{\varepsilon''}} = (1 + j)\sqrt{\frac{\pi f\mu}{\sigma}} = (1 + j)\frac{\alpha}{\sigma} \quad (\Omega). \qquad (7.77c)$$

In Eq. (7.77c), we used the relation given by Eq. (1.53): $\sqrt{j} = (1 + j)/\sqrt{2}$. For a perfect conductor with $\sigma = \infty$, these expressions give $\alpha = \beta = \infty$, and $\eta_c = 0$. A perfect conductor is equivalent to a short circuit.

Expressions for the propagation parameters in various types of media are summarized in Table 7-1 for easy reference.

Example 7-4   Plane Wave in Seawater

A uniform plane wave is traveling downward in the $+z$-direction in seawater, with the x–y plane denoting the sea surface and $z = 0$ denoting a point just below the surface. The constitutive parameters of seawater are $\varepsilon_r = 80$, $\mu_r = 1$, and $\sigma = 4$ S/m. If the magnetic field at $z = 0$ is given by $\mathbf{H}(0, t) = \hat{\mathbf{y}} 100 \cos(2\pi \times 10^3 t + 15°)$ (mA/m),

(a) obtain expressions for $\mathbf{E}(z, t)$ and $\mathbf{H}(z, t)$, and

(b) determine the depth at which the amplitude of $\mathbf{E}$ is 1% of its value at $z = 0$.

Solution: (a) Since $\mathbf{H}$ is along $\hat{\mathbf{y}}$ and the propagation direction is $\hat{\mathbf{z}}$, $\mathbf{E}$ must be along $\hat{\mathbf{x}}$. Hence, the general expressions for the phasor fields are

$$\widetilde{\mathbf{E}}(z) = \hat{\mathbf{x}} E_{x0} e^{-\alpha z} e^{-j\beta z}, \qquad (7.78a)$$

$$\widetilde{\mathbf{H}}(z) = \hat{\mathbf{y}} \frac{E_{x0}}{\eta_c} e^{-\alpha z} e^{-j\beta z}. \qquad (7.78b)$$

Table 7-1: Expressions for α, β, η_{c}, u_{p}, and λ for various types of media.

	Any Medium	Lossless Medium $(\sigma = 0)$	Low-loss Medium $(\varepsilon''/\varepsilon' \ll 1)$	Good Conductor $(\varepsilon''/\varepsilon' \gg 1)$	Units
$\alpha =$	$\omega \left[\dfrac{\mu\varepsilon'}{2} \left[\sqrt{1 + \left(\dfrac{\varepsilon''}{\varepsilon'}\right)^2} - 1 \right] \right]^{1/2}$	0	$\dfrac{\sigma}{2}\sqrt{\dfrac{\mu}{\varepsilon}}$	$\sqrt{\pi f \mu \sigma}$	(Np/m)
$\beta =$	$\omega \left[\dfrac{\mu\varepsilon'}{2} \left[\sqrt{1 + \left(\dfrac{\varepsilon''}{\varepsilon'}\right)^2} + 1 \right] \right]^{1/2}$	$\omega\sqrt{\mu\varepsilon}$	$\omega\sqrt{\mu\varepsilon}$	$\sqrt{\pi f \mu \sigma}$	(rad/m)
$\eta_{\mathrm{c}} =$	$\sqrt{\dfrac{\mu}{\varepsilon'}}\left(1 - j\dfrac{\varepsilon''}{\varepsilon'}\right)^{-1/2}$	$\sqrt{\dfrac{\mu}{\varepsilon}}$	$\sqrt{\dfrac{\mu}{\varepsilon}}$	$(1+j)\dfrac{\alpha}{\sigma}$	(Ω)
$u_{\mathrm{p}} =$	ω/β	$1/\sqrt{\mu\varepsilon}$	$1/\sqrt{\mu\varepsilon}$	$\sqrt{4\pi f/\mu\sigma}$	(m/s)
$\lambda =$	$2\pi/\beta = u_{\mathrm{p}}/f$	u_{p}/f	u_{p}/f	u_{p}/f	(m)

Notes: $\varepsilon' = \varepsilon$; $\varepsilon'' = \sigma/\omega$; in free space, $\varepsilon = \varepsilon_0$, $\mu = \mu_0$; in practice, a material is considered a low-loss medium if $\varepsilon''/\varepsilon' = \sigma/\omega\varepsilon < 0.01$ and a good conducting medium if $\varepsilon''/\varepsilon' > 100$.

To determine α, β, and η_{c} for seawater, we begin by evaluating the ratio $\varepsilon''/\varepsilon'$. From the argument of the cosine function of $\mathbf{H}(0, t)$, we deduce that $\omega = 2\pi \times 10^3$ (rad/s), and therefore $f = 1$ kHz. Hence,

$$\frac{\varepsilon''}{\varepsilon'} = \frac{\sigma}{\omega\varepsilon} = \frac{\sigma}{\omega\varepsilon_{\mathrm{r}}\varepsilon_0} = \frac{4}{2\pi \times 10^3 \times 80 \times (10^{-9}/36\pi)}$$

$$= 9 \times 10^5.$$

Since $\varepsilon''/\varepsilon' \gg 1$, seawater is a good conductor at 1 kHz. This allows us to use the good-conductor expressions given in Table 7-1:

$$\alpha = \sqrt{\pi f \mu \sigma}$$

$$= \sqrt{\pi \times 10^3 \times 4\pi \times 10^{-7} \times 4}$$

$$= 0.126 \quad \text{(Np/m)}, \tag{7.79a}$$

$$\beta = \alpha = 0.126 \quad \text{(rad/m)}, \tag{7.79b}$$

$$\eta_{\mathrm{c}} = (1 + j)\frac{\alpha}{\sigma}$$

$$= (\sqrt{2}\, e^{j\pi/4})\frac{0.126}{4} = 0.044 e^{j\pi/4} \quad (\Omega). \tag{7.79c}$$

As no explicit information has been given about the electric field amplitude E_{x0}, we should assume it to be complex; that is, $E_{x0} = |E_{x0}|e^{j\phi_0}$. The wave's instantaneous electric and magnetic fields are then given by

$$\mathbf{E}(z, t) = \Re\mathfrak{e}\left[\hat{\mathbf{x}}|E_{x0}|e^{j\phi_0}e^{-\alpha z}e^{-j\beta z}e^{j\omega t}\right]$$

$$= \hat{\mathbf{x}}|E_{x0}|e^{-0.126z}\cos(2\pi \times 10^3 t - 0.126z + \phi_0)$$

$$\text{(V/m)}, \tag{7.80a}$$

$$\mathbf{H}(z, t) = \Re\mathfrak{e}\left[\hat{\mathbf{y}}\frac{|E_{x0}|e^{j\phi_0}}{0.044 e^{j\pi/4}}e^{-\alpha z}e^{-j\beta z}e^{j\omega t}\right]$$

$$= \hat{\mathbf{y}}22.5|E_{x0}|e^{-0.126z}\cos(2\pi \times 10^3 t$$

$$- 0.126z + \phi_0 - 45°) \quad \text{(A/m)}. \tag{7.80b}$$

At $z = 0$,

$$\mathbf{H}(0, t) = \hat{\mathbf{y}}\, 22.5|E_{x0}| \cos(2\pi \times 10^3 t + \phi_0 - 45°) \quad \text{(A/m)}. \tag{7.81}$$

By comparing Eq. (7.81) with the expression given in the problem statement,

$$\mathbf{H}(0, t) = \hat{\mathbf{y}}\, 100 \cos(2\pi \times 10^3 t + 15°) \quad \text{(mA/m)},$$

we deduce that

$$22.5|E_{x0}| = 100 \times 10^{-3}$$

or

$$|E_{x0}| = 4.44 \quad \text{(mV/m)},$$

and

$$\phi_0 - 45° = 15° \quad \text{or} \quad \phi_0 = 60°.$$

Hence, the final expressions for $\mathbf{E}(z, t)$ and $\mathbf{H}(z, t)$ are

$$\mathbf{E}(z, t) = \hat{\mathbf{x}}\, 4.44 e^{-0.126z} \cos(2\pi \times 10^3 t - 0.126z + 60°)$$

$$\text{(mV/m)}, \tag{7.82a}$$

$$\mathbf{H}(z, t) = \hat{\mathbf{y}}\, 100 e^{-0.126z} \cos(2\pi \times 10^3 t - 0.126z + 15°)$$

$$\text{(mA/m)}. \tag{7.82b}$$

(b) The depth at which the amplitude of $\mathbf{E}$ has decreased to 1% of its initial value at $z = 0$ is obtained from

$$0.01 = e^{-0.126z} \quad \text{or} \quad z = \frac{\ln(0.01)}{-0.126} = 36 \text{ m}. \quad \blacksquare$$

EXERCISE 7.7　The constitutive parameters of copper are $\mu = \mu_0 = 4\pi \times 10^{-7}$ (H/m), $\varepsilon = \varepsilon_0 \simeq (1/36\pi) \times 10^{-9}$ (F/m), and $\sigma = 5.8 \times 10^7$ (S/m). Assuming that these parameters are frequency independent, over what frequency range of the electromagnetic spectrum [see Fig. 1-15] is copper a good conductor?

Ans.　$f < 1.04 \times 10^{16}$ Hz, which includes the radio spectrum, the infrared and visible regions, and part of the ultraviolet region.

EXERCISE 7.8　Over what frequency range may dry soil, with $\varepsilon_r = 3$, $\mu_r = 1$, and $\sigma = 10^{-4}$ (S/m), be regarded as a low-loss dielectric medium?

Ans.　$f > 60$ MHz.

EXERCISE 7.9　For a wave traveling in a medium with a skin depth δ_s, what is the amplitude of $\mathbf{E}$ at a distance of $3\delta_s$ compared with its initial value?

Ans.　$e^{-3} \approx 0.05$ or 5%.

7-5　Current Flow in a Good Conductor

When a d-c voltage is connected across the ends of a conducting wire, the current flowing through the wire has a uniform current density $\mathbf{J}$ over the wire's cross section. That is, $\mathbf{J}$ has the same value along the axis of the wire as along its outer perimeter [Fig. 7-14(a)]. This is not true in the a-c case. As we will see shortly, the current density

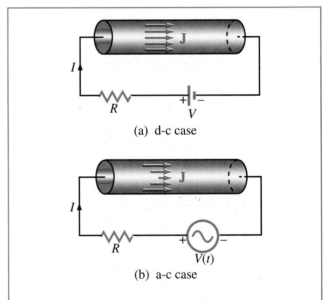

(a) d-c case

(b) a-c case

Figure 7-14:　Current density $\mathbf{J}$ in a conducting wire is (a) uniform across its cross section in the d-c case, but (b) in the a-c case, $\mathbf{J}$ is highest along the wire's perimeter.

in the time-varying case is maximum along the perimeter of the wire and decreases exponentially as a function of distance toward the axis of the wire [Fig. 7-14(b)]. In fact, at very high frequencies most of the current flows through a thin outer layer of the wire, and if the wire material is a perfect conductor, the current flows entirely along the surface of the wire.

Before we deal with the situation for a wire with circular cross section, let us consider the simpler geometry of a semi-infinite solid, as shown in Fig. 7-15(a). The conducting solid is infinite in depth and has a planar surface coincident with the x–y plane. If at $z = 0^-$ (just above the surface), an x-polarized EM field with $\widetilde{\mathbf{E}} = \hat{\mathbf{x}}E_0$ and $\widetilde{\mathbf{H}} = \hat{\mathbf{y}}E_0/\eta$ exists in the medium above the conductor, a similarly polarized EM field will be induced in the conducting medium and will constitute a plane wave that travels along the $+z$-direction. As a consequence of the boundary condition mandating that the tangential component of $\mathbf{E}$ be continuous across the boundary between any two contiguous media, at $z = 0^+$ (just below the boundary) the electric field of the wave is $\widetilde{\mathbf{E}}(0) = \hat{\mathbf{x}}E_0$. The fields of the EM wave at any depth z in the conductor are then given by

$$\widetilde{\mathbf{E}}(z) = \hat{\mathbf{x}}E_0 e^{-\alpha z} e^{-j\beta z}, \qquad (7.83\text{a})$$

$$\widetilde{\mathbf{H}}(z) = \hat{\mathbf{y}}\frac{E_0}{\eta_c} e^{-\alpha z} e^{-j\beta z}. \qquad (7.83\text{b})$$

From $\mathbf{J} = \sigma \mathbf{E}$, the current flows in the x-direction, and its density is

$$\widetilde{\mathbf{J}}(z) = \hat{\mathbf{x}}\,\widetilde{J}_x(z), \qquad (7.84)$$

with

$$\widetilde{J}_x(z) = \sigma E_0 e^{-\alpha z} e^{-j\beta z} = J_0 e^{-\alpha z} e^{-j\beta z}, \qquad (7.85)$$

where $J_0 = \sigma E_0$ is the magnitude of the current density at the surface. In terms of the skin depth $\delta_s = 1/\alpha$ defined by Eq. (7.72) and in view of Eq. (7.77b), which states that $\alpha = \beta$ for a good conductor, Eq. (7.85) can be written as

$$\widetilde{J}_x(z) = J_0 e^{-(1+j)z/\delta_s} \qquad (\text{A/m}^2). \qquad (7.86)$$

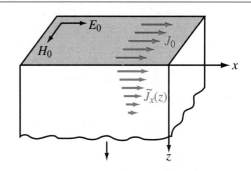

(a) Exponentially decaying $\widetilde{J}_x(z)$

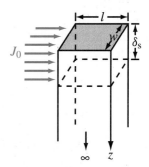

(b) Equivalent J_0 over skin depth δ_s

Figure 7-15: Exponential decay of current density $\widetilde{J}_x(z)$ with z in a solid conductor. The total current flowing through (a) a section of width w extending between $z = 0$ and $z = \infty$ is equivalent to (b) a constant current density J_0 flowing through a section of depth δ_s.

The current flowing through a rectangular strip extending between zero and ∞ in the z-direction and of width w in the y-direction is

$$\widetilde{I} = w \int_0^\infty \widetilde{J}_x(z)\, dz$$

$$= w \int_0^\infty J_0 e^{-(1+j)z/\delta_s}\, dz = \frac{J_0 w \delta_s}{(1+j)} \qquad (\text{A}). \quad (7.87)$$

The numerator of Eq. (7.87) is equivalent to a uniform

current density J_0 flowing through a thin surface section of width w and depth δ_s. Because $\widetilde{J}_x(z)$ decreases exponentially with depth z, a conductor with a finite thickness d may be treated in practice as infinitely deep as long as d is several skin depths in extent. If $d = 3\delta_s$ [instead of ∞ in the integral of Eq. (7.87)], the error incurred in using the result on the right-hand side of Eq. (7.87) is less than 5%; and if $d = 5\delta_s$, the error is less than 1%.

The voltage across a length l at the surface [Fig. 7-15(b)] is given by

$$\widetilde{V} = E_0 l = \frac{J_0}{\sigma}l. \qquad (7.88)$$

Hence, the impedance of a slab of width w, length l, and depth $d = \infty$ (or, in practice, $d > 5\delta_s$) is

$$Z = \frac{\widetilde{V}}{\widetilde{I}} = \frac{1+j}{\sigma\delta_s}\frac{l}{w} \qquad (\Omega). \qquad (7.89)$$

It is customary to represent Z as

$$Z = Z_s \frac{l}{w}, \qquad (7.90)$$

where Z_s, which is called the *internal* or *surface impedance* of the conductor, is defined as the impedance Z for a unit length $l = 1$ m and a unit width $w = 1$ m. Thus,

$$\boxed{Z_s = \frac{1+j}{\sigma\delta_s} \qquad (\Omega). \qquad (7.91)}$$

Since the reactive part of Z_s is positive, Z_s can be defined as

$$Z_s = R_s + j\omega L_s$$

with

$$R_s = \frac{1}{\sigma\delta_s} = \sqrt{\frac{\pi f \mu}{\sigma}} \qquad (\Omega), \qquad (7.92a)$$

$$L_s = \frac{1}{\omega\sigma\delta_s} = \frac{1}{2}\sqrt{\frac{\mu}{\pi f \sigma}} \qquad (H), \qquad (7.92b)$$

where we used the relation $\delta_s = 1/\alpha \simeq 1/\sqrt{\pi f \mu \sigma}$ given by Eq. (7.77a). In terms of the *surface resistance* R_s, the *a-c resistance* of a slab of width w and length l is

$$R = R_s \frac{l}{w} = \frac{l}{\sigma\delta_s w} \qquad (\Omega). \qquad (7.93)$$

The expression for the a-c resistance R is equivalent to the d-c resistance of a plane conductor of length l and cross section $A = \delta_s w$.

The results obtained for the planar conductor will now be extended to the coaxial cable shown in Fig. 7-16(a). If the conductors are made of copper with $\sigma = 5.8 \times 10^7$ S/m, the skin depth $\delta_s = 1/\sqrt{\pi f \mu \sigma} = 0.066$ mm at 1 MHz, and since δ_s varies as $1/\sqrt{f}$, it becomes smaller at higher frequencies. For the inner conductor, as long as its radius a is greater than $5\delta_s$, or 0.33 mm at 1 MHz, its "depth" may be regarded as semiinfinite, and a similar criterion applies to the thickness of the outer conductor. The current flowing

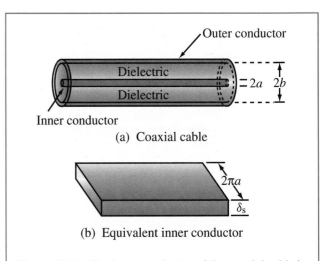

(a) Coaxial cable

(b) Equivalent inner conductor

Figure 7-16: The inner conductor of the coaxial cable in (a) is represented in (b) by a planar conductor of width $2\pi a$ and depth δ_s, as if its skin has been cut along its length on the bottom side and then unfurled into a planar geometry.

through the inner conductor is concentrated on its outer surface and is approximately equivalent to a uniform current flowing through a thin layer of depth δ_s and circumference $2\pi a$. This is equivalent to a planar conductor of width $w = 2\pi a$, as shown in Fig. 7-16(b). The corresponding resistance per unit length is obtained by setting $w = 2\pi a$ and dividing by l in Eq. (7.93):

$$R_1' = \frac{R}{l} = \frac{R_s}{2\pi a} \quad (\Omega/\text{m}). \qquad (7.94)$$

Similarly, for the outer conductor, the current is concentrated within a thin layer on the inside surface of the conductor adjacent to the insulating medium between the two conductors, which is where the EM fields exist. The resistance per unit length for the outer conductor with radius b is

$$R_2' = \frac{R_s}{2\pi b} \quad (\Omega/\text{m}), \qquad (7.95)$$

and the total a-c resistance per unit length is

$$R' = R_1' + R_2' = \frac{R_s}{2\pi}\left(\frac{1}{a} + \frac{1}{b}\right) \quad (\Omega/\text{m}). \qquad (7.96)$$

This expression was used in Chapter 2 for characterizing the resistance per unit length of a coaxial transmission line.

REVIEW QUESTIONS

Q7.6 How does β of a low-loss dielectric medium compare to that of a lossless medium?

Q7.7 In a good conductor, does the phase of **H** lead or lag that of **E** and by how much?

Q7.8 Attenuation means that a wave loses energy as it propagates in a lossy medium. What happens to the lost energy?

Q7.9 Is a conducting medium dispersive or dispersionless? Explain.

Q7.10 Compare the flow of current through a wire in the d-c and a-c cases. Compare the corresponding d-c and a-c resistances of the wire.

7-6 Electromagnetic Power Density

This section deals with the flow of power carried by an electromagnetic wave. For any wave with an electric field **E** and magnetic field **H**, the *Poynting vector* **S** is defined as

$$\mathbf{S} = \mathbf{E} \times \mathbf{H} \quad (\text{W/m}^2). \qquad (7.97)$$

The unit of **S** is $(\text{V/m}) \times (\text{A/m}) = (\text{W/m}^2)$, and the direction of **S** is along the propagation direction of the wave, $\hat{\mathbf{k}}$. Thus, **S** represents the power per unit area (power density) carried by the wave, and if the wave is incident upon an aperture of area A with outward surface unit vector $\hat{\mathbf{n}}$ as shown in Fig. 7-17, then the total power that flows through or is intercepted by the aperture is

$$P = \int_A \mathbf{S} \cdot \hat{\mathbf{n}} \, dA \quad (\text{W}). \qquad (7.98)$$

For a uniform plane wave propagating in a direction $\hat{\mathbf{k}}$ that makes an angle θ with $\hat{\mathbf{n}}$, $P = SA\cos\theta$, where $S = |\mathbf{S}|$.

Except for the fact that the units of **S** are per unit area, Eq. (7.97) is the vector analogue of the scalar expression

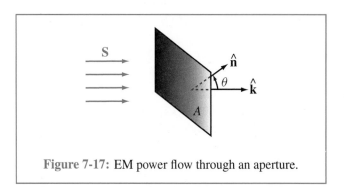

Figure 7-17: EM power flow through an aperture.

for the instantaneous power $P(z, t)$ flowing through a transmission line; that is,

$$P(z, t) = v(z, t) \, i(z, t), \qquad (7.99)$$

where $v(z, t)$ and $i(z, t)$ are the instantaneous voltage and current on the line.

Since both **E** and **H** are functions of time, so is the Poynting vector **S**. In practice, however, the quantity of greater interest is the *average power density* of the wave, $\mathbf{S}_{av}$, which is the time-average value of **S**. For a transmission line excited by a time-harmonic (sinusoidal) source, we showed in Chapter 2 that the time-average power flowing toward the load can be computed by applying Eq. (2.87):

$$P_{av}(z) = \tfrac{1}{2} \, \mathfrak{Re} \left[\tilde{V}(z) \, \tilde{I}^*(z) \right], \qquad (7.100)$$

where $\tilde{V}(z)$ and $\tilde{I}(z)$ are the phasors corresponding to $v(z, t)$ and $i(z, t)$, respectively. The analogous expression for an electromagnetic wave is

$$\mathbf{S}_{av} = \tfrac{1}{2} \, \mathfrak{Re} \left[\tilde{\mathbf{E}} \times \tilde{\mathbf{H}}^* \right] \qquad (\text{W/m}^2). \qquad (7.101)$$

7-6.1 Plane Wave in a Lossless Medium

The general expression for the phasor electric field of a uniform plane wave with arbitrary polarization, traveling in the $+z$-direction, is given by

$$\begin{aligned}
\tilde{\mathbf{E}}(z) &= \hat{\mathbf{x}} \, \tilde{E}_x(z) + \hat{\mathbf{y}} \, \tilde{E}_y(z) \\
&= (\hat{\mathbf{x}} \, E_{x0} + \hat{\mathbf{y}} \, E_{y0}) e^{-jkz}, \qquad (7.102)
\end{aligned}$$

where, in the general case, E_{x0} and E_{y0} may be complex quantities. The magnitude of $\tilde{\mathbf{E}}$ is

$$|\tilde{\mathbf{E}}| = (\tilde{\mathbf{E}} \cdot \tilde{\mathbf{E}}^*)^{1/2} = [|E_{x0}|^2 + |E_{y0}|^2]^{1/2}. \qquad (7.103)$$

The phasor magnetic field associated with $\tilde{\mathbf{E}}$ is obtained by applying Eq. (7.39a):

$$\tilde{\mathbf{H}}(z) = \frac{1}{\eta} \hat{\mathbf{z}} \times \tilde{\mathbf{E}} = \frac{1}{\eta} (-\hat{\mathbf{x}} E_{y0} + \hat{\mathbf{y}} E_{x0}) e^{-jkz}. \quad (7.104)$$

The wave can be considered to be the sum of two waves, one with $(\tilde{E}_x, \tilde{H}_y)$ and another with $(\tilde{E}_y, \tilde{H}_x)$. Using Eqs. (7.102) and (7.104) in Eq. (7.101) leads to

$$\begin{aligned}
\mathbf{S}_{av} &= \hat{\mathbf{z}} \frac{1}{2\eta} (|E_{x0}|^2 + |E_{y0}|^2) \\
&= \hat{\mathbf{z}} \frac{|\tilde{\mathbf{E}}|^2}{2\eta} \qquad (\text{W/m}^2), \qquad (7.105)
\end{aligned}$$

which states that the power flow is in the z-direction and the average power density is equal to the sum of the average power densities of the $(\tilde{E}_x, \tilde{H}_y)$ wave and the $(\tilde{E}_y, \tilde{H}_x)$ wave. Note that, because $\mathbf{S}_{av}$ depends only on η and $|\tilde{\mathbf{E}}|$, waves characterized by different polarizations carry the same amount of average power if their electric fields have the same magnitudes.

Example 7-5   Solar Power

If solar illumination is characterized by a power density of 1 kW/m² at Earth's surface, find (a) the total power radiated by the sun, (b) the total power intercepted by Earth, and (c) the electric field of the power density incident upon Earth's surface, assuming that all the solar illumination is at a single frequency. The radius of Earth's orbit around the sun, R_s, is approximately 1.5×10^8 km, and Earth's mean radius R_e is 6,380 km.

Solution: (a) Assuming that the sun radiates isotropically (equally in all directions), the total power it radiates is equal to $S_{av} A_{sph}$, where A_{sph} is the area of a spherical shell of radius R_s [Fig. 7-18(a)]. Thus,

$$\begin{aligned}
P_{sun} &= S_{av}(4\pi R_s^2) = 1 \times 10^3 \times 4\pi \times (1.5 \times 10^{11})^2 \\
&= 2.8 \times 10^{26} \text{ W.}
\end{aligned}$$

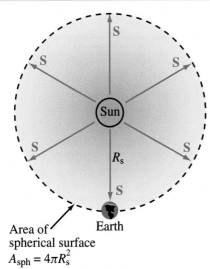

Area of spherical surface $A_{sph} = 4\pi R_s^2$

(a) Radiated solar power

$A_e = \pi R_e^2$ Earth

(b) Earth intercepted power

Figure 7-18: Solar radiation intercepted by (a) a spherical surface of radius R_s, and (b) Earth's surface (Example 7-5).

(b) With reference to Fig. 7-18(b), the power intercepted by Earth's cross section $A_e = \pi R_e^2$ is

$$P_{int} = S_{av}(\pi R_e^2) = 1 \times 10^3 \times \pi \times (6.38 \times 10^6)^2$$
$$= 1.28 \times 10^{17} \text{ W}.$$

(c) The power density S_{av} is related to the magnitude of the electric field $|E_0|$ by

$$S_{av} = \frac{|E_0|^2}{2\eta_0},$$

where $\eta_0 = 377$ (Ω) for air. Hence,

$$|E_0| = \sqrt{2\eta_0 S_{av}} = \sqrt{2 \times 377 \times 10^3}$$
$$= 870 \quad \text{(V/m)}. \quad \blacksquare$$

7-6.2 Plane Wave in a Lossy Medium

The expressions given by Eqs. (7.68) and (7.69) characterize the electric and magnetic fields of a z-directed, x-polarized plane wave propagating in a lossy medium with a propagation constant $\gamma = \alpha + j\beta$. By extending these expressions to the more general case of a wave with components along both x and y, we have

$$\widetilde{\mathbf{E}}(z) = \hat{\mathbf{x}}\widetilde{E}_x(z) + \hat{\mathbf{y}}\widetilde{E}_y(z)$$
$$= (\hat{\mathbf{x}}E_{x0} + \hat{\mathbf{y}}E_{y0})e^{-\alpha z}e^{-j\beta z}, \quad (7.106a)$$

$$\widetilde{\mathbf{H}}(z) = \frac{1}{\eta_c}(-\hat{\mathbf{x}}E_{y0} + \hat{\mathbf{y}}E_{x0})e^{-\alpha z}e^{-j\beta z}, \quad (7.106b)$$

where η_c is the intrinsic impedance of the lossy medium. Application of Eq. (7.101) gives

$$\mathbf{S}_{av}(z) = \frac{1}{2}\mathfrak{Re}\left[\widetilde{\mathbf{E}} \times \widetilde{\mathbf{H}}^*\right]$$
$$= \frac{\hat{\mathbf{z}}(|E_{x0}|^2 + |E_{y0}|^2)}{2}e^{-2\alpha z}\mathfrak{Re}\left(\frac{1}{\eta_c^*}\right). \quad (7.107)$$

By expressing η_c in polar form as

$$\eta_c = |\eta_c|e^{j\theta_\eta}, \quad (7.108)$$

Eq. (7.107) can be rewritten as

$$\boxed{\mathbf{S}_{av}(z) = \hat{\mathbf{z}}\frac{|E_0|^2}{2|\eta_c|}e^{-2\alpha z}\cos\theta_\eta \quad (\text{W/m}^2), \quad (7.109)}$$

where $|E_0| = [|E_{x0}|^2 + |E_{y0}|^2]^{1/2}$ is the magnitude of $\widetilde{\mathbf{E}}(z)$ at $z = 0$. Whereas the fields $\widetilde{\mathbf{E}}(z)$ and $\widetilde{\mathbf{H}}(z)$ decay with z as $e^{-\alpha z}$, the power density $\mathbf{S}_{av}$ decreases as $e^{-2\alpha z}$. When a wave propagates through a distance $z = \delta_s = 1/\alpha$, the magnitudes of its electric and magnetic fields decrease to $e^{-1} \approx 37\%$ of their initial values, and its average power density decreases to $e^{-2} \approx 14\%$ of its initial value.

7-6.3   Decibel Scale for Power Ratios

The unit for power P is watts (W). In many engineering problems, the quantity of interest is the ratio of two power levels, P_1 and P_2, such as the incident and reflected powers on a transmission line, and often the ratio P_1/P_2 may vary over several orders of magnitude. The decibel (dB) scale is logarithmic, thereby providing a convenient representation of the power ratio, particularly when numerical values of P_1/P_2 are plotted against some variable of interest. If

$$G = \frac{P_1}{P_2}, \qquad (7.110)$$

then

$$G \text{ [dB]} \triangleq 10 \log G = 10 \log \left(\frac{P_1}{P_2} \right) \quad \text{(dB).} \quad (7.111)$$

Table 7-2 provides a comparison between some values of G and the corresponding values of G [dB]. Even though decibels are defined for power ratios, they can sometimes be used to represent other quantities. For example, if $P_1 = V_1^2/R$ is the power dissipated in a resistor R with voltage V_1 across it at time t_1, and $P_2 = V_2^2/R$ is the power dissipated in the same resistor at time t_2, then

$$\begin{aligned} G \text{ [dB]} &= 10 \log \left(\frac{P_1}{P_2} \right) \\ &= 10 \log \left(\frac{V_1^2/R}{V_2^2/R} \right) \\ &= 20 \log \left(\frac{V_1}{V_2} \right) \\ &= 20 \log(g) \triangleq g \text{ [dB]}, \qquad (7.112) \end{aligned}$$

where $g = V_1/V_2$ is the voltage ratio. *Note that for voltage (or current) ratios the scale factor is 20 rather than 10,* which results in G [dB] $= g$ [dB].

Table 7-2: Power ratios in natural numbers and in decibels.

G	G [dB]
10^x	$10x$ dB
4	6 dB
2	3 dB
1	0 dB
0.5	-3 dB
0.25	-6 dB
0.1	-10 dB
10^{-3}	-30 dB

The *attenuation rate*, representing the rate of decrease of the magnitude of $\mathbf{S}_{av}(z)$ as a function of propagation distance, is defined as

$$\begin{aligned} A &= 10 \log \left[\frac{S_{av}(z)}{S_{av}(0)} \right] \\ &= 10 \log(e^{-2\alpha z}) \\ &= -20\alpha z \log e \\ &= -8.68 \alpha z = -\alpha \text{ [dB/m] } z \quad \text{(dB),} \quad (7.113) \end{aligned}$$

where

$$\alpha \text{ [dB/m]} \triangleq 8.68\alpha \text{ [Np/m].} \qquad (7.114)$$

We also note that, since $\mathbf{S}_{av}(z)$ is directly proportional to $|\mathbf{E}(z)|^2$,

$$A = 10 \log \left[\frac{|E(z)|^2}{|E(0)|^2} \right] = 20 \log \left[\frac{|E(z)|}{|E(0)|} \right] \quad \text{(dB).} \tag{7.115}$$

Example 7-6   Power Received by a Submarine Antenna

A submarine at a depth of 200 m uses a wire antenna to receive signal transmissions at 1 kHz. Determine the power density incident upon the submarine antenna due to the EM wave of Example 7-4.

Solution: From Example 7-4, $|E_0| = |E_{x0}| = 4.44$ (mV/m), $\alpha = 0.126$ (Np/m), and $\eta_c = 0.044\underline{/45°}$ (Ω). Application of Eq. (7.109) gives

$$\mathbf{S}_{av}(z) = \hat{\mathbf{z}}\frac{|E_0|^2}{2|\eta_c|}e^{-2\alpha z}\cos\theta_\eta$$

$$= \hat{\mathbf{z}}\frac{(4.44 \times 10^{-3})^2}{2 \times 0.044}e^{-0.252z}\cos 45°$$

$$= \hat{\mathbf{z}}\,0.16e^{-0.252z} \quad (\text{mW/m}^2).$$

At $z = 200$ m, the incident power density is

$$\mathbf{S}_{av} = \hat{\mathbf{z}}(0.16 \times 10^{-3}e^{-0.252 \times 200})$$

$$= 2.1 \times 10^{-26} \quad (\text{W/m}^2). \quad \blacksquare$$

EXERCISE 7.10 Convert the following values of the power ratio G from natural numbers to decibels: (a) 2.3, (b) 4×10^3, (c) 3×10^{-2}.

Ans. (a) 3.6 dB, (b) 36 dB, (c) -15.2 dB.

EXERCISE 7.11 Find the voltage ratio g in natural units corresponding to the following decibel values of the power ratio G: (a) 23 dB, (b) -14 dB, (c) -3.6 dB.

Ans. (a) 14.13, (b) 0.2, (c) 0.66.

CHAPTER HIGHLIGHTS

- A spherical wave radiated by a source becomes approximately a uniform plane wave at large distances from the source.

- The electric and magnetic fields of a transverse electromagnetic (TEM) wave are orthogonal to each other, and both are perpendicular to the direction of wave travel.

- The magnitudes of the electric and magnetic fields of a TEM wave are related by the intrinsic impedance of the medium.

- Wave polarization describes the shape and locus of the tip of the **E** vector at a given point in space as a function of time. The polarization state, which may be linear, circular, or elliptical, is governed by the ratio of the magnitudes of and the difference in phase between the two orthogonal components of the electric field vector.

- Media are classified as lossless, low-loss, quasi-conducting, or good conducting on the basis of the ratio $\varepsilon''/\varepsilon' = \sigma/\omega\varepsilon$.

- Unlike the d-c case, wherein the current flowing through a wire is distributed uniformly across its cross section, in the a-c case most of the current is concentrated along the outer perimeter of the wire.

- Power density carried by a plane EM wave traveling in an unbounded medium is akin to the power carried by the voltage/current wave on a transmission line.

GLOSSARY OF IMPORTANT TERMS

Provide definitions or explain the meaning of the following terms:

guided medium
unbounded media
spherical wave
uniform plane wave
complex permittivity ε_c
wavenumber k
TEM wave
intrinsic impedance η
wave polarization
elliptical polarization
circular polarization
linear polarization
LHC and RHC polarizations
attenuation constant α
phase constant β
skin depth δ_s
low-loss dielectric
quasi-conductor
good conductor

internal or surface impedance
surface resistance R_s
d-c and a-c resistances
Poynting vector **S**
average power density $\mathbf{S}_{av}$

PROBLEMS

Section 7-2: Propagation in Lossless Media

7.1* The magnetic field of a wave propagating through a certain nonmagnetic material is given by

$$\mathbf{H} = \hat{\mathbf{z}}\, 50 \cos(10^9 t - 5y) \quad \text{(mA/m)}$$

Find the following:
(a) The direction of wave propagation.
(b) The phase velocity.
(c) The wavelength in the material.
(d) The relative permittivity of the material.
(e) The electric field phasor.

7.2 Write general expressions for the electric and magnetic fields of a 1-GHz sinusoidal plane wave traveling in the $+y$-direction in a lossless nonmagnetic medium with relative permittivity $\varepsilon_r = 9$. The electric field is polarized along the x-direction, its peak value is 3 V/m, and its intensity is 2 V/m at $t = 0$ and $y = 2$ cm.

7.3* The electric field phasor of a uniform plane wave is given by $\widetilde{\mathbf{E}} = \hat{\mathbf{y}}\, 10 e^{j0.2z}$ (V/m). If the phase velocity of the wave is 1.5×10^8 m/s and the relative permeability of the medium is $\mu_r = 2.4$, find the following:
(a) The wavelength.
(b) The frequency f of the wave.
(c) The relative permittivity of the medium.
(d) The magnetic field $\mathbf{H}(z, t)$.

---

*Answer(s) available in Appendix D.
 Solution available in CD-ROM.

7.4 The electric field of a plane wave propagating in a nonmagnetic material is given by

$$\mathbf{E} = [\hat{\mathbf{y}}\, 3 \sin(2\pi \times 10^7 t - 0.4\pi x)$$
$$+ \hat{\mathbf{z}}\, 4 \cos(2\pi \times 10^7 t - 0.4\pi x)] \quad \text{(V/m)}$$

Determine:
(a) The wavelength.
(b) ε_r.
(c) **H**.

7.5* A wave radiated by a source in air is incident upon a soil surface, whereupon a part of the wave is transmitted into the soil medium. If the wavelength of the wave is 30 cm in air and 15 cm in the soil medium, what is the soil's relative permittivity? Assume the soil to be a very low-loss medium.

7.6 The electric field of a plane wave propagating in a lossless, nonmagnetic, dielectric material with $\varepsilon_r = 2.56$ is given by

$$\mathbf{E} = \hat{\mathbf{y}}\, 20 \cos(8\pi \times 10^9 t - kz) \quad \text{(V/m)}$$

Determine:
(a) $f, u_p, \lambda, k,$ and η.
(b) The magnetic field **H**.

Section 7-3: Wave Polarization

7.7* An RHC-polarized wave with a modulus of 2 (V/m) is traveling in free space in the negative z-direction. Write the expression for the wave's electric field vector, given that the wavelength is 6 cm.

7.8 For a wave characterized by the electric field

$$\mathbf{E}(z, t) = \hat{\mathbf{x}} a_x \cos(\omega t - kz) + \hat{\mathbf{y}} a_y \cos(\omega t - kz + \delta)$$

identify the polarization state, determine the polarization angles (γ, χ), and sketch the locus of $\mathbf{E}(0, t)$ for each of the following cases:
(a) $a_x = 3$ V/m, $a_y = 4$ V/m, and $\delta = 0$
(b) $a_x = 3$ V/m, $a_y = 4$ V/m, and $\delta = 180°$
(c) $a_x = 3$ V/m, $a_y = 3$ V/m, and $\delta = 45°$
(d) $a_x = 3$ V/m, $a_y = 4$ V/m, and $\delta = -135°$

7.9* The electric field of a uniform plane wave propagating in free space is given by

$$\widetilde{\mathbf{E}} = (\hat{\mathbf{x}} + j\hat{\mathbf{y}})20e^{-j\pi z/6} \quad \text{(V/m)}$$

Specify the modulus and direction of the electric field intensity at the $z = 0$ plane at $t = 0$, 5, and 10 ns.

7.10 A linearly polarized plane wave of the form $\widetilde{\mathbf{E}} = \hat{\mathbf{x}}a_x e^{-jkz}$ can be expressed as the sum of an RHC polarized wave with magnitude a_R, and an LHC polarized wave with magnitude a_L. Prove this statement by finding expressions for a_R and a_L in terms of a_x.

7.11* The electric field of an elliptically polarized plane wave is given by

$$\mathbf{E}(z, t) = [-\hat{\mathbf{x}}\, 10 \sin(\omega t - kz - 60°)$$
$$+ \hat{\mathbf{y}}\, 20 \cos(\omega t - kz)] \quad \text{(V/m)}$$

Determine the following:
(a) The polarization angles (γ, χ).
(b) The direction of rotation.

7.12 Compare the polarization states of each of the following pairs of plane waves:
(a) Wave 1: $\mathbf{E}_1 = \hat{\mathbf{x}}2 \cos(\omega t - kz) + \hat{\mathbf{y}}2 \sin(\omega t - kz)$.
    Wave 2: $\mathbf{E}_2 = \hat{\mathbf{x}}2 \cos(\omega t + kz) + \hat{\mathbf{y}}2 \sin(\omega t + kz)$.
(b) Wave 1: $\mathbf{E}_1 = \hat{\mathbf{x}}2 \cos(\omega t - kz) - \hat{\mathbf{y}}2 \sin(\omega t - kz)$.
    Wave 2: $\mathbf{E}_2 = \hat{\mathbf{x}}2 \cos(\omega t + kz) - \hat{\mathbf{y}}2 \sin(\omega t + kz)$.

7.13 Plot the locus of $\mathbf{E}(0, t)$ for a plane wave with

$$\mathbf{E}(z, t) = \hat{\mathbf{x}} \sin(\omega t + kz) + \hat{\mathbf{y}}\, 2 \cos(\omega t + kz)$$

Determine the polarization state from your plot.

Sections 7-4: Propagation in a Lossy Medium

7.14 For each of the following combinations of parameters, determine if the material is a low-loss dielectric, a quasi-conductor, or a good conductor, and then calculate α, β, λ, u_p, and η_c:
(a) Glass with $\mu_r = 1$, $\varepsilon_r = 5$, and $\sigma = 10^{-12}$ S/m at 10 GHz.
(b) Animal tissue with $\mu_r = 1$, $\varepsilon_r = 12$, and $\sigma = 0.3$ S/m at 100 MHz.
(c) Wood with $\mu_r = 1$, $\varepsilon_r = 3$, and $\sigma = 10^{-4}$ S/m at 1 kHz.

7.15 Dry soil is characterized by $\varepsilon_r = 2.5$, $\mu_r = 1$, and $\sigma = 10^{-4}$ (S/m). At each of the following frequencies, determine if dry soil may be considered a good conductor, a quasi-conductor, or a low-loss dielectric, and then calculate α, β, λ, μ_p, and η_c:
(a) 60 Hz
(b) 1 kHz
(c) 1 MHz
(d) 1 GHz

7.16 In a medium characterized by $\varepsilon_r = 9$, $\mu_r = 1$, and $\sigma = 0.1$ S/m, determine the phase angle by which the magnetic field leads the electric field at 100 MHz.

7.17 Generate a plot for the skin depth δ_s versus frequency for seawater for the range from 1 kHz to 10 GHz (use log–log scales). The constitutive parameters of seawater are $\mu_r = 1$, $\varepsilon_r = 80$, and $\sigma = 4$ S/m.

7.18 Ignoring reflection at the air–soil boundary, if the amplitude of a 2-GHz incident wave is 10 V/m at the surface of a wet soil medium, at what depth will it be down to 1 mV/m? Wet soil is characterized by $\mu_r = 1$, $\varepsilon_r = 16$, and $\sigma = 5 \times 10^{-4}$ S/m.

7.19* The skin depth of a certain nonmagnetic conducting material is 2 μm at 5 GHz. Determine the phase velocity in the material.

7.20 Based on wave attenuation and reflection measurements conducted at 1 MHz, it was determined that the intrinsic impedance of a certain medium is $28.1\underline{/45°}$ (Ω) and the skin depth is 5 m. Determine the following:

(a) The conductivity of the material.
(b) The wavelength in the medium.
(c) The phase velocity.

7.21* The electric field of a plane wave propagating in a nonmagnetic medium is given by

$$\mathbf{E} = \hat{\mathbf{z}}\, 25 e^{-30x} \cos(2\pi \times 10^9 t - 40x) \qquad \text{(V/m)}$$

Obtain the corresponding expression for $\mathbf{H}$.

Section 7-5: Current Flow in Conductors

7.22 In a nonmagnetic, lossy, dielectric medium, a 300-MHz plane wave is characterized by the magnetic field phasor

$$\tilde{\mathbf{H}} = (\hat{\mathbf{x}} - j4\hat{\mathbf{z}}) e^{-2y} e^{-j9y} \qquad \text{(A/m)}$$

Obtain time-domain expressions for the electric and magnetic field vectors.

7.23* A rectangular copper block is 30 cm in height (along z). In response to a wave incident upon the block from above, a current is induced in the block in the positive x-direction. Determine the ratio of the a-c resistance of the block to its d-c resistance at 1 kHz. The relevant properties of copper are given in Appendix B.

7.24 The inner and outer conductors of a coaxial cable have radii of 0.5 cm and 1 cm, respectively. The conductors are made of copper with $\varepsilon_r = 1$, $\mu_r = 1$, and $\sigma = 5.8 \times 10^7$ S/m, and the outer conductor is 0.1 cm thick. At 10 MHz:
(a) Are the conductors thick enough to be considered infinitely thick as far as the flow of current through them is concerned?
(b) Determine the surface resistance R_s.
(c) Determine the a-c resistance per unit length of the cable.

Section 7-6: EM Power Density

7.25* The magnetic field of a plane wave traveling in air is given by $\mathbf{H} = \hat{\mathbf{x}}\, 25 \sin(2\pi \times 10^7 t - ky)$ (mA/m). Determine the average power density carried by the wave.

7.26 A wave traveling in a nonmagnetic medium with $\varepsilon_r = 9$ is characterized by an electric field given by

$$\mathbf{E} = [\hat{\mathbf{y}}\, 3 \cos(\pi \times 10^7 t + kx)$$
$$- \hat{\mathbf{z}}\, 2 \cos(\pi \times 10^7 t + kx)] \qquad \text{(V/m)}$$

Determine the direction of wave travel and average power density carried by the wave.

7.27* The electric-field phasor of a uniform plane wave traveling downward in water is given by

$$\tilde{\mathbf{E}} = \hat{\mathbf{x}}\, 10 e^{-0.2z} e^{-j0.2z} \qquad \text{(V/m)}$$

where $\hat{\mathbf{z}}$ is the downward direction and $z = 0$ is the water surface. If $\sigma = 4$ S/m,
(a) Obtain an expression for the average power density.
(b) Determine the attenuation rate.
(c) Determine the depth at which the power density has been reduced by 40 dB.

7.28 The amplitudes of an elliptically polarized plane wave traveling in a lossless, nonmagnetic medium with $\varepsilon_r = 4$ are $H_{y0} = 6$ (mA/m) and $H_{z0} = 8$ (mA/m). Determine the average power flowing through an aperture in the y–z plane if its area is 20 m^2.

7.29* A wave traveling in a lossless, nonmagnetic medium has an electric field amplitude of 24.56 V/m and an average power density of 4 W/m^2. Determine the phase velocity of the wave.

7.30 At microwave frequencies, the power density considered safe for human exposure is 1 (mW/cm^2). A radar radiates a wave with an electric field amplitude E that decays with distance as $E(R) = (3{,}000/R)$ (V/m), where R is the distance in meters. What is the radius of the unsafe region?

7.31 Consider the imaginary rectangular box shown in Fig. 7-19.

(a) Determine the net power flux $P(t)$ entering the box due to a plane wave in air given by

$$\mathbf{E} = \hat{\mathbf{x}} E_0 \cos(\omega t - ky) \quad \text{(V/m)}$$

(b) Determine the net time-average power entering the box.

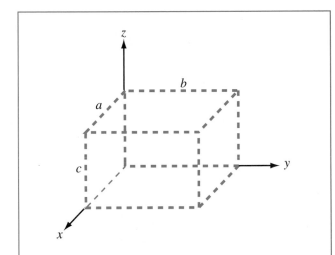

Figure 7-19: Imaginary rectangular box of Problems 7.31 and 7.32.

7.32 Repeat Problem 7.31 for a wave traveling in a lossy medium in which

$$\mathbf{E} = \hat{\mathbf{x}}\, 100 e^{-30y} \cos(2\pi \times 10^9 t - 40y) \quad \text{(V/m)}$$

$$\mathbf{H} = -\hat{\mathbf{z}}\, 0.64 e^{-30y} \cos(2\pi \times 10^9 t - 40y - 36.85°)$$

$$\text{(A/m)}$$

The box has dimensions $A = 1$ cm, $b = 2$ cm, and $c = 0.5$ cm.

7.33 Given a wave with

$$\mathbf{E} = \hat{\mathbf{x}} E_0 \cos(\omega t - kz)$$

calculate:

(a) The time-average electric energy density

$$(w_e)_{av} = \frac{1}{T} \int_0^T w_e \, dt = \frac{1}{2T} \int_0^T \varepsilon E^2 \, dt$$

(b) The time-average magnetic energy density

$$(w_m)_{av} = \frac{1}{T} \int_0^T w_m \, dt = \frac{1}{2T} \int_0^T \mu H^2 \, dt$$

(c) Show that $(w_e)_{av} = (w_m)_{av}$.

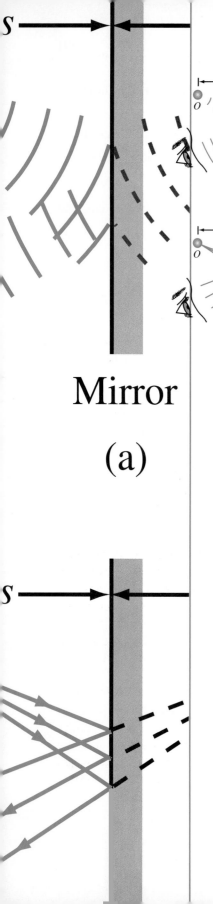

Mirror

(a)

CHAPTER 8

Wave Reflection and Transmission, and Geometric Optics

EM Waves at Boundaries

The sketch in Fig. 8-1 depicts the propagation path that a signal travels between a shipboard transmitter and a receiver on a submerged submarine. Let us use this communication system to examine the wave-related processes that take place along the signal path. Starting with the transmitter (denoted Tx for short in Fig. 8-1), the signal travels along a transmission line to the antenna. The relationship between the transmitter (generator) output power, P_t, and the power supplied to the antenna is governed by the transmission-line equations given in Chapter 2. If the transmission line is approximately lossless and if it is properly matched to the transmitter antenna, then all of P_t is delivered to the antenna. The next wave-related process is that of radiation; that is, converting the guided wave provided to the antenna by the transmission line into a spherical wave radiated outward into space. The radiation process is the subject of Chapter 9. From point 1, denoting the location of the shipboard antenna, to point 2, denoting the point of incidence of the wave onto the water's surface, the signal is governed by the equations characterizing wave propagation in a lossless medium, which we covered in Chapter 7. As the wave impinges upon the air–water boundary, a part of it gets reflected by the surface and another part gets transmitted across the boundary into the water medium. The transmitted part undergoes refraction, wherein the direction of wave travel moves closer toward the vertical. The reflection and transmission processes are treated in this chapter. Wave travel from point 3, representing a point just below the water surface, to point 4, denoting the location of the submarine antenna, is subject to the laws of wave propagation in a lossy medium, which also were treated in Chapter 7. The final step involves intercepting the wave incident

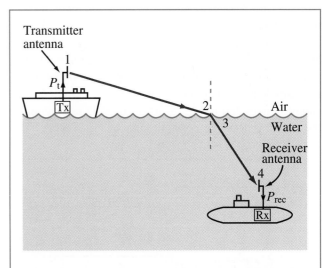

Figure 8-1: Signal path between a shipboard transmitter (Tx) and a submarine receiver (Rx).

upon the receiver antenna and converting its power into a received power, P_{rec}, for delivery via a transmission line to the receiver. The receiving properties of antennas are covered in Chapter 9. In summary, then, each wave-related aspect of the transmission process depicted in Fig. 8-1, starting with the transmitter and ending up with the receiver, is treated in some section or chapter in this book.

This chapter begins with examinations of the reflection and transmission properties of plane waves incident upon planar boundaries and concludes with sections on geometric optics, in which we discuss the imaging properties of mirrors and lenses. The in-between topics include discussions of power relations, and fiber optics.

8-1 Wave Reflection and Transmission at Normal Incidence

We know from Chapter 2 that when a guided wave traveling along a transmission line encounters an impedance discontinuity, such as that shown in Fig. 8-2(a) at the boundary between two lines with different characteristic impedances, the incident wave is partly reflected back toward the source and partly transmitted across the boundary into the second line. A similar process applies to a uniform plane wave propagating in an *unbounded* medium when it encounters a boundary. In fact, the situation depicted in Fig. 8-2(b) is exactly analogous to the transmission-line configuration of Fig. 8-2(a). The

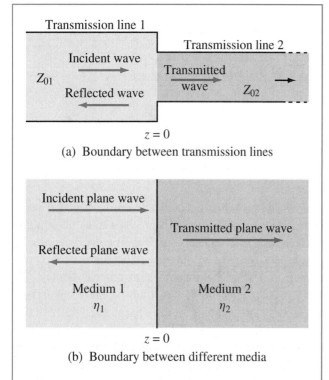

(a) Boundary between transmission lines

(b) Boundary between different media

Figure 8-2: Discontinuity between two different transmission lines is analogous to that between two dissimilar media.

boundary conditions governing the relationships between the electric and magnetic fields of the incident, reflected, and transmitted waves in Fig. 8-2(b) are similar to those we developed in Chapter 2 for the voltages and currents of the corresponding waves on the transmission line.

For convenience, we divide our treatment of wave reflection by and transmission through planar boundaries into two parts: in this section we confine our discussion to the normal-incidence case depicted in Fig. 8-3(a), and in Sections 8-2 to 8-4 we will examine the more general oblique-incidence situation depicted in Fig. 8-3(b). We will show the basis for the analogy between the transmission-line and plane-wave configurations so that we may use transmission-line equivalent models for solving plane-wave problems.

Before we proceed with our treatment, however, we should explain the relationship between rays and wavefronts, as both will be used to represent the propagation of electromagnetic waves. A *ray* is a line drawn to represent the direction of flow of electromagnetic energy carried by the wave, and therefore it is parallel to the propagation unit vector $\hat{\mathbf{k}}$ and orthogonal to the *wavefront*. The ray representation of wave incidence, reflection, and transmission shown in Fig. 8-3(b) is equivalent to the wavefront representation depicted in Fig. 8-3(c). The two representations are complimentary; the ray representation is easier to use in graphical illustrations, whereas the wavefront representation provides greater physical insight when examining what happens to a wave when it encounters a discontinuous boundary. Both representations will be used in our forthcoming discussions.

8-1.1 Boundary between Lossless Media

The planar boundary located at $z = 0$ in Fig. 8-4(a) separates two lossless, homogeneous, dielectric media. Medium 1, defined for $z \leq 0$, is characterized by (ε_1, μ_1), and medium 2, defined for $z \geq 0$, is characterized by (ε_2, μ_2). In medium 1, an incident x-polarized plane wave with fields $(\mathbf{E}^i, \mathbf{H}^i)$ is traveling in direction

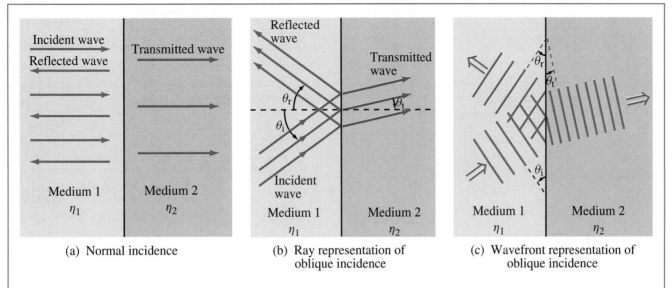

Figure 8-3: Ray representation of wave reflection and transmission at (a) normal incidence and (b) oblique incidence, and (c) wavefront representation of oblique incidence.

$\hat{\mathbf{k}}_i = \hat{\mathbf{z}}$ toward medium 2. Reflection and transmission at the discontinuous boundary result in a reflected wave $(\mathbf{E}^r, \mathbf{H}^r)$ with $\hat{\mathbf{k}}_r = -\hat{\mathbf{z}}$ in medium 1 and a transmitted wave $(\mathbf{E}^t, \mathbf{H}^t)$ with $\hat{\mathbf{k}}_t = \hat{\mathbf{z}}$ in medium 2. On the basis of the formulations developed in Sections 7-2 and 7-3 for characterizing the fields of a TEM wave, the three waves can be described in phasor form by:

Incident Wave

$$\widetilde{\mathbf{E}}^i(z) = \hat{\mathbf{x}} E_0^i e^{-jk_1 z}, \tag{8.1a}$$

$$\widetilde{\mathbf{H}}^i(z) = \hat{\mathbf{z}} \times \frac{\widetilde{\mathbf{E}}^i(z)}{\eta_1} = \hat{\mathbf{y}} \frac{E_0^i}{\eta_1} e^{-jk_1 z}. \tag{8.1b}$$

Reflected Wave

$$\widetilde{\mathbf{E}}^r(z) = \hat{\mathbf{x}} E_0^r e^{jk_1 z}, \tag{8.2a}$$

$$\widetilde{\mathbf{H}}^r(z) = (-\hat{\mathbf{z}}) \times \frac{\widetilde{\mathbf{E}}^r(z)}{\eta_1} = -\hat{\mathbf{y}} \frac{E_0^r}{\eta_1} e^{jk_1 z}. \tag{8.2b}$$

Transmitted Wave

$$\widetilde{\mathbf{E}}^t(z) = \hat{\mathbf{x}} E_0^t e^{-jk_2 z}, \tag{8.3a}$$

$$\widetilde{\mathbf{H}}^t(z) = \hat{\mathbf{z}} \times \frac{\widetilde{\mathbf{E}}^t(z)}{\eta_2} = \hat{\mathbf{y}} \frac{E_0^t}{\eta_2} e^{-jk_2 z}. \tag{8.3b}$$

The quantities E_0^i, E_0^r, and E_0^t are, respectively, the amplitudes of the incident, reflected, and transmitted electric fields, all specified at $t = 0$ and $z = 0$ (the boundary between the two media). The wavenumber and intrinsic impedance of medium 1 are $k_1 = \omega\sqrt{\mu_1 \varepsilon_1}$ and $\eta_1 = \sqrt{\mu_1/\varepsilon_1}$ and, similarly, $k_2 = \omega\sqrt{\mu_2 \varepsilon_2}$ and $\eta_2 = \sqrt{\mu_2/\varepsilon_2}$ for medium 2.

The amplitude E_0^i is related to the source responsible for generating the incident wave, and therefore it is assumed to be a known quantity. Our goal is to relate E_0^r and E_0^t each to E_0^i. We do so by applying boundary conditions for $\widetilde{\mathbf{E}}$ and $\widetilde{\mathbf{H}}$ at $z = 0$. According to Table 6-2, the tangential component of $\mathbf{E}$ is always continuous across a boundary between two contiguous media, and in the

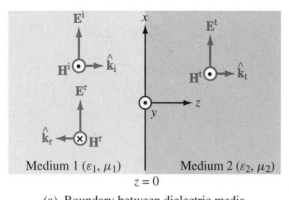

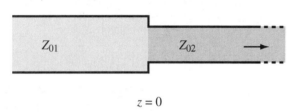

Figure 8-4: The two dielectric media separated by the x–y plane in (a) can be represented by the transmission-line analogue in (b).

governed by the values of the intrinsic impedances of the two media, η_1 and η_2.

The total electric field $\widetilde{\mathbf{E}}_1(z)$ in medium 1 is the sum of the electric fields of the incident and reflected waves, and a similar statement applies to the magnetic field $\widetilde{\mathbf{H}}_1(z)$. Hence,

Medium 1

$$\widetilde{\mathbf{E}}_1(z) = \widetilde{\mathbf{E}}^{\mathrm{i}}(z) + \widetilde{\mathbf{E}}^{\mathrm{r}}(z)$$
$$= \hat{\mathbf{x}}(E_0^{\mathrm{i}} e^{-jk_1 z} + E_0^{\mathrm{r}} e^{jk_1 z}), \qquad (8.4a)$$
$$\widetilde{\mathbf{H}}_1(z) = \widetilde{\mathbf{H}}^{\mathrm{i}}(z) + \widetilde{\mathbf{H}}^{\mathrm{r}}(z)$$
$$= \hat{\mathbf{y}} \frac{1}{\eta_1} (E_0^{\mathrm{i}} e^{-jk_1 z} - E_0^{\mathrm{r}} e^{jk_1 z}). \qquad (8.4b)$$

With only the transmitted wave present in medium 2, the fields are

Medium 2

$$\widetilde{\mathbf{E}}_2(z) = \widetilde{\mathbf{E}}^{\mathrm{t}}(z) = \hat{\mathbf{x}} E_0^{\mathrm{t}} e^{-jk_2 z}, \qquad (8.5a)$$
$$\widetilde{\mathbf{H}}_2(z) = \widetilde{\mathbf{H}}^{\mathrm{t}}(z) = \hat{\mathbf{y}} \frac{E_0^{\mathrm{t}}}{\eta_2} e^{-jk_2 z}. \qquad (8.5b)$$

At the boundary ($z = 0$), the tangential components of the electric and magnetic fields are continuous. Hence,

$$\widetilde{\mathbf{E}}_1(0) = \widetilde{\mathbf{E}}_2(0) \quad \text{or} \quad E_0^{\mathrm{i}} + E_0^{\mathrm{r}} = E_0^{\mathrm{t}}, \qquad (8.6a)$$
$$\widetilde{\mathbf{H}}_1(0) = \widetilde{\mathbf{H}}_2(0) \quad \text{or} \quad \frac{E_0^{\mathrm{i}}}{\eta_1} - \frac{E_0^{\mathrm{r}}}{\eta_1} = \frac{E_0^{\mathrm{t}}}{\eta_2}. \qquad (8.6b)$$

Simultaneous solutions for E_0^{r} and E_0^{t} in terms of E_0^{i} give

$$E_0^{\mathrm{r}} = \left(\frac{\eta_2 - \eta_1}{\eta_2 + \eta_1} \right) E_0^{\mathrm{i}} = \Gamma E_0^{\mathrm{i}}, \qquad (8.7a)$$
$$E_0^{\mathrm{t}} = \left(\frac{2\eta_2}{\eta_2 + \eta_1} \right) E_0^{\mathrm{i}} = \tau E_0^{\mathrm{i}}, \qquad (8.7b)$$

absence of current sources at the boundary, the tangential component of **H** also is continuous across the boundary. In the present case, both **E** and **H** of the normally incident wave are tangential to the boundary. Consequently, since no free charges or currents exist at the boundary, the fields of the reflected and transmitted waves will have tangential components only. In Fig. 8-4(a) and correspondingly in Eqs. (8.2a) and (8.3a), we arbitrarily chose the directions of $\widetilde{\mathbf{E}}^{\mathrm{r}}$ and $\widetilde{\mathbf{E}}^{\mathrm{t}}$ to coincide with the direction of $\widetilde{\mathbf{E}}^{\mathrm{i}}$ along the positive x-direction. Their true directions, relative to the assumed directions, will be determined by the polarities of the amplitudes E_0^{r} and E_0^{t}. As we will see shortly, both the magnitudes and polarities of these two amplitudes are

where

$$\Gamma = \frac{E_0^r}{E_0^i} = \frac{\eta_2 - \eta_1}{\eta_2 + \eta_1} \quad \text{(normal incidence)}, \quad (8.8a)$$

$$\tau = \frac{E_0^t}{E_0^i} = \frac{2\eta_2}{\eta_2 + \eta_1} \quad \text{(normal incidence)}. \quad (8.8b)$$

The quantities Γ and τ are called the *reflection coefficient* and *transmission coefficient*, respectively. For lossless dielectric media, η_1 and η_2 are real quantities; consequently, both Γ and τ are real also. As we will see in Section 8-1.4, the expressions given by Eqs. (8.8a) and (8.8b) are equally applicable when the media are conductive, but in that case η_1 and η_2 may be complex, and hence Γ and τ may be complex as well. From Eqs. (8.8a) and (8.8b), it can be easily shown that Γ and τ are interrelated by the simple formula

$$\tau = 1 + \Gamma \quad \text{(normal incidence)}. \quad (8.9)$$

For nonmagnetic media,

$$\eta_1 = \frac{\eta_0}{\sqrt{\varepsilon_{r_1}}} ,$$

$$\eta_2 = \frac{\eta_0}{\sqrt{\varepsilon_{r_2}}} ,$$

where η_0 is the intrinsic impedance of free space, in which case Eq. (8.8a) may be rewritten as

$$\Gamma = \frac{\sqrt{\varepsilon_{r_1}} - \sqrt{\varepsilon_{r_2}}}{\sqrt{\varepsilon_{r_1}} + \sqrt{\varepsilon_{r_2}}} \quad \text{(nonmagnetic media)}. \quad (8.10)$$

8-1.2 Transmission-Line Analogue

The transmission-line configuration shown in Fig. 8-4(b) consists of a lossless transmission line with characteristic impedance Z_{01}, connected at $z = 0$ to an infinitely long lossless transmission line with characteristic impedance Z_{02}. The input impedance of an infinitely long line is equal to its characteristic impedance. Hence, at $z = 0$, the voltage reflection coefficient (looking toward the boundary from the vantage point of the first line) is

$$\Gamma = \frac{Z_{02} - Z_{01}}{Z_{02} + Z_{01}} ,$$

which is identical in form to Eq. (8.8a). To show the basis for the analogy between the plane-wave and transmission-line situations, the expressions for the two cases are given in Table 8-1. Comparison of the two columns shows that there is a one-to-one correspondence between the transmission-line parameters ($\widetilde{V}, \widetilde{I}, \beta, Z_0$) and the plane-wave parameters ($\widetilde{\mathbf{E}}, \widetilde{\mathbf{H}}, k, \eta$). This correspondence allows us to use the techniques we developed in Chapter 2, including the Smith-chart method for calculating impedance transformations, to solve plane-wave propagation problems.

Simultaneous presence of incident and reflected waves in a medium, such as medium 1 in Fig. 8-4(a), gives rise to a standing-wave pattern. By analogy with the transmission-line case, the *standing-wave ratio* in medium 1 is given by

$$S = \frac{|\widetilde{E}_1|_{max}}{|\widetilde{E}_1|_{min}} = \frac{1 + |\Gamma|}{1 - |\Gamma|} . \quad (8.15)$$

If the two media have equal impedances ($\eta_1 = \eta_2$), then $\Gamma = 0$ and $S = 1$, and if medium 2 is a perfect conductor with $\eta_2 = 0$ (which is equivalent to a short-circuited transmission line), then $\Gamma = -1$ and $S = \infty$. The distances from the boundary to where the magnitude of the electric field intensity in medium 1 is at a maximum, denoted l_{max}, are described by the same expression as

Table 8-1: Analogy between plane-wave equations for normal incidence and transmission-line equations, both under lossless conditions.

Plane Wave [Fig. 8-4(a)]		Transmission Line [Fig. 8-4(b)]	
$\widetilde{\mathbf{E}}_1(z) = \hat{\mathbf{x}} E_0^{\mathrm{i}}(e^{-jk_1z} + \Gamma e^{jk_1z})$	(8.11a)	$\widetilde{V}_1(z) = V_0^+(e^{-j\beta_1 z} + \Gamma e^{j\beta_1 z})$	(8.11b)
$\widetilde{\mathbf{H}}_1(z) = \hat{\mathbf{y}} \dfrac{E_0^{\mathrm{i}}}{\eta_1}(e^{-jk_1z} - \Gamma e^{jk_1z})$	(8.12a)	$\widetilde{I}_1(z) = \dfrac{V_0^+}{Z_{01}}(e^{-j\beta_1 z} - \Gamma e^{j\beta_1 z})$	(8.12b)
$\widetilde{\mathbf{E}}_2(z) = \hat{\mathbf{x}} \tau E_0^{\mathrm{i}} e^{-jk_2z}$	(8.13a)	$\widetilde{V}_2(z) = \tau V_0^+ e^{-j\beta_2 z}$	(8.13b)
$\widetilde{\mathbf{H}}_2(z) = \hat{\mathbf{y}} \tau \dfrac{E_0^{\mathrm{i}}}{\eta_2} e^{-jk_2z}$	(8.14a)	$\widetilde{I}_2(z) = \tau \dfrac{V_0^+}{Z_{02}} e^{-j\beta_2 z}$	(8.14b)
$\Gamma = (\eta_2 - \eta_1)/(\eta_2 + \eta_1)$		$\Gamma = (Z_{02} - Z_{01})/(Z_{02} + Z_{01})$	
$\tau = 1 + \Gamma$		$\tau = 1 + \Gamma$	
$k_1 = \omega\sqrt{\mu_1 \varepsilon_1}, \qquad k_2 = \omega\sqrt{\mu_2 \varepsilon_2}$		$\beta_1 = \omega\sqrt{\mu_1 \varepsilon_1}, \qquad \beta_2 = \omega\sqrt{\mu_2 \varepsilon_2}$	
$\eta_1 = \sqrt{\mu_1/\varepsilon_1}, \qquad \eta_2 = \sqrt{\mu_2/\varepsilon_2}$		Z_{01} and Z_{02} depend on transmission-line parameters	

that given by Eq. (2.56) for the voltage maxima on a transmission line:

$$-z = l_{\max} = \frac{\theta_{\mathrm{r}} + 2n\pi}{2k_1} = \frac{\theta_{\mathrm{r}}\lambda_1}{4\pi} + \frac{n\lambda_1}{2},$$

$$\begin{cases} n = 1, 2, \ldots, & \text{if } \theta_{\mathrm{r}} < 0, \\ n = 0, 1, 2, \ldots, & \text{if } \theta_{\mathrm{r}} \geq 0, \end{cases} \qquad (8.16)$$

where $\lambda_1 = 2\pi/k_1$ and θ_{r} is the phase angle of Γ (i.e., $\Gamma = |\Gamma|e^{j\theta_{\mathrm{r}}}$, and θ_{r} is bounded in the range $-\pi < \theta_{\mathrm{r}} \leq \pi$). The spacing between adjacent maxima is $\lambda_1/2$, and the spacing between a maximum and the nearest minimum is $\lambda_1/4$. The electric-field minima occur at

$$l_{\min} = \begin{cases} l_{\max} + \lambda_1/4, & \text{if } l_{\max} < \lambda_1/4, \\ l_{\max} - \lambda_1/4, & \text{if } l_{\max} \geq \lambda_1/4. \end{cases} \qquad (8.17)$$

The expressions for $l_{\max}$ and $l_{\min}$ are valid provided that the medium containing the standing-wave pattern is either lossless or a low-loss dielectric, but no restrictions are imposed on the nature of the reflecting medium.

8-1.3   Power Flow in Lossless Media

Medium 1 of Fig. 8-4(a) contains an incident wave and a reflected wave, which together produce the electric and magnetic fields $\widetilde{\mathbf{E}}_1(z)$ and $\widetilde{\mathbf{H}}_1(z)$ given by Eqs. (8.11a) and (8.12a) of Table 8-1. Using Eq. (7.101), the net average power density flowing in medium 1 is

$$\mathbf{S}_{\mathrm{av}_1}(z) = \tfrac{1}{2}\mathfrak{Re}[\widetilde{\mathbf{E}}_1(z) \times \widetilde{\mathbf{H}}_1^*(z)]$$

$$= \tfrac{1}{2}\mathfrak{Re}\Big[\hat{\mathbf{x}} E_0^{\mathrm{i}}(e^{-jk_1z} + \Gamma e^{jk_1z})$$

$$\times \hat{\mathbf{y}} \frac{E_0^{\mathrm{i}*}}{\eta_1}(e^{jk_1z} - \Gamma^* e^{-jk_1z}) \Big]$$

$$= \hat{\mathbf{z}} \frac{|E_0^{\mathrm{i}}|^2}{2\eta_1}(1 - |\Gamma|^2), \qquad (8.18)$$

which is analogous to Eq. (2.86) for the lossless trans-mission-line case. The first term in Eq. (8.18) represents the average power density of the incident wave, and the second term (proportional to $|\Gamma|^2$) represents the average power density of the reflected wave. Thus,

$$\mathbf{S}_{av_1} = \mathbf{S}_{av}^i + \mathbf{S}_{av}^r, \tag{8.19a}$$

with

$$\mathbf{S}_{av}^i = \hat{\mathbf{z}} \frac{|E_0^i|^2}{2\eta_1}, \tag{8.19b}$$

$$\mathbf{S}_{av}^r = -\hat{\mathbf{z}}|\Gamma|^2 \frac{|E_0^i|^2}{2\eta_1} = -|\Gamma|^2 \mathbf{S}_{av}^i. \tag{8.19c}$$

Even though Γ is purely real when both media are lossless dielectrics, we chose to treat it as complex, thereby providing in Eq. (8.19c) an expression that is also valid when medium 2 is conducting.

The average power density of the transmitted wave in medium 2 is

$$\mathbf{S}_{av_2}(z) = \tfrac{1}{2}\mathfrak{Re}[\widetilde{\mathbf{E}}_2(z) \times \widetilde{\mathbf{H}}_2^*(z)]$$

$$= \tfrac{1}{2}\mathfrak{Re}\left[\hat{\mathbf{x}}\tau E_0^i e^{-jk_2 z} \times \hat{\mathbf{y}}\tau^* \frac{E_0^{i*}}{\eta_2} e^{jk_2 z}\right]$$

$$= \hat{\mathbf{z}}|\tau|^2 \frac{|E_0^i|^2}{2\eta_2}. \tag{8.20}$$

This expression is applicable when both media are lossless, as well as when medium 1 is conducting and only medium 2 is lossless.

Through the use of Eqs. (8.8a) and (8.8b), it can be easily shown that for lossless media (for which Γ and τ are real)

$$\boxed{\frac{\tau^2}{\eta_2} = \frac{1 - \Gamma^2}{\eta_1}} \quad \text{(lossless media)}, \tag{8.21}$$

which leads to

$$\mathbf{S}_{av_1} = \mathbf{S}_{av_2}.$$

This result is as expected from considerations of power conservation.

Example 8-1 Radar Radome Design

A 10-GHz aircraft radar uses a narrow-beam scanning antenna mounted on a gimbal behind a dielectric radome, as shown in Fig. 8-5. Even though the radome shape is far from planar, it is approximately planar over the narrow extent of the radar beam. If the radome material is a lossless dielectric with $\mu_r = 1$ and $\varepsilon_r = 9$, choose its thickness d such that the radome appears transparent to the radar beam. Mechanical integrity requires d to be greater than 2.3 cm.

Solution: The propagation problem is shown in Fig. 8-6(a) at an expanded scale. The incident wave is approximated as a plane wave propagating in medium 1 (air) with intrinsic impedance η_0, the radome (medium 2) is of thickness d and intrinsic impedance η_r, and medium 3 is semi-infinite with intrinsic impedance η_0. Figure 8-6(b) is an equivalent transmission-line model with $z = 0$ selected to coincide with the outside surface of the radome, and the load impedance $Z_L = \eta_0$ represents the input impedance of the semi-infinite medium.

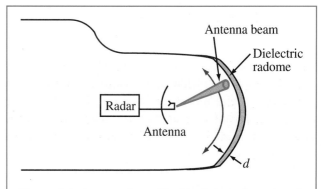

Figure 8-5: Antenna beam "looking" through an aircraft radome of thickness d (Example 8-1).

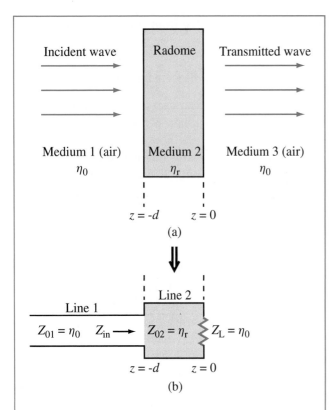

Figure 8-6: (a) Planar section of the radome of Fig. 8-5 at an expanded scale and (b) its transmission-line equivalent model (Example 8-1).

Requiring the radome to "appear" transparent to the incident wave simply means that the reflection coefficient must be zero at $z = -d$, thereby achieving total transmission of the incident power into medium 3. Since $Z_L = \eta_0$ in Fig. 8-6(b), no reflection will take place at $z = -d$ if $Z_{in} = \eta_0$, which can be realized by choosing $d = n\lambda_2/2$ [see Section 2-7.4], where λ_2 is the wavelength in medium 2 and n is a positive integer. At 10 GHz, the wavelength in air is $\lambda_0 = c/f = 3$ cm, and in the radome material

$$\lambda_2 = \frac{\lambda_0}{\sqrt{\varepsilon_r}} = \frac{3 \text{ cm}}{3} = 1 \text{ cm}.$$

Hence, if we choose $d = 5\lambda_2/2 = 2.5$ cm, we will satisfy both the no-reflection and the mechanical integrity requirements. ■

Example 8-2   Yellow Light Incident upon a Glass Surface

A beam of yellow light with wavelength of 0.6 μm is normally incident in air upon a glass surface. If the surface is situated in the plane $z = 0$ and the relative permittivity of glass is 2.25, determine:

(a) the locations of the electric field maxima in medium 1 (air),

(b) the standing-wave ratio, and

(c) the fraction of the incident power transmitted into the glass medium.

Solution: (a) We begin by determining the value of η_1, η_2, and Γ:

$$\eta_1 = \sqrt{\frac{\mu_1}{\varepsilon_1}} = \sqrt{\frac{\mu_0}{\varepsilon_0}} \simeq 120\pi \ (\Omega),$$

$$\eta_2 = \sqrt{\frac{\mu_2}{\varepsilon_2}} = \sqrt{\frac{\mu_0}{\varepsilon_0}} \cdot \frac{1}{\sqrt{\varepsilon_r}} \simeq \frac{120\pi}{\sqrt{2.25}} = 80\pi \ (\Omega),$$

$$\Gamma = \frac{\eta_2 - \eta_1}{\eta_2 + \eta_1} = \frac{80\pi - 120\pi}{80\pi + 120\pi} = -0.2.$$

Hence, $|\Gamma| = 0.2$ and $\theta_r = \pi$. From Eq. (8.16), the electric-field magnitude is a maximum at

$$l_{max} = \frac{\theta_r \lambda_1}{4\pi} + n\frac{\lambda_1}{2}$$

$$= \frac{\lambda_1}{4} + n\frac{\lambda_1}{2} \quad (n = 0, 1, 2, \ldots)$$

with $\lambda_1 = 0.6 \ \mu$m.

(b)

$$S = \frac{1 + |\Gamma|}{1 - |\Gamma|} = \frac{1 + 0.2}{1 - 0.2} = 1.5.$$

(c) The fraction of the incident power transmitted into the glass medium is equal to the ratio of the transmitted power density, given by Eq. (8.20), to the incident power density, $S_{av}^i = |E_0^i|^2/2\eta_1$:

$$\frac{S_{av_2}}{S_{av}^i} = \tau^2 \frac{|E_0^i|^2}{2\eta_2} \Big/ \left[\frac{|E_0^i|^2}{2\eta_1}\right] = \tau^2 \frac{\eta_1}{\eta_2}.$$

In view of Eq. (8.21),

$$\frac{S_{av_2}}{S_{av}^i} = 1 - |\Gamma|^2 = 1 - (0.2)^2 = 0.96, \text{ or } 96\%. \quad \blacksquare$$

8-1.4 Boundary between Lossy Media

In Section 8-1.1 we considered a plane wave in a lossless medium incident normally on a planar boundary of another lossless medium. We will now generalize our expressions to lossy media. In a medium with constitutive parameters $(\varepsilon, \mu, \sigma)$, the propagation parameters of interest are the propagation constant $\gamma = \alpha + j\beta$ and the complex intrinsic impedance η_c. The general expressions for α, β, and η_c are given by Eqs. (7.66a), (7.66b), and (7.70), respectively, and approximate expressions are given in Table 7-1 for the special cases of low-loss media and good conducting media. If medium 1 is characterized by $(\varepsilon_1, \mu_1, \sigma_1)$ and medium 2 by $(\varepsilon_2, \mu_2, \sigma_2)$, as shown in Fig. 8-7, the expressions for the electric and magnetic fields in media 1 and 2 can be obtained from Eqs. (8.11a) through (8.14a) of Table 8-1 by replacing jk with γ and η with η_c everywhere. Thus,

Medium 1

$$\widetilde{\mathbf{E}}_1(z) = \hat{\mathbf{x}} E_0^i (e^{-\gamma_1 z} + \Gamma e^{\gamma_1 z}), \qquad (8.22a)$$

$$\widetilde{\mathbf{H}}_1(z) = \hat{\mathbf{y}} \frac{E_0^i}{\eta_{c_1}} (e^{-\gamma_1 z} - \Gamma e^{\gamma_1 z}), \qquad (8.22b)$$

Medium 2

$$\widetilde{\mathbf{E}}_2(z) = \hat{\mathbf{x}} \tau E_0^i e^{-\gamma_2 z}, \qquad (8.23a)$$

$$\widetilde{\mathbf{H}}_2(z) = \hat{\mathbf{y}} \tau \frac{E_0^i}{\eta_{c_2}} e^{-\gamma_2 z}, \qquad (8.23b)$$

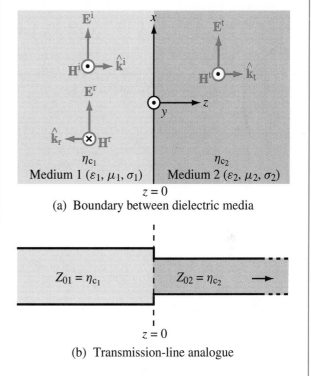

(a) Boundary between dielectric media

(b) Transmission-line analogue

Figure 8-7: Normal incidence at a planar boundary between two lossy media.

where $\gamma_1 = \alpha_1 + j\beta_1$, $\gamma_2 = \alpha_2 + j\beta_2$, and

$$\Gamma = \frac{\eta_{c_2} - \eta_{c_1}}{\eta_{c_2} + \eta_{c_1}}, \qquad (8.24a)$$

$$\tau = 1 + \Gamma = \frac{2\eta_{c_2}}{\eta_{c_2} + \eta_{c_1}}. \qquad (8.24b)$$

Because η_{c_1} and η_{c_2} are, in general, complex, Γ and τ may be complex as well.

Example 8-3 Normal Incidence on a Metal Surface

A 1-GHz x-polarized TEM wave traveling in the $+z$-direction is incident in air upon a metal surface coincident with the x–y plane at $z = 0$. If the amplitude of the electric field of the incident wave is 12 (mV/m) and the metal surface is made of copper with $\mu_r = 1$, $\varepsilon_r = 1$, and $\sigma = 5.8 \times 10^7$ (S/m), obtain expressions for the instantaneous electric and magnetic fields in the air medium. Assume the metal surface to be several skin depths deep.

Solution: In medium 1 (air), $\alpha = 0$,

$$\beta = k_1 = \frac{\omega}{c} = \frac{2\pi \times 10^9}{3 \times 10^8} = \frac{20\pi}{3} \quad \text{(rad/m)},$$

$$\eta_1 = \eta_0 = 377 \ (\Omega), \qquad \lambda = \frac{2\pi}{k_1} = 0.3 \text{ m}.$$

At $f = 1$ GHz, copper is an excellent conductor because

$$\frac{\varepsilon''}{\varepsilon'} = \frac{\sigma}{\omega \varepsilon_r \varepsilon_0} = \frac{5.8 \times 10^7}{2\pi \times 10^9 \times (10^{-9}/36\pi)} = 1 \times 10^9 \gg 1.$$

Use of Eq. (7.77c) gives

$$\eta_{c_2} = (1 + j)\sqrt{\frac{\pi f \mu}{\sigma}}$$

$$= (1 + j)\left[\frac{\pi \times 10^9 \times 4\pi \times 10^{-7}}{5.8 \times 10^7}\right]^{1/2}$$

$$= 8.25(1 + j) \quad (\text{m}\Omega).$$

Since η_{c_2} is so small compared to $\eta_0 = 377 \ (\Omega)$ for air, the copper surface acts, in effect, like a short circuit. Hence,

$$\Gamma = \frac{\eta_{c_2} - \eta_0}{\eta_{c_2} + \eta_0} \simeq -1.$$

Upon setting $\Gamma = -1$ in Eqs. (8.11a) and (8.12a) of Table 8-1, we have

$$\widetilde{\mathbf{E}}_1(z) = \hat{\mathbf{x}} E_0^i (e^{-jk_1 z} - e^{jk_1 z})$$

$$= -\hat{\mathbf{x}} j 2 E_0^i \sin k_1 z, \qquad (8.25a)$$

$$\widetilde{\mathbf{H}}_1(z) = \hat{\mathbf{y}} \frac{E_0^i}{\eta_1} (e^{-jk_1 z} + e^{jk_1 z})$$

$$= \hat{\mathbf{y}} 2 \frac{E_0^i}{\eta_1} \cos k_1 z. \qquad (8.25b)$$

With $E_0^i = 12$ (mV/m), the instantaneous fields corresponding to these phasors are

$$\mathbf{E}_1(z, t) = \mathfrak{Re}[\widetilde{\mathbf{E}}_1(z) \, e^{j\omega t}]$$

$$= \hat{\mathbf{x}} 2 E_0^i \sin k_1 z \sin \omega t$$

$$= \hat{\mathbf{x}} 24 \sin(20\pi z/3) \sin(2\pi \times 10^9 t) \quad \text{(mV/m)},$$

$$\mathbf{H}_1(z, t) = \mathfrak{Re}[\widetilde{\mathbf{H}}_1(z) \, e^{j\omega t}]$$

$$= \hat{\mathbf{y}} 2 \frac{E_0^i}{\eta_1} \cos k_1 z \cos \omega t$$

$$= \hat{\mathbf{y}} 64 \cos(20\pi z/3) \cos(2\pi \times 10^9 t) \quad (\mu\text{A/m}).$$

Plots of the magnitude of $\mathbf{E}_1(z, t)$ and $\mathbf{H}_1(z, t)$ are shown in Fig. 8-8 as a function of negative z at various values of ωt. The standing-wave patterns exhibit a repetition period of $\lambda/2$, and E and H are in phase quadrature (90° phase shift) in both space and time. This behavior is identical with that of the standing-wave patterns for voltage and current on a shorted transmission line. ■

REVIEW QUESTIONS

Q8.1 What boundary conditions were used in the derivations of the expressions for Γ and τ?

Q8.2 In the radar radome design of Example 8-1, all the incident energy in medium 1 ends up getting transmitted into medium 3, and vice versa. Does this imply that no reflections take place within medium 2? Explain.

Q8.3 Explain on the basis of boundary conditions why it is necessary that $\Gamma = -1$ at the boundary between a dielectric and a perfect conductor.

EXERCISE 8.1 To eliminate wave reflections, a dielectric slab of thickness d and relative permittivity ε_{r_2} is to be inserted between two semi-infinite media with relative permittivities $\varepsilon_{r_1} = 1$ and $\varepsilon_{r_3} = 16$. Use the quarter-wave

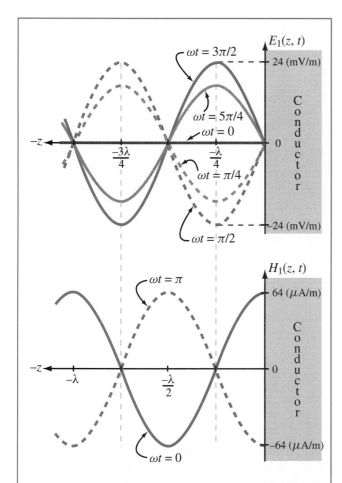

Figure 8-8: Standing-wave patterns for fields $E_1(z, t)$ and $H_1(z, t)$ of Example 8-3.

transformer technique to select d and ε_{r_2} for normally incident waves at 3 GHz.

Ans. $\varepsilon_{r_2} = 4$ and $d = (1.25 + 2.5n)$ (cm), with $n = 0, 1, 2, \ldots$.

EXERCISE 8.2 Express the normal-incidence reflection coefficient at the boundary between two nonmagnetic, conducting media in terms of their complex permittivities.

Ans. For incidence in medium 1 $(\varepsilon_1, \mu_0, \sigma_1)$ onto medium 2 $(\varepsilon_2, \mu_0, \sigma_2)$,

$$\Gamma = \frac{\sqrt{\varepsilon_{c_1}} - \sqrt{\varepsilon_{c_2}}}{\sqrt{\varepsilon_{c_1}} + \sqrt{\varepsilon_{c_2}}},$$

with $\varepsilon_{c_1} = (\varepsilon_1 - j\sigma_1/\omega)$ and $\varepsilon_{c_2} = (\varepsilon_2 - j\sigma_2/\omega)$.

EXERCISE 8.3 Obtain expressions for the average power densities in media 1 and 2 for the fields described by Eqs. (8.22a) through (8.23b), assuming medium 1 is slightly lossy with η_{c_1} approximately real.

Ans.

$$\mathbf{S}_{av_1} = \hat{\mathbf{z}}\frac{|E_0^i|^2}{2\eta_{c_1}} \left(e^{-2\alpha_1 z} - |\Gamma|^2 e^{2\alpha_1 z}\right),$$

$$\mathbf{S}_{av_2} = \hat{\mathbf{z}}|\tau|^2 \frac{|E_0^i|^2}{2} e^{-2\alpha_2 z} \, \mathfrak{Re}\left(\frac{1}{\eta_{c_2}^*}\right).$$

8-2 Snell's Laws

In the preceding sections we examined the reflection and transmission properties of plane waves when incident normally upon a planar interface between two different media. We now consider the oblique-incidence case depicted in Fig. 8-9. The $z = 0$ plane is the boundary between two dielectric media characterized by (ε_1, μ_1) for medium 1 and (ε_2, μ_2) for medium 2. The two lines with direction $\hat{\mathbf{k}}_i$ represent rays drawn normal to the wavefront of the incident wave and, similarly, those along $\hat{\mathbf{k}}_r$ and $\hat{\mathbf{k}}_t$ represent the reflected and transmitted waves, respectively. Defined with respect to the normal to the boundary (the z-axis), the *angles of incidence, reflection*, and *transmission* (or *refraction*) are, respectively, θ_i, θ_r, and θ_t. These three angles are interrelated by *Snell's laws*, which we will derive by considering the propagation of the wavefronts of the three waves. The incident wave intersects the boundary at O and O'. The constant-phase wavefront of the incident wave is A_iO, and the wavefronts of the reflected and transmitted waves are

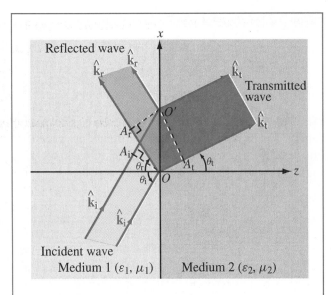

Figure 8-9: Wave reflection and refraction at a planar boundary between different media.

$A_r O'$ and $A_t O'$, as shown in Fig. 8-9. The incident and reflected waves propagate in medium 1 with the same phase velocity $u_{p_1} = 1/\sqrt{\mu_1 \varepsilon_1}$, and the transmitted wave in medium 2 propagates with a velocity $u_{p_2} = 1/\sqrt{\mu_2 \varepsilon_2}$. The time it takes the incident wave to travel from A_i to O' is the same as the time it takes the reflected wave to travel from O to A_r and also the time it takes the transmitted wave to travel from O to A_t. Since time is equal to distance divided by velocity, it follows that

$$\frac{\overline{A_i O'}}{u_{p_1}} = \frac{\overline{O A_r}}{u_{p_1}} = \frac{\overline{O A_t}}{u_{p_2}}. \qquad (8.26)$$

From the geometries of the three right triangles in Fig. 8-9, we deduce that

$$\overline{A_i O'} = \overline{O O'} \sin \theta_i, \qquad (8.27a)$$

$$\overline{O A_r} = \overline{O O'} \sin \theta_r, \qquad (8.27b)$$

$$\overline{O A_t} = \overline{O O'} \sin \theta_t. \qquad (8.27c)$$

Use of these expressions in Eq. (8.26) leads to

$$\theta_i = \theta_r \quad \text{(Snell's law of reflection)}, \qquad (8.28a)$$

$$\frac{\sin \theta_t}{\sin \theta_i} = \frac{u_{p_2}}{u_{p_1}} = \sqrt{\frac{\mu_1 \varepsilon_1}{\mu_2 \varepsilon_2}}$$

$$\text{(Snell's law of refraction)}. \qquad (8.28b)$$

Snell's law of reflection states that the angle of reflection is equal to the angle of incidence, and *Snell's law of refraction* provides a relation between $\sin \theta_t$ and $\sin \theta_i$ in terms of the ratio of the phase velocities.

The *index of refraction* of a medium, n, is defined as the ratio of the phase velocity in free space (i.e., the speed of light c) to the phase velocity in the medium. Thus,

$$n = \frac{c}{u_p} = \sqrt{\frac{\mu \varepsilon}{\mu_0 \varepsilon_0}} = \sqrt{\mu_r \varepsilon_r}. \qquad (8.29)$$

In view of Eq. (8.29), Eq. (8.28b) may be rewritten as

$$\frac{\sin \theta_t}{\sin \theta_i} = \frac{n_1}{n_2} = \sqrt{\frac{\mu_{r_1} \varepsilon_{r_1}}{\mu_{r_2} \varepsilon_{r_2}}}. \qquad (8.30)$$

For nonmagnetic materials, $\mu_{r_1} = \mu_{r_2} = 1$, in which case

$$\frac{\sin \theta_t}{\sin \theta_i} = \frac{n_1}{n_2} = \sqrt{\frac{\varepsilon_{r_1}}{\varepsilon_{r_2}}} = \frac{\eta_2}{\eta_1} \quad \text{(for } \mu_1 = \mu_2 \text{)}, \qquad (8.31)$$

where $\eta = \sqrt{\mu/\varepsilon}$ is the intrinsic impedance of a dielectric medium. Usually, materials with higher densities have higher permittivities. Air, with $\mu_r = \varepsilon_r = 1$, has an index of refraction $n_0 = 1$. Since for nonmagnetic materials $n = \sqrt{\varepsilon_r}$, *a material is often referred to as more dense than a second material if the index of refraction of the first material is greater than that of the second.*

At normal incidence ($\theta_i = 0$), Eq. (8.31) gives $\theta_t = 0$, as expected, and at oblique incidence $\theta_t < \theta_i$ when $n_2 > n_1$ and $\theta_t > \theta_i$ when $n_2 < n_1$. That is, *if the wave*

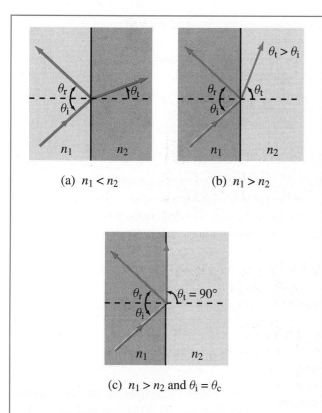

(a) $n_1 < n_2$ (b) $n_1 > n_2$

(c) $n_1 > n_2$ and $\theta_i = \theta_c$

Figure 8-10: Snell's laws state that $\theta_r = \theta_i$ and $\sin \theta_t = (n_1/n_2) \sin \theta_i$. Refraction is (a) inward if $n_1 < n_2$ and (b) outward if $n_1 > n_2$; and (c) the refraction angle is $90°$ if $n_1 > n_2$ and θ_i is equal to or greater than the critical angle $\theta_c = \sin^{-1}(n_2/n_1)$.

in medium 1 is incident on a more dense medium, as in Fig. 8-10(a), the transmitted wave refracts inwardly toward the z-axis, and the opposite is true if the wave is incident on a less dense medium [Fig. 8-10(b)]. A case of particular interest is when $\theta_t = \pi/2$, as shown in Fig. 8-10(c); in this case, the refracted wave flows along the surface and no energy is transmitted into medium 2. The value of the angle of incidence θ_i corresponding to $\theta_t = \pi/2$ is called the *critical angle* θ_c and is obtained

from Eq. (8.30) as

$$\sin \theta_c = \frac{n_2}{n_1} \sin \theta_t \Big|_{\theta_t = \pi/2} = \frac{n_2}{n_1} \quad (8.32a)$$

$$= \sqrt{\frac{\varepsilon_{r_2}}{\varepsilon_{r_1}}} \quad (\text{for } \mu_1 = \mu_2). \quad (8.32b)$$

If θ_i exceeds θ_c, the incident wave is totally reflected, and the refracted wave becomes a nonuniform *surface wave* that travels along the boundary between the two media. This wave behavior is called *total internal reflection*.

Example 8-4   Light Beam Passing through a Slab

A dielectric slab with index of refraction n_2 is surrounded by a medium with index of refraction n_1, as shown in Fig. 8-11. If $\theta_i < \theta_c$, show that the emerging beam is parallel to the incident beam.

Solution: At the slab's upper surface, Snell's law gives

$$\sin \theta_2 = \frac{n_1}{n_2} \sin \theta_1 \quad (8.33)$$

and, similarly, at the slab's lower surface,

$$\sin \theta_3 = \frac{n_2}{n_3} \sin \theta_2 = \frac{n_2}{n_1} \sin \theta_2. \quad (8.34)$$

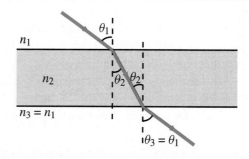

Figure 8-11: The exit angle θ_3 is equal to the incidence angle θ_1 if the dielectric slab has parallel boundaries and is surrounded by the same index of refraction on both sides (Example 8-4).

Substituting Eq. (8.33) into Eq. (8.34) gives

$$\sin \theta_3 = \left(\frac{n_2}{n_1}\right)\left(\frac{n_1}{n_2}\right) \sin \theta_1 = \sin \theta_1.$$

Hence, $\theta_3 = \theta_1$. The slab displaces the beam's position, but the beam's direction remains unchanged. ∎

EXERCISE 8.4   In the visible part of the electromagnetic spectrum, the index of refraction of water is 1.33. What is the critical angle for light waves generated by an upward-looking underwater light source?

Ans. $\theta_c = 48.8°$.

EXERCISE 8.5   If the light source of Exercise 8.4 is situated at a depth of 1 m below the water surface and if its beam is isotropic (radiates in all directions), how large a circle would it illuminate when observed from above?

Ans.   Circle's diameter $= 2.28$ m.

8-3   Fiber Optics

By successive total internal reflections, as indicated in Fig. 8-12(a), light can be guided through thin dielectric rods made of glass or transparent plastic, known as *optical fibers*. Because the light is confined to traveling within the rod, the only loss in power is due to reflections at the sending and receiving ends of the fiber and absorption by the fiber material (because it is not a perfect dielectric). Fiber optics is useful for the transmission of wide-bandwidth signals and in a wide range of imaging applications.

An optical fiber usually consists of a cylindrical *fiber core* with an index of refraction n_f, surrounded by another cylinder of lower index of refraction, n_c, called a *cladding*, as shown in Fig. 8-12(b). The cladding layer serves to optically isolate the fiber from adjacent fibers when a large number of fibers are packed in close

proximity, thereby avoiding the leakage of light from one fiber to another. To satisfy the condition of total internal reflection, the incident angle θ_3 in the fiber core must be equal to or greater than the critical angle θ_c for a wave in the fiber medium (with n_f) incident upon the cladding medium (with n_c). From Eq. (8.32a), we have

$$\sin \theta_c = \frac{n_c}{n_f}. \qquad (8.35)$$

To meet the total-reflection requirement that $\theta_3 \geq \theta_c$, it is then necessary that $\sin \theta_3 \geq n_c/n_f$. The angle θ_2 is the complement of angle θ_3, and $\cos \theta_2 = \sin \theta_3$. Hence, the necessary condition may be written as

$$\cos \theta_2 \geq \frac{n_c}{n_f}. \qquad (8.36)$$

Moreover, θ_2 is related to the incidence angle on the face of the fiber, θ_i, by Snell's law:

$$\sin \theta_2 = \frac{n_0}{n_f} \sin \theta_i, \qquad (8.37)$$

where n_0 is the index of refraction of the medium surrounding the fiber ($n_0 = 1$ for air and $n_0 = 1.33$ if the fiber is in water), or

$$\cos \theta_2 = \left[1 - \left(\frac{n_0}{n_f}\right)^2 \sin^2 \theta_i\right]^{1/2}. \qquad (8.38)$$

Using Eq. (8.38) in the left-hand side of Eq. (8.36) and then solving for $\sin \theta_i$ gives

$$\sin \theta_i \leq \frac{1}{n_0} (n_f^2 - n_c^2)^{1/2}. \qquad (8.39)$$

The *acceptance angle* θ_a is defined as the maximum value of θ_i for which the condition of total internal reflection remains satisfied:

$$\boxed{\sin \theta_a = \frac{1}{n_0} (n_f^2 - n_c^2)^{1/2}. \qquad (8.40)}$$

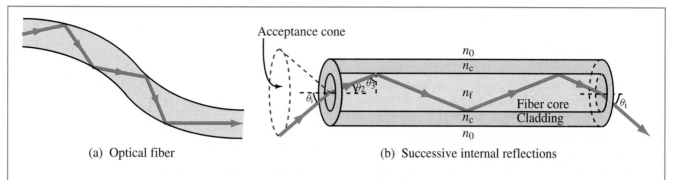

(a) Optical fiber   (b) Successive internal reflections

Figure 8-12: Waves can be guided along optical fibers as long as the reflection angles exceed the critical angle for total internal reflection.

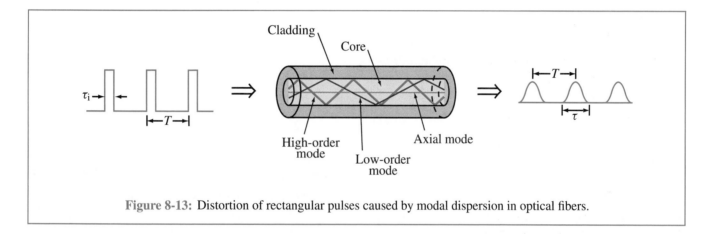

Figure 8-13: Distortion of rectangular pulses caused by modal dispersion in optical fibers.

The angle θ_a is equal to half the angle of the acceptance cone of the fiber. Any ray of light incident upon the face of the core fiber at an incidence angle within the acceptance cone can propagate down the core. This means that there can be a larger number of ray paths, called *modes*, by which light energy can travel in the core. Higher-angle rays travel longer paths than rays that propagate along the axis of the fiber, as illustrated by the three modes shown in Fig. 8-13. Consequently, different modes have different transit times between the two ends of the fiber. This property of optical fibers is called *modal dispersion* and has the undesirable effect of changing the shape of pulses used for the transmission of

digital data. When a rectangular pulse of light incident upon the face of the fiber gets broken up into many modes and the different modes do not arrive at the other end of the fiber at the same time, the pulse shape gets distorted, both in shape and length. In the example shown in Fig. 8-13, the narrow rectangular pulses at the input side of the optical fiber are of width τ_i separated by a time duration T. After propagating through the fiber core, modal dispersion causes the pulses to look more like spread-out sine waves with spread-out width τ. If the output pulses spread out so much that $\tau > T$, the output signals will smear out, making it impossible to read the transmitted message. Hence, to ensure that the transmitted pulses remain distinguishable at the output side of the fiber, it is necessary that τ be shorter than T. As a safety margin, it is common practice to require that $T \geq 2\tau$.

The spread-out width τ is equal to the time delay Δt between the arrival of the slowest ray and the fastest ray. The slowest ray is the one traveling the longest distance and corresponds to the ray incident upon the input face of the fiber at the acceptance angle θ_a. From the geometry of Fig. 8-12(b) and Eq. (8.36), this ray corresponds to $\cos \theta_2 = n_c/n_f$. For an optical fiber of length l, the length of the path traveled by such a ray is

$$l_{max} = \frac{l}{\cos \theta_2} = l\frac{n_f}{n_c}, \tag{8.41}$$

and its travel time in the fiber at the velocity $u_p = c/n_f$ is

$$t_{max} = \frac{l_{max}}{u_p} = \frac{ln_f^2}{cn_c}. \tag{8.42}$$

The minimum time of travel is realized by the axial ray and is given by

$$t_{min} = \frac{l}{u_p} = \frac{l}{c} n_f. \tag{8.43}$$

The total time delay is therefore

$$\tau = \Delta t = t_{max} - t_{min} = \frac{ln_f}{c}\left(\frac{n_f}{n_c} - 1\right) \quad \text{(s).} \tag{8.44}$$

As we stated before, to retrieve the desired information from the transmitted signals, it is advisable that T, the interpulse period of the input train of pulses, be no shorter than 2τ. This, in turn, means that the data rate (in bits per second), or equivalently the number of pulses per second, that can be transmitted through the fiber is limited to

$$f_p = \frac{1}{T} = \frac{1}{2\tau} = \frac{cn_c}{2ln_f(n_f - n_c)} \quad \text{(bits/s).} \tag{8.45}$$

Example 8-5 **Transmission Data Rate on Optical Fibers**

A 1-km-long optical fiber (in air) is made of a fiber core with an index of refraction of 1.52 and a cladding with an index of refraction of 1.49. Determine
(a) the acceptance angle θ_a, and
(b) the maximum usable data rate that can be transmitted through the fiber.

Solution: (a) From Eq. (8.40),

$$\sin \theta_a = \frac{1}{n_0}(n_f^2 - n_c^2)^{1/2} = [(1.52)^2 - (1.49)^2]^{1/2} = 0.3,$$

which corresponds to $\theta_a = 17.5°$.

(b) From Eq. (8.45),

$$f_p = \frac{cn_c}{2ln_f(n_f - n_c)}$$

$$= \frac{3 \times 10^8 \times 1.49}{2 \times 10^3 \times 1.52(1.52 - 1.49)} = 4.9 \text{ (Mb/s).} \quad \blacksquare$$

EXERCISE 8.6 If the index of refraction of the cladding material in Example 8-5 is increased to 1.50, what would be the new maximum usable data rate?

Ans. 7.4 (Mb/s).

8-4 Wave Reflection and Transmission at Oblique Incidence

For normal incidence, the reflection coefficient Γ and transmission coefficient τ of a boundary between two different media is independent of the polarization of the incident wave, because the electric and magnetic fields of a normally incident plane wave are both always tangential to the boundary regardless of the wave polarization. This is not the case for oblique incidence at an angle

$\theta_i \neq 0$. A wave with any specified polarization may be described as the superposition of two orthogonally polarized waves, one with its electric field parallel to the plane of incidence—and it is called *parallel polarization*—and another with its electric field perpendicular to the plane of incidence—and it is called *perpendicular polarization*. *The plane of incidence is defined as the plane containing the normal to the boundary and the direction of propagation of the incident wave.* These two polarization configurations are shown in Fig. 8-14, in which the plane of incidence is coincident with the x–z plane. Polarization with **E** perpendicular to the plane of incidence is also called *transverse electric* (TE) polarization because **E** is perpendicular to the plane of incidence, and that with **E** parallel to the plane of incidence is called *transverse magnetic* (TM) polarization because in this case it is the magnetic field that is perpendicular to the plane of incidence.

Instead of solving the reflection and transmission problems for the general case of a wave with an arbitrary polarization, it is more convenient in practice to first decompose the incident wave $(\mathbf{E}^i, \mathbf{H}^i)$ into a perpendicularly polarized component $(\mathbf{E}^i_\perp, \mathbf{H}^i_\perp)$ and a parallel polarized component $(\mathbf{E}^i_\parallel, \mathbf{H}^i_\parallel)$, and then after determining the reflected waves $(\mathbf{E}^r_\perp, \mathbf{H}^r_\perp)$ and $(\mathbf{E}^r_\parallel, \mathbf{H}^r_\parallel)$ due to the two incident components, the reflected waves can be added together to give the total reflected wave corresponding to the original incident wave. A similar process applies to the transmitted wave.

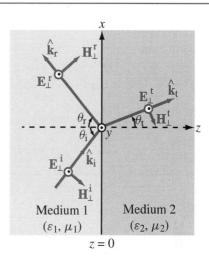

(a) Perpendicular polarization

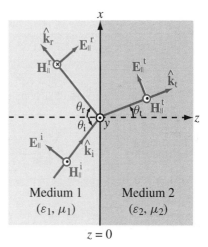

(b) Parallel polarization

Figure 8-14: The plane of incidence is the plane containing the direction of wave travel, $\hat{\mathbf{k}}$, and the surface normal to the boundary, which in the present case is the plane of the paper. A wave is (a) perpendicularly polarized when its **E** is perpendicular to the plane of incidence and (b) parallel polarized when its **E** lies in the plane of incidence.

8-4.1 Perpendicular Polarization

In Fig. 8-15, we show a perpendicularly polarized incident plane wave propagating along the x_i-direction in dielectric medium 1. The electric field phasor $\widetilde{\mathbf{E}}^i_\perp$ points along the y-direction, and the associated magnetic field phasor $\widetilde{\mathbf{H}}^i_\perp$ is along the y_i-axis. The directions of $\widetilde{\mathbf{E}}^i_\perp$ and $\widetilde{\mathbf{H}}^i_\perp$ satisfy the condition that $\widetilde{\mathbf{E}}^i_\perp \times \widetilde{\mathbf{H}}^i_\perp$ points along the propagation direction, $\hat{\mathbf{x}}_i$. The expressions for such a

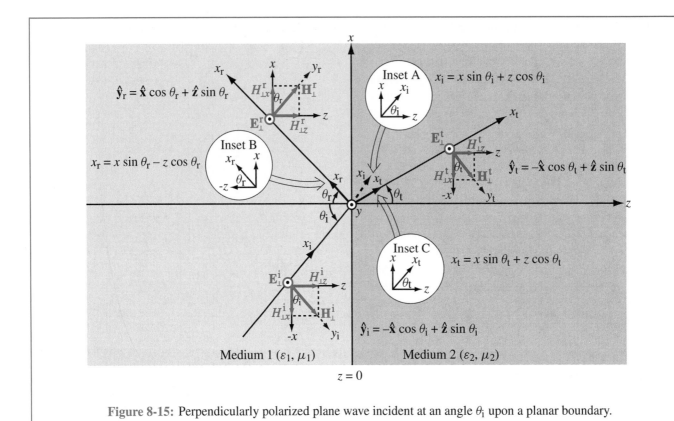

Figure 8-15: Perpendicularly polarized plane wave incident at an angle θ_i upon a planar boundary.

plane wave are given by

$$\widetilde{\mathbf{E}}^i_\perp = \hat{\mathbf{y}} E^i_{\perp 0} e^{-jk_1 x_i}, \tag{8.46a}$$

$$\widetilde{\mathbf{H}}^i_\perp = \hat{\mathbf{y}}_i \frac{E^i_{\perp 0}}{\eta_1} e^{-jk_1 x_i}, \tag{8.46b}$$

where $E^i_{\perp 0}$ is the amplitude of the electric field phasor at $x_i = 0$, $k_1 = \omega\sqrt{\mu_1 \varepsilon_1}$ is the wavenumber, and $\eta_1 = \sqrt{\mu_1/\varepsilon_1}$ is the intrinsic impedance, both for medium 1. From Fig. 8-15, the distance x_i and the unit vector $\hat{\mathbf{y}}_i$ may be expressed in terms of the (x, y, z) global coordinate system as follows:

$$x_i = x \sin\theta_i + z \cos\theta_i, \tag{8.47a}$$

$$\hat{\mathbf{y}}_i = -\hat{\mathbf{x}}\cos\theta_i + \hat{\mathbf{z}}\sin\theta_i. \tag{8.47b}$$

Substituting Eqs. (8.47a) and (8.47b) into Eqs. (8.46a) and (8.46b) gives:

Incident Wave

$$\widetilde{\mathbf{E}}^i_\perp = \hat{\mathbf{y}} E^i_{\perp 0} e^{-jk_1(x \sin\theta_i + z \cos\theta_i)}, \tag{8.48a}$$

$$\widetilde{\mathbf{H}}^i_\perp = (-\hat{\mathbf{x}}\cos\theta_i + \hat{\mathbf{z}}\sin\theta_i)$$
$$\times \frac{E^i_{\perp 0}}{\eta_1} e^{-jk_1(x \sin\theta_i + z \cos\theta_i)}. \tag{8.48b}$$

With the aid of the directional relationships given in Fig. 8-15 for the reflected and transmitted waves, the fields are given by:

Reflected Wave

$$\widetilde{\mathbf{E}}_\perp^r = \hat{\mathbf{y}} E_{\perp 0}^r e^{-jk_1 x_r}$$
$$= \hat{\mathbf{y}} E_{\perp 0}^r e^{-jk_1(x \sin\theta_r - z \cos\theta_r)}, \qquad (8.49a)$$

$$\widetilde{\mathbf{H}}_\perp^r = \hat{\mathbf{y}}_r \frac{E_{\perp 0}^r}{\eta_1} e^{-jk_1 x_r}$$
$$= (\hat{\mathbf{x}} \cos\theta_r + \hat{\mathbf{z}} \sin\theta_r)$$
$$\times \frac{E_{\perp 0}^r}{\eta_1} e^{-jk_1(x \sin\theta_r - z \cos\theta_r)}, \qquad (8.49b)$$

Transmitted Wave

$$\widetilde{\mathbf{E}}_\perp^t = \hat{\mathbf{y}} E_{\perp 0}^t e^{-jk_2 x_t}$$
$$= \hat{\mathbf{y}} E_{\perp 0}^t e^{-jk_2(x \sin\theta_t + z \cos\theta_t)}, \qquad (8.49c)$$

$$\widetilde{\mathbf{H}}_\perp^t = \hat{\mathbf{y}}_t \frac{E_{\perp 0}^t}{\eta_2} e^{-jk_2 x_t}$$
$$= (-\hat{\mathbf{x}} \cos\theta_t + \hat{\mathbf{z}} \sin\theta_t)$$
$$\times \frac{E_{\perp 0}^t}{\eta_2} e^{-jk_2(x \sin\theta_t + z \cos\theta_t)}, \qquad (8.49d)$$

where θ_r and θ_t are the reflection and transmission angles shown in Fig. 8-15 and k_2 and η_2 are the wavenumber and intrinsic impedance of medium 2. Our goal is to characterize the reflected and transmitted fields in terms of the parameters of the incident wave, which include the incidence angle θ_i and the amplitude $E_{\perp 0}^i$. The four expressions given by Eqs. (8.49a) through (8.49d) contain four unknown quantities: $E_{\perp 0}^r$, $E_{\perp 0}^t$, θ_r, and θ_t. Angles θ_r and θ_t can be related to θ_i by Snell's laws, Eqs. (8.28a) and (8.28b), but we choose to keep them as unknowns for the time being, because we intend to show that Snell's laws can also be derived by applying the fields' boundary conditions at $z = 0$. The total electric field in medium 1 is the sum of the incident and reflected electric fields: $\widetilde{\mathbf{E}}_\perp^1 = \widetilde{\mathbf{E}}_\perp^i + \widetilde{\mathbf{E}}_\perp^r$; and the same is true for the total magnetic field in medium 1: $\widetilde{\mathbf{H}}_\perp^1 = \widetilde{\mathbf{H}}_\perp^i + \widetilde{\mathbf{H}}_\perp^r$. Boundary conditions state that the tangential components of $\widetilde{\mathbf{E}}$ and $\widetilde{\mathbf{H}}$ must each be continuous across the boundary between the two media. Field components tangential to the boundary are those along $\hat{\mathbf{x}}$ and $\hat{\mathbf{y}}$. Since the electric fields in media 1 and 2 have $\hat{\mathbf{y}}$-components only, the boundary condition for $\widetilde{\mathbf{E}}$ is

$$(\widetilde{E}_{\perp y}^i + \widetilde{E}_{\perp y}^r)\big|_{z=0} = \widetilde{E}_{\perp y}^t\big|_{z=0}. \qquad (8.50)$$

Upon using Eqs. (8.48a), (8.49a), and (8.49c) in Eq. (8.50) and then setting $z = 0$, we have

$$E_{\perp 0}^i e^{-jk_1 x \sin\theta_i} + E_{\perp 0}^r e^{-jk_1 x \sin\theta_r} = E_{\perp 0}^t e^{-jk_2 x \sin\theta_t}. \qquad (8.51)$$

The boundary condition for the tangential component of the magnetic field (i.e., the x-component) is

$$(\widetilde{H}_{\perp x}^i + \widetilde{H}_{\perp x}^r)\big|_{z=0} = \widetilde{H}_{\perp x}^t\big|_{z=0}, \qquad (8.52)$$

or

$$-\frac{E_{\perp 0}^i}{\eta_1} \cos\theta_i \, e^{-jk_1 x \sin\theta_i} + \frac{E_{\perp 0}^r}{\eta_1} \cos\theta_r \, e^{-jk_1 x \sin\theta_r}$$
$$= -\frac{E_{\perp 0}^t}{\eta_2} \cos\theta_t \, e^{-jk_2 x \sin\theta_t}. \qquad (8.53)$$

To satisfy Eqs. (8.51) and (8.53) for all possible values of x (all along the boundary), it follows that all three exponential arguments must be equal. That is,

$$k_1 \sin\theta_i = k_1 \sin\theta_r = k_2 \sin\theta_t, \qquad (8.54)$$

which is known as the ***phase-matching condition***. The first equality in Eq. (8.54) leads to

$$\boxed{\theta_r = \theta_i \qquad \text{(Snell's law of reflection)}, \qquad (8.55)}$$

and the second equality leads to

$$\boxed{\begin{array}{c} \dfrac{\sin\theta_t}{\sin\theta_i} = \dfrac{k_1}{k_2} = \dfrac{\omega\sqrt{\mu_1\varepsilon_1}}{\omega\sqrt{\mu_2\varepsilon_2}} = \dfrac{n_1}{n_2} \\ \text{(Snell's law of refraction)}. \qquad (8.56) \end{array}}$$

The results expressed by Eqs. (8.55) and (8.56) are identical with those derived previously in Section 8-2 through consideration of the ray path traversed by the incident, reflected, and transmitted wavefronts.

In view of Eq. (8.54), the boundary conditions given by Eqs. (8.51) and (8.53) reduce to

$$E_{\perp 0}^i + E_{\perp 0}^r = E_{\perp 0}^t, \qquad (8.57a)$$

$$\frac{\cos \theta_i}{\eta_1}(-E_{\perp 0}^i + E_{\perp 0}^r) = -\frac{\cos \theta_t}{\eta_2} E_{\perp 0}^t. \qquad (8.57b)$$

These two equations can be solved simultaneously to yield the following expressions for the reflection and transmission coefficients in the perpendicular polarization case:

$$\Gamma_{\perp} = \frac{E_{\perp 0}^r}{E_{\perp 0}^i} = \frac{\eta_2 \cos \theta_i - \eta_1 \cos \theta_t}{\eta_2 \cos \theta_i + \eta_1 \cos \theta_t}, \qquad (8.58a)$$

$$\tau_{\perp} = \frac{E_{\perp 0}^t}{E_{\perp 0}^i} = \frac{2\eta_2 \cos \theta_i}{\eta_2 \cos \theta_i + \eta_1 \cos \theta_t}. \qquad (8.58b)$$

These two coefficients, which formally are known as the *Fresnel reflection and transmission coefficients for perpendicular polarization*, are related by

$$\tau_{\perp} = 1 + \Gamma_{\perp}. \qquad (8.59)$$

If medium 2 is a perfect conductor ($\eta_2 = 0$), Eqs. (8.58a) and (8.58b) reduce to $\Gamma_{\perp} = -1$ and $\tau_{\perp} = 0$, respectively, which means that the incident wave is totally reflected by the conducting medium.

For nonmagnetic dielectrics with $\mu_1 = \mu_2 = \mu_0$ and with the help of Eq. (8.56), the expression for $\Gamma_{\perp}$ can be written as

$$\Gamma_{\perp} = \frac{\cos \theta_i - \sqrt{(\varepsilon_2/\varepsilon_1) - \sin^2 \theta_i}}{\cos \theta_i + \sqrt{(\varepsilon_2/\varepsilon_1) - \sin^2 \theta_i}} \quad \text{(for } \mu_1 = \mu_2\text{)}.$$

$$(8.60)$$

Since $(\varepsilon_2/\varepsilon_1) = (n_2/n_1)^2$, this expression can also be written in terms of the indices of refraction n_1 and n_2.

Example 8-6 Wave Incident Obliquely on a Soil Surface

Using the coordinate system of Fig. 8-15, a plane wave radiated by a distant antenna is incident in air upon a plane soil surface at $z = 0$. The electric field of the incident wave is given by

$$\mathbf{E}^i = \hat{\mathbf{y}}100 \cos(\omega t - \pi x - 1.73\pi z) \quad \text{(V/m)}, \quad (8.61)$$

and the soil medium may be assumed to be a lossless dielectric with a relative permittivity of 4.

(a) Determine k_1, k_2, and the incidence angle θ_i.
(b) Obtain expressions for the total electric fields in air and in the soil medium.
(c) Determine the average power density carried by the wave traveling in the soil medium.

Solution: (a) We begin by converting Eq. (8.61) into phasor form, akin to the expression given by Eq. (8.46a):

$$\widetilde{\mathbf{E}}^i = \hat{\mathbf{y}}100 e^{-j\pi x - j1.73\pi z}$$

$$= \hat{\mathbf{y}}100 e^{-jk_1 x_i} \quad \text{(V/m)}, \qquad (8.62)$$

where x_i is the axis along which the wave is traveling, and

$$k_1 x_i = \pi x + 1.73\pi z. \qquad (8.63)$$

Using Eq. (8.47a), we have

$$k_1 x_i = k_1 x \sin \theta_i + k_1 z \cos \theta_i. \qquad (8.64)$$

Hence,

$$k_1 \sin \theta_i = \pi,$$

$$k_1 \cos \theta_i = 1.73\pi,$$

which together give

$$k_1 = \sqrt{\pi^2 + (1.73\pi)^2} = 2\pi \quad \text{(rad/m)},$$

$$\theta_i = \tan^{-1}\left(\frac{\pi}{1.73\pi}\right) = 30°.$$

The wavelength in medium 1 (air) is

$$\lambda_1 = \frac{2\pi}{k_1} = 1 \text{ m,}$$

and the wavelength in medium 2 (soil) is

$$\lambda_2 = \frac{\lambda_1}{\sqrt{\varepsilon_{r_2}}} = \frac{1}{\sqrt{4}} = 0.5 \text{ m.}$$

The corresponding wave number in medium 2 is

$$k_2 = \frac{2\pi}{\lambda_2} = 4\pi \quad \text{(rad/m)}.$$

Since $\widetilde{\mathbf{E}}^i$ is along $\hat{\mathbf{y}}$, it is perpendicularly polarized ($\hat{\mathbf{y}}$ is perpendicular to the plane of incidence containing the surface normal $\hat{\mathbf{z}}$ and the propagation direction $\hat{\mathbf{x}}_i$).

(b) Corresponding to $\theta_i = 30°$, the transmission angle θ_t is obtained with the help of Eq. (8.56):

$$\sin\theta_t = \frac{k_1}{k_2}\sin\theta_i = \frac{2\pi}{4\pi}\sin 30° = 0.25$$

or

$$\theta_t = 14.5°.$$

With $\varepsilon_1 = \varepsilon_0$ and $\varepsilon_2 = \varepsilon_{r_2}\varepsilon_0 = 4\varepsilon_0$, the reflection and transmission coefficients for perpendicular polarization are determined with the help of Eqs. (8.59) and (8.60),

$$\Gamma_\perp = \frac{\cos\theta_i - \sqrt{(\varepsilon_2/\varepsilon_1) - \sin^2\theta_i}}{\cos\theta_i + \sqrt{(\varepsilon_2/\varepsilon_1) - \sin^2\theta_i}} = -0.38,$$

$$\tau_\perp = 1 + \Gamma_\perp = 0.62.$$

Using Eqs. (8.48a) and (8.49a) with $E^i_{\perp 0} = 100$ V/m and $\theta_i = \theta_r$, the total electric field phasor in medium 1 is

$$\widetilde{\mathbf{E}}^1_\perp = \widetilde{\mathbf{E}}^i_\perp + \widetilde{\mathbf{E}}^r_\perp$$

$$= \hat{\mathbf{y}} E^i_{\perp 0} e^{-jk_1(x\sin\theta_i + z\cos\theta_i)}$$

$$+ \hat{\mathbf{y}}\Gamma E^i_{\perp 0} e^{-jk_1(x\sin\theta_i - z\cos\theta_i)}$$

$$= \hat{\mathbf{y}}100 e^{-j(\pi x + 1.73\pi z)} - \hat{\mathbf{y}}38 e^{-j(\pi x - 1.73\pi z)},$$

and the corresponding instantaneous electric field in medium 1 is

$$\mathbf{E}^1_\perp(x, z, t) = \mathfrak{Re}\left[\widetilde{\mathbf{E}}^1_\perp e^{j\omega t}\right]$$

$$= \hat{\mathbf{y}}[100\cos(\omega t - \pi x - 1.73\pi z)$$

$$- 38\cos(\omega t - \pi x + 1.73\pi z)] \quad \text{(V/m)}.$$

In medium 2, using Eq. (8.49c) with $E^t_{\perp 0} = \tau_\perp E^i_{\perp 0}$ gives

$$\widetilde{\mathbf{E}}^t_\perp = \hat{\mathbf{y}}\tau E^i_{\perp 0} e^{-jk_2(x\sin\theta_t + z\cos\theta_t)}$$

$$= \hat{\mathbf{y}}62 e^{-j(\pi x + 3.87\pi z)}$$

and, correspondingly,

$$\mathbf{E}^t_\perp(x, z, t) = \mathfrak{Re}\left[\widetilde{\mathbf{E}}^t_\perp e^{j\omega t}\right]$$

$$= \hat{\mathbf{y}}62\cos(\omega t - \pi x - 3.87\pi z) \quad \text{(V/m)}.$$

(c) In medium 2, $\eta_2 = \eta_0/\sqrt{\varepsilon_{r_2}} \simeq 120\pi/\sqrt{4} = 60\pi$ (Ω), and the average power density carried by the wave is

$$S^t_{av} = \frac{|E^t_{\perp 0}|^2}{2\eta_2} = \frac{(62)^2}{2 \times 60\pi} = 10.2 \quad \text{(W/m}^2\text{)}. \quad \blacksquare$$

8-4.2 Parallel Polarization

If we interchange **E** and **H** of the perpendicular polarization situation, while keeping in mind the requirement that relates the directions of **E** and **H** to the direction of propagation for each of the incident, reflected, and transmitted waves, we end up with the geometry shown in Fig. 8-16 for parallel polarization. Now the electric fields lie in the plane of incidence, and the associated magnetic fields are perpendicular to the plane of incidence. With reference to the directions indicated in Fig. 8-16, the fields of the incident, reflected, and transmitted waves are given by

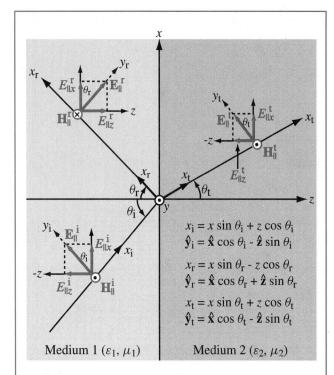

$$x_i = x \sin \theta_i + z \cos \theta_i$$
$$\hat{\mathbf{y}}_i = \hat{\mathbf{x}} \cos \theta_i - \hat{\mathbf{z}} \sin \theta_i$$

$$x_r = x \sin \theta_r - z \cos \theta_r$$
$$\hat{\mathbf{y}}_r = \hat{\mathbf{x}} \cos \theta_r + \hat{\mathbf{z}} \sin \theta_r$$

$$x_t = x \sin \theta_t + z \cos \theta_t$$
$$\hat{\mathbf{y}}_t = \hat{\mathbf{x}} \cos \theta_t - \hat{\mathbf{z}} \sin \theta_t$$

Medium 1 (ε_1, μ_1) Medium 2 (ε_2, μ_2)

Figure 8-16: Parallel polarized plane wave incident at an angle θ_i upon a planar boundary.

Incident Wave

$$\widetilde{\mathbf{E}}_\parallel^i = \hat{\mathbf{y}}_i E_{\parallel 0}^i e^{-jk_1 x_i}$$

$$= (\hat{\mathbf{x}} \cos \theta_i - \hat{\mathbf{z}} \sin \theta_i) E_{\parallel 0}^i e^{-jk_1(x \sin \theta_i + z \cos \theta_i)}, \quad (8.65a)$$

$$\widetilde{\mathbf{H}}_\parallel^i = \hat{\mathbf{y}} \frac{E_{\parallel 0}^i}{\eta_1} e^{-jk_1 x_i} = \hat{\mathbf{y}} \frac{E_{\parallel 0}^i}{\eta_1} e^{-jk_1(x \sin \theta_i + z \cos \theta_i)}, \quad (8.65b)$$

Reflected Wave

$$\widetilde{\mathbf{E}}_\parallel^r = \hat{\mathbf{y}}_r E_{\parallel 0}^r e^{-jk_1 x_r}$$

$$= (\hat{\mathbf{x}} \cos \theta_r + \hat{\mathbf{z}} \sin \theta_r) E_{\parallel 0}^r e^{-jk_1(x \sin \theta_r - z \cos \theta_r)}, \quad (8.65c)$$

$$\widetilde{\mathbf{H}}_\parallel^r = -\hat{\mathbf{y}} \frac{E_{\parallel 0}^r}{\eta_1} e^{-jk_1 x_r}$$

$$= -\hat{\mathbf{y}} \frac{E_{\parallel 0}^r}{\eta_1} e^{-jk_1(x \sin \theta_r - z \cos \theta_r)}, \quad (8.65d)$$

Transmitted Wave

$$\widetilde{\mathbf{E}}_\parallel^t = \hat{\mathbf{y}}_t E_{\parallel 0}^t e^{-jk_2 x_t}$$

$$= (\hat{\mathbf{x}} \cos \theta_t - \hat{\mathbf{z}} \sin \theta_t) E_{\parallel 0}^t e^{-jk_2(x \sin \theta_t + z \cos \theta_t)}, \quad (8.65e)$$

$$\widetilde{\mathbf{H}}_\parallel^t = \hat{\mathbf{y}} \frac{E_{\parallel 0}^t}{\eta_2} e^{-jk_2 x_t} = \hat{\mathbf{y}} \frac{E_{\parallel 0}^t}{\eta_2} e^{-jk_2(x \sin \theta_t + z \cos \theta_t)}. \quad (8.65f)$$

By matching the tangential components of $\widetilde{\mathbf{E}}$ and $\widetilde{\mathbf{H}}$ in the two media at $z = 0$, as we did previously in the perpendicular-polarization case, we again obtain the relations defining Snell's laws, as well as the following expressions for the *Fresnel reflection and transmission coefficients for parallel polarization*:

$$\Gamma_\parallel = \frac{E_{\parallel 0}^r}{E_{\parallel 0}^i} = \frac{\eta_2 \cos \theta_t - \eta_1 \cos \theta_i}{\eta_2 \cos \theta_t + \eta_1 \cos \theta_i}, \quad (8.66a)$$

$$\tau_\parallel = \frac{E_{\parallel 0}^t}{E_{\parallel 0}^i} = \frac{2\eta_2 \cos \theta_i}{\eta_2 \cos \theta_t + \eta_1 \cos \theta_i}. \quad (8.66b)$$

The preceding expressions can be shown to yield the relation

$$\tau_\parallel = (1 + \Gamma_\parallel) \frac{\cos \theta_i}{\cos \theta_t}. \quad (8.67)$$

We noted earlier in connection with the perpendicular-polarization case that, when the second medium is a perfect conductor with $\eta_2 = 0$, the incident wave gets totally reflected by the boundary. The same is true for parallel polarization; setting $\eta_2 = 0$ in Eqs. (8.66a) and (8.66b) gives $\Gamma_\parallel = -1$ and $\tau_\parallel = 0$.

For nonmagnetic materials, Eq. (8.66a) becomes

$$\Gamma_\parallel = \frac{-(\varepsilon_2/\varepsilon_1) \cos \theta_i + \sqrt{(\varepsilon_2/\varepsilon_1) - \sin^2 \theta_i}}{(\varepsilon_2/\varepsilon_1) \cos \theta_i + \sqrt{(\varepsilon_2/\varepsilon_1) - \sin^2 \theta_i}}$$

$$\text{(for } \mu_1 = \mu_2\text{).} \quad (8.68)$$

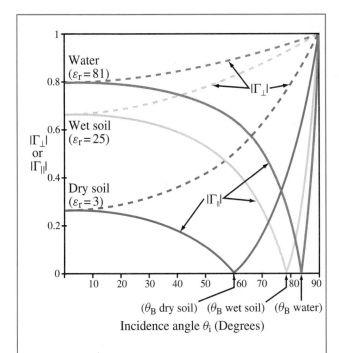

Figure 8-17: Plots for $|\Gamma_\perp|$ and $|\Gamma_\parallel|$ as a function of θ_i for a dry soil surface, a wet-soil surface, and a water surface. For each surface, $|\Gamma_\parallel| = 0$ at the Brewster angle.

To illustrate the angular variations of the magnitudes of $\Gamma_\perp$ and $\Gamma_\parallel$, in Fig. 8-17 we show plots for waves incident in air onto three different types of dielectric surfaces: dry soil ($\varepsilon_r = 3$), wet soil ($\varepsilon_r = 25$), and water ($\varepsilon_r = 81$). For each of the surfaces, (1) $\Gamma_\perp = \Gamma_\parallel$ at normal incidence ($\theta_i = 0$), as expected, (2) $|\Gamma_\perp| = |\Gamma_\parallel| = 1$ at *grazing* incidence ($\theta_i = 90°$), and (3) $\Gamma_\parallel$ goes to zero at an angle labeled the ***Brewster angle*** in Fig. 8-17. For nonmagnetic materials, the Brewster angle exists only for parallel polarization, and its value depends on the ratio ($\varepsilon_2/\varepsilon_1$), as we will see shortly. *At the Brewster angle, the parallel-polarized component of the incident wave is totally transmitted into medium 2.*

8-4.3 Brewster Angle

The Brewster angle θ_B is defined as the incidence angle θ_i at which the Fresnel reflection coefficient $\Gamma = 0$.

Perpendicular Polarization

For perpendicular polarization, the Brewster angle $\theta_{B\perp}$ can be obtained by setting the numerator of the expression for $\Gamma_\perp$, given by Eq. (8.58a), equal to zero or, equivalently, when

$$\eta_2 \cos \theta_i = \eta_1 \cos \theta_t. \qquad (8.69)$$

After (1) squaring both sides of Eq. (8.69), (2) using Eq. (8.56), (3) solving for θ_i, and then denoting θ_i as $\theta_{B\perp}$, we have

$$\sin \theta_{B\perp} = \sqrt{\frac{1 - (\mu_1\varepsilon_2/\mu_2\varepsilon_1)}{1 - (\mu_1/\mu_2)^2}}. \qquad (8.70)$$

Because the denominator of Eq. (8.70) goes to zero when $\mu_1 = \mu_2$, $\theta_{B\perp}$ *does not exist for nonmagnetic materials.*

Parallel Polarization

The value of θ_i, denoted $\theta_{B\parallel}$, at which $\Gamma_\parallel = 0$ can be found by setting the numerator of Eq. (8.66a) equal to zero. The result is identical with Eq. (8.70), but with μ and ε interchanged. That is,

$$\sin \theta_{B\parallel} = \sqrt{\frac{1 - (\varepsilon_1\mu_2/\varepsilon_2\mu_1)}{1 - (\varepsilon_1/\varepsilon_2)^2}}. \qquad (8.71)$$

For nonmagnetic materials,

$$\boxed{\begin{aligned} \theta_{B\parallel} &= \sin^{-1}\sqrt{\frac{1}{1 + (\varepsilon_1/\varepsilon_2)}} \\ &= \tan^{-1}\sqrt{\frac{\varepsilon_2}{\varepsilon_1}} \quad (\text{for } \mu_1 = \mu_2). \quad (8.72) \end{aligned}}$$

The Brewster angle is also called the *polarizing angle*. This is because, if a wave composed of both perpendicular and parallel polarization components is incident upon a nonmagnetic surface at the Brewster angle $\theta_{B\parallel}$, the parallel polarized component is totally transmitted into the second medium, and only the perpendicularly polarized component is reflected by the surface. Natural light, including sunlight and light generated by most manufactured sources, is considered unpolarized because the direction of the electric field of the light waves varies randomly in angle over the plane perpendicular to the direction of propagation. Thus, on average half of the intensity of natural light is perpendicularly polarized and the other half is parallel polarized. When unpolarized light is incident upon a surface at the Brewster angle, the reflected wave is strictly perpendicularly polarized. Hence, the reflection process acts as a polarizer.

REVIEW QUESTIONS

Q8.4 Can total internal reflection take place for a wave incident in medium 1 (with n_1) onto medium 2 (with n_2) when $n_2 > n_1$?

Q8.5 What is the difference between the boundary conditions applied in Section 8-1.1 for normal incidence and those applied in Section 8-4.1 for oblique incidence with perpendicular polarization?

Q8.6 Why is the Brewster angle also called the polarizing angle?

Q8.7 At the boundary, the vector sum of the tangential components of the incident and reflected electric fields has to be equal to the tangential component of the transmitted electric field. For $\varepsilon_{r_1} = 1$ and $\varepsilon_{r_2} = 16$, determine the Brewster angle and then verify the validity of the preceding statement by sketching to scale the tangential components of the three electric fields at the Brewster angle.

EXERCISE 8.7 A wave in air is incident upon the flat boundary of a soil medium with $\varepsilon_r = 4$ and $\mu_r = 1$ at $\theta_i = 50°$. Determine $\Gamma_\perp$, $\tau_\perp$, $\Gamma_\parallel$, and $\tau_\parallel$,

Ans. $\Gamma_\perp = -0.48, \tau_\perp = 0.52, \Gamma_\parallel = -0.16, \tau_\parallel = 0.58.$

EXERCISE 8.8 Determine the Brewster angle for the boundary of Exercise 8.7.

Ans. $\theta_B = 63.4°.$

EXERCISE 8.9 Show that the incident, reflected, and transmitted electric and magnetic fields given by Eqs. (8.65a) through (8.65f) all have the same exponential phase function along the x-direction.

Ans. With the help of Eqs. (8.55) and (8.56), all six fields are shown to vary as $e^{-jk_1 x \sin\theta_i}$.

8-5 Reflectivity and Transmissivity

The reflection and transmission coefficients represent the ratios of the reflected and transmitted electric field amplitudes to the amplitude of the incident wave. We now examine power ratios, and we start the process by considering the perpendicular polarization case. Figure 8-18 shows a circular beam of electromagnetic energy incident upon the boundary between two contiguous, lossless media. The area of the spot illuminated by the beam is A, and the incident, reflected, and transmitted beams have electric-field amplitudes $E^i_{\perp 0}$, $E^r_{\perp 0}$, and $E^t_{\perp 0}$, respectively. The average power densities carried by the incident, reflected, and transmitted beams are

$$S^i_\perp = \frac{|E^i_{\perp 0}|^2}{2\eta_1}, \tag{8.73a}$$

$$S^r_\perp = \frac{|E^r_{\perp 0}|^2}{2\eta_1}, \tag{8.73b}$$

$$S^t_\perp = \frac{|E^t_{\perp 0}|^2}{2\eta_2}, \tag{8.73c}$$

where η_1 and η_2 are the intrinsic impedances of media 1 and 2, respectively. The cross-sectional areas of the

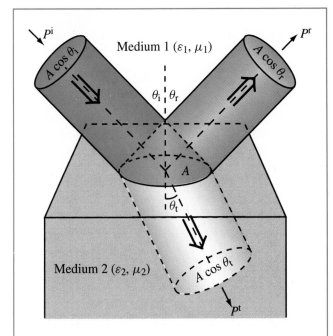

Figure 8-18: Reflection and transmission of an incident circular beam illuminating a spot of size A on the interface.

incident, reflected, and transmitted beams are

$$A_i = A \cos \theta_i, \qquad (8.74a)$$

$$A_r = A \cos \theta_r, \qquad (8.74b)$$

$$A_t = A \cos \theta_t, \qquad (8.74c)$$

and the corresponding average powers carried by the beams are

$$P_\perp^i = S_\perp^i A_i = \frac{|E_{\perp 0}^i|^2}{2\eta_1} A \cos \theta_i, \qquad (8.75a)$$

$$P_\perp^r = S_\perp^r A_r = \frac{|E_{\perp 0}^r|^2}{2\eta_1} A \cos \theta_r, \qquad (8.75b)$$

$$P_\perp^t = S_\perp^t A_t = \frac{|E_{\perp 0}^t|^2}{2\eta_2} A \cos \theta_t. \qquad (8.75c)$$

The *reflectivity R* (also called *reflectance* in optics) is defined as the ratio of the reflected power to the incident power. The reflectivity for perpendicular polarization is then

$$R_\perp = \frac{P_\perp^r}{P_\perp^i} = \frac{|E_{\perp 0}^r|^2 \cos \theta_r}{|E_{\perp 0}^i|^2 \cos \theta_i} = \left| \frac{E_{\perp 0}^r}{E_{\perp 0}^i} \right|^2, \qquad (8.76)$$

where we used the fact that $\theta_r = \theta_i$, in accordance with Snell's law of reflection. The ratio of the reflected to incident electric field amplitudes, $|E_{\perp 0}^r / E_{\perp 0}^i|$, is simply equal to the magnitude of the reflection coefficient $\Gamma_\perp$. Hence,

$$\boxed{R_\perp = |\Gamma_\perp|^2, \qquad (8.77)}$$

and, similarly, for parallel polarization

$$\boxed{R_\parallel = \frac{P_\parallel^r}{P_\parallel^i} = |\Gamma_\parallel|^2. \qquad (8.78)}$$

The *transmissivity T* (or *transmittance* in optics) is defined as the ratio of the transmitted power to incident power:

$$\boxed{\begin{aligned} T_\perp &= \frac{P_\perp^t}{P_\perp^i} = \frac{|E_{\perp 0}^t|^2}{|E_{\perp 0}^i|^2} \frac{\eta_1}{\eta_2} \frac{A \cos \theta_t}{A \cos \theta_i} \\ &= |\tau_\perp|^2 \left(\frac{\eta_1 \cos \theta_t}{\eta_2 \cos \theta_i} \right), \qquad (8.79a) \\ T_\parallel &= \frac{P_\parallel^t}{P_\parallel^i} = |\tau_\parallel|^2 \left(\frac{\eta_1 \cos \theta_t}{\eta_2 \cos \theta_i} \right). \qquad (8.79b) \end{aligned}}$$

The incident, reflected, and transmitted waves do not have to obey any such laws as conservation of electric field, conservation of magnetic field, or conservation of power density, but they do have to obey the law of conservation of power. In fact, in many cases the transmitted electric field is larger than the incident electric field.

Table 8-2: Expressions for Γ, τ, R, and T for wave incidence from a medium with intrinsic impedance η_1 onto a medium with intrinsic impedance η_2. Angles θ_i and θ_t are the angles of incidence and transmission, respectively.

Property	Normal Incidence $\theta_i = \theta_t = 0$	Perpendicular Polarization	Parallel Polarization
Reflection coefficient	$\Gamma = \dfrac{\eta_2 - \eta_1}{\eta_2 + \eta_1}$	$\Gamma_\perp = \dfrac{\eta_2 \cos\theta_i - \eta_1 \cos\theta_t}{\eta_2 \cos\theta_i + \eta_1 \cos\theta_t}$	$\Gamma_\parallel = \dfrac{\eta_2 \cos\theta_t - \eta_1 \cos\theta_i}{\eta_2 \cos\theta_t + \eta_1 \cos\theta_i}$
Transmission coefficient	$\tau = \dfrac{2\eta_2}{\eta_2 + \eta_1}$	$\tau_\perp = \dfrac{2\eta_2 \cos\theta_i}{\eta_2 \cos\theta_i + \eta_1 \cos\theta_t}$	$\tau_\parallel = \dfrac{2\eta_2 \cos\theta_i}{\eta_2 \cos\theta_t + \eta_1 \cos\theta_i}$
Relation of Γ to τ	$\tau = 1 + \Gamma$	$\tau_\perp = 1 + \Gamma_\perp$	$\tau_\parallel = (1 + \Gamma_\parallel)\dfrac{\cos\theta_i}{\cos\theta_t}$
Reflectivity	$R = |\Gamma|^2$	$R_\perp = |\Gamma_\perp|^2$	$R_\parallel = |\Gamma_\parallel|^2$
Transmissivity	$T = |\tau|^2 \left(\dfrac{\eta_1}{\eta_2}\right)$	$T_\perp = |\tau_\perp|^2 \dfrac{\eta_1 \cos\theta_t}{\eta_2 \cos\theta_i}$	$T_\parallel = |\tau_\parallel|^2 \dfrac{\eta_1 \cos\theta_t}{\eta_2 \cos\theta_i}$
Relation of R to T	$T = 1 - R$	$T_\perp = 1 - R_\perp$	$T_\parallel = 1 - R_\parallel$

Notes: (1) $\sin\theta_t = \sqrt{\mu_1\varepsilon_1/\mu_2\varepsilon_2}\,\sin\theta_i$; (2) $\eta_1 = \sqrt{\mu_1/\varepsilon_1}$; (3) $\eta_2 = \sqrt{\mu_2/\varepsilon_2}$; (4) for nonmagnetic media, $\eta_2/\eta_1 = n_1/n_2$.

Conservation of power requires that the incident power be equal to the sum of the reflected and transmitted powers. That is, for perpendicular polarization, for example,

$$P_\perp^i = P_\perp^r + P_\perp^t, \qquad (8.80)$$

or

$$\frac{|E_{\perp 0}^i|^2}{2\eta_1} A \cos\theta_i = \frac{|E_{\perp 0}^r|^2}{2\eta_1} A \cos\theta_r$$
$$+ \frac{|E_{\perp 0}^t|^2}{2\eta_2} A \cos\theta_t. \qquad (8.81)$$

Use of Eqs. (8.76), (8.79a), and (8.79b) leads to

$$R_\perp + T_\perp = 1, \qquad (8.82a)$$
$$R_\parallel + T_\parallel = 1, \qquad (8.82b)$$

or

$$|\Gamma_\perp|^2 + |\tau_\perp|^2 \left(\frac{\eta_1 \cos\theta_t}{\eta_2 \cos\theta_i}\right) = 1, \qquad (8.83a)$$

$$|\Gamma_\parallel|^2 + |\tau_\parallel|^2 \left(\frac{\eta_1 \cos\theta_t}{\eta_2 \cos\theta_i}\right) = 1. \qquad (8.83b)$$

Figure 8-19 shows plots for $(R_\parallel, T_\parallel)$ as a function of θ_i for an air–glass interface. Note that the sum of $R_\parallel$ and $T_\parallel$ is always equal to 1, as mandated by Eq. (8.82b). We also note that, at the Brewster angle θ_B, $R_\parallel = 0$ and $T_\parallel = 1$.

Table 8-2 provides a summary of the general expressions for Γ, τ, R, and T for both normal incidence and oblique incidence.

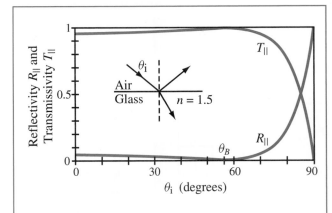

Figure 8-19: Angular plots for $(R_\parallel, T_\parallel)$ for an air–glass interface.

Example 8-7   Beam of Light

A 5-W beam of light with circular cross section is incident in air upon a plane boundary of a dielectric medium with an index of refraction of 5. If the angle of incidence is 60° and the incident wave is parallel polarized, determine the transmission angle and the powers contained in the reflected and transmitted beams.

Solution: From Eq. (8.56),

$$\sin \theta_t = \frac{n_1}{n_2} \sin \theta_i = \frac{1}{5} \sin 60° = 0.17$$

or

$$\theta_t = 10°.$$

With $\varepsilon_2/\varepsilon_1 = n_2^2/n_1^2 = (5)^2 = 25$, the reflection coefficients for parallel polarization can be computed by applying Eq. (8.68) as follows:

$$
\Gamma_\parallel = \frac{-(\varepsilon_2/\varepsilon_1) \cos \theta_i + \sqrt{(\varepsilon_2/\varepsilon_1) - \sin^2 \theta_i}}{(\varepsilon_2/\varepsilon_1) \cos \theta_i + \sqrt{(\varepsilon_2/\varepsilon_1) - \sin^2 \theta_i}}
$$

$$
= \frac{-25 \cos 60° + \sqrt{25 - \sin^2 60°}}{25 \cos 60° + \sqrt{25 - \sin^2 60°}} = -0.435.
$$

The reflected and transmitted powers are then

$$P_\parallel^r = P_\parallel^i |\Gamma_\parallel|^2 = 5(0.435)^2 = 0.95 \text{ W},$$

$$P_\parallel^t = P_\parallel^i - P_\parallel^r = 5 - 0.95 = 4.05 \text{ W}. \quad \blacksquare$$

8-6 Geometric Optics

In the foregoing material, we examined the reflection and refraction behavior of *uniform, plane* electromagnetic waves when incident upon *planar* boundaries between dissimilar materials. In practice, wave interaction with matter may involve waves that are neither uniform nor plane. Also, the objects that the waves interact with may be of any shape and size. How then do we put the results we obtained in the preceding sections to use when dealing with practical situations? The answer depends in large part on the size of the object and the degree of curvature of its surfaces relative to the wavelength of the incident wave. We have available to us two fundamental approaches for describing wave interaction with matter; these are the physical-optics method and the geometric-optics method. In *physical optics*, also known as *wave optics*, a wave is characterized mathematically by all its attributes: its amplitude, phase factor, and polarization vector. In contrast, in *geometric optics*, otherwise known as *ray optics*, a wave is represented by a ray denoting the direction of travel of the wave's energy. Neither the phase nor the polarization of the wave is accounted for explicitly in geometric optics. Thus, geometric optics is a ray approximation of physical optics. Although it is mathematically superior and its formulation is traceable back to Maxwell's equations, the physical-optics method is mathematically more complicated to apply than the geometric-optics method, and therefore physical optics is used whenever geometric optics is inapplicable or when a greater degree of accuracy is desired than that used in geometric optics. *In general, the geometric-optics method provides fairly accurate results whenever the size of the object illuminated by an electromagnetic wave is much larger than the wavelength* λ. There is also the

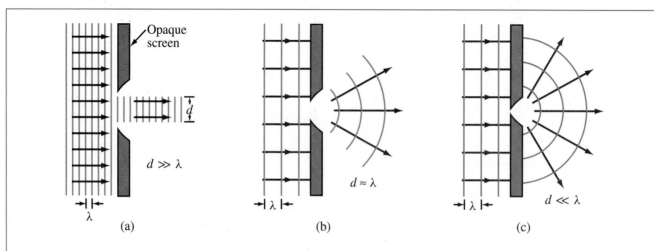

Figure 8-20: A plane wave of wavelength λ incident upon an opaque screen with opening of diameter d. The wave emerging from the opening (a) maintains its original direction if $d \gg \lambda$, (b) gets diffracted by the opening when $d \approx \lambda$, and (c) behaves like a spherical wave emanating from a point source when $d \ll \lambda$.

additional secondary requirement that the curvatures of the object's surfaces be relatively smooth relative to λ.

By treating waves as rays of energy, geometric optics does not account for the relative phases of multiple waves arriving at a given point in space. Consequently, *interference effects* due to phase addition and subtraction cannot be predicted on the basis of geometric optics. Another wave phenomenon ignored by geometric optics is *diffraction*, which is the deviation of wave propagation from the straight-line representation used in geometric optics. Diffraction accounts for the "bending" of waves around the edges of objects. To explain what we mean by diffraction, let us consider the three situations depicted in Fig. 8-20. In part (a), a plane wave is incident upon an opaque barrier with a circular opening whose diameter d is very large relative to the wavelength λ. Apart from some small edge effects, the wave emerging from the opening continues to move in a straight line. This situation adheres to the condition of geometric optics, that the dimensions of reflecting boundaries (in a transparent medium) or, conversely, the dimensions of openings in

opaque boundaries be much larger than λ. The situations in Fig. 8-20(b) and (c) are for openings with diameters comparable with and smaller than λ, respectively. In both cases, the incident wave is said to be *diffracted* by the opening; as a consequence, the emerging wave is no longer a plane wave. To account for wave interference and diffraction, it is necessary to use the physical-optics representation of electromagnetic waves. We should note that the term "optics" in the names "geometric optics" and "physical optics" is historically based and does not imply that these methods are limited to optical systems; the methods are applicable across the entire electromagnetic spectrum.

The next two sections are intended to provide an overview of geometric optics, with particular emphasis on how mirrors and lenses are used as imaging devices.

8-7 Images Formed by Mirrors

Mirrors are used throughout the electromagnetic spectrum, including the x-ray, ultraviolet, visible, infrared, and radio regions. A mirror is usually made by coating a dielectric substrate, such as glass, with a thin layer of silver or aluminum. A perfect mirror is a perfectly conducting surface with a reflectivity of 1, and it is analogous to a short circuit at the end of a transmission line. In this section, we will examine how images are formed by both plane mirrors and spherical mirrors.

8-7.1 Images Formed by Plane Mirrors

The point source at point O in Fig. 8-21 is at a distance s, called the *object distance*, in front of a plane mirror.

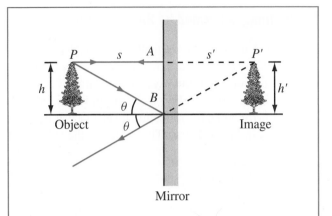

Figure 8-22: Image location and orientation are determined by ray optics and Snell's law of reflection.

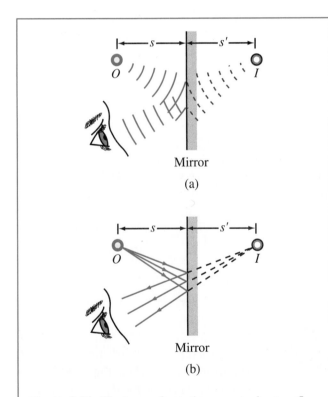

Figure 8-21: The image-formation process due to reflection from a plane mirror may be represented (a) in terms of spherical wavefronts or (b) in terms of rays of light emanating from the source at point O.

The source radiates spherical waves, which may be represented by wavefronts, as shown in Fig. 8-21(a), or, equivalently, the waves may be represented by light rays, as in Fig. 8-21(b). To a viewer at the location shown in the figure, the reflected rays appear to come from point I located a distance s', called the *image distance*, behind the mirror. The image is called a *real image* if the rays of reflected light actually pass through the image point, and it is called a *virtual image* if the rays do not actually pass through the image point but appear to diverge from that point. The image formed by the plane mirror in this case is a virtual image.

By considering a self-luminous or externally illuminated object to consist of a large number of point sources, the point-source imaging process can be extended to generate the image of the object shown in Fig. 8-22. To determine where an image is formed, it is necessary to follow at least two rays of light as they are reflected by the mirror. Point P', the image of point P, is at the intersection of the horizontal ray to the mirror, which reflects back on itself, and the oblique ray that reflects at point B on the mirror. Because triangles PAB and $P'AB$ are congruent, $s = s'$ and $h = h'$.

8-7.2 Images Formed by Spherical Mirrors

A spherical mirror is a segment of a sphere painted or covered with a highly reflective material. It is called a *concave* mirror when the rays of light are reflected by its inner surface, as in Fig. 8-23(a), and it is called a *convex* mirror when the reflection takes place at its outer surface, as in Fig. 8-23(b). The center of the sphere (of

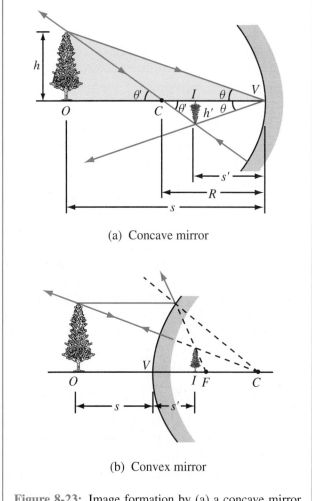

(a) Concave mirror

(b) Convex mirror

Figure 8-23: Image formation by (a) a concave mirror and (b) a convex mirror.

which the spherical mirror is a part) is called the *center of curvature* and is denoted by point C, and the radius R is called the *radius of curvature*. The *principal axis* of the mirror is the line connecting point C to point V, the center (or vertex) of the spherical segment. In Fig. 8-23(a), the object of height h and located at object distance s is imaged as an inverted real image of height h' at image distance s'. From the large triangle with included angle θ, we have $\tan \theta = h/s$, and the small triangle with angle θ gives $\tan \theta = -h'/s'$, where the negative sign signifies that the image is inverted because h' is negative. The *lateral magnification* of the mirror is defined as the ratio of h' to h:

$$M = \frac{h'}{h} = -\frac{s'}{s}. \qquad (8.84)$$

From the two similar triangles involving the angles θ',

$$\tan \theta' = \frac{h}{s - R} = -\frac{h'}{R - s'}. \qquad (8.85)$$

Upon combining Eqs. (8.84) and (8.85), we obtain the *mirror equation*

$$\frac{1}{s} + \frac{1}{s'} = \frac{2}{R}. \qquad (8.86)$$

When the object is very far away from the mirror compared to R, we can regard it as being at infinity, in which case $1/s \simeq 0$ and $s' \simeq R/2$. Thus, the image point is halfway between the center of curvature C and the center of the mirror V, as shown in Fig. 8-24. In this case, the image point is called the *focal point* F, and the image distance is called the *focal length* of the mirror, f. That is,

$$f = \frac{R}{2}, \qquad (8.87)$$

Table 8-3: Sign convention for spherical mirrors.

Quantity	Sign	Condition
s	+	if object is in front side of mirror (real object)
s	−	if object is in back side of mirror (virtual object)
s'	+	if image is in front side of mirror (real image)
s'	−	if image is in back side of mirror (virtual image)
f and R	+	if C is in front side of mirror (concave)
f and R	−	if C is in back side of mirror (convex)

Notes: (1) The front side of a mirror is defined as the side from which the light approaches the mirror, regardless of the mirror's shape. (2) An object cannot be virtual in single-mirror reflection, but its image, if virtual, can become a virtual object for reflections by additional mirrors.

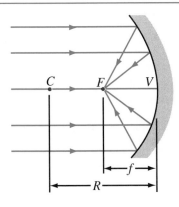

Figure 8-24: A source point at infinity is imaged at the focal point F of the mirror. The focal length $f = R/2$.

and the mirror equation can be rewritten in terms of f as

$$\frac{1}{s} + \frac{1}{s'} = \frac{1}{f}. \qquad (8.88)$$

An object at infinity is imaged at the focal point and, conversely, an object at the focal point is imaged at infinity.

Equations (8.84) and (8.88) can be used for both concave and convex mirrors, as long as we adhere to the sign convention given in Table 8-3. The sign convention refers to two regions: (1) the *front side* of the mirror, which is the region in which the light rays actually exist, and (2) the *back side*, which is the region where virtual images of real objects are formed. For both mirrors in Figs. 8-23(a) and (b), the front sides are the sides to the left of the mirrors. Note that f is positive for concave mirrors and negative for convex mirrors.

The magnification and orientation (erect or inverted) of the image depend on the type of mirror (concave or convex) and the location of the object relative to the mirror center. A summary of these properties is provided in Table 8-4 for various object locations.

Example 8-8   Image Formed by a Concave Mirror

A flower is placed at a distance of 20 cm in front of a concave mirror with a 15-cm focal length. Locate and describe the image.

Solution: Inserting $s = 20$ cm and $f = 15$ cm in Eq. (8.88) gives

$$\frac{1}{20} + \frac{1}{s'} = \frac{1}{15},$$

Table 8-4: Images of real objects formed by spherical mirrors.[a]

Concave				
Object	Image			
Location	Type	Location	Orientation	Relative size
$2f < s < \infty$	Real	$2f > s' > f$	Inverted	Minified
$s = 2f$	Real	$s' = 2f$	Inverted	Same size
$f < s < 2f$	Real	$\infty > s' > 2f$	Inverted	Magnified
$s = f$		$s' = \infty$		
$s < f$	Virtual	$|s'| > s$	Erect	Magnified
Convex				
Object	Image			
Location	Type	Location	Orientation	Relative size
Anywhere	Virtual	$|s'| < |f|$, $s > |s'|$	Erect	Minified

[a]Adapted from E. Hecht, *Optics*, second edition, Addison-Wesley Publishing Co., 1990.

which leads to $s' = 60$ cm. According to Eq. (8.84), the lateral magnification is

$$M = -\frac{s'}{s} = -\frac{60}{15} = -4.$$

The magnitude of M means that the image is magnified by a factor of 4, and the negative sign means that the image is inverted. ∎

8-8 Images Formed by Spherical Lenses

Lenses are commonly used to form images in optical instruments, such as cameras, telescopes, and microscopes, but their use also extends to nonimaging applications and to many parts of the electromagnetic spectrum. A lens is characterized in terms of three features: its size, shape, and index of refraction. For the present treatment, we will limit our discussion to *thin* lenses with spherical or planar surfaces, such as those shown in Fig. 8-25. For each lens, the signs of the radii of curvature R_1 and R_2 of its two surfaces are governed by the sign convention discussed

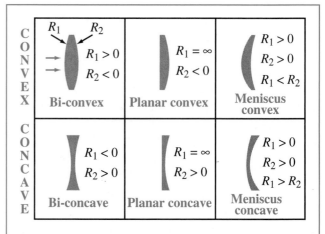

Figure 8-25: Convex and concave lenses of various shapes. The indicated radii of curvature are for wave illumination from the left side of the lens. Convex lenses have positive focal lengths, whereas concave lenses have negative focal lengths.

later and summarized in Table 8-5. Convex lenses, which also are called *converging* or *positive lenses*, are thicker

Table 8-5: Sign convention for spherical refracting surfaces and thin lenses.

Quantity	Sign	Condition
s	$+$	if object is in front side of surface (real object)
s	$-$	if object is in back side of surface (virtual object)
s'	$+$	if image is in back side of surface (real image)
s'	$-$	if image is in front side of surface (virtual image)
R	$+$	if C is in back side of surface
R	$-$	if C is in front side of surface
f	$+$	convex lens
f	$-$	concave lens

Notes: (1) The front side of a surface is defined as the side from which the light approaches the surface. (2) An object cannot be virtual with refraction by a single surface, but its image, if virtual, can become a virtual object for refraction by another surface.

at the center than at the ends. In contrast, concave lenses, otherwise known as *diverging* or *negative lenses*, are narrower at the center than at the ends.

The imaging process performed by a lens involves refraction at both of its surfaces. As an initial step toward understanding the process, we begin by considering the refraction at a spherical boundary between a medium with index of refraction n_1 and a second medium with index of refraction n_2, as shown in Fig. 8-26. Spherical waves radiated by a point source at point O are represented by rays emanating radially from the source in the direction of energy flow. When a ray reaches a point on the spherical boundary, it reflects and refracts in accordance with Snell's laws as if the surface is part of an infinite plane tangent to the surface at that point. Of the total energy arriving at point I, the bulk of it is due to rays incident upon the central region of the lens. Rays originating at the source and refracting at points within the central region of the spherical surface are called *paraxial rays*. These rays form an image of the source at a single point, called the image point. The imaging process can be demonstrated by considering any one of these paraxial rays. One such ray is the ray at angle ϕ relative to the horizontal in

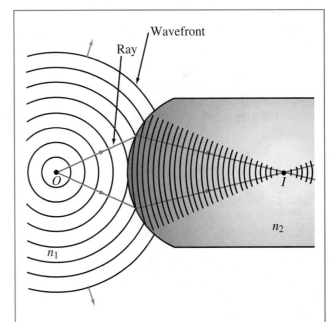

Figure 8-26: Spherical waves radiated by a source at O are refracted by the spherical boundary of medium 2 (with $n_2 > n_1$).

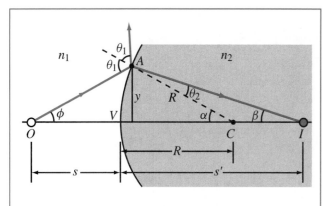

Figure 8-27: Ray refraction at a spherical boundary.

The paraxial approximation noted earlier can again be invoked to allow us to use the approximation $\tan\theta \simeq \theta$ for angles ϕ, α, and β. Hence,

$$\phi \simeq \tan\phi = \frac{y}{s}, \qquad (8.93a)$$

$$\alpha \simeq \tan\alpha = \frac{y}{R}, \qquad (8.93b)$$

$$\beta \simeq \tan\beta = \frac{y}{s'}, \qquad (8.93c)$$

where s and s' are the source (or object) and image distances and R is the radius of curvature of the spherical surface. Upon substituting these relations in Eq. (8.92) and dividing all terms by y, we have

$$\boxed{\frac{n_1}{s} + \frac{n_2}{s'} = \frac{n_2 - n_1}{R}} \quad \text{(single spherical boundary)}.$$
$$(8.94)$$

Since for a fixed source distance s, the image s' is independent of the angle ϕ that the ray makes with the axis, all paraxial rays focus at the image point I.

In contrast to mirrors, where real images are formed on the same side as the object, with a refracting surface, real images are formed on the side opposite the one from which the source of energy originates. The sign convention for spherical refracting surfaces is given in Table 8-5, which will also be used for thin lenses later.

We will now extend our results to a lens with two spherical surfaces of radii of curvature R_1 and R_2, as in Fig. 8-28. The lens has an index of refraction n_l and is surrounded by a medium with index of refraction n_m. The imaging process can be analyzed in two different ways, which we shall call the actual imaging process and the equivalent imaging process. If we have an object or source at point O in Fig. 8-28, the rays emanating from it will refract at the first surface of the lens, propagate to the second surface, and then refract again at the second surface. All paraxial rays end up intersecting at the image point I. Although this is what actually happens, it is not

Fig. 8-27. At point A, Snell's law of refraction gives

$$n_1 \sin\theta_1 = n_2 \sin\theta_2, \qquad (8.89)$$

where θ_1 is the angle between $\overline{OA}$ and the normal to the spherical surface at point A, which originates at C, the center of curvature of the spherical surface. Over the central region of the spherical surface, angles θ_1 and θ_2 are sufficiently small to allow use of the approximations $\sin\theta_1 \simeq \theta_1$ and $\sin\theta_2 \simeq \theta_2$. Hence, Eq. (8.89) may be rewritten as

$$n_1\theta_1 \simeq n_2\theta_2. \qquad (8.90)$$

From the geometry of triangles OAC and AIC, we know that the exterior angle θ_1 is equal to the sum of angles ϕ and α and, similarly, angle α is equal to the sum of angles θ_2 and β. Thus,

$$\theta_1 = \phi + \alpha, \qquad (8.91a)$$

$$\alpha = \theta_2 + \beta. \qquad (8.91b)$$

Upon using these relations to replace θ_1 and θ_2 in Eq. (8.90), we have

$$n_1\phi + n_2\beta = (n_2 - n_1)\alpha. \qquad (8.92)$$

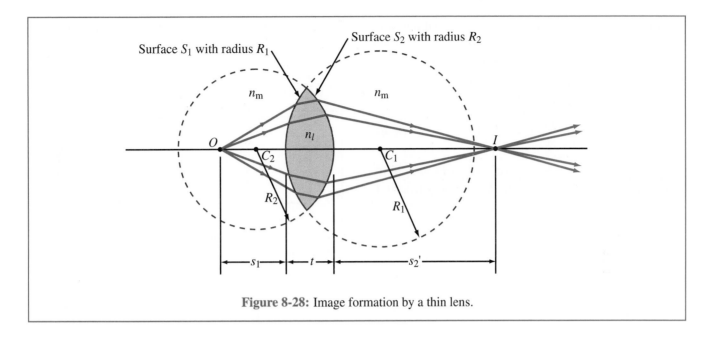

Figure 8-28: Image formation by a thin lens.

immediately obvious why the rays do intersect at point I or how to easily relate the image distance s_2' to the object distance s_1.

These questions can be addressed by using an equivalent imaging process that divides the refractive effects of the two surfaces of the lens into two sequential steps. Considering the first surface alone and treating it as if it were the boundary between the medium containing the object and an extended medium of index of refraction n_l (i.e., the same as the boundary we treated earlier in Fig. 8-27), the object located at point O in Fig. 8-29 produces an image I_1, located a distance s_1' from surface S_1. Because of the particular choice we made for the location of the object in Fig. 8-29, the image I_1 is virtual. The relationship governing the distances s_1 and s_1' is obtained from Eq. (8.94) upon replacing n_1 with n_m and n_2 with n_l:

$$\frac{n_\mathrm{m}}{s_1} + \frac{n_l}{s_1'} = \frac{n_l - n_\mathrm{m}}{R_1} \, , \qquad (8.95)$$

where R_1 is the radius of curvature of surface S_1. We now repeat the process with respect to surface S_2 by treating

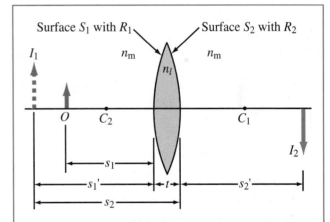

Figure 8-29: Image I_2 of object O can be determined by first locating image I_1 formed by surface S_1 of the lens and then using I_1 as the object for surface S_2.

the image I_1 as a virtual object at a distance s_2 from surface S_2. In this case, we regard S_2 as the boundary between a medium with index of refraction n_l to its left (and containing its object I_1) and a medium with index of refraction n_m to its right. Hence, we need to reverse the assignments of n_l and n_m when using Eq. (8.94). Accordingly, for surface S_2, the final image I_2 at distance s_2' is related to s_2 by

$$\frac{n_l}{s_2} + \frac{n_m}{s_2'} = \frac{n_m - n_l}{R_2}. \qquad (8.96)$$

As far as surface S_2 is concerned, its object I_1 is in its front side (the side containing the rays coming from I_1). Hence, by our sign convention, the distance s_2 must be positive; that is, $s_2 = |s_2|$. When we apply the sign convention to s_1', the image distance due to surface S_1, we find that s_1' must be negative because the image was formed on the same side as the object O. Hence, $s_1' = -|s_1'|$. From Fig. 8-29,

$$|s_2| = |s_1'| + t, \qquad (8.97)$$

where t is the thickness of the lens. In view of the fact that s_1' is negative and s_2 is positive, it follows that

$$s_2 = -s_1' + t. \qquad (8.98)$$

Upon using Eq. (8.98) in Eq. (8.96) and then adding Eqs. (8.95) and (8.96), we have

$$\frac{n_m}{s_1} + \frac{n_m}{s_2'} = (n_l - n_m)\left(\frac{1}{R_1} - \frac{1}{R_2}\right) + \frac{n_l t}{s_1'(s_1' - t)}. \qquad (8.99)$$

For a thin lens with thickness t sufficiently small as to allow us to ignore the last term, we obtain the ***thin-lens equation*** given by

$$\frac{n_m}{s_1} + \frac{n_m}{s_2'} = (n_l - n_m)\left(\frac{1}{R_1} - \frac{1}{R_2}\right). \qquad (8.100)$$

We note that, by our sign convention, $R_1 > 0$ and $R_2 < 0$ for the bi-convex lens we used in Fig. 8-29. The radii of curvature for lenses with various combinations of concave, convex, and planar surfaces are given in Fig. 8-25.

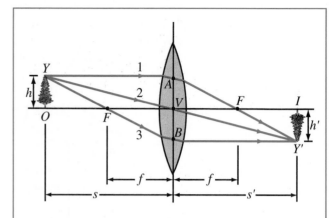

Figure 8-30: Rotation between object and image for a thin lens.

If the surrounding medium is air ($n_m = 1$), Eq. (8.100) reduces to the ***lensmaker's formula***:

$$\frac{1}{s} + \frac{1}{s'} = (n-1)\left(\frac{1}{R_1} - \frac{1}{R_2}\right) \quad \text{(thin lens)}, (8.101)$$

where we have omitted the subscripts on s_1 and s_2' and n_l in Eq. (8.100). Since t is negligibly small, we now refer to s and s' as the object and image distances, respectively, and we measure them from the lens center, as shown in Fig. 8-30.

In Section 8-7.2, we defined the focal length f of a reflecting mirror as the image distance when the object is moved out to infinity. Applying the same definition for the lens leads to

$$\frac{1}{f} = (n-1)\left(\frac{1}{R_1} - \frac{1}{R_2}\right), \qquad (8.102)$$

in which case Eq. (8.101) can be rewritten as

$$\frac{1}{s} + \frac{1}{s'} = \frac{1}{f} \quad \text{(Gaussian lens formula)}, \qquad (8.103)$$

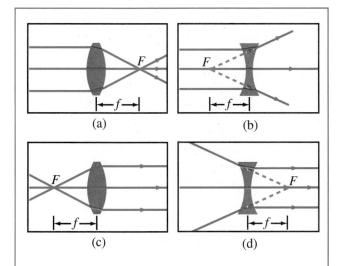

Figure 8-31: An object at infinity is imaged at the focal point, as in (a) and (b), and if the object is at the focal point, as in (c) and (d), it is imaged at infinity.

which is known as the *Gaussian lens formula* and is identical in form to Eq. (8.88) for mirrors. The sign convention for thin lenses is the same as that outlined in Table 8-5 for spherical refracting surfaces. *Convex lenses have positive focal lengths, and concave lenses have negative focal lengths.* According to Eq. (8.103), if the object is at infinity ($s = \infty$), then $s' = f$; this means that rays coming from infinity (and therefore parallel to the optic axis) are refracted by the lens and intersect at the focal point F, as in Fig. 8-31(a) for the positive-focal-length convex lens and in Fig. 8-31(b) for the negative-focal-length concave lens. The convex lens is a converging lens, whereas the concave lens is a diverging lens.

The Gaussian lens formula of Eq. (8.103) also states that if the object is at the focal point ($s = f$) then $s' = \infty$; the corresponding ray representation is illustrated in Figs. 8-31(c) and (d). These two properties will prove useful in locating the image of an object and in relating the size of the image to that of the object. In Fig. 8-30, the object at location O is of height h and its tip is denoted

by point Y. The image at location I is inverted, it has a height h' (which is negative in this case), and its tip is at point Y'. Point Y' is at the intersection of rays 1, 2, and 3. Ray 1 arrives at the lens in a direction parallel to the optic axis, whereupon it is refracted by the lens and emerges in a direction that passes through the focal point F. (In the case of a diverging lens, ray 1 would appear to come from the focal point to the left of the lens.) Ray 3, drawn from point Y through the focal point F to the left of the lens, is refracted by the lens and emerges in a direction parallel to the optic axis. A third useful property of thin lenses pertains to the ray drawn through the center of the lens. Such a ray, labeled 2 in Fig. 8-30, emerges from the lens undeviated, meaning that points Y and Y' are joined by a straight line that passes through the center of the lens.

For rays emanating from an object to the left of the lens, the ray properties of thin lenses may be summarized as follows:

(1) For a converging lens, the ray arriving at the lens in a direction parallel to the optic axis, such as ray 1 in Fig. 8-30, gets diffracted by the lens and emerges in a direction that passes through the focal point F to the right of the lens. In the case of a diverging lens, the refracted ray would appear to come from the focal point to the left of the lens, as shown in Fig. 8-31(b).

(2) For both converging and diverging lenses, the ray directed toward the center of the lens, such as ray 2 in Fig. 8-30, emerges from the lens undeviated in direction.

(3) For a converging lens, the ray arriving at the lens through the focal point to the left of the lens, such as ray 3 in Fig. 8-30, gets refracted by the lens and emerges in a direction parallel to the optic axis. For a diverging lens, the ray traveling in the direction of the focal point to the right of the lens, such as the rays shown in Fig. 8-31(d), gets diffracted by the lens and appears to come from the left side of the lens along a direction parallel to the optic axis.

The *lateral magnification* of a thin lens is defined as the ratio of the image height h' to the object height h.

Table 8-6: Images of real objects formed by thin lenses.[a]

Convex				
Object	**Image**			
Location	Type	Location	Orientation	Relative size
$2f < s < \infty$	Real	$2f > s' > f$	Inverted	Minified
$s = 2f$	Real	$s' = 2f$	Inverted	Same size
$f < s < 2f$	Real	$\infty > s' > 2f$	Inverted	Magnified
$s = f$		$s' = \infty$		
$s < f$	Virtual	$|s'| > s$	Erect	Magnified
Concave				
Object	**Image**			
Location	Type	Location	Orientation	Relative size
Anywhere	Virtual	$|s'| < |f|,$ $s > |s'|$	Erect	Minified

[a]Adapted from E. Hecht, *Optics*, second edition, Addison-Wesley Publishing Co., 1990.

From Fig. 8-30,

$$M = \frac{h'}{h} = -\frac{s'}{s} . \quad (8.104)$$

For the example shown in Fig. 8-30, M is negative because h' is negative. By our sign convention, both s and s' are positive. Hence, in order for M to be positive (image is erect), s' has to be negative (assuming that the object is real with positive s), which corresponds to a virtual image. Thus, *a lens can generate a real inverted image to the right of the lens or an erect virtual image to its left.* Image properties are summarized in Table 8-6 for both convex and concave lenses.

The image-construction process is extendible to imaging systems consisting of multiple lenses. In a two-lens system, for example, the image of an object formed by the first lens serves as the object for the second lens.

Example 8-9   Image Formed by a Bi-convex Lens

A 5-cm-tall pencil is placed at a distance of 25 cm in front of a bi-convex lens with radii of curvature of 50 cm and 12.5 cm. If the index of refraction of the lens is 1.5, determine the location of the image and describe its properties.

Solution: A bi-convex lens has a positive focal length f, corresponding to radii $R_1 > 0$ and $R_2 < 0$ [see Fig. 8-25]. We can choose $R_1 = 50$ cm and $R_2 = -12.5$ cm or $R_1 = 12.5$ cm and $R_2 = -50$ cm . Either choice gives the same value for f when applied to Eq. (8.102). The first choice gives

$$\frac{1}{f} = (1.5 - 1) \left(\frac{1}{0.5} - \frac{1}{-0.125} \right),$$

from which we determine that $f = 0.2$ m. From Eq. (8.103),

$$\frac{1}{0.25} + \frac{1}{s'} = \frac{1}{0.2} ,$$

which gives $s' = 1$ m.

The lateral magnification is

$$M = \frac{h'}{h} = -\frac{s'}{s} = -\frac{1}{0.25} = -4.$$

Hence, the pencil's image is inverted (M is negative), and its height $h' = -4 \times 5$ cm $= -20$ cm. ■

REVIEW QUESTIONS

Q8.8 Under what conditions is geometric optics applicable?

Q8.9 What are paraxial rays and what approximation do they allow us to make?

Q8.10 What is the thin-lens approximation?

EXERCISE 8.10 Show that for a plane refracting surface Eq. (8.94) reduces to $s' = -(n_2/n_1)s$.

EXERCISE 8.11 Calculate the focal length of a planar concave lens with a radius of curvature of 10 cm and an index of refraction of 1.5.

Ans. $f = -20$ cm.

CHAPTER HIGHLIGHTS

- The relations describing the reflection and transmission behavior of a plane EM wave at the boundary between two different media are the consequence of satisfying the conditions of continuity of the tangential components of **E** and **H** across the boundary.

- Snell's laws state that $\theta_i = \theta_r$ and

$$\sin \theta_t = (n_1/n_2) \sin \theta_i.$$

For media such that $n_2 < n_1$, the incident wave is reflected totally by the boundary when $\theta_i \geq \theta_c$, where θ_c is the critical angle given by $\theta_c = \sin^{-1}(n_2/n_1)$.

- By successive multiple reflections, light can be guided through optical fibers. The maximum data rate of digital pulses that can be transmitted along optical fibers is dictated by modal dispersion.

- At the Brewster angle for a given polarization, the incident wave is transmitted totally across the boundary. For nonmagnetic materials, the Brewster angle exists for parallel polarization only.

- Any plane wave incident on a plane boundary can be synthesized as the sum of a perpendicularly polarized wave and a parallel polarized wave.

- Transmission-line equivalent models can be used to characterize wave propagation, reflection by, and transmission through boundaries between different media.

- In geometric optics, waves are represented by rays that travel in straight lines. Geometric optics, which can be used for constructing images formed by mirrors and lenses, is valid as long as the objects with which the waves interact are much larger in size than the wavelength and their surfaces are relatively smooth relative to λ.

GLOSSARY OF IMPORTANT TERMS

Provide definitions or explain the meaning of the following terms:

rays
wavefront
reflection coefficient Γ
transmission coefficient τ
standing-wave ratio S
Snell's laws

PROBLEMS

Section 8-1: Reflection and Transmission at Normal Incidence

8.1* A plane wave in air with an electric field amplitude of 10 V/m is incident normally upon the surface of a lossless, nonmagnetic medium with $\varepsilon_r = 25$. Determine the following:

(a) The reflection and transmission coefficients.

(b) The standing-wave ratio in the air medium.

(c) The average power densities of the incident, reflected, and transmitted waves.

8.2 A plane wave traveling in medium 1 with $\epsilon_{r1} = 2.25$ is normally incident upon medium 2 with $\epsilon_{r2} = 4$. Both media are made of nonmagnetic, non-conducting materials. If the electric field of the incident wave is given by

$$\mathbf{E}^i = \hat{\mathbf{y}}4\cos(6\pi \times 10^9 t - 30\pi x) \quad \text{(V/m)}$$

(a) Obtain time-domain expressions for the electric and magnetic fields in each of the two media.

(b) Determine the average power densities of the incident, reflected and transmitted waves.

8.3 A plane wave traveling in a medium with $\varepsilon_{r_1} = 9$ is normally incident upon a second medium with $\varepsilon_{r_2} = 4$. Both media are made of nonmagnetic, non-conducting materials. If the magnetic field of the incident plane wave is given by

$$\mathbf{H}^i = \hat{\mathbf{z}}2\cos(2\pi \times 10^9 t - ky) \quad \text{(A/m)}$$

(a) Obtain time-domain expressions for the electric and magnetic fields in each of the two media.

(b) Determine the average power densities of the incident, reflected, and transmitted waves.

---

*Answer(s) available in Appendix D.

Solution available in CD-ROM.

8.4 A 200-MHz, left-hand circularly polarized plane wave with an electric field modulus of 10 V/m is normally incident in air upon a dielectric medium with $\varepsilon_r = 4$, and occupies the region defined by $z \geq 0$.

(a) Write an expression for the electric field phasor of the incident wave, given that the field is a positive maximum at $z = 0$ and $t = 0$.

(b) Calculate the reflection and transmission coefficients.

(c) Write expressions for the electric field phasors of the reflected wave, the transmitted wave, and the total field in the region $z \leq 0$.

(d) Determine the percentages of the incident average power reflected by the boundary and transmitted into the second medium.

8.5* Repeat Problem 8.4, but replace the dielectric medium with a poor conductor characterized by $\varepsilon_r = 2.25$, $\mu_r = 1$, and $\sigma = 10^{-4}$ S/m.

8.6 A 50-MHz plane wave with electric field amplitude of 30 V/m is normally incident in air onto a semi-infinite, perfect dielectric medium with $\varepsilon_r = 36$. Determine the following:

(a) Γ

(b) The average power densities of the incident and reflected waves.

(c) The distance in the air medium from the boundary to the nearest minimum of the electric field intensity, $|\mathbf{E}|$.

8.7* What is the maximum amplitude of the total electric field in the air medium of Problem 8.6, and at what nearest distance from the boundary does it occur?

8.8 Repeat Problem 8.6, but replace the dielectric medium with a conductor with $\varepsilon_r = 1$, $\mu_r = 1$, and $\sigma = 2.78 \times 10^{-3}$ S/m.

8.9* The three regions shown in Fig. 8-32 contain perfect dielectrics. For a wave in medium 1, incident normally upon the boundary at $z = -d$, what combination of ε_{r_2} and d produces no reflection? Express your answers in terms of ε_{r_1}, ε_{r_3} and the oscillation frequency of the wave, f.

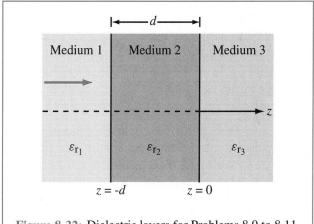

Figure 8-32: Dielectric layers for Problems 8.9 to 8.11.

8.10 For the configuration shown in Fig. 8-32, use transmission-line equations (or the Smith chart) to calculate the input impedance at $z = -d$ for $\varepsilon_{r_1} = 1$, $\varepsilon_{r_2} = 9$, $\varepsilon_{r_3} = 4$, $d = 1.2$ m, and $f = 50$ MHz. Also determine the fraction of the incident average power density reflected by the structure. Assume all media are lossless and nonmagnetic.

8.11* Repeat Problem 8.10, but interchange ε_{r_1} and ε_{r_3}.

8.12 Orange light of wavelength 0.61 μm in air enters a block of glass with $\varepsilon_r = 2.25$. What color would it appear to a sensor embedded in the glass? The wavelength ranges of colors are violet (0.39 to 0.45 μm), blue (0.45 to 0.49 μm), green (0.49 to 0.58 μm), yellow (0.58 to 0.60 μm), orange (0.60 to 0.62 μm), and red (0.62 to 0.78 μm).

8.13* A plane wave of unknown frequency is normally incident in air upon the surface of a perfect conductor. Using an electric-field meter, it was determined that the total electric field in the air medium is always zero when measured at a distance of 2.5 m from the conductor surface. Moreover, no such nulls were observed at distances closer to the conductor. What is the frequency of the incident wave?

8.14 Consider a thin film of soap in air under illumination by yellow light with $\lambda = 0.6$ μm in vacuum. If the film is treated as a planar dielectric slab with $\varepsilon_r = 1.72$, surrounded on both sides by air, what film thickness would produce strong reflection of the yellow light at normal incidence?

8.15* A 5-MHz plane wave with electric field amplitude of 20 (V/m) is normally incident in air onto the plane surface of a semi-infinite conducting material with $\varepsilon_r = 4$, $\mu_r = 1$, and $\sigma = 100$ (S/m). Determine the average power dissipated (lost) per unit cross-sectional area in a 2-mm penetration of the conducting medium.

8.16 A 0.5-MHz antenna carried by an airplane flying over the ocean surface generates a wave that approaches the water surface in the form of a normally incident plane wave with an electric-field amplitude of 3,000 (V/m). Seawater is characterized by $\varepsilon_r = 72$, $\mu_r = 1$, and $\sigma = 4$ (S/m). The plane is trying to communicate a message to a submarine submerged at a depth d below the water surface. If the submarine's receiver requires a minimum signal amplitude of 0.1 (μV/m), what is the maximum depth d to which successful communication is still possible?

Sections 8-2 and 8-3: Snell's Laws and Fiber Optics

8.17* A light ray is incident on a prism in air at an angle θ as shown in Fig. 8-33. The ray is refracted at the first surface and again at the second surface. In terms of the apex angle ϕ of the prism and its index of refraction n, determine the smallest value of θ for which the ray will emerge from the other side. Find this minimum θ for $n = 1.5$ and $\phi = 60°$.

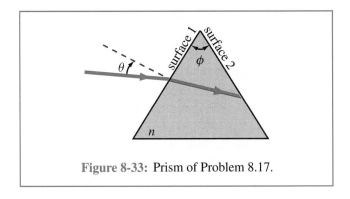

Figure 8-33: Prism of Problem 8.17.

8.18 For some types of glass, the index of refraction varies with wavelength. A prism made of a material with

$$n = 1.71 - \frac{4}{30}\lambda_0, \qquad (\lambda_0 \text{ in } \mu\text{m})$$

where λ_0 is the wavelength in vacuum, was used to disperse white light as shown in Fig. 8-34. The white light is incident at an angle of 50°, the wavelength λ_0 of red light is 0.7 μm, and that of violet light is 0.4 μm. Determine the angular dispersion in degrees.

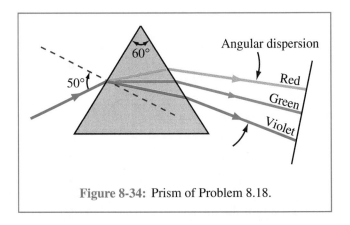

Figure 8-34: Prism of Problem 8.18.

8.19* The two prisms in Fig. 8-35 are made of glass with $n = 1.52$. What fraction of the power density carried by the ray incident upon the top prism emerges from the bottom prism? Neglect multiple internal reflections.

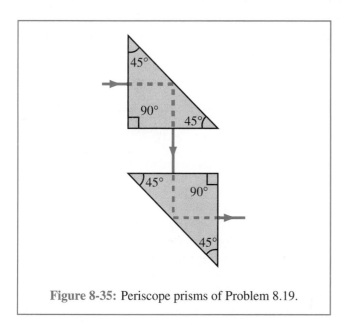

Figure 8-35: Periscope prisms of Problem 8.19.

8.20 A light ray incident at 45° passes through two dielectric materials with the indices of refraction and thicknesses given in Fig. 8-36. If the ray strikes the surface of the first dielectric at a height of 2 cm, at what height will it strike the screen?

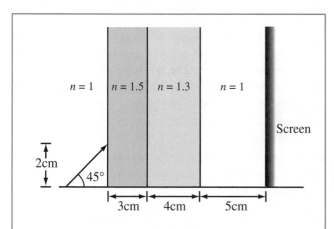

Figure 8-36: Light incident on a screen through a multi-layered dielectric (Problem 8.20).

8.21* Figure 8-37 depicts a beaker containing a block of glass on the bottom and water over it. The glass block contains a small air bubble at an unknown depth below the water surface. When viewed from above at an angle of 60°, the air bubble appears at a depth of 6.81 cm. What is the true depth of the air bubble?

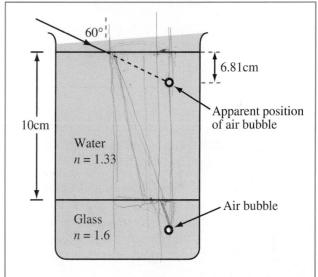

Figure 8-37: Apparent position of the air bubble in Problem 8.21.

8.22 A glass semicylinder with $n = 1.5$ is positioned such that its flat face is horizontal, as shown in Fig. 8-38, and its horizontal surface supports a drop of oil, as also shown. When light is directed radially toward the oil, total internal reflection occurs if θ exceeds 60°. What is the index of refraction of the oil?

8.23* A penny lies at the bottom of a water fountain at a depth of 30 cm. Determine the diameter of a piece of paper which, if placed to float on the surface of the water directly above the penny, would totally obscure the penny from view. Treat the penny as a point and assume that $n = 1.33$ for water.

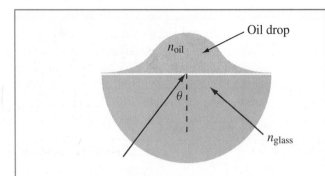

Figure 8-38: Oil drop on the flat surface of a glass semicylinder (Problem 8.22).

8.24 Suppose that the optical fiber of Example 8-5 is submerged in water (with $n = 1.33$) instead of air. Determine θ_a and f_p in that case.

8.25* Equation (8.45) was derived for the case where the light incident upon the sending end of the optical fiber extends over the entire acceptance cone shown in Fig. 8-12(b). Suppose the incident light is constrained to a narrower range extending between normal incidence and θ', where $\theta' < \theta_a$.

(a) Obtain an expression for the maximum data rate f_p in terms of θ'.

(b) Evaluate f_p for the fiber of Example 8-5 when $\theta' = 3°$.

Sections 8-4 and 8-5: Reflection and Transmission at Oblique Incidence

8.26 A plane wave in air with

$$\widetilde{\mathbf{E}}^i = \hat{\mathbf{y}}\, 10 e^{-j(3x+4z)} \quad \text{(V/m)}$$

is incident upon the planar surface of a dielectric material, with $\varepsilon_r = 4$, occupying the half-space $z \geq 0$. Determine:

(a) The polarization of the incident wave.

(b) The angle of incidence.

(c) The time-domain expressions for the reflected electric and magnetic fields.

(d) The time-domain expressions for the transmitted electric and magnetic fields.

(e) The average power density carried by the wave in the dielectric medium.

8.27 Repeat Problem 8.26 for a wave in air with

$$\widetilde{\mathbf{H}}^i = \hat{\mathbf{y}}\, 2 \times 10^{-2} e^{-j(8x+6z)} \quad \text{(A/m)}$$

incident upon the planar boundary of a dielectric medium ($z \geq 0$) with $\varepsilon_r = 9$.

8.28 Natural light is randomly polarized, which means that, on average, half the light energy is polarized along any given direction (in the plane orthogonal to the direction of propagation) and the other half of the energy is polarized along the direction orthogonal to the first polarization direction. Hence, when treating natural light incident upon a planar boundary, we can consider half of its energy to be in the form of parallel-polarized waves and the other half as perpendicularly polarized waves. Determine the fraction of the incident power reflected by the planar surface of a piece of glass with $n = 1.5$ when illuminated by natural light at $70°$.

8.29* A parallel-polarized plane wave is incident from air onto a dielectric medium with $\varepsilon_r = 9$ at the Brewster angle. What is the refraction angle?

8.30 A perpendicularly polarized wave in air is obliquely incident upon a planar glass–air interface at an incidence angle of $30°$. The wave frequency is 600 THz (1 THz $= 10^{12}$ Hz), which corresponds to green light, and the index of refraction of the glass is 1.6. If the electric field amplitude of the incident wave is 50 V/m, determine the following:

(a) The reflection and transmission coefficients.

(b) The instantaneous expressions for $\mathbf{E}$ and $\mathbf{H}$ in the glass medium.

8.31 Show that the reflection coefficient $\Gamma_\perp$ can be written in the following form:

$$\Gamma_\perp = \frac{\sin(\theta_t - \theta_i)}{\sin(\theta_t + \theta_i)}$$

8.32 Show that for nonmagnetic media, the reflection coefficient $\Gamma_\parallel$ can be written in the following form:

$$\Gamma_\parallel = \frac{\tan(\theta_t - \theta_i)}{\tan(\theta_t + \theta_i)}$$

8.33* A parallel-polarized beam of light with an electric field amplitude of 20 (V/m) is incident in air on polystyrene with $\mu_r = 1$ and $\varepsilon_r = 2.6$. If the incidence angle at the air–polystyrene planar boundary is 50°, determine the following:
(a) The reflectivity and transmissivity.
(b) The power carried by the incident, reflected, and transmitted beams if the spot on the boundary illuminated by the incident beam is 1 m² in area.

Sections 8-6 to 8-8: Geometric Optics

8.34 A man is 2 m tall. How long should a mirror be, when placed in front of him at a distance d, for him to have a full-length view of his reflected image? Does the answer depend on d?

8.35* A gas flame is 2 m from a screen, as shown in Fig. 8-39. A concave mirror is used to produce a real image of the flame on the screen, magnified four times. Determine s and the focal length of the mirror.

8.36 A dentist uses a concave mirror with a radius of curvature of 6 cm to view a filling in a tooth. If the mirror is placed 2 cm from the tooth, by how many times will the image of the filling be magnified?

8.37* A flower is 20 cm from a concave spherical mirror. If the image of the flower is formed at a distance of 80 cm from the mirror, determine the following:
(a) The mirror's radius of curvature.
(b) The lateral magnification.

8.38 A candle in air is placed at a distance of 25 cm to the left of a glass medium with a concave spherical boundary characterized by a radius of curvature of 50 cm. If the index of refraction of glass is 1.5, determine the location of the candle's image.

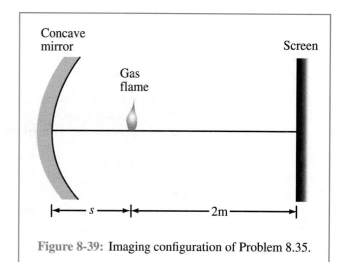

Figure 8-39: Imaging configuration of Problem 8.35.

8.39* A thin bi-convex lens with an index of refraction of 1.6 has radii of curvature of 20 cm and 30 cm. Determine the location, magnification, and orientation of an object placed 20 cm from the lens.

8.40 Repeat Problem 8.39 for an object placed 50 cm from a planar-concave lens with a radius of curvature of 15 cm.

8.41 The curved face of a planar convex lens has a radius of curvature of 3 cm and the index of refraction of the lens material is 1.5. Determine the focal length of the lens when:
(a) The planar surface of the lens faces the light.
(b) The convex surface of the lens faces the light.

8.42 An image formed by a convex lens was observed to move by a distance of 1 cm when the object was moved from far away (essentially at infinity) to only 42 cm from the lens. What is the focal length of the lens?

8.43* Equation (8.102) defines the focal length of a thin lens in air.
(a) Derive an expression for f for the general case of a thin lens with index of refraction n_l immersed in a medium with index of refraction n_m.

(b) Determine the focal length of a bi-concave lens with radii of curvature of 10 cm and 15 cm and an index of refraction of 1.5 when placed (i) in air, and (ii) in water (with an index of refraction of 1.33).

8.44 A positive lens is used to image an object placed 50 cm in front of it on a screen located 1 m behind the lens. What is the focal length of the lens?

8.45* An imaging system consists of two thin positive lenses with focal lengths of 15 cm for the first lens and 5 cm for the second. If the two lenses are separated by a distance of 50 cm, locate the image of an object placed 25 cm in front of the first lens, relative to the location of the second lens.

8.46 Two thin lenses with focal lengths f_1 and f_2 are placed in contact with each other, as shown in Fig. 8-40. Show that F, the focal length of the combination, is given by

$$\frac{1}{F} = \frac{1}{f_1} + \frac{1}{f_2}$$

8.47* Two thin lenses of focal lengths $f_1 = 5$ cm and $f_2 = -10$ cm are separated by a distance of 5 cm, as shown in Fig. 8-41. Determine the location of the object.

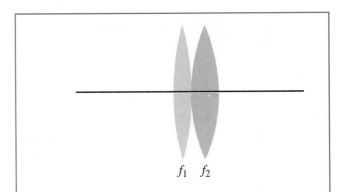

Figure 8-40: Two lenses in contact. The lenses are so thin that they may be considered to be at the same location (Problem 8.46).

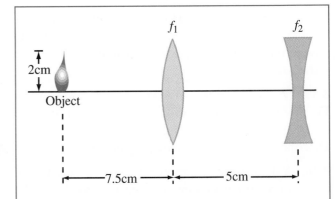

Figure 8-41: Imaging configuration of Problem 8.47.

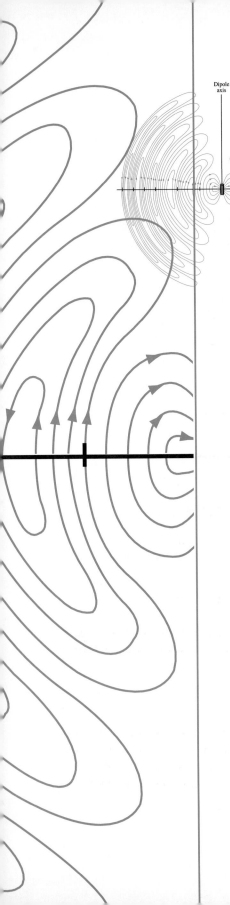

Dipole
axis

C H A P T E R 9

Radiation
and Antennas

Overview

OVERVIEW

An *antenna* may be defined as a transducer between a guided wave propagating in a transmission line and an electromagnetic wave propagating in an unbounded medium (usually free space), or vice versa. Figure 9-1 shows how a wave is launched by a hornlike antenna, with the horn acting as a transition between the waveguide and free space. Although any conducting or dielectric structure can serve this function, an antenna is designed to radiate or receive electromagnetic energy with directional and polarization properties suitable for the intended application. Also, to minimize reflection at the transmission line–antenna juncture, it is important to know the impedance of the antenna and to match it to the transmission line.

Antennas are made in various shapes and sizes [Fig. 9-2] and are used in radio and television broadcasting and reception, radio-wave communication systems, cellular telephones, radar systems, and anticollision automobile sensors, among many other applications. The radiation and impedance properties of an antenna are governed by its shape and size and the material of which it is made. The dimensions of an antenna are usually measured in units of the wavelength of the wave it is launching or receiving; a 1-m-long dipole antenna operating at a wavelength of 2 m exhibits the same properties as a 1-cm-long dipole operating at $\lambda = 2$ cm. Hence, in most of our discussions in this chapter, we will refer to antenna dimensions in wavelength units.

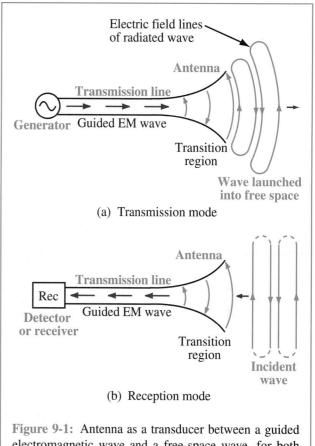

(a) Transmission mode

(b) Reception mode

Figure 9-1: Antenna as a transducer between a guided electromagnetic wave and a free-space wave, for both transmission and reception.

Reciprocity

The directional function characterizing the *relative* distribution of power radiated by an antenna is known as the *antenna radiation pattern*, or simply *the antenna pattern*. An *isotropic* antenna is a hypothetical antenna that radiates equally in all directions, and it is often used as a reference radiator when describing the radiation properties of real antennas. Most antennas are *recip-*

341

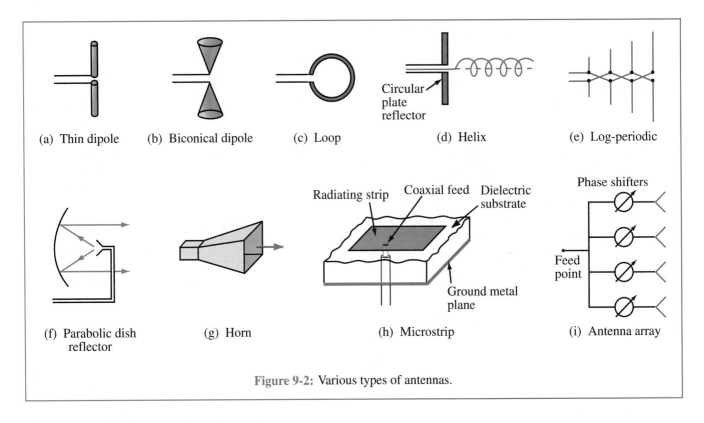

Figure 9-2: Various types of antennas.

rocal devices, exhibiting the same radiation pattern for transmission as for reception. Reciprocity means that, if in the transmission mode a given antenna transmits in direction A 100 times the power it transmits in direction B, then when used in the reception mode it will be 100 times more sensitive to electromagnetic radiation incident from direction A than B. All the antennas shown in Fig. 9-2 obey the reciprocity law, but not all antennas are reciprocal devices. Reciprocity may not hold for some solid-state antennas composed of nonlinear semiconductors or ferrite materials. Such nonreciprocal antennas are beyond the scope of this chapter, and hence reciprocity will be assumed throughout. The reciprocity property is very convenient because it allows us to compute the radiation pattern of an antenna in the transmission mode, even when the antenna is intended to operate as a receiver.

Antenna performance consists of two aspects: its radiation properties and its impedance. The radiation properties include its directional radiation pattern and the associated polarization state of the radiated wave when the antenna is used in the transmission mode. This polarization state is called *antenna polarization*. Being a reciprocal device, an antenna, when operating in the receiving mode, can extract from an incident wave only that component of the wave whose electric field is parallel to that of the antenna polarization direction. The second aspect, the *antenna impedance*, pertains to the transfer of power from a generator to the antenna when the antenna is used as a transmitter and, conversely, the transfer of power from the antenna to a load when the antenna is used as a receiver, as will be discussed later in Section 9-5. It should be noted that throughout our discussions in this chapter it will be assumed that the antenna is properly matched to the transmission line connected to its terminals, thereby avoiding reflections and their associated problems.

Radiation Sources

Radiation sources fall into two groups: currents and aperture fields. The dipole and loop antennas [Fig. 9-2(a) and (c)] are examples of current sources; the time-varying currents flowing in the conducting wires give rise to the radiated electromagnetic field. A horn antenna [Fig. 9-2(g)] is an example of the second group because the electric and magnetic fields across the horn's aperture serve as the sources of the radiated field. The aperture fields are themselves induced by time-varying currents on the surfaces of the horn's walls, and therefore ultimately all radiation is due to time-varying currents. The choice of currents or apertures as the sources is merely a computational convenience dictated by the structure of the antenna. We will examine the basic methodology used for computing the radiation pattern for each of these two types of sources.

Far-Field Region

The wave radiated by a point source is spherical in shape, with the wavefront expanding outward at a rate equal to the phase velocity u_p (or the velocity of light c if the medium is free space). If R, the distance between the transmitting antenna and the receiving antenna, is sufficiently large that the wavefront across the receiving aperture may be considered a plane wave [Fig. 9-3], the receiving aperture is said to be in the *far-field* (or *far-zone*) region of the transmitting point source. This region is of particular significance because, for most applications, the observation region of interest is indeed in the far-field region of the antenna. The far-field plane-wave approximation allows the use of certain mathematical approximations that simplify the computation of the radiated field and, conversely, provide convenient techniques for synthesizing the appropriate

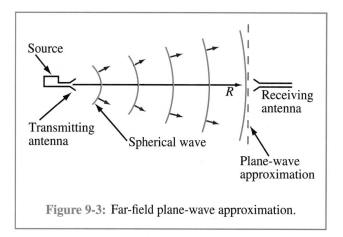

Figure 9-3: Far-field plane-wave approximation.

antenna structure that would give rise to the desired far-field antenna pattern.

Antenna Arrays

When several antennas are connected together, the combination is called an *antenna array* [Fig. 9-2(i)], and the array as a whole behaves as if it were a single antenna. By controlling the magnitude and phase of the signal feeding each antenna individually, it is possible to *shape the radiation pattern* of the array and to *steer the direction of the beam electronically*. These topics are treated in Sections 9-9 to 9-11.

9-1 The Short Dipole

By regarding a linear antenna as consisting of a series of a large number of infinitesimally short conducting elements, each of which is so short that current may be considered uniform over its length, the field of the entire antenna may be obtained by integrating the fields from all these differential antennas, with the proper magnitudes and phases taken into account. We shall first examine the radiation properties of such a differential antenna, known as a *short dipole*, and then in Section 9-3 we will extend the results to compute the fields radiated by a half-wave

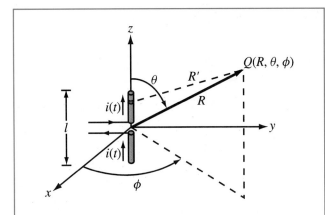

Figure 9-4: Short dipole placed at the origin of a spherical coordinate system.

dipole, which is commonly used as a standard antenna for many applications.

A short dipole, also called a *Hertzian dipole*, is a thin, linear conductor whose length l is very short compared with the wavelength λ; to satisfy the uniform-current assumption, l should not exceed $\lambda/50$. The wire, which is oriented along the z-direction as shown in Fig. 9-4, carries a sinusoidally varying current given by

$$i(t) = I_0 \cos \omega t = \Re[I_0 e^{j\omega t}] \quad \text{(A)}, \qquad (9.1)$$

where I_0 is the current amplitude. From Eq. (9.1), the phasor current $\tilde{I} = I_0$. Even though the current has to go to zero at the two ends of the dipole, we shall treat it as constant across its entire length.

The customary approach for finding the electric and magnetic fields at a point Q in space [Fig. 9-4] due to radiation by a current source is through the retarded vector potential **A**. From Eq. (6.84), the phasor retarded vector potential $\tilde{\mathbf{A}}(R)$ at a distance vector **R** from a volume v' containing a phasor current distribution $\tilde{\mathbf{J}}$ is given by

$$\tilde{\mathbf{A}}(\mathbf{R}) = \frac{\mu_0}{4\pi} \int_{v'} \frac{\tilde{\mathbf{J}} e^{-jkR'}}{R'} \, dv', \qquad (9.2)$$

where μ_0 is the magnetic permeability of free space (because the observation point is in air) and $k = \omega/c = 2\pi/\lambda$ is the wavenumber. For the dipole, the current density is simply $\tilde{\mathbf{J}} = \hat{\mathbf{z}}(I_0/s)$, where s is the cross-sectional area of the dipole wire, $dv' = s\, dz$, and the limits of integration are from $z = -l/2$ to $z = l/2$. In Fig. 9-4, the distance R' between the observation point and a given point along the dipole is not the same as the distance to its center, R, but because we are dealing with a very short dipole, we can set $R' \simeq R$. Hence,

$$\tilde{\mathbf{A}} = \frac{\mu_0}{4\pi} \frac{e^{-jkR}}{R} \int_{-l/2}^{l/2} \hat{\mathbf{z}} I_0 \, dz$$

$$= \hat{\mathbf{z}} \frac{\mu_0}{4\pi} I_0 l \left(\frac{e^{-jkR}}{R} \right), \qquad (9.3)$$

The function (e^{-jkR}/R), called the *spherical propagation factor*, accounts for the $1/R$ decay of the magnitude with distance as well as for the phase change represented by e^{-jkR}. The direction of $\tilde{\mathbf{A}}$ is dictated by the direction of flow of the current (z-direction).

Because our objective is to characterize the directional character of the radiated power at a fixed distance R from the antenna, the spherical coordinate system shown in Fig. 9-5 is deemed appropriate for the presentation of antenna pattern plots, with the variables R, θ, and ϕ referred to as the *range, zenith angle*, and *azimuth angle*, respectively. To this end, we need to write $\tilde{\mathbf{A}}$ in terms of its spherical coordinate components, which is realized [with the help of Eq. (3.65c)] by expressing $\hat{\mathbf{z}}$ in terms of the unit vectors of the spherical coordinate system:

$$\hat{\mathbf{z}} = \hat{\mathbf{R}} \cos \theta - \hat{\boldsymbol{\theta}} \sin \theta. \qquad (9.4)$$

Upon substituting Eq. (9.4) into Eq. (9.3), we have

$$\tilde{\mathbf{A}} = (\hat{\mathbf{R}} \cos \theta - \hat{\boldsymbol{\theta}} \sin \theta) \frac{\mu_0 I_0 l}{4\pi} \left(\frac{e^{-jkR}}{R} \right)$$

$$= \hat{\mathbf{R}} \tilde{A}_R + \hat{\boldsymbol{\theta}} \tilde{A}_\theta + \hat{\boldsymbol{\phi}} \tilde{A}_\phi, \qquad (9.5)$$

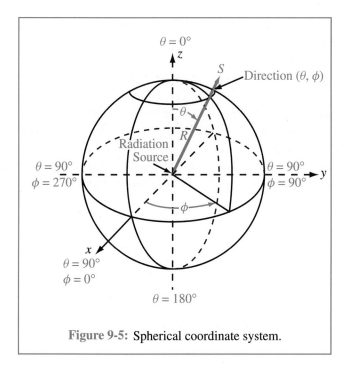

Figure 9-5: Spherical coordinate system.

with

$$\tilde{A}_R = \frac{\mu_0 I_0 l}{4\pi} \cos\theta \left(\frac{e^{-jkR}}{R}\right), \qquad (9.6a)$$

$$\tilde{A}_\theta = -\frac{\mu_0 I_0 l}{4\pi} \sin\theta \left(\frac{e^{-jkR}}{R}\right), \qquad (9.6b)$$

$$\tilde{A}_\phi = 0.$$

With the spherical coordinate components of $\tilde{\mathbf{A}}$ known, the next step is straightforward; we simply apply the free-space relationships given by Eqs. (6.85) and (6.86),

$$\tilde{\mathbf{H}} = \frac{1}{\mu_0} \nabla \times \tilde{\mathbf{A}}, \qquad (9.7a)$$

$$\tilde{\mathbf{E}} = \frac{1}{j\omega\varepsilon_0} \nabla \times \tilde{\mathbf{H}}, \qquad (9.7b)$$

to obtain the expressions

$$\tilde{H}_\phi = \frac{I_0 l k^2}{4\pi} e^{-jkR} \left[\frac{j}{kR} + \frac{1}{(kR)^2}\right] \sin\theta \qquad (9.8a)$$

$$\tilde{E}_R = \frac{2I_0 l k^2}{4\pi} \eta_0 e^{-jkR} \left[\frac{1}{(kR)^2} - \frac{j}{(kR)^3}\right] \cos\theta \quad (9.8b)$$

$$\tilde{E}_\theta = \frac{I_0 l k^2}{4\pi} \eta_0 e^{-jkR} \left[\frac{j}{kR} + \frac{1}{(kR)^2} - \frac{j}{(kR)^3}\right] \sin\theta, \qquad (9.8c)$$

where $\eta_0 = \sqrt{\mu_0/\varepsilon_0} \cong 120\pi$ (Ω) is the intrinsic impedance of free space. The remaining components ($\tilde{H}_R$, $\tilde{H}_\theta$, and $\tilde{E}_\phi$) are everywhere zero. Figure 9-6 depicts the electric field lines of the wave radiated by the short dipole.

9-1.1 Far-Field Approximation

As was stated earlier, in most antenna applications, we are primarily interested in the radiation pattern of the antenna at great distances from the source. For the electrical dipole, this corresponds to distances R such that $R \gg \lambda$ or, equivalently, $kR = 2\pi R/\lambda \gg 1$. This condition allows us to neglect the terms varying as $1/(kR)^2$ and $1/(kR)^3$ in Eqs. (9.8a) to (9.8c) in favor of the terms varying as $1/kR$, which yields the far-field expressions

$$\tilde{E}_\theta = \frac{jI_0 l k \eta_0}{4\pi} \left(\frac{e^{-jkR}}{R}\right) \sin\theta \quad \text{(V/m)}, \qquad (9.9a)$$

$$\tilde{H}_\phi = \frac{\tilde{E}_\theta}{\eta_0} \quad \text{(A/m)}, \qquad (9.9b)$$

and $\tilde{E}_R$ is negligible. At the observation point Q [Fig. 9-4], the wave now appears to be similar to a uniform plane wave with its electric and magnetic fields in time phase, related by the intrinsic impedance of the medium

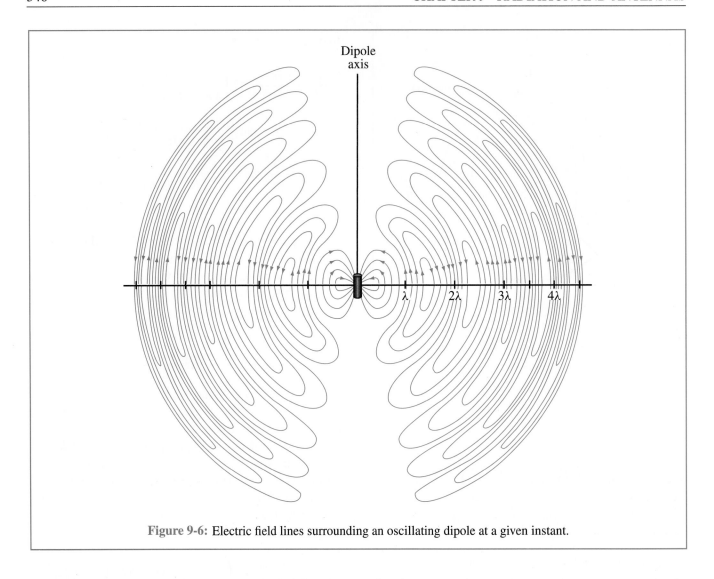

Figure 9-6: Electric field lines surrounding an oscillating dipole at a given instant.

η_0, and orthogonal to each other and to the direction of propagation ($\hat{\mathbf{R}}$). Both fields are proportional to $\sin\theta$ and independent of ϕ (which is as expected from symmetry considerations).

9-1.2 Power Density

Given $\widetilde{\mathbf{E}}$ and $\widetilde{\mathbf{H}}$ in phasor form, the *time-average Poynting vector* of the radiated wave, which is also called the *power density*, can be obtained by applying Eq. (7.101), that is,

$$\mathbf{S}_{\text{av}} = \tfrac{1}{2}\mathfrak{Re}\left(\widetilde{\mathbf{E}} \times \widetilde{\mathbf{H}}^*\right) \quad \text{(W/m}^2\text{).} \quad (9.10)$$

For the short dipole, use of Eqs. (9.9a) and (9.9b) gives

$$\mathbf{S}_{\text{av}} = \hat{\mathbf{R}}\, S(R, \theta), \quad (9.11)$$

with

$$S(R, \theta) = \left(\frac{\eta_0 k^2 I_0^2 l^2}{32\pi^2 R^2} \right) \sin^2 \theta$$

$$= S_0 \sin^2 \theta \quad (\text{W/m}^2). \quad (9.12)$$

The directional pattern of any antenna is described in terms of the *normalized radiation intensity* $F(\theta, \phi)$, defined as the ratio of the power density $S(R, \theta, \phi)$ to S_{max}, the maximum value of $S(R, \theta, \phi)$, at a specified range R:

$$F(\theta, \phi) = \frac{S(R, \theta, \phi)}{S_{\text{max}}} \quad (\text{dimensionless}). \quad (9.13)$$

For the Hertzian dipole, the $\sin^2 \theta$ dependence in Eq. (9.12) indicates that the radiation is maximum in the broadside direction ($\theta = \pi/2$), corresponding to the azimuth plane, and is given by

$$S_{\text{max}} = S_0 = \frac{\eta_0 k^2 I_0^2 l^2}{32\pi^2 R^2}$$

$$= \frac{15\pi I_0^2}{R^2} \left(\frac{l}{\lambda} \right)^2 \quad (\text{W/m}^2), \quad (9.14)$$

where use was made of the relations $k = 2\pi/\lambda$ and $\eta_0 \simeq 120\pi$. We observe that S_{max} is directly proportional to I_0^2 and to l^2 (with l measured in wavelengths), and it decreases with distance as $1/R^2$.

From the definition of the normalized radiation intensity given by Eq. (9.13), it follows that

$$F(\theta, \phi) = F(\theta) = \sin^2 \theta. \quad (9.15)$$

Plots of $F(\theta)$ are shown in Fig. 9-7 in both the elevation plane (also called the θ-plane) and the azimuth plane (ϕ-plane). No energy is radiated by the dipole along the direction of the dipole axis and maximum radiation ($F = 1$) occurs in the broadside direction. Since $F(\theta)$ is independent of ϕ, the pattern is doughnut-shaped in θ–ϕ space.

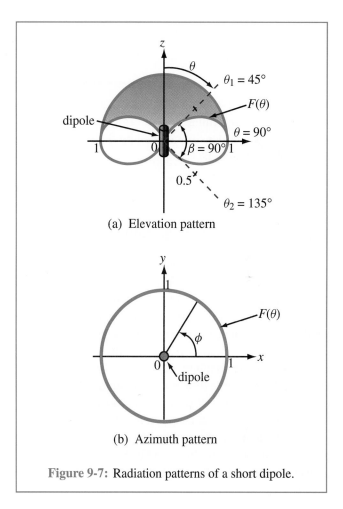

(a) Elevation pattern

(b) Azimuth pattern

Figure 9-7: Radiation patterns of a short dipole.

REVIEW QUESTIONS

Q9.1 When is an antenna a reciprocal device?

Q9.2 What is the radiated wave like in the far-field region of the antenna?

Q9.3 How short should the length of a wire antenna be so that it may be considered a Hertzian dipole? What is the underlying assumption about the current flowing through the wire?

Q9.4 Outline the basic steps used in order to relate the current in the wire to the radiated power density.

EXERCISE 9.1 A 1-m-long dipole is excited by a 5-MHz current with an amplitude of 5 A. At a distance of 2 km, what is the power density radiated by the antenna along its broadside direction?

Ans. $S_0 = 8.2 \times 10^{-8}$ W/m^2.

9-2 Antenna Radiation Characteristics

An *antenna pattern* describes the far-field directional properties of an antenna when measured at a fixed distance from the antenna. In general, the antenna pattern is a three-dimensional plot that displays the strength of the radiated field or power density as a function of direction, with direction being specified by the zenith angle θ and the azimuth angle ϕ. By virtue of the reciprocity theorem, *a receiving antenna has the same directional antenna pattern as the pattern that it exhibits when operated in the transmission mode.*

Consider a transmitting antenna placed at the origin of the observation sphere shown in Fig. 9-8. The differential power radiated by the antenna through an elemental area dA is

$$dP_{rad} = \mathbf{S}_{av} \cdot d\mathbf{A} = \mathbf{S}_{av} \cdot \hat{\mathbf{R}}\, dA = S\, dA \quad \text{(W)}, \quad (9.16)$$

where S is the radial component of the time-average Poynting vector $\mathbf{S}_{av}$. In the far-field region of any antenna, $\mathbf{S}_{av}$ is always in the radial direction. In a spherical coordinate system,

$$dA = R^2 \sin\theta\, d\theta\, d\phi, \quad (9.17)$$

and the *solid angle* $d\Omega$ associated with dA, defined as the subtended area divided by R^2, is given by

$$d\Omega = \frac{dA}{R^2} = \sin\theta\, d\theta\, d\phi \quad \text{(sr).} \quad (9.18)$$

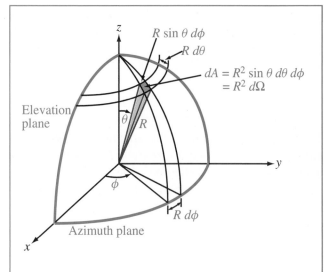

Figure 9-8: Definition of solid angle $d\Omega = \sin\theta\, d\theta\, d\phi$.

Note that, whereas a planar angle is measured in radians and the angular measure of a complete circle is 2π (rad), a solid angle is measured in **steradians** (sr), and the angular measure for a spherical surface is $\Omega = (4\pi R^2)/R^2 = 4\pi$ (sr). The solid angle of a hemisphere is 2π (sr).

Using the relation $dA = R^2\, d\Omega$, dP_{rad} can be rewritten as

$$dP_{rad} = R^2\, S(R, \theta, \phi)\, d\Omega. \quad (9.19)$$

The total power radiated by an antenna through a spherical surface at a fixed distance R is obtained by integrating Eq. (9.19):

$$
\begin{aligned}
P_{rad} &= R^2 \int_{\phi=0}^{2\pi} \int_{\theta=0}^{\pi} S(R, \theta, \phi) \sin\theta\, d\theta\, d\phi \\
&= R^2 S_{max} \int_{\phi=0}^{2\pi} \int_{\theta=0}^{\pi} F(\theta, \phi) \sin\theta\, d\theta\, d\phi \\
&= R^2 S_{max} \iint_{4\pi} F(\theta, \phi)\, d\Omega \quad \text{(W)}, \quad (9.20)
\end{aligned}
$$

where $F(\theta, \phi)$ is the normalized radiation intensity defined by Eq. (9.13). The 4π symbol under the integral sign is used as an abbreviation for the indicated limits on θ and ϕ. Formally, P_{rad} is called the *total radiated power*.

9-2.1 Antenna Pattern

Each specific combination of the zenith angle θ and the azimuth angle ϕ denotes a specific direction in the spherical coordinate system of Fig. 9-8. The normalized radiation intensity $F(\theta, \phi)$ characterizes the directional pattern of the energy radiated by an antenna, and a plot of $F(\theta, \phi)$ as a function of both θ and ϕ constitutes a three-dimensional pattern, an example of which is shown in Fig. 9-9.

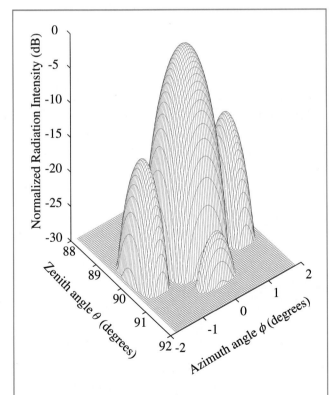

Figure 9-9: Three-dimensional pattern of a narrow-beam antenna.

Often, it is of interest to characterize the variation of $F(\theta, \phi)$ in the form of two-dimensional plots in specific planes in the spherical coordinate system. The two planes most commonly specified for this purpose are the elevation and azimuth planes. The *elevation plane*, also called the θ plane, is the plane corresponding to a constant value of ϕ. For example, $\phi = 0$ defines the x–z plane and $\phi = 90°$ defines the y–z plane, both of which are elevation planes (Fig. 9-8). A plot of $F(\theta, \phi)$ versus θ in either of these planes constitutes a two-dimensional pattern in the elevation plane. This is not to imply, however, that the elevation-plane pattern is necessarily the same in all elevation planes. The *azimuth plane*, also called the ϕ plane, is specified by $\theta = 90°$ and corresponds to the x–y plane. The elevation and azimuth planes are often called the two *principal planes* of the spherical coordinate system.

Some antennas exhibit highly directive patterns with narrow beams, in which case it is found convenient to plot the antenna pattern on a decibel scale by expressing F in decibels:

$$F \text{ (dB)} = 10 \log F.$$

As an example, the antenna pattern shown in Fig. 9-10(a) is plotted on a decibel scale in polar coordinates, with intensity as the radial variable. This format permits a convenient visual interpretation of the directional distribution of the *radiation lobes*. Another format commonly used for inspecting the pattern of a narrow-beam antenna is the rectangular display shown in Fig. 9-10(b), which permits the pattern to be easily expanded by changing the scale of the horizontal axis. These plots represent the variation in only one plane in the observation sphere, the $\phi = 0$ plane. Unless the pattern is symmetrical in ϕ, additional patterns are required to define the variation of $F(\theta, \phi)$ with θ and ϕ.

Strictly speaking, the polar angle θ is always positive, being defined over the range from $0°$ (z-direction) to $180°$ ($-z$-direction), and yet the θ-axis in Fig. 9-10(b) is shown to have both positive and negative values. This

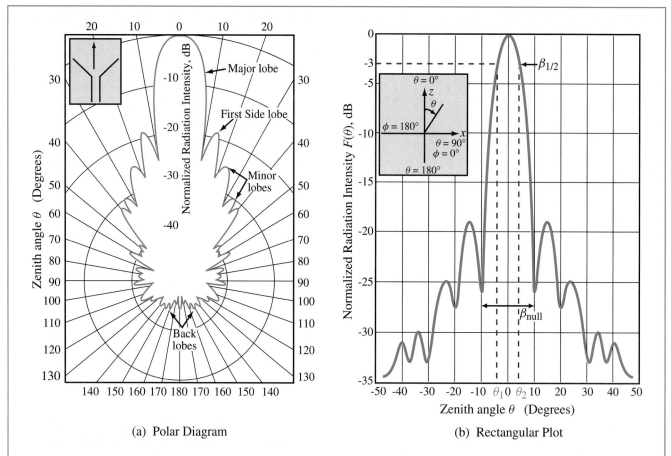

(a) Polar Diagram   (b) Rectangular Plot

Figure 9-10: Representative plots of the normalized radiation pattern of a microwave antenna in (a) polar form and (b) rectangular form.

is not a contradiction, but rather a different form of plotting antenna patterns. The right-hand half of the plot represents the variation of F (dB) with θ as θ is increased in a clockwise direction in the x–z plane [see inset in Fig. 9-10(b)], corresponding to $\phi = 0$, whereas the left-hand half of the plot represents the variation of F (dB) with θ as θ is increased in a counterclockwise direction at $\phi = 180°$. Thus, a negative θ value simply

denotes that the direction (θ, ϕ) is in the left-hand half of the x–z plane.

The pattern shown in Fig. 9-10(a) indicates that the antenna is fairly directive, since most of the energy is radiated through a narrow range called the *main lobe*. In addition to the main lobe, the pattern exhibits several *side lobes* and *back lobes* as well. For most applications, these extra lobes are considered undesirable because they represent wasted energy for transmitting antennas and potential interference directions for receiving antennas.

9-2.2 Beam Dimensions

For an antenna with a single main lobe, the *pattern solid angle* Ω_p describes the equivalent width of the main lobe of the antenna pattern [Fig. 9-11]. It is defined as the integral of the normalized radiation intensity $F(\theta, \phi)$ over a sphere:

$$\Omega_p = \iint_{4\pi} F(\theta, \phi)\, d\Omega \quad \text{(sr)}. \quad (9.21)$$

For an isotropic antenna with $F(\theta, \phi) = 1$ in all directions, $\Omega_p = 4\pi$ (sr).

The pattern solid angle characterizes the directional properties of the three-dimensional radiation pattern. To characterize the width of the main lobe in a given plane, the term used is *beamwidth*. The *half-power beamwidth*, or simply the beamwidth β, is defined as the angular width of the main lobe between the two

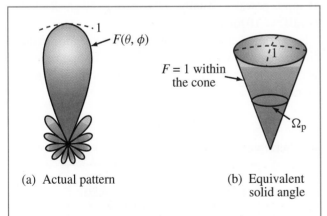

(a) Actual pattern

(b) Equivalent solid angle

Figure 9-11: The pattern solid angle Ω_p defines an equivalent cone over which all the radiation of the actual antenna is concentrated with equal intensity equal to the maximum of the actual pattern.

angles at which the magnitude of $F(\theta, \phi)$ is equal to half of its peak value (or -3 dB on a decibel scale). For example, for the pattern displayed in Fig. 9-10(b), β is given by

$$\beta = \theta_2 - \theta_1, \quad (9.22)$$

where θ_1 and θ_2 are the *half-power angles* at which $F(\theta, 0) = 0.5$ (with θ_2 denoting the larger value and θ_1 denoting the smaller one, as shown in the figure). If the pattern is symmetrical and the peak value of $F(\theta, \phi)$ is at $\theta = 0$, then $\beta = 2\theta_2$. For the short-dipole pattern shown earlier in Fig. 9-7(a), $F(\theta)$ is maximum at $\theta = 90°$, θ_2 is at $135°$, and θ_1 is at $45°$. In this case, $\beta = 135° - 45° = 90°$. The beamwidth β is also known as the *3-dB beamwidth*. In addition to the half-power beamwidth, other beam dimensions may be of interest for certain applications, such as the *null beamwidth* β_{null}, which is the width of the spacing between the first nulls on the two sides of the peak [Fig. 9-10(b)].

9-2.3 Antenna Directivity

The *directivity D* of an antenna is defined as the ratio of its maximum normalized radiation intensity, $F_{\max}$ (which by definition is equal to 1), to the average value of $F(\theta, \phi)$ over 4π space:

$$D = \frac{F_{\max}}{F_{\text{av}}}$$

$$= \frac{1}{\dfrac{1}{4\pi} \displaystyle\iint_{4\pi} F(\theta, \phi)\, d\Omega}$$

$$= \frac{4\pi}{\Omega_p} \quad \text{(dimensionless)}, \quad (9.23)$$

where Ω_p is the pattern solid angle defined by Eq. (9.21). Thus, the narrower Ω_p of an antenna pattern is, the greater is the antenna directivity. For an isotropic antenna, $\Omega_p = 4\pi$; hence, its directivity $D_{\text{iso}} = 1$.

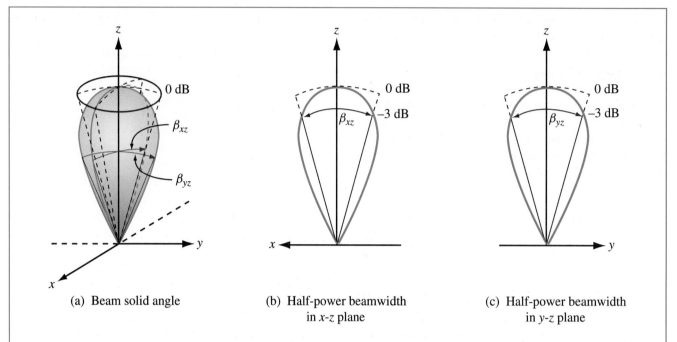

(a) Beam solid angle

(b) Half-power beamwidth in x-z plane

(c) Half-power beamwidth in y-z plane

Figure 9-12: The solid angle of a unidirectional radiation pattern is approximately equal to the product of the half-power beamwidths in the two principal planes; that is, $\Omega_{\mathrm{p}} \simeq \beta_{xz}\beta_{yz}$.

By using Eq. (9.20) in Eq. (9.23), D can be expressed as

$$D = \frac{4\pi R^2 S_{\max}}{P_{\mathrm{rad}}} = \frac{S_{\max}}{S_{\mathrm{av}}}, \qquad (9.24)$$

where $S_{\mathrm{av}} = P_{\mathrm{rad}}/(4\pi R^2)$ is the average value of the radiated power density and is equal to the total power radiated by the antenna, P_{rad}, divided by the surface area of a sphere of radius R. Since $S_{\mathrm{av}} = S_{\mathrm{iso}}$, where S_{iso} is the power density radiated by an isotropic antenna, D represents the ratio of the maximum power density radiated by the antenna under consideration to the power density radiated by an isotropic antenna, both measured at the same range R and excited by the same amount of input power. Usually, D is expressed in decibels:[*] $D \,(\mathrm{dB}) = 10 \log D$.

For an antenna with a single main lobe pointing in the z-direction as shown in Fig. 9-12, Ω_{p} may be approximated as the product of the half-power beamwidths β_{xz} and β_{yz} (in radians):

$$\Omega_{\mathrm{p}} \simeq \beta_{xz}\beta_{yz}, \qquad (9.25)$$

and therefore

$$D = \frac{4\pi}{\Omega_{\mathrm{p}}} \simeq \frac{4\pi}{\beta_{xz}\beta_{yz}}. \qquad (9.26)$$

---

[*]A note of caution: even though we often express certain dimensionless quantities in decibels, we should always convert their decibel values to natural values before using them in the relations given in this chapter.

Although approximate, this relation provides a useful method for estimating the antenna directivity from measurements of the beamwidths in the two orthogonal planes whose intersection is the axis of the main lobe.

Example 9-1   Antenna Radiation Properties

Determine (a) the direction of maximum radiation, (b) pattern solid angle, (c) directivity, and (d) half-power beamwidth in the y–z plane for an antenna that radiates into only the upper hemisphere and its normalized radiation intensity is given by $F(\theta, \phi) = \cos^2 \theta$.

Solution: Mathematically, the statement that the antenna radiates along directions covering only the upper hemisphere can be written as

$$F(\theta, \phi) = F(\theta) = \begin{cases} \cos^2 \theta, & \text{for } 0 \le \theta \le \pi/2 \\ & \text{and } 0 \le \phi \le 2\pi, \\ 0, & \text{elsewhere.} \end{cases}$$

(a) The function $F(\theta) = \cos^2 \theta$ is independent of ϕ and is maximum when $\theta = 0°$. A polar plot of $F(\theta)$ is shown in Fig. 9-13.

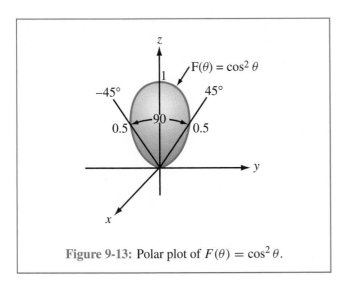

Figure 9-13: Polar plot of $F(\theta) = \cos^2 \theta$.

(b) From Eq. (9.21), the pattern solid angle Ω_p is given by

$$\Omega_p = \iint_{4\pi} F(\theta, \phi) \, d\Omega$$

$$= \int_{\phi=0}^{2\pi} \left[\int_{\theta=0}^{\pi/2} \cos^2 \theta \sin \theta \, d\theta \right] d\phi$$

$$= \int_{\phi=0}^{2\pi} \left[-\frac{\cos^3 \theta}{3} \right]_0^{\pi/2} d\phi$$

$$= \int_0^{2\pi} \frac{1}{3} \, d\phi = \frac{2\pi}{3} \quad \text{(sr).}$$

(c) Application of Eq. (9.23) gives

$$D = \frac{4\pi}{\Omega_p} = 4\pi \left(\frac{3}{2\pi} \right) = 6,$$

which corresponds to D (dB) $= 10 \log 6 = 7.78$ dB.

(d) The half-power beamwidth β is obtained by setting $F(\theta) = 0.5$. That is,

$$F(\theta) = \cos^2 \theta = 0.5,$$

which gives the half-power angles $\theta_1 = -45°$ and $\theta_2 = 45°$. Hence,

$$\beta = \theta_2 - \theta_1 = 90°. \quad \blacksquare$$

Example 9-2   Directivity of a Hertzian Dipole

Calculate the directivity of a Hertzian dipole.

Solution: Application of Eq. (9.23) with $F(\theta) = \sin^2 \theta$ [from Eq. (9.15)] gives

$$D = \frac{4\pi}{\displaystyle\iint_{4\pi} F(\theta, \phi) \sin \theta \, d\theta \, d\phi}$$

$$= \frac{4\pi}{\displaystyle\int_{\phi=0}^{2\pi} \int_{\theta=0}^{\pi} \sin^3 \theta \, d\theta \, d\phi} = \frac{4\pi}{8\pi/3} = 1.5$$

or, equivalently, 1.76 dB. $\quad \blacksquare$

9-2.4 Antenna Gain

Of the total power P_t (transmitter power) supplied to the antenna, a part, P_{rad}, is radiated out into space, and the remainder, P_{loss}, is dissipated as heat loss in the antenna structure. The *radiation efficiency* ξ is defined as the ratio of P_{rad} to P_t:

$$\xi = \frac{P_{rad}}{P_t} \quad \text{(dimensionless).} \quad (9.27)$$

The *gain* of an antenna is defined as

$$G = \frac{4\pi R^2 S_{max}}{P_t}, \quad (9.28)$$

which is similar in form to the expression given by Eq. (9.24) for the directivity D except that it is referred to the input power supplied to the antenna, P_t, rather than to the radiated power P_{rad}. In view of Eq. (9.27),

$$G = \xi D \quad \text{(dimensionless).} \quad (9.29)$$

The gain accounts for ohmic losses in the antenna material, whereas the directivity does not. For a lossless antenna, $\xi = 1$.

9-2.5 Radiation Resistance

To the transmission line connected to its terminal, an antenna is merely an impedance. If the transmission line is matched to the antenna impedance, part of P_t, the power supplied by the generator, is radiated out into space, and the remainder is dissipated as heat in the antenna. The resistance part of the antenna impedance may be defined as consisting of a *radiation resistance* R_{rad} and a *loss resistance* R_{loss}. The corresponding time-average radiated power P_{rad} and dissipated power P_{loss} are

$$P_{rad} = \frac{1}{2}I_0^2 R_{rad}, \quad (9.30a)$$

$$P_{loss} = \frac{1}{2}I_0^2 R_{loss}, \quad (9.30b)$$

where I_0 is the amplitude of the sinusoidal current exciting the antenna. As defined earlier, the radiation efficiency is the ratio of P_{rad} to P_t,

$$\xi = \frac{P_{rad}}{P_t} = \frac{P_{rad}}{P_{rad} + P_{loss}} = \frac{R_{rad}}{R_{rad} + R_{loss}}. \quad (9.31)$$

The radiation resistance R_{rad} can be calculated by integrating the far-field power density over a sphere to obtain P_{rad} and then equating the result to Eq. (9.30a).

Example 9-3   Radiation Resistance and Efficiency of a Hertzian Dipole

A 4-cm-long center-fed dipole is used as an antenna at 75 MHz. The antenna wire is made of copper and has a radius $a = 0.4$ mm. From Eqs. (7.92a) and (7.94), the loss resistance of a circular wire of length l is given by

$$R_{loss} = \frac{l}{2\pi a}\sqrt{\frac{\pi f \mu_c}{\sigma_c}}, \quad (9.32)$$

where μ_c and σ_c are the magnetic permeability and conductivity of the wire, respectively. Calculate the radiation resistance and the radiation efficiency of the dipole antenna.

Solution: At 75 MHz,

$$\lambda = \frac{c}{f} = \frac{3 \times 10^8}{7.5 \times 10^7} = 4 \text{ m}.$$

The length to wavelength ratio is $l/\lambda = 4$ cm/4 m $= 10^{-2}$. Hence, this is a short dipole. From Eq. (9.24), P_{rad} is given by

$$P_{rad} = \frac{4\pi R^2}{D} S_{max}. \quad (9.33)$$

For the Hertzian dipole, S_{max} is given by Eq. (9.14), and from Example 9-2 we established that $D = 1.5$. Hence,

$$P_{rad} = \frac{4\pi R^2}{1.5} \times \frac{15\pi I_0^2}{R^2}\left(\frac{l}{\lambda}\right)^2$$

$$= 40\pi^2 I_0^2 \left(\frac{l}{\lambda}\right)^2. \qquad (9.34)$$

Equating this result to Eq. (9.30a) and then solving for the radiation resistance R_{rad} gives

$$\boxed{R_{rad} = 80\pi^2(l/\lambda)^2 \quad (\Omega). \qquad (9.35)}$$

For $l/\lambda = 10^{-2}$, $R_{rad} = 0.08\ \Omega$.

Next, we will find the loss resistance R_{loss}. For copper, Appendix B gives $\mu_c \simeq \mu_0 = 4\pi \times 10^{-7}$ H/m and $\sigma_c = 5.8 \times 10^7$ S/m. Hence,

$$R_{loss} = \frac{l}{2\pi a}\sqrt{\frac{\pi f \mu_c}{\sigma_c}}$$

$$= \frac{4 \times 10^{-2}}{2\pi \times 4 \times 10^{-4}}\left[\frac{\pi \times 75 \times 10^6 \times 4\pi \times 10^{-7}}{5.8 \times 10^7}\right]^{1/2}$$

$$= 0.036\ \Omega,$$

and therefore the radiation efficiency is

$$\xi = \frac{R_{rad}}{R_{rad} + R_{loss}} = \frac{0.08}{0.08 + 0.036} = 0.69,$$

or 69% efficient. ∎

REVIEW QUESTIONS

Q9.5 What does the pattern solid angle represent?

Q9.6 What is the magnitude of the directivity of an isotropic antenna?

Q9.7 What physical and material properties affect the radiation efficiency of a fixed-length Hertzian dipole antenna?

EXERCISE 9.2 An antenna has a conical radiation pattern with a normalized radiation intensity $F(\theta) = 1$ for θ between $0°$ and $45°$ and zero intensity for θ between $45°$ and $180°$. The pattern is independent of the azimuth angle ϕ. Find (a) the pattern solid angle and (b) the directivity.

Ans. (a) $\Omega_p = 1.84$ sr, (b) $D = 6.83$ or, equivalently, 8.3 dB.

EXERCISE 9.3 The maximum power density radiated by a short dipole at a distance of 1 km is 60 (nW/m^2). If $I_0 = 10$ A, find the radiation resistance.

Ans. $R_{rad} = 10$ mΩ.

9-3 Half-Wave Dipole Antenna

In Section 9-1 we developed expressions for the electric and magnetic fields radiated by a short dipole of length much shorter than λ. We shall now use these expressions as building blocks to obtain expressions for the fields radiated by a half-wave dipole antenna, so named because its length $l = \lambda/2$. As shown in Fig. 9-14, the half-wave dipole consists of a thin wire fed at its center by a generator connected to the antenna terminals via a transmission line. The current flowing through the wire has a symmetrical distribution with respect to the center of the dipole, and the current has to be zero at the ends. Mathematically, $i(t)$ is given by

$$i(t) = I_0 \cos\omega t \cos kz = \mathfrak{Re}\left[I_0 \cos kz\, e^{j\omega t}\right], \quad (9.36)$$

whose current phasor is

$$\tilde{I}(z) = I_0 \cos kz, \qquad \frac{-\lambda}{4} \le z \le \frac{\lambda}{4}, \qquad (9.37)$$

and $k = 2\pi/\lambda$. Equation (9.9a) gives an expression for $\tilde{E}_\theta$, the far field (at distance R) radiated by a short dipole of length l when excited by a current I_0. Let us adapt that expression to an infinitesimal dipole segment

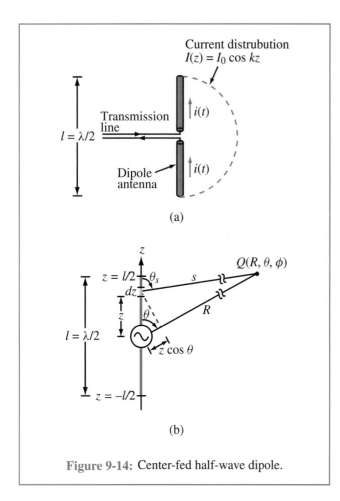

Figure 9-14: Center-fed half-wave dipole.

of length dz, excited by a current $\widetilde{I}(z)$ and located at a distance s from the observation point [Fig. 9-14(b)]. Thus,

$$d\widetilde{E}_\theta(z) = \frac{jk\eta_0}{4\pi}\,\widetilde{I}(z)\,dz\left(\frac{e^{-jks}}{s}\right)\sin\theta_s, \qquad (9.38a)$$

and the associated magnetic field is

$$d\widetilde{H}_\phi(z) = \frac{d\widetilde{E}_\theta(z)}{\eta_0}. \qquad (9.38b)$$

The far field due to radiation by the entire antenna is obtained by integrating the fields from all the Hertzian

dipoles making up the antenna:

$$\widetilde{E}_\theta = \int_{z=-\lambda/4}^{\lambda/4} d\widetilde{E}_\theta. \qquad (9.39)$$

Before we calculate this integral, we shall make the following two approximations. The first approximation relates to the magnitude part of the spherical propagation factor, $1/s$. In Fig. 9-14(b), the distance s between the current element and the observation point Q is considered so large in comparison to the length of the dipole that the difference between s and R may be neglected in terms of its effect on $1/s$. Hence, we may set $1/s \simeq 1/R$, and by the same argument we set $\theta_s \simeq \theta$. The error Δ between s and R is a maximum when the observation point is along the z-axis and it is equal to $\lambda/4$ (corresponding to half of the antenna length). If $R \gg \lambda$, this error will have an insignificant effect on $1/s$. For the phase factor e^{-jks}, such an error in distance corresponds to an error in phase of $k\Delta = (2\pi/\lambda)(\lambda/4) = \pi/2$. As a rule of thumb, a phase error greater than $\pi/8$ is considered unacceptable because it may lead to a significant error in the computed value of the field $\widetilde{E}_\theta$. Hence, the approximation $s \simeq R$ is too crude for the phase factor and cannot be used. A more tolerable option is to use the parallel-ray approximation given by

$$s \simeq R - z\cos\theta, \qquad (9.40)$$

as illustrated in Fig. 9-14(b).

Substituting Eq. (9.40) for s in the phase factor of Eq. (9.38a) and replacing s with R and θ_s with θ elsewhere in the expression, we have

$$d\widetilde{E}_\theta = \frac{jk\eta_0}{4\pi}\,\widetilde{I}(z)\,dz\left(\frac{e^{-jkR}}{R}\right)\sin\theta\,e^{jkz\cos\theta}. \qquad (9.41)$$

After (1) inserting Eq. (9.41) into Eq. (9.39), (2) using the expression for $\widetilde{I}(z)$ given by Eq. (9.37), and (3) carrying out the integration, the following expressions are obtained:

$$\widetilde{E}_\theta = j\,60 I_0 \left[\frac{\cos[(\pi/2)\cos\theta]}{\sin\theta} \right]\left(\frac{e^{-jkR}}{R} \right), \quad (9.42a)$$

$$\widetilde{H}_\phi = \frac{\widetilde{E}_\theta}{\eta_0}, \quad\quad (9.42b)$$

and the corresponding time-average power density is

$$S(R,\theta) = \frac{|\widetilde{E}_\theta|^2}{2\eta_0}$$

$$= \frac{15 I_0^2}{\pi R^2} \left[\frac{\cos^2[(\pi/2)\cos\theta]}{\sin^2\theta} \right],$$

$$= S_0 \left[\frac{\cos^2[(\pi/2)\cos\theta]}{\sin^2\theta} \right] \quad (W/m^2). \quad (9.43)$$

Examination of Eq. (9.43) reveals that $S(R,\theta)$ is maximum at $\theta = \pi/2$, and its value is

$$S_{max} = S_0 = \frac{15 I_0^2}{(\pi R^2)}.$$

Hence, the normalized radiation intensity is

$$F(\theta) = \frac{S(R,\theta)}{S_0} = \left[\frac{\cos[(\pi/2)\cos\theta]}{\sin\theta} \right]^2. \quad (9.44)$$

The radiation pattern of the half-wave dipole exhibits roughly the same doughnutlike shape shown earlier in Fig. 9-7 for the short dipole. Its directivity is slightly larger (1.64 compared to 1.5 for the short dipole), but its radiation resistance is 73 Ω (as will be shown later in Section 9-3.2), which is orders of magnitude larger than that of a short dipole.

9-3.1 Directivity of $\lambda/2$ Dipole

To evaluate both the directivity D and the radiation resistance R_{rad} of the half-wave dipole, we first need to calculate the total radiated power P_{rad} by applying Eq. (9.20):

$$P_{rad} = R^2 \iint_{4\pi} S(R,\theta)\,d\Omega$$

$$= \frac{15 I_0^2}{\pi} \int_0^{2\pi}\int_0^\pi \left[\frac{\cos[(\pi/2)\cos\theta]}{\sin\theta} \right]^2 \sin\theta\,d\theta\,d\phi.$$

$$(9.45)$$

The integration over ϕ is equal to 2π, and numerical evaluation of the integration over θ gives the value 1.22. Consequently,

$$P_{rad} = 36.6\,I_0^2 \quad (W). \quad (9.46)$$

From Eq. (9.43), we found that $S_{max} = 15 I_0^2/(\pi R^2)$. Using this in Eq. (9.24) gives the following result for the directivity D of the half-wave dipole:

$$D = \frac{4\pi R^2 S_{max}}{P_{rad}} = \frac{4\pi R^2}{36.6 I_0^2}\left(\frac{15 I_0^2}{\pi R^2} \right) = 1.64 \quad (9.47)$$

or, equivalently, 2.15 dB.

9-3.2 Radiation Resistance of $\lambda/2$ Dipole

Using Eq. (9.30a) to relate the radiation resistance R_{rad} to the total radiated power P_{rad}, we have

$$R_{rad} = \frac{2 P_{rad}}{I_0^2} = \frac{2 \times 36.6 I_0^2}{I_0^2} \simeq 73\ \Omega. \quad (9.48)$$

As was noted earlier in Example 9-3, because the radiation resistance of a short dipole is comparable in magnitude to that of its loss resistance R_{loss}, its radiation efficiency ξ is quite small. For the 4-cm-long dipole of Example 9-3, $R_{rad} = 0.08\ \Omega$ (at 75 MHz) and $R_{loss} = 0.036\ \Omega$. If we keep the frequency the same and increase the length of the dipole to 2 m ($\lambda = 4$ m at $f = 75$ MHz), R_{rad} becomes 73 Ω and R_{loss} increases to 1.8 Ω. The radiation efficiency increases from 69% for the short dipole to 98% for the half-wave dipole. More significant is the fact that it is practically impossible to match a transmission line to an antenna with a resistance on the order of 0.1 Ω, while it is quite easy to do so when $R_{\text{rad}} = 73$ Ω.

Another interesting feature of the half-wave dipole pertains to its reactance. As seen by a transmission line, an antenna may be represented by an *input impedance* connected to the line at the antenna terminals. The input impedance Z_{in} consists of a real part R_{in} and an imaginary part X_{in}:

$$Z_{\text{in}} = R_{\text{in}} + j X_{\text{in}}, \qquad (9.49)$$

where R_{in} is the sum of the radiation resistance R_{rad} and the loss resistance R_{loss},

$$R_{\text{in}} = R_{\text{rad}} + R_{\text{loss}}. \qquad (9.50)$$

For the half-wave dipole, R_{loss} is much smaller than R_{rad} and can be ignored. Hence,

$$Z_{\text{in}} \cong R_{\text{rad}} + j X_{\text{in}}. \qquad (9.51)$$

Deriving an expression for X_{in} for the half-wave dipole is fairly complicated and beyond the scope of this book. However, it is significant to note that X_{in} is a strong function of l/λ, and it decreases from 42 Ω at $l/\lambda = 0.5$ down to zero at $l/\lambda = 0.48$, whereas R_{rad} remains approximately unchanged. Hence, by reducing the length of the half-wave dipole by 4%, Z_{in} becomes purely real and equal to 73 Ω, thereby making it possible to match the dipole to a 75-Ω transmission line without resorting to the use of a matching network.

9-3.3 Quarter-Wave Monopole Antenna

When placed over a conducting ground plane, a quarter-wave monopole antenna excited by a source at its base [Fig. 9-15(a)] exhibits the same radiation pattern in the region above the ground plane as a half-wave dipole in free space. This is because, from image theory [Section 4-12], the conducting plane can be replaced with the image of the $\lambda/4$ monopole, as illustrated in Fig. 9-15(b). Thus, the $\lambda/4$ monopole will radiate an electric field identical to that given by Eq. (9.42a), and its normalized radiation intensity is given by Eq. (9.44);

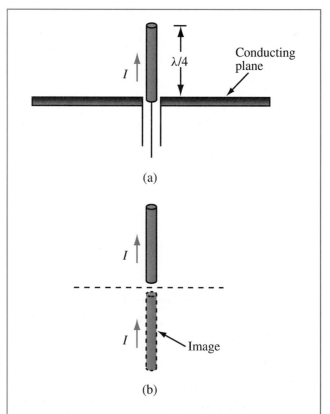

(a)

(b)

Figure 9-15: A quarter-wave monopole above a conducting plane is equivalent to a full half-wave dipole in free space.

but the radiation is limited to the upper half-space defined by $0 \leq \theta \leq \pi/2$. Hence, a monopole radiates only half as much power as the dipole. Consequently, for a $\lambda/4$ monopole, $P_{\mathrm{rad}} = 18.3I_0^2$ and its radiation resistance is $R_{\mathrm{rad}} = 36.5\ \Omega$.

The approach used for the quarter-wave monopole is also valid for any vertical wire antenna placed above a conducting plane, including a Hertzian monopole.

REVIEW QUESTIONS

Q9.8 What is the physical length of a half-wave dipole operating at (a) 1 MHz (in the AM broadcast band), (b) 100 MHz (FM broadcast band), and (c) 10 GHz (microwave band)?

Q9.9 How does the radiation pattern of a half-wave dipole compare with that of a Hertzian dipole? How do their directivities, radiation resistances, and radiation efficiencies compare?

Q9.10 How does the radiation efficiency of a quarter-wave monopole compare with that of a half-wave dipole, assuming that both are made of the same material and have the same cross section?

EXERCISE 9.4 For the half-wave dipole antenna, evaluate $F(\theta)$ versus θ in order to determine the half-power beamwidth in the elevation plane (the plane containing the dipole axis).

Ans. $\beta = 78°$.

EXERCISE 9.5 If the maximum power density radiated by a half-wave dipole is 50 (μW/m^2) at a range of 1 km, what is the current amplitude I_0?

Ans. $I_0 = 3.24$ A.

9-4 Dipole of Arbitrary Length

So far, we have examined the radiation properties of the short Hertzian dipole and of the half-wave dipole. We will now consider the more general case of a linear dipole of any arbitrary length l, relative to λ. For a center-fed dipole, as depicted in Fig. 9-16, the currents flowing through its two halves are symmetrical and must go to zero at its ends. Hence, the current phasor $\tilde{I}(z)$ can be expressed as a sine function with an argument that goes to zero at $z = \pm l/2$:

$$\tilde{I}(z) = \begin{cases} I_0 \sin\left[k\left(l/2 - z\right)\right], & \text{for } 0 \leq z \leq l/2, \\ I_0 \sin\left[k\left(l/2 + z\right)\right], & \text{for } -l/2 \leq z < 0, \end{cases} \tag{9.52}$$

where I_0 is the current amplitude. The procedure for calculating the electric and magnetic fields and the

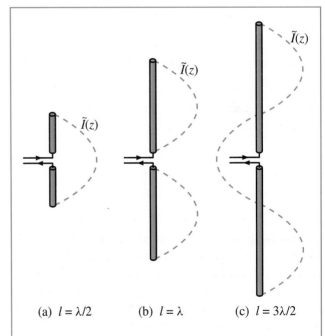

(a) $l = \lambda/2$ (b) $l = \lambda$ (c) $l = 3\lambda/2$

Figure 9-16: Current distribution for three center-fed dipoles.

associated power density of the wave radiated by such an antenna is basically the same as that used previously in connection with the half-wave dipole antenna. The only difference is the current distribution $\widetilde{I}(z)$. If we insert the expression for $\widetilde{I}(z)$ given by Eq. (9.52) in Eq. (9.41), we obtain the following expression for the differential electric field $d\widetilde{E}_\theta$ of the wave radiated by an elemental length dz at location z along the dipole:

$$d\widetilde{E}_\theta = \frac{jk\eta_0 I_0}{4\pi} \left(\frac{e^{-jkR}}{R} \right) \sin\theta \, e^{jkz\cos\theta} \, dz$$

$$\times \begin{cases} \sin\left[k\left(l/2 - z \right) \right], & \text{for } 0 \le z \le l/2, \\ \sin\left[k\left(l/2 + z \right) \right], & \text{for } -l/2 \le z < 0. \end{cases}$$

$$(9.53)$$

The total field radiated by the dipole is

$$\widetilde{E}_\theta = \int_{-l/2}^{l/2} d\widetilde{E}_\theta$$

$$= \int_0^{l/2} d\widetilde{E}_\theta + \int_{-l/2}^0 d\widetilde{E}_\theta$$

$$= \frac{jk\eta_0 I_0}{4\pi} \left(\frac{e^{-jkR}}{R} \right) \sin\theta$$

$$\times \left[\int_0^{l/2} e^{jkz\cos\theta} \sin[k(l/2 - z)] \, dz \right.$$

$$\left. + \int_{-l/2}^0 e^{jkz\cos\theta} \sin[k(l/2 + z)] \, dz \right]. \quad (9.54)$$

If we apply Euler's identity to express $e^{jkz\cos\theta}$ as $[\cos(kz\cos\theta) + j\sin(kz\cos\theta)]$, we can integrate the two integrals and obtain the result

$$\widetilde{E}_\theta = j60 I_0 \left(\frac{e^{-jkR}}{R} \right)$$

$$\times \left[\frac{\cos\left(\frac{kl}{2}\cos\theta \right) - \cos\left(\frac{kl}{2} \right)}{\sin\theta} \right]. \quad (9.55)$$

The corresponding time-average power density radiated by the dipole antenna is given by

$$S(\theta) = \frac{|\widetilde{E}_\theta|^2}{2\eta_0}$$

$$= \frac{15 I_0^2}{\pi R^2} \left[\frac{\cos\left(\frac{\pi l}{\lambda}\cos\theta \right) - \cos\left(\frac{\pi l}{\lambda} \right)}{\sin\theta} \right]^2, \quad (9.56)$$

where we have used the relations $\eta_0 \simeq 120\pi$ (Ω) and $k = 2\pi/\lambda$. For $l = \lambda/2$, Eq. (9.56) reduces to the expression given by Eq. (9.43) for the half-wave dipole. Plots of the normalized radiation intensity, $F(\theta) = S(R,\theta)/S_{\max}$, are shown in Fig. 9-17 for dipoles with lengths of $\lambda/2$, λ, and $3\lambda/2$. The dipoles with $l = \lambda/2$ and $l = \lambda$ have similar radiation patterns with maxima along $\theta = 90°$, but the half-power beamwidth of the wavelength-long dipole is narrower than that of the half-wave dipole, and $S_{\max} = 60 I_0^2/(\pi R^2)$ for the wavelength-long dipole, which is four times that for the half-wave dipole. The pattern of the dipole with length $l = 3\lambda/2$ exhibits a multiple lobe structure, and its direction of maximum radiation is not along $\theta = 90°$.

9-5   Effective Area of a Receiving Antenna

So far, we have examined the radiation characteristics of antennas by treating them as radiators of the energy supplied by a source. Now we shall consider the reverse process, that is, how a receiving antenna extracts energy from an incident wave and delivers it to a load. The ability of an antenna to capture energy from an incident wave of power density S_i (W/m^2) and to convert it into an *intercepted power* P_{int} (W) for delivery to a matched load is characterized by the *effective area* A_e:

$$\boxed{A_e = \frac{P_{\text{int}}}{S_i} \quad (\text{m}^2). \quad (9.57)}$$

Other commonly used names for A_e include *effective aperture* and *receiving cross section*. The antenna

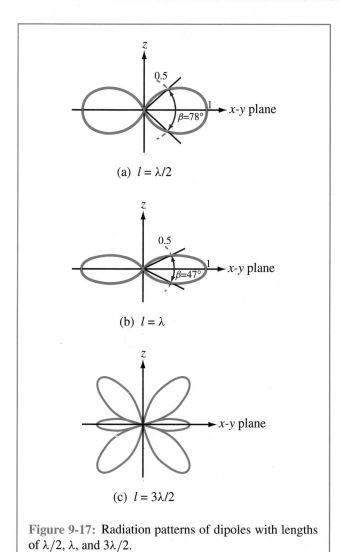

(a) $l = \lambda/2$

(b) $l = \lambda$

(c) $l = 3\lambda/2$

Figure 9-17: Radiation patterns of dipoles with lengths of $\lambda/2$, λ, and $3\lambda/2$.

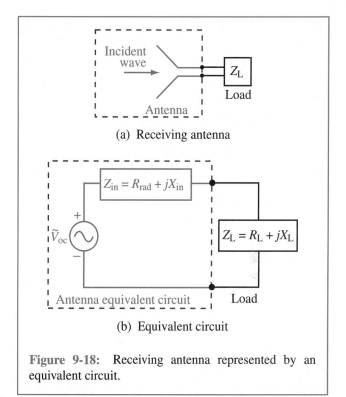

(a) Receiving antenna

(b) Equivalent circuit

Figure 9-18: Receiving antenna represented by an equivalent circuit.

receiving process may be modeled in the form of a Thévenin equivalent circuit as shown in Fig. 9-18, where $\tilde{V}_{oc}$ is the open-circuit voltage phasor induced in the receiving antenna by the incident wave, Z_{in} is the antenna impedance, and Z_L is the impedance of the load for which the received power is intended. In general, both Z_{in} and Z_L are complex:

$$Z_{in} = R_{rad} + j X_{in}, \qquad (9.58a)$$

$$Z_L = R_L + j X_L, \qquad (9.58b)$$

where R_{rad} denotes the radiation resistance of the antenna, and it is assumed that the antenna loss resistance is much smaller than R_{rad} and can be ignored. As we will see later in Example 9-4, for maximum power transfer, the load impedance must be chosen such that $Z_L = Z_{in}^*$, or $R_L = R_{rad}$ and $X_L = -X_{in}$, in which case the circuit reduces to a source $\tilde{V}_{oc}$ connected across a resistance equal to $2R_{rad}$. Since $\tilde{V}_{oc}$ is a sinusoidal voltage phasor, the time-average power delivered to the load is

$$P_L = \frac{1}{2}|\tilde{I}_L|^2 R_{rad} = \frac{1}{2}\left[\frac{|\tilde{V}_{oc}|}{2R_{rad}}\right]^2 R_{rad} = \frac{|\tilde{V}_{oc}|^2}{8R_{rad}},$$

$$(9.59)$$

where $\widetilde{I}_L = \widetilde{V}_{oc}/(2R_{rad})$ is the phasor current flowing through the circuit. Since the antenna is lossless, all the intercepted power P_{int} ends up in the load resistance R_L. Hence,

$$P_{int} = P_L = \frac{|\widetilde{V}_{oc}|^2}{8R_{rad}}. \tag{9.60}$$

For an incident wave with electric field $\widetilde{E}_i$ parallel to the antenna polarization direction, the power density carried by the wave is

$$S_i = \frac{|\widetilde{E}_i|^2}{2\eta_0} = \frac{|\widetilde{E}_i|^2}{240\pi}. \tag{9.61}$$

The ratio of the results provided by Eqs. (9.60) and (9.61) gives

$$A_e = \frac{P_{int}}{S_i} = \frac{30\pi|\widetilde{V}_{oc}|^2}{R_{rad}|\widetilde{E}_i|^2}. \tag{9.62}$$

The open-circuit voltage $\widetilde{V}_{oc}$ induced in the receiving antenna is due to the incident field $\widetilde{E}_i$, but the relation between them depends on the specific antenna under consideration. By way of illustration, let us consider the case of the short-dipole antenna of Section 9-1. Because the length l of the short dipole is small compared to λ, the current induced by the incident field will be uniform across its length, and the open-circuit voltage will simply be $\widetilde{V}_{oc} = \widetilde{E}_i l$. Noting that $R_{rad} = 80\pi^2(l/\lambda)^2$ for the short dipole [see Eq. (9.35)] and using $\widetilde{V}_{oc} = \widetilde{E}_i l$, Eq. (9.62) simplifies to

$$A_e = \frac{3\lambda^2}{8\pi} \quad (m^2) \quad \text{(short dipole).} \tag{9.63}$$

In Example 9-2 it was shown that for the short dipole the directivity $D = 1.5$. In terms of D, Eq. (9.63) can be rewritten in the form

$$A_e = \frac{\lambda^2 D}{4\pi} \quad (m^2) \quad \text{(any antenna).} \tag{9.64}$$

Despite the fact that the relation between A_e and D given by Eq. (9.64) was derived for a short dipole, it can be shown that it is also valid *for any antenna* under matched-impedance conditions.

Example 9-4 Maximum Power Transfer

The circuit shown in Fig. 9-18(b) consists of a generator comprised of a voltage phasor $\widetilde{V}_{oc}$ and an internal impedance Z_{in} connected to a load Z_L. If

$$Z_{in} = R_{rad} + jX_{in},$$
$$Z_L = R_L + jX_L,$$

what values of R_L and X_L yield the maximum transfer of power from the generator into the load?

Solution: The current phasor flowing through the circuit is

$$\widetilde{I}_L = \frac{\widetilde{V}_{oc}}{Z_{in} + Z_L} = \frac{\widetilde{V}_{oc}}{(R_{rad} + R_L) + j(X_{in} + X_L)}, \tag{9.65}$$

and the time-average power dissipated in the load resistance R_L is

$$P_L = \frac{1}{2}|\widetilde{I}_L|^2 R_L$$
$$= \frac{1}{2}|\widetilde{V}_{oc}|^2 \frac{R_L}{(R_{rad} + R_L)^2 + (X_{in} + X_L)^2}. \tag{9.66}$$

Our task is to specify the load parameters, R_L and X_L, that maximize P_L. This is achieved by differentiating the expression for P_L with respect to R_L, equating it to zero, and then repeating the process with respect to X_L. The first of these two steps gives

$$\frac{\partial P_L}{\partial R_L} = \frac{1}{2}|\widetilde{V}_{oc}|^2 \left[\frac{1}{(R_{rad} + R_L)^2 + (X_{in} + X_L)^2} \right.$$
$$\left. + \frac{-2R_L(R_{rad} + R_L)}{\left[(R_{rad} + R_L)^2 + (X_{in} + X_L)^2\right]^2} \right] = 0,$$

which yields the result

$$(X_{in} + X_L)^2 - R_L^2 + R_{rad}^2 = 0. \qquad (9.67)$$

Application of the second step gives

$$\frac{\partial P_L}{\partial X_L} = \frac{1}{2}|\tilde{V}_{oc}|^2 \left[\frac{-2R_L(X_{in} + X_L)}{[(R_{rad} + R_L)^2 + (X_{in} + X_L)^2]^2} \right] = 0,$$

which yields a second condition,

$$R_L(X_{in} + X_L) = 0. \qquad (9.68)$$

To satisfy this result, we have to choose either $R_L = 0$, which is not an acceptable result because it means that $P_L = 0$, or

$$\boxed{X_L = -X_{in}. \qquad (9.69a)}$$

Using Eq. (9.69a) in Eq. (9.67) gives

$$\boxed{R_L = R_{rad}. \qquad (9.69b)}$$

These two results can be combined into

$$\boxed{Z_L = Z_{in}^*, \qquad (9.70)}$$

where Z_{in}^* is the complex conjugate of Z_{in}. Thus, *for maximum power transfer, the load impedance should be equal to the complex conjugate of the generator impedance.* ■

EXERCISE 9.6 The effective area of an antenna is 9 m². What is its directivity in decibels at 3 GHz?

Ans. $D = 40.53$ dB.

EXERCISE 9.7 At 100 MHz, the pattern solid angle of an antenna is 1.3 sr. Find (a) the antenna directivity D and (b) its effective area A_e.

Ans. (a) $D = 9.67$, (b) $A_e = 6.92$ m².

9-6 Friis Transmission Formula

The two antennas shown in Fig. 9-19 are part of a free-space communication link, with the separation between the antennas, R, being large enough for each antenna to be in the far-field region of the other. The transmitting and receiving antennas have effective areas A_t and A_r and radiation efficiencies ξ_t and ξ_r, respectively. Our objective is to find a relationship between P_t, the transmitter power supplied to the transmitting antenna, and P_{rec}, the power delivered to the receiver by the receiving antenna. As always, we assume that both antennas are impedance matched to their respective transmission lines. Initially, we shall consider the case where the two antennas are oriented such that the peak of the radiation pattern of each antenna points in the direction of the other.

Let us start by treating the transmitting antenna as a lossless isotropic radiator. In this case, the power density incident upon the receiving antenna at a distance R from an isotropic transmitting antenna is simply equal to the transmitter power P_t divided by the surface area of a sphere of radius R:

$$S_{iso} = \frac{P_t}{4\pi R^2}. \qquad (9.71)$$

The real transmitting antenna is neither lossless nor isotropic. Hence, the power density S_r due to the real

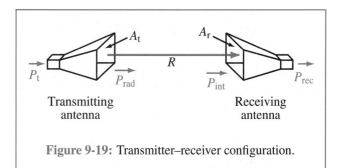

Figure 9-19: Transmitter–receiver configuration.

antenna is given by

$$S_r = G_t S_{iso} = \xi_t D_t S_{iso} = \frac{\xi_t D_t P_t}{4\pi R^2}, \qquad (9.72)$$

where, through the gain $G_t = \xi_t D_t$, ξ_t accounts for the fact that only part of the power P_t supplied to the antenna actually is radiated out into space, and D_t accounts for the directivity of the transmitting antenna (in the direction of the receiving antenna). Using Eq. (9.64), Eq. (9.72) can be expressed in terms of the effective area A_t of the transmitting antenna:

$$S_r = \frac{\xi_t A_t P_t}{\lambda^2 R^2}. \qquad (9.73)$$

On the receiving-antenna side, the power intercepted by the receiving antenna is equal to the product of the incident power density S_r and the effective area A_r:

$$P_{int} = S_r A_r = \frac{\xi_t A_t A_r P_t}{\lambda^2 R^2}. \qquad (9.74)$$

The received power P_{rec} delivered to the receiver is equal to the intercepted power P_{int} multiplied by the radiation efficiency of the receiving antenna, ξ_r. Thus, $P_{rec} = \xi_r P_{int}$, which leads to the result

$$\boxed{\frac{P_{rec}}{P_t} = \frac{\xi_t \xi_r A_t A_r}{\lambda^2 R^2} = G_t G_r \left(\frac{\lambda}{4\pi R}\right)^2. \qquad (9.75)}$$

This relation is known as the *Friis transmission formula*, and P_{rec}/P_t is sometimes called the *power transfer ratio*. In deriving the form of the Friis formula involving the gains G_t and G_r of the transmitting and receiving antennas, we used the fact that $\xi_t A_t = \xi_t D_t \lambda^2/4\pi = G_t \lambda^2/4\pi$, and similar relations apply to the receiving antenna.

If the two antennas are not oriented in the direction of maximum power transfer, Eq. (9.75) can be rewritten in the general form

$$\frac{P_{rec}}{P_t} = G_t G_r \left(\frac{\lambda}{4\pi R}\right)^2 F_t(\theta_t, \phi_t) \, F_r(\theta_r, \phi_r), \qquad (9.76)$$

where $F_t(\theta_t, \phi_t)$ is the normalized radiation intensity of the transmitting antenna, evaluated at the direction (θ_t, ϕ_t) corresponding to the direction of the receiving antenna (as seen by the antenna pattern of the transmitting antenna), and a similar definition applies to $F_r(\theta_r, \phi_r)$ for the receiving antenna.

Example 9-5   Satellite Communication System

A 6-GHz direct-broadcast TV satellite system transmits 100 W through a 2-m-diameter parabolic dish antenna from a distance of approximately 40,000 km above Earth's surface. Each TV channel occupies a bandwidth of 5 MHz. Due to electromagnetic noise picked up by the antenna as well as noise generated by the receiver electronics, a ground home receiving TV station has a noise level given by

$$P_n = K T_{sys} B \qquad (\text{W}), \qquad (9.77)$$

where T_{sys} [measured in kelvins (K)] is a figure of merit, called the *system noise temperature*, that characterizes the noise performance of the receiver–antenna combination, K is Boltzmann's constant [1.38×10^{-23} (J/K)], and B is the receiver bandwidth in Hz.

The *signal-to-noise ratio* S_n (which should not be confused with the power density S) is defined as the ratio of P_{rec}, the signal power received from the transmitter, to P_n:

$$S_n = P_{rec}/P_n \qquad (\text{dimensionless}). \qquad (9.78)$$

For a receiver with $T_{sys} = 580$ K, what minimum diameter of a parabolic dish receiving antenna is required for high-quality TV reception with $S_n = 40$ dB? The satellite and ground receiving antennas may be assumed lossless, and their effective areas may be assumed equal to their physical apertures.

Solution: The following quantities are given:

$$P_t = 100 \text{ W}, \quad f = 6 \text{ GHz} = 6 \times 10^9 \text{ Hz}, \quad S_n = 10^4,$$

$$\text{Transmit antenna diameter } d_t = 2 \text{ m},$$

$$T_{sys} = 580 \text{ K},$$

$$R = 40,000 \text{ km} = 4 \times 10^7 \text{ m},$$

$$B = 5 \text{ MHz} = 5 \times 10^6 \text{ Hz}.$$

The wavelength $\lambda = c/f = 5 \times 10^{-2}$ m, and the area of the transmitting satellite antenna is $A_t = (\pi d_t^2/4) = \pi$ (m^2). From Eq. (9.77), the receiver noise power is

$$P_n = KT_{sys}B = 1.38 \times 10^{-23} \times 580 \times 5 \times 10^6$$

$$= 4 \times 10^{-14} \text{ W}.$$

Using Eq. (9.75) for lossless antennas ($\xi_t = \xi_r = 1$), the received signal power is

$$P_{rec} = \frac{P_t A_t A_r}{\lambda^2 R^2} = \frac{100\pi A_r}{(5 \times 10^{-2})^2 (4 \times 10^7)^2}$$

$$= 7.85 \times 10^{-11} A_r.$$

The area of the receiving antenna, A_r, can now be determined by equating the ratio P_{rec}/P_n to $S_n = 10^4$:

$$10^4 = \frac{7.85 \times 10^{-11} A_r}{4 \times 10^{-14}},$$

which yields the value $A_r = 5.1$ m^2. The required minimum diameter is $d_r = \sqrt{4A_r/\pi} = 2.55$ m. ∎

REVIEW QUESTIONS

Q9.11 For an a-c generator connected to a load, what is the condition for maximum transfer of power from the generator to the load?

Q9.12 If the two antennas of a communication system have constant radiation efficiencies and effective areas, how does the power transfer ratio vary with λ? Explain in terms of how the antenna beams vary with λ.

EXERCISE 9.8 Suppose that the operating frequency of the communication system described in Example 9-5 were to be doubled to 12 GHz. What would then be the minimum required diameter of a home receiving TV antenna?

Ans. $d_r = 1.27$ m.

EXERCISE 9.9 A 3-GHz microwave link consists of two identical antennas each with a gain of 30 dB. If the transmitter output power is 1 kW and the two antennas are 10 km apart, find the received power.

Ans. $P_{rec} = 6.33 \times 10^{-4}$ W.

EXERCISE 9.10 The effective area of a parabolic dish antenna is approximately equal to its physical aperture. If the directivity of a dish antenna is 30 dB at 10 GHz, what is its effective area? If the frequency is increased to 30 GHz, what will be its new directivity?

Ans. $A_e = 0.07$ m^2, $D = 39.54$ dB.

9-7 Radiation by Large-Aperture Antennas

For wire antennas, the sources of radiation are the infinitesimal current elements comprising the current distribution along the wire, and the total radiated field at a given point in space is equal to the sum, or integral, of the fields radiated by all the elements. A parallel arrangement applies to aperture antennas, except that now the source of radiation is the electric-field distribution across the aperture. Consider the horn antenna shown in Fig. 9-20, which is connected to a source through a coaxial transmission line, with the outer conductor of the coaxial line being connected to the metal body of the horn and the inside conductor made to protrude, through a small hole, partially into the throat end of the horn. The protruding conductor acts as a monopole antenna, generating waves that radiate outwardly toward

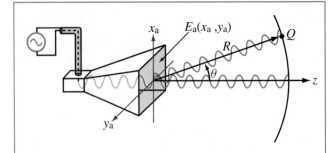

Figure 9-20: A horn antenna with aperture field distribution $E_a(x_a, y_a)$.

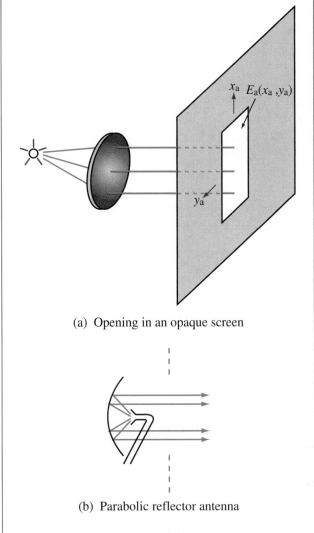

(a) Opening in an opaque screen

(b) Parabolic reflector antenna

Figure 9-21: Radiation by apertures: (a) an opening in an opaque screen illuminated by a light source through a collimating lens and (b) a parabolic dish reflector illuminated by a small horn antenna.

the horn's aperture. The electric field of the energy arriving at the aperture, which may vary as a function of x_a and y_a over the horn's aperture, is called the electric-field aperture distribution, $E_a(x_a, y_a)$. Inside the horn, wave propagation is guided by the horn's geometry; but as the wave transitions from a guided wave into an unbounded wave, every point of its wavefront serves as a source of spherical secondary wavelets. The aperture may then be represented as a distribution of isotropic radiators, each with an intensity proportional to the electric field $E_a(x_a, y_a)$ at its location (x_a, y_a). At a distant point Q, the combination of all the waves arriving from all of these radiators constitutes the total wave that would be observed by a receiver placed at that point.

The radiation process described for the horn antenna is equally applicable to any aperture upon which an electromagnetic wave is incident. For example, if a light source is used to illuminate an opening in an opaque screen through a collimating lens, as shown in Fig. 9-21(a), the opening becomes a source of secondary spherical wavelets, much like the aperture of the horn antenna. In the case of the parabolic reflector shown in Fig. 9-21(b), it can be described in terms of an imaginary aperture representing the electric-field distribution across a plane in front of the reflector.

Two types of mathematical formulations are available for computing the electromagnetic fields of waves radi-

ated by apertures. The first type is a *scalar formulation* based on Kirchhoff's work and is used commonly in optical diffraction problems and for large-aperture antennas. The second type is a *vector formulation* based on Maxwell's equations. Although theoretically superior, the vector approach generally is more difficult to apply. Hence, it is used mostly in connection with antenna apertures whose dimensions are comparable to or smaller than a wavelength, because in this case the scalar approach is inapplicable.

In this section, we shall limit our presentation to the scalar diffraction approach, in part because of its inherent simplicity and also because it is applicable to a wide range of practical applications. *The key requirement for the validity of the scalar formulation is that the antenna aperture be at least several wavelengths long along each of its principal dimensions.* A distinctive feature of such an antenna is its high directivity and correspondingly narrow beam, which makes it attractive for radar and free-space microwave communication systems. The frequency range commonly used for these applications is the 1- to 30-GHz microwave band. Because the corresponding wavelength range is 30 to 1 cm, respectively, it is quite practical to construct and use antennas (in this frequency range) with aperture dimensions that are many wavelengths in size.

The x_a–y_a plane in Fig. 9-22, denoted plane A, contains an aperture with an electric field distribution $E_a(x_a, y_a)$, usually called the *aperture illumination*. For the sake of convenience, the opening has been chosen to be rectangular in shape, with dimensions l_x along x_a and l_y along y_a, even though the formulation we are about to discuss is general enough to accommodate any two-dimensional aperture distribution, including those associated with circular and elliptical apertures. At a distance z from the aperture plane A in Fig. 9-22, we have an observation plane O with axes (x, y). The two planes have parallel axes and are separated by a distance z that is sufficiently large so that any point Q in the observation plane is in the far-field region of the aperture. To satisfy the far-field condition, it is necessary that

$$R \geq 2d^2/\lambda, \quad (9.79)$$

where d is the longest linear dimension of the radiating aperture.

The position of observation point Q is specified by the range R between the center of the aperture and point Q and by the angles θ and ϕ shown in Fig. 9-22, which together define the direction of the observation point relative to the coordinate system of the aperture. The electric field phasor of the wave incident upon point Q is denoted $\widetilde{E}(R, \theta, \phi)$. Kirchhoff's scalar diffraction theory provides the following relationship between the radiated field $\widetilde{E}(R, \theta, \phi)$ and the aperture illumination $\widetilde{E}_a(x_a, y_a)$:

$$\widetilde{E}(R, \theta, \phi) = \frac{j}{\lambda} \left(\frac{e^{-jkR}}{R} \right) \widetilde{h}(\theta, \phi), \quad (9.80)$$

where

$$\widetilde{h}(\theta, \phi) = \iint_{-\infty}^{\infty} \widetilde{E}_a(x_a, y_a)$$
$$\cdot \exp\left[jk \sin\theta (x_a \cos\phi + y_a \sin\phi) \right] dx_a \, dy_a.$$
$$(9.81)$$

We shall refer to $\widetilde{h}(\theta, \phi)$ as the *form factor* of $\widetilde{E}(R, \theta, \phi)$. Its integral is written with infinite limits, with the understanding that $\widetilde{E}_a(x_a, y_a)$ is identically zero outside the aperture. The spherical propagation factor (e^{-jkR}/R) accounts for wave propagation between the center of the aperture and the observation point, and $\widetilde{h}(\theta, \phi)$ represents an integration of the exciting field $\widetilde{E}_a(x_a, y_a)$ over the extent of the aperture, taking into account [through the exponential function in Eq. (9.81)] the approximate deviation in distance between R and s, where s is the distance to any point (x_a, y_a) in the aperture plane [see Fig. 9-22]. *The polarization direction of the radiated field $\widetilde{E}(R, \theta, \phi)$ is the same as that of the aperture field*

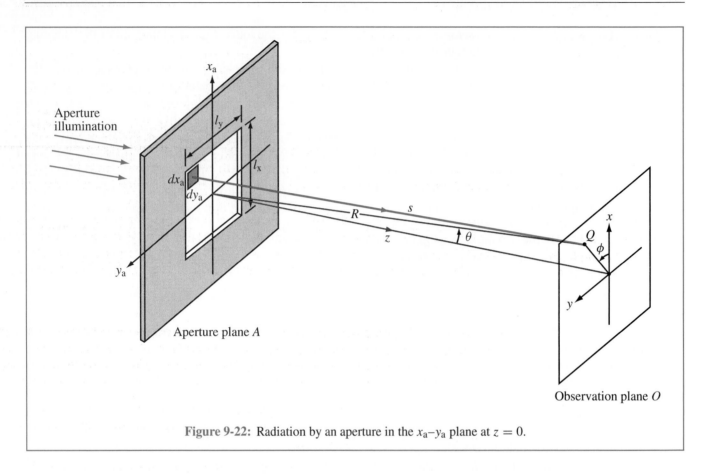

Figure 9-22: Radiation by an aperture in the x_a–y_a plane at $z = 0$.

$\widetilde{E}_a(x_a, y_a)$, and the power density of the radiated wave is given by

$$S(R, \theta, \phi) = \frac{|\widetilde{E}(R, \theta, \phi)|^2}{2\eta_0} = \frac{|\widetilde{h}(\theta, \phi)|^2}{2\eta_0\lambda^2 R^2} . \quad (9.82)$$

9-8 Rectangular Aperture with Uniform Aperture Distribution

By way of illustrating the scalar diffraction technique, let us consider a rectangular aperture of height l_x and width l_y, both at least a few wavelengths long. The

aperture is excited by a uniform field distribution given by

$$\widetilde{E}_a(x_a, y_a) = \begin{cases} E_0, & \text{for } -l_x/2 \leq x_a \leq l_x/2 \\ & \text{and } -l_y/2 \leq y_a \leq l_y/2, \\ 0, & \text{otherwise.} \end{cases}$$
$$(9.83)$$

To keep the mathematics simple, let us confine our examination to the radiation pattern at a fixed range R in the x–z plane, corresponding to $\phi = 0$. In this case, Eq. (9.81) simplifies to

$$\widetilde{h}(\theta) = \int_{y_a=-l_y/2}^{l_y/2} \int_{x_a=-l_x/2}^{l_x/2} E_0 \exp[jkx_a \sin \theta] \, dx_a \, dy_a.$$
$$(9.84)$$

In preparation for performing the integration in Eq. (9.84), we introduce the intermediate variable u defined as

$$u = k \sin \theta = \frac{2\pi \sin \theta}{\lambda}, \qquad (9.85)$$

in which case Eq. (9.84) becomes

$$\tilde{h}(\theta) = E_0 \int_{-l_x/2}^{l_x/2} e^{jux_a} \, dx_a \cdot \int_{-l_y/2}^{l_y/2} dy_a$$

$$= E_0 \left[\frac{e^{jul_x/2} - e^{-jul_x/2}}{ju} \right] \cdot l_y$$

$$= \frac{2E_0 l_y}{u} \left[\frac{e^{jul_x/2} - e^{-jul_x/2}}{2j} \right]$$

$$= \frac{2E_0 l_y}{u} \sin(ul_x/2). \qquad (9.86)$$

Upon replacing u with its defining expression, we have

$$\tilde{h}(\theta) = \frac{2E_0 l_y}{\left(\dfrac{2\pi}{\lambda} \sin \theta \right)} \sin(\pi l_x \sin \theta/\lambda)$$

$$= E_0 l_x l_y \frac{\sin(\pi l_x \sin \theta/\lambda)}{\pi l_x \sin \theta/\lambda}$$

$$= E_0 A_{\mathrm{p}} \operatorname{sinc}(\pi l_x \sin \theta/\lambda), \qquad (9.87)$$

where $A_{\mathrm{p}} = l_x l_y$ is the physical area of the aperture and where we have used the standard definition of the sinc function, which, for any argument t, is defined as

$$\operatorname{sinc} t = \frac{\sin t}{t}. \qquad (9.88)$$

Using Eq. (9.82), we obtain the following expression for the power density of the radiated wave at the observation point:

$$\boxed{S(R, \theta) = S_0 \operatorname{sinc}^2(\pi l_x \sin \theta/\lambda) \quad (x\text{–}z \text{ plane}), \quad (9.89)}$$

where $S_0 = E_0^2 A_{\mathrm{p}}^2/(2\eta_0 \lambda^2 R^2)$. The sinc function is at a maximum when its argument is equal to zero and its value is equal to unity. This occurs when $\theta = 0$. Thus, at a fixed range R, $S_{\max} = S(\theta = 0) = S_0$. The normalized radiation intensity is then given by

$$F(\theta) = \frac{S(R, \theta)}{S_{\max}}$$

$$= \operatorname{sinc}^2(\pi l_x \sin \theta/\lambda)$$

$$= \operatorname{sinc}^2(\pi \gamma) \quad (x\text{–}z \text{ plane}), \qquad (9.90)$$

and it is plotted (on a decibel scale) in Fig. 9-23 as a function of the intermediate variable $\gamma = (l_x/\lambda) \sin \theta$. The pattern exhibits nulls at nonzero integer values of γ.

9-8.1 Beamwidth

The normalized radiation intensity $F(\theta)$ is symmetrical in the x–z plane, and its maximum is along the boresight direction ($\theta = 0$, in this case). Its half-power beamwidth $\beta_{xz} = \theta_2 - \theta_1$, where θ_1 and θ_2 are the values of θ at which $F(\theta, 0) = 0.5$ (or -3 dB on a decibel scale), as shown in Fig. 9-23. Since the pattern is symmetrical with respect to $\theta = 0$, $\theta_1 = -\theta_2$ and $\beta_{xz} = 2\theta_2$. The angle θ_2 can be obtained from a solution of

$$F(\theta_2) = \operatorname{sinc}^2(\pi l_x \sin \theta/\lambda) = 0.5. \qquad (9.91)$$

From tabulated values of the sinc function, it is found that Eq. (9.91) yields the result

$$\frac{\pi l_x}{\lambda} \sin \theta_2 = 1.39, \qquad (9.92)$$

or

$$\sin \theta_2 = 0.44 \frac{\lambda}{l_x}. \qquad (9.93)$$

Because $\lambda/l_x \ll 1$ (a fundamental condition of scalar diffraction theory is that the aperture dimensions be much

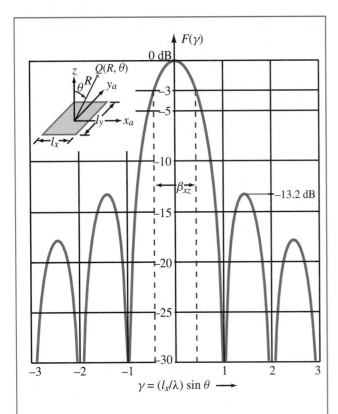

Figure 9-23: Normalized radiation pattern of a uniformly illuminated rectangular aperture in the x–z plane ($\phi = 0$).

larger than the wavelength λ), θ_2 is a small angle, in which case we can use the approximation $\sin \theta_2 \simeq \theta_2$. Hence,

$$\beta_{xz} = 2\theta_2 \simeq 2 \sin \theta_2 = 0.88 \frac{\lambda}{l_x} \quad \text{(rad).} \quad (9.94a)$$

A similar solution for the y–z plane ($\phi = \pi/2$) gives

$$\beta_{yz} = 0.88 \frac{\lambda}{l_y} \quad \text{(rad).} \quad (9.94b)$$

It is worth noting that the uniform aperture distribution ($\widetilde{E}_a = E_0$ across the aperture) gives a far-field

pattern with the narrowest possible beamwidth. The first sidelobe level is 13.2 dB below the peak value [see Fig. 9-23], which is equivalent to 4.8% of the peak value. If the intended application calls for a pattern with a lower sidelobe level (to avoid interference with signals from sources along directions outside the main beam of the antenna pattern), this can be accomplished by using a *tapered aperture distribution*, one that is a maximum at the center of the aperture and decreases toward the edges. A tapered distribution provides a pattern with lower side lobes, but the main lobe becomes wider. The steeper the taper, the lower are the side lobes and the wider is the main lobe. In general, the beamwidth in a given plane, say the x–z plane, is given by

$$\beta_{xz} = k_x \frac{\lambda}{l_x}, \quad (9.95)$$

where k_x is a constant related to the steepness of the taper. For a uniform distribution with no taper, $k_x = 0.88$, and for a highly tapered distribution, $k_x \simeq 2$. In the typical case, $k_x \simeq 1$.

To illustrate the relationship between the antenna dimensions and the corresponding beam shape, we show in Fig. 9-24 the radiation patterns of a circular reflector and a cylindrical reflector. The circular reflector has a circularly symmetric pattern, whereas the pattern of the cylindrical reflector has a narrow beam in the azimuth plane corresponding to its long dimension and a wide beam in the elevation plane corresponding to its narrow dimension. For a circularly symmetric antenna pattern, the beamwidth β is related to the diameter d by the approximate relation $\beta \simeq \lambda/d$.

9-8.2 Directivity and Effective Area

In Section 9-2.3, we derived an approximate expression [Eq. (9.26)] for the antenna directivity D in terms of the half-power beamwidths β_{xz} and β_{yz} for antennas

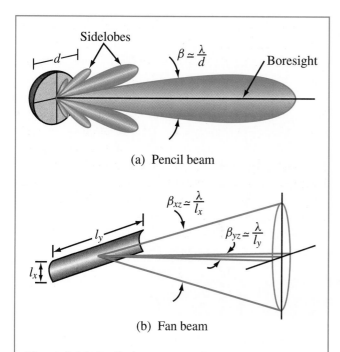

(a) Pencil beam

(b) Fan beam

Figure 9-24: Radiation patterns of (a) a circular reflector and (b) a cylindrical reflector (side lobes not shown).

Hence, *for aperture antennas, their effective apertures are approximately equal to their physical apertures;* **that** is, $A_e \simeq A_p$.

EXERCISE 9.11 Verify that Eq. (9.92) is a solution of Eq. (9.91) by calculating $\mathrm{sinc}^2 t$ for $t = 1.39$.

EXERCISE 9.12 With its boresight direction along z, a square aperture was observed to have half-power beamwidths of 3° in both the x–z and y–z planes. Determine its directivity in decibels.

Ans. $D = 4{,}583.66 = 36.61$ dB.

EXERCISE 9.13 What condition must be satisfied in order to use scalar diffraction to compute the field radiated by an aperture antenna? Can we use it to compute the directional pattern of the eye's pupil ($d \cong 0.2$ cm) in the visible part of the spectrum ($\lambda = 0.35$ to 0.7 μm)? What would the beamwidth of the eye's directional pattern be at $\lambda = 0.5$ μm if we assume that the pupil is uniformly illuminated?

Ans. $\beta = 0.88\,\lambda/d = 2.2 \times 10^{-4}$ rad $= 0.76'$ (arc minute, with $60' = 1°$).

---

characterized by a single major lobe whose boresight is along the z-direction:

$$D \simeq \frac{4\pi}{\beta_{xz}\beta_{yz}}. \qquad (9.96)$$

If we use the approximate relations $\beta_{xz} \simeq \lambda/l_x$ and $\beta_{yz} \simeq \lambda/l_y$, we have

$$D \simeq \frac{4\pi l_x l_y}{\lambda^2} = \frac{4\pi A_p}{\lambda^2}. \qquad (9.97)$$

For any antenna, its directivity is related to its effective area A_e by Eq. (9.64):

$$D = \frac{4\pi A_e}{\lambda^2}. \qquad (9.98)$$

9-9 Antenna Arrays

AM broadcast services operate in the 535- to 1605-kHz band. The antennas they use are basically vertical dipoles mounted along tall towers. The antennas range in height from $\lambda/6$ to $5\lambda/8$, depending on the operating characteristics desired and other considerations. Their physical heights vary from 46 m (150 ft) to 274 m (900 ft); the wavelength at 1 MHz, approximately in the middle of the AM band, is 300 m. Because the field radiated by a single dipole is uniform in the horizontal plane (as discussed in Sections 9-1 and 9-3), it is necessary to use more than one antenna tower to direct the horizontal antenna pattern along directions of interest (such as a city)

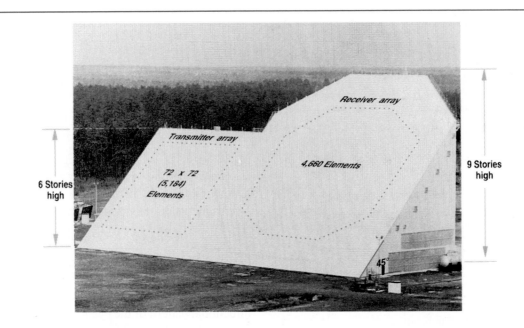

Figure 9-25: The AN/FPS-85 Phased Array Radar Facility in the Florida panhandle, near the city of Freeport. A several-mile no-fly zone surrounds the radar installation as a safety concern for electroexplosive devices, such as ejection seats and munitions, carried on military aircraft.

and to minimize it along directions of low population density or a direction corresponding to an area serviced by another station operating at the same frequency (to avoid undesirable interference effects). When two or more antennas are used together, the combination is called an *antenna array*.

The AM broadcast antenna array is only one application example of antenna arrays; they are used extensively in numerous communication systems and radar applications as well. Antenna arrays provide the antenna designer the flexibility to obtain high directivity, narrow beams, low side lobes, steerable beams, and shaped antenna patterns. Figure 9-25 shows a photograph of a very large radar system consisting of a transmitter array

composed of 5,184 individual dipole antenna elements and a receiver array composed of 4,660 elements. The radar system, part of the Space Surveillance Network operated by the U.S. Air Force, operates at 442 MHz and transmits a combined peak power of 30 MW!

Although an array need not consist of similar radiating elements, most arrays usually use identical elements, such as dipoles, slots, horn antennas, or parabolic dishes, excited by the same type of current or field distribution. The antenna elements comprising an array may be arranged in various configurations, but the most common are the linear one-dimensional configuration, wherein the elements are arranged along a straight line, and the two-dimensional lattice configuration, wherein the elements form a rectangular grid. The desired shape of the far-field radiation pattern of the array can be synthesized by controlling the relative amplitudes of the array elements. Also, through the use of electronically controlled solid-state phase shifters, the beam direction of the antenna array can be steered electronically by controlling the relative phases of the array elements. This flexibility of the array antenna has led to numerous applications, including *electronic steering* and *multiple-beam generation*.

The purpose of this and the next two sections is to introduce the reader to the basic principles of array theory and to the design techniques used in shaping the antenna pattern and steering the main lobe. The presentation will be confined to the one-dimensional linear array with equal spacing between adjacent elements.

A linear array of N identical radiators is arranged along the z-axis as shown in Fig. 9-26. The radiators are fed by a common oscillator through a branching network. In each branch, an attenuator (or amplifier) and phase shifter are inserted in series to control the amplitude and phase of the signal feeding the antenna element in that branch.

In the far-field region of any radiating element, the phasor electric-field intensity $\widetilde{E}_e(R, \theta, \phi)$ may be expressed as a product of two functions, the spherical propagation factor e^{-jkR}/R, which accounts for the dependence on the range R, and $\widetilde{f}_e(\theta, \phi)$, which accounts for the directional dependence of the element's electric field. Thus, for an isolated element, the radiated field is

$$\widetilde{E}_e(R, \theta, \phi) = \frac{e^{-jkR}}{R}\, \widetilde{f}_e(\theta, \phi), \qquad (9.99)$$

and the corresponding power density S_e is

$$S_e(R, \theta, \phi) = \frac{1}{2\eta_0}|\widetilde{E}_e(R, \theta, \phi)|^2$$

$$= \frac{1}{2\eta_0 R^2}|\widetilde{f}_e(\theta, \phi)|^2. \qquad (9.100)$$

With regard to the elements in the array shown in Fig. 9-26(b), the far-zone field due to element i at range R_i

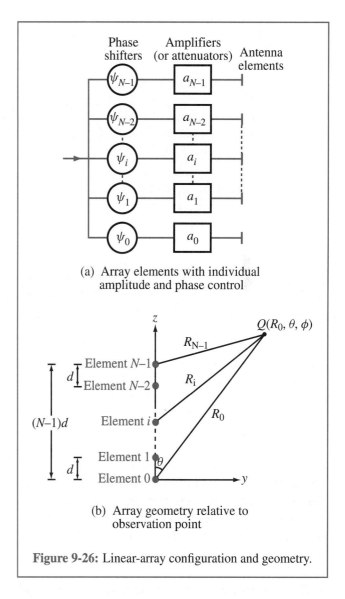

(a) Array elements with individual amplitude and phase control

(b) Array geometry relative to observation point

Figure 9-26: Linear-array configuration and geometry.

from the observation point Q may then be written in the form

$$\widetilde{E}_i(R_i, \theta, \phi) = A_i \frac{e^{-jkR_i}}{R_i}\, \widetilde{f}_e(\theta, \phi), \qquad (9.101)$$

where $A_i = a_i e^{j\psi_i}$ is a *complex feeding coefficient* representing the amplitude a_i and phase ψ_i of the excitation giving rise to $\widetilde{E}_i$, relative to a reference excitation. In practice, the excitation of one of the elements is used as reference. Note that R_i and A_i may be different for the different elements in the array, but $\widetilde{f}_e(\theta, \phi)$ is the same for all the elements since they are assumed to be identical and hence exhibit identical directional patterns.

The total field at the observation point $Q(R_0, \theta, \phi)$ is the sum of the fields due to the N elements:

$$\widetilde{E}(R_0, \theta, \phi) = \sum_{i=0}^{N-1} \widetilde{E}_i(R_i, \theta, \phi)$$

$$= \left[\sum_{i=0}^{N-1} A_i \frac{e^{-jkR_i}}{R_i} \right] \widetilde{f}_e(\theta, \phi), \quad (9.102)$$

where R_0 denotes the range of Q from the center of the coordinate system, chosen to be at the location of the zeroth element. To satisfy the far-field condition given by Eq. (9.79) for an array of length $l = (N-1)d$, where d is the interelement spacing, the range R_0 should be sufficiently large that

$$R_0 \geq \frac{2l^2}{\lambda} = \frac{2(N-1)^2 d^2}{\lambda}. \quad (9.103)$$

This condition allows us to ignore differences in the distances from Q to the individual elements as far as the magnitude of the radiated field is concerned. Thus, we can set $R_i = R_0$ in the denominator in Eq. (9.102) for all i. With regard to the phase part of the propagation factor, we can use the parallel-ray approximation given by

$$R_i \simeq R_0 - z_i \cos\theta = R_0 - id\cos\theta, \quad (9.104)$$

where $z_i = id$ is the distance between the ith element and the zeroth element [Fig. 9-27]. Employing these two approximations in Eq. (9.102) leads to

$$\widetilde{E}(R_0, \theta, \phi) = \widetilde{f}_e(\theta, \phi) \left(\frac{e^{-jkR_0}}{R_0} \right)$$

$$\times \left[\sum_{i=0}^{N-1} A_i e^{jikd\cos\theta} \right], \quad (9.105)$$

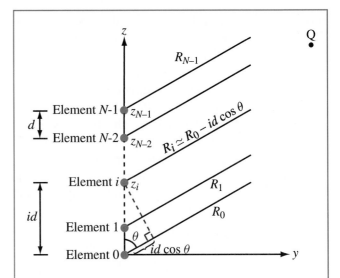

Figure 9-27: The rays between the elements and a faraway observation point are approximately parallel lines. Hence, the distance $R_i \simeq R_0 - id\cos\theta$.

and the corresponding array-antenna power density is given by

$$S(R_0, \theta, \phi) = \frac{1}{2\eta_0} |\widetilde{E}(R_0, \theta, \phi)|^2$$

$$= \frac{1}{2\eta_0 R_0^2} |\widetilde{f}_e(\theta, \phi)|^2 \left| \sum_{i=0}^{N-1} A_i e^{jikd\cos\theta} \right|^2$$

$$= S_e(R_0, \theta, \phi) \left| \sum_{i=0}^{N-1} A_i e^{jikd\cos\theta} \right|^2, \quad (9.106)$$

where use has been made of Eq. (9.100). This expression is a product of two factors. The first factor, $S_e(R_0, \theta, \phi)$, is the power density of the energy radiated by an individual element, and the second, usually called the *array factor*, is a function of the positions of the individual elements and their feeding coefficients, but not a function of the specific type of radiators used. *The array factor represents the far-field radiation intensity of the*

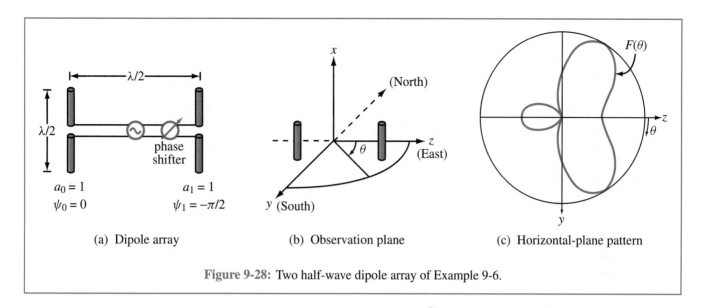

Figure 9-28: Two half-wave dipole array of Example 9-6.

(a) Dipole array   (b) Observation plane   (c) Horizontal-plane pattern

N elements, had the elements been isotropic radiators. Denoting the array factor by

$$F_a(\theta) = \left| \sum_{i=0}^{N-1} A_i e^{jikd\cos\theta} \right|^2, \qquad (9.107)$$

the power density of the antenna array is then written as

$$S(R_0, \theta, \phi) = S_e(R_0, \theta, \phi)\, F_a(\theta). \quad (9.108)$$

This equation is called the *pattern multiplication principle*. It allows us to find the far-field power density of the antenna array by first computing the far-field power pattern with the array elements replaced with isotropic radiators, which yields the array factor $F_a(\theta)$, and then multiplying the result by $S_e(R_0, \theta, \phi)$, the power density for a single element (which should be the same for all the elements).

The feeding coefficient A_i is, in general, a complex amplitude consisting of an amplitude factor a_i and a phase factor ψ_i:

$$A_i = a_i e^{j\psi_i}. \qquad (9.109)$$

Insertion of Eq. (9.109) into Eq. (9.107) leads to

$$F_a(\theta) = \left| \sum_{i=0}^{N-1} a_i e^{j\psi_i} e^{jikd\cos\theta} \right|^2. \quad (9.110)$$

The array factor is governed by two input functions: the *array amplitude distribution* given by the a_i's and the *array phase distribution* given by the ψ_i's. The amplitude distribution serves to control the shape of the array radiation pattern, while the phase distribution can be used to steer its direction.

Example 9-6   Array of Two Vertical Dipoles

An AM radio station uses two vertically oriented half-wave dipoles separated by a distance of $\lambda/2$, as shown in Fig. 9-28(a). The vector from the location of the first dipole to the location of the second dipole points toward the east. The two dipoles are fed with equal-amplitude excitations, and the dipole farther east is excited with a phase shift of $-\pi/2$ relative to the other one. Find and plot the antenna pattern of the antenna array in the horizontal plane.

Solution: The array factor given by Eq. (9.110) was derived for radiators arranged along the z-axis. To keep the coordinate system the same, we choose the easterly direction to be the z-axis as shown in Fig. 9-28(b), and we place the first dipole at $z = -\lambda/4$ and the second at $z = \lambda/4$. A half-wave dipole radiates uniformly in the plane perpendicular to its axis, which in this case is the horizontal plane. Hence, $S_e = S_0$ for all angles θ in Fig. 9-28(b), where S_0 is the maximum value of the power density radiated by each dipole individually. Consequently, the power density radiated by the two-dipole array is

$$S(R, \theta) = S_0\, F_a(\theta).$$

For two elements separated by $d = \lambda/2$ and excited with equal amplitudes ($a_0 = a_1 = 1$) and with phase angles $\psi_0 = 0$ and $\psi_1 = -\pi/2$, Eq. (9.110) gives

$$F_a(\theta) = \left| \sum_{i=0}^{1} a_i e^{j\psi_i} e^{jikd\cos\theta} \right|^2$$

$$= |1 + e^{-j\pi/2} e^{j(2\pi/\lambda)(\lambda/2)\cos\theta}|^2$$

$$= \left| 1 + e^{j(\pi\cos\theta - \pi/2)} \right|^2.$$

A function of the form $|1 + e^{jx}|^2$ can be evaluated by factoring out $e^{jx/2}$ from both terms:

$$|1 + e^{jx}|^2 = |e^{jx/2}(e^{-jx/2} + e^{jx/2})|^2$$

$$= |e^{jx/2}|^2\, |e^{-jx/2} + e^{jx/2}|^2$$

$$= |e^{jx/2}|^2 \left| 2\frac{[e^{-jx/2} + e^{jx/2}]}{2} \right|^2.$$

The absolute value of $e^{jx/2}$ is equal to 1, and we recognize the function inside the square bracket as $\cos(x/2)$. Hence,

$$|1 + e^{jx}|^2 = 4\cos^2\left(\frac{x}{2}\right).$$

Applying this result to the expression for $F_a(\theta)$, we have

$$F_a(\theta) = 4\cos^2\left(\frac{\pi}{2}\cos\theta - \frac{\pi}{4}\right).$$

The power density radiated by the array is then

$$S(R, \theta) = S_0 F_a(\theta) = 4S_0 \cos^2\left(\frac{\pi}{2}\cos\theta - \frac{\pi}{4}\right).$$

This function has a maximum value $S_{max} = 4S_0$, and it occurs when the argument of the cosine function is equal to zero. Thus,

$$\frac{\pi}{2}\cos\theta - \frac{\pi}{4} = 0,$$

which leads to the solution: $\theta = 60°$. Upon normalizing $S(R, \theta)$ by its maximum value, we obtain the normalized radiation intensity given by

$$F(\theta) = \frac{S(R, \theta)}{S_{max}} = \cos^2\left(\frac{\pi}{2}\cos\theta - \frac{\pi}{4}\right).$$

The pattern of $F(\theta)$ is shown in Fig. 9-28(c). ∎

Example 9-7   Pattern Synthesis

In Example 9-6, we were given the array parameters a_0, a_1, ψ_0, ψ_1, and d, and we were then asked to determine the pattern of the two-element dipole array. We will now consider the reverse process; we will be given specifications on the desired pattern, and we are then asked to specify the array parameters to meet those specifications.

Given two vertical dipoles, as depicted in Fig. 9-28(b), specify the array parameters such that the array exhibits maximum radiation toward the east and no radiation toward the north or south.

Solution: From Example 9-6, we established that because each dipole radiates equally along all directions in the y–z plane, the radiation pattern of the two-dipole array in that plane is governed solely by the array factor $F_a(\theta)$. The shape of the pattern of the array factor depends on three parameters: the amplitude ratio a_1/a_0, the phase difference $\psi_1 - \psi_0$, and the spacing d

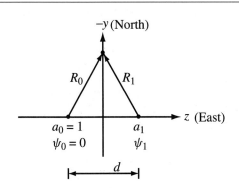

−y (North)

R_0 R_1

z (East)

$a_0 = 1$ a_1
$\psi_0 = 0$ ψ_1

d

(a) Array arrangement

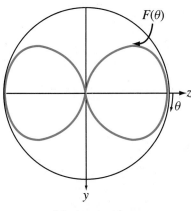

$F(\theta)$

z

θ

y

(b) Array pattern

Figure 9-29: (a) Two vertical dipoles separated by a distance d along the z-axis; (b) normalized array pattern in the y–z plane for $a_0 = a_1 = 1$, $\psi_1 = \psi_0 = -\pi$, and $d = \lambda/2$.

[Fig. 9-29(a)]. For convenience, we choose $a_0 = 1$ and $\psi_0 = 0$. Accordingly, Eq. (9.107) gives

$$F_a(\theta) = \left| \sum_{i=0}^{1} a_i e^{j\psi_i} e^{jikd\cos\theta} \right|^2$$

$$= |1 + a_1 e^{j\psi_1} e^{j(2\pi d/\lambda)\cos\theta}|^2.$$

Let us now consider the given specification requiring that F_a be equal to zero when $\theta = 90°$ (north and south directions in Fig. 9-29(a)). For any observation point on the y-axis, the ranges R_0 and R_1 shown in Fig. 9-29(a) are equal, which means that the propagation phases associated with the time travel of the waves radiated by the two dipoles to that point are identical. Hence, to satisfy the stated condition, we need to choose $a_1 = a_0$ and $\psi_1 = \pm\pi$. With these choices, the signals radiated by the two dipoles will have equal amplitudes and opposite phases, thereby interfering destructively. This conclusion can be ascertained by evaluating the array factor at $\theta = 90°$, with $a_0 = a_1 = 1$ and $\psi_1 = \pm\pi$:

$$F_a(\theta = 90°) = |1 + 1e^{\pm j\pi}|^2 = |1 - 1| = 0.$$

The two values of ψ_1, π and $-\pi$, lead to the same solution for the spacing d when we consider the specification that the array radiation pattern should have a maximum toward the east, corresponding to $\theta = 0°$. Let us choose $\psi_1 = -\pi$ and examine the array factor at $\theta = 0°$:

$$F_a(\theta = 0) = |1 + e^{-j\pi} e^{j2\pi d/\lambda}|^2$$

$$= |1 + e^{j(-\pi + 2\pi d/\lambda)}|^2.$$

For $F_a(\theta = 0)$ to be a maximum, we require the phase angle of the second term to be zero or a multiple of 2π:

$$-\pi + \frac{2\pi d}{\lambda} = 2n\pi,$$

or

$$d = (2n + 1)\frac{\lambda}{2}, \qquad n = 0, 1, 2, \ldots$$

In summary, the two-dipole array will meet the given specifications if $a_0 = a_1$, $\psi_1 - \psi_0 = -\pi$, and $d = (2n + 1)\lambda/2$.

For $d = \lambda/2$, the array factor at any angle θ is given by

$$F_a(\theta) = |1 + e^{-j\pi} e^{j\pi \cos\theta}|^2$$
$$= |1 - e^{j\pi \cos\theta}|^2$$
$$= \left| 2je^{-j(\pi/2)\cos\theta} \left[\frac{e^{j(\pi/2)\cos\theta} - e^{-j(\pi/2)\cos\theta}}{2j} \right] \right|^2$$
$$= 4\sin^2\left(\frac{\pi}{2}\cos\theta\right).$$

The array factor has a maximum value of 4, which is the maximum level attainable from a two-element array with unit amplitudes. The directions along which $F_a(\theta)$ is a maximum are those corresponding to $\theta = 0$ (east) and $\theta = 180°$ (west), as shown in Fig. 9-29(b). ∎

EXERCISE 9.14 Derive an expression for the array factor of a two-element array excited in phase with $a_0 = 1$ and $a_1 = 3$. The elements are positioned along the z-axis and are separated by $\lambda/2$.

Ans. $F_a(\theta) = [10 + 6\cos(\pi\cos\theta)]$.

EXERCISE 9.15 An equally spaced N-element array arranged along the z-axis is fed with equal amplitudes and phases; that is, $A_i = 1$ for $i = 0, 1, \ldots, N - 1$. What is the magnitude of the array factor in the broadside direction?

Ans. $F_a(\theta = 90°) = N^2$.

9-10 N-Element Array with Uniform Phase Distribution

We now consider an array of N elements with equal spacing d and equal-phase excitations; that is, $\psi_i = \psi_0$ for $i = 1, 2, \ldots, N - 1$. Such an array of in-phase elements is sometimes referred to as a *broadside array* because the main beam of the radiation pattern of its array

factor is always in the direction broadside to the array axis. From Eq. (9.110), its array factor is given by

$$F_a(\theta) = \left| e^{j\psi_0} \sum_{i=0}^{N-1} a_i e^{jikd\cos\theta} \right|^2$$
$$= |e^{j\psi_0}|^2 \left| \sum_{i=0}^{N-1} a_i e^{jikd\cos\theta} \right|^2$$
$$= \left| \sum_{i=0}^{N-1} a_i e^{jikd\cos\theta} \right|^2. \qquad (9.111)$$

The phase difference between the fields radiated by adjacent elements is

$$\gamma = kd\cos\theta = \frac{2\pi d}{\lambda}\cos\theta. \qquad (9.112)$$

In terms of γ, Eq. (9.111) takes the compact form

$$\boxed{F_a(\gamma) = \left| \sum_{i=0}^{N-1} a_i e^{ji\gamma} \right|^2 \qquad \text{(uniform phase).} \qquad (9.113)}$$

For a uniform amplitude distribution with $a_i = 1$ for $i = 0, 1, \ldots, N - 1$, Eq. (9.113) becomes

$$F_a(\gamma) = |1 + e^{j\gamma} + e^{j2\gamma} + \cdots + e^{j(N-1)\gamma}|^2. \qquad (9.114)$$

This geometric series can be rewritten in a more compact form by applying the following procedure. First, we define

$$F_a(\gamma) = |f_a(\gamma)|^2, \qquad (9.115)$$

with

$$f_a(\gamma) = [1 + e^{j\gamma} + e^{j2\gamma} + \cdots + e^{j(N-1)\gamma}]. \qquad (9.116)$$

Next, we multiply $f_a(\gamma)$ by $e^{j\gamma}$ to obtain

$$f_a(\gamma)e^{j\gamma} = (e^{j\gamma} + e^{j2\gamma} + \cdots + e^{jN\gamma}). \qquad (9.117)$$

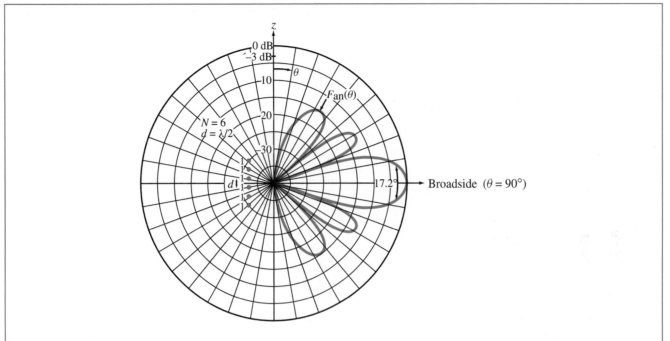

Figure 9-30: Normalized array pattern of a uniformly excited six-element array with interelement spacing $d = \lambda/2$.

Subtracting Eq. (9.117) from Eq. (9.116) gives

$$f_a(\gamma)\,(1 - e^{j\gamma}) = 1 - e^{jN\gamma}, \qquad (9.118)$$

which, in turn, gives

$$
\begin{aligned}
f_a(\gamma) &= \frac{1 - e^{jN\gamma}}{1 - e^{j\gamma}} \\
&= \frac{e^{jN\gamma/2}}{e^{j\gamma/2}}\,\frac{(e^{-jN\gamma/2} - e^{jN\gamma/2})}{(e^{-j\gamma/2} - e^{j\gamma/2})} \\
&= e^{j(N-1)\gamma/2}\,\frac{\sin(N\gamma/2)}{\sin(\gamma/2)}. \qquad (9.119)
\end{aligned}
$$

After multiplying $f_a(\gamma)$ by its complex conjugate, we obtain the result:

$$F_a(\gamma) = \frac{\sin^2(N\gamma/2)}{\sin^2(\gamma/2)}$$

(uniform amplitude and phase).   (9.120)

The maximum value of $F_a(\gamma)$ can be shown to occur at $\gamma = 0$ (or $\theta = \pi/2$) and is equal to N^2, which is easy to verify by evaluating Eq. (9.114) for $\gamma = 0$. Hence, the normalized array factor is given by

$$
\begin{aligned}
F_{an}(\gamma) &= \frac{F_a(\gamma)}{F_{a,\max}} = \frac{\sin^2(N\gamma/2)}{N^2 \sin^2(\gamma/2)} \\
&= \frac{\sin^2\!\left[\dfrac{N\pi d}{\lambda}\cos\theta\right]}{N^2 \sin^2\!\left[\dfrac{\pi d}{\lambda}\cos\theta\right]}. \quad (9.121)
\end{aligned}
$$

A polar plot of $F_{an}(\theta)$ with $N = 6$ is shown in Fig. 9-30 for $d = \lambda/2$. The reader is reminded that this is a plot of the radiation pattern of the array factor alone; the pattern for the antenna array is equal to the product of this pattern and that of a single element, as discussed earlier in connection with the pattern multiplication principle.

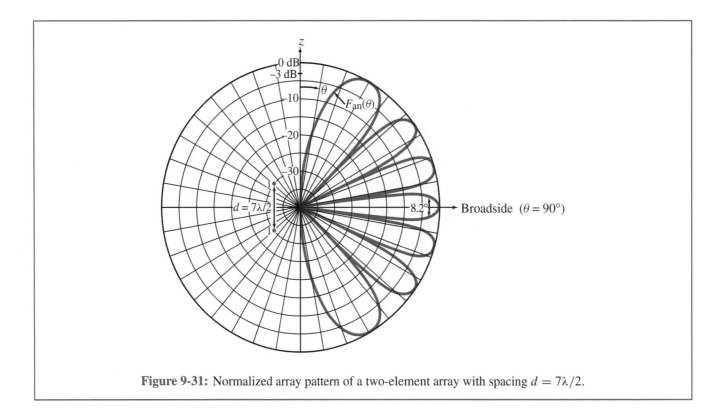

Figure 9-31: Normalized array pattern of a two-element array with spacing $d = 7\lambda/2$.

Example 9-8   Multiple-Beam Array

Obtain an expression for the array factor of a two-element array with equal excitation and a separation $d = 7\lambda/2$, and then plot the array pattern.

Solution: The array factor for a two-element array ($N = 2$) with equal excitation ($a_0 = a_1 = 1$) is given by

$$F_a(\gamma) = \left| \sum_{i=0}^{1} a_i e^{ji\gamma} \right|^2$$
$$= |1 + e^{j\gamma}|^2,$$
$$= |e^{j\gamma/2}(e^{-j\gamma/2} + e^{j\gamma/2})|^2$$
$$= |e^{j\gamma/2}|^2 |e^{-j\gamma/2} + e^{j\gamma/2}|^2$$
$$= 4\cos^2(\gamma/2),$$

where $\gamma = (2\pi d/\lambda)\cos\theta$. The normalized array pattern, shown in Fig. 9-31, consists of seven beams, all with the same peak value, but not the same angular width. The number of beams in the angular range between $\theta = 0$ and $\theta = \pi$ is equal to the separation between the array elements, d, measured in units of $\lambda/2$. ■

9-11 Electronic Scanning of Arrays

The discussion in the preceding section was concerned with uniform-phase arrays, for which the phases of the feeding coefficients, ψ_0 to ψ_{N-1}, are all equal. In this section, we examine the use of phase delay between adjacent elements as a tool for *electronically steering* the direction of the array-antenna beam from broadside at $\theta = 90°$ to any desired angle θ_0. In addition to

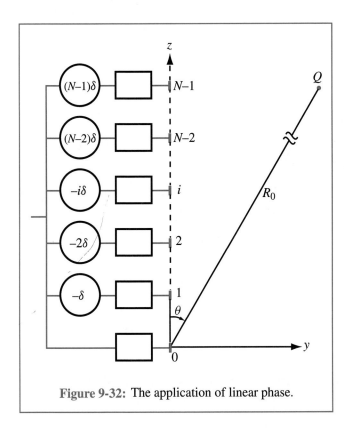

Figure 9-32: The application of linear phase.

eliminating the need to mechanically steer an antenna to change its beam's direction, electronic steering allows beam scanning at very fast rates.

Electronic steering is achieved by applying linearly progressive phase delays from element to element across the array, as shown in Fig. 9-32. Relative to the phase of the zeroth element, the phase of the ith element is

$$\psi_i = -i\delta, \qquad (9.122)$$

where δ is the *incremental phase delay* between adjacent elements. Use of Eq. (9.122) in Eq. (9.110) leads to

$$F_a(\theta) = \left| \sum_{i=0}^{N-1} a_i e^{-ji\delta} e^{jikd\cos\theta} \right|^2$$

$$= \left| \sum_{i=0}^{N-1} a_i e^{ji(kd\cos\theta - \delta)} \right|^2$$

$$= \left| \sum_{i=0}^{N-1} a_i e^{ji\gamma'} \right|^2 \triangleq F_a(\gamma'), \qquad (9.123)$$

where

$$\gamma' = kd\cos\theta - \delta. \qquad (9.124)$$

For reasons that will become clear later, let us define the phase shift δ in terms of an angle θ_0, which we shall call the *scan angle*, as follows:

$$\delta = kd\cos\theta_0. \qquad (9.125)$$

In this case, γ' becomes

$$\boxed{\gamma' = kd(\cos\theta - \cos\theta_0). \qquad (9.126)}$$

The array factor given by Eq. (9.123) has the same functional form as the array factor developed earlier for the case of the uniform-phase array [see Eq. (9.113)], except that γ is replaced with γ'. Hence, *for any amplitude distribution across the array, the array factor of an array excited by a linear-phase distribution may be obtained from the expression developed for the array assuming a uniform-phase distribution simply by replacing γ with γ'.* For an amplitude distribution symmetrical with respect to the array center, the array factor $F_a(\gamma')$ is maximum when its argument $\gamma' = 0$. When the phase is uniform ($\delta = 0$), this condition corresponds to the direction $\theta = 90°$, which is why the uniform-phase case is called a broadside array. According to Eq. (9.126), in the more general case of a linearly phased array, $\gamma' = 0$ corresponds to $\theta = \theta_0$. Thus, by applying linear phase across the array, the array pattern is shifted along the $\cos\theta$-axis by an amount $\cos\theta_0$, and the direction of maximum radiation is *steered* from the broadside direction ($\theta = 90°$) to the direction $\theta = \theta_0$. To steer the beam all the way to the *end-fire* direction ($\theta = 0$), the incremental phase shift δ should be equal to kd radians.

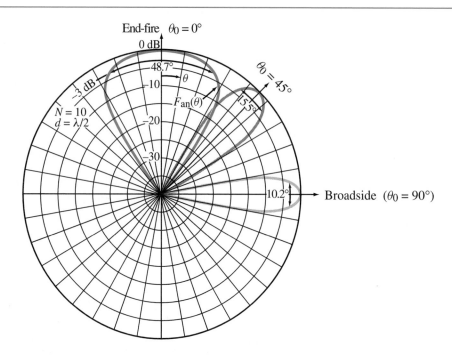

Figure 9-33: Normalized array pattern of a 10-element array with $\lambda/2$ spacing between adjacent elements. All elements are excited with equal amplitude. Through the application of linear phase across the array, the main beam can be steered from the broadside direction ($\theta_0 = 90°$) to any scan angle θ_0. Equiphase excitation corresponds to $\theta_0 = 90°$.

9-11.1 Uniform-Amplitude Excitation

To illustrate the process with an example, let us consider the case of the N-element array excited by a uniform-amplitude distribution. Its normalized array factor is given by Eq. (9.121). Upon replacing γ with γ', we have

$$F_{an}(\gamma') = \frac{\sin^2(N\gamma'/2)}{N^2 \sin^2(\gamma'/2)}, \qquad (9.127)$$

with γ' defined by Eq. (9.126). The expression given by Eq. (9.127) now characterizes the array factor of an array excited with equal amplitudes ($a_i = 1$ for $i = 0, 1, \ldots, N-1$) and with incremental phases ($\psi_i = -i\delta$).

For $N = 10$ and $d = \lambda/2$, plots of the main lobe of $F_{an}(\theta)$ are shown in Fig. 9-33 for $\theta_0 = 0°$, $45°$, and $90°$. We note that the half-power beamwidth increases as the array beam is steered from broadside to end fire.

9-11.2 Array Feeding

According to the foregoing discussion, to steer the antenna beam to an angle θ_0, two conditions must be satisfied: (1) the phase distribution must be linear across the array, and (2) the magnitude of the incremental phase delay δ must satisfy Eq. (9.125). The combination of these two conditions provides the necessary translation of the phase front from $\theta = 90°$ (broadside) to $\theta = \theta_0$. This can be accomplished

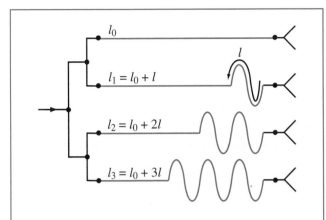

Figure 9-34: An example of a feeding arrangement for frequency-scanned arrays.

by controlling the excitation of each radiating element individually through the use of electronically controlled phase shifters. Alternatively, a technique known as *frequency scanning* can be used to provide control of the phases of all the elements simultaneously. Figure 9-34 shows an example of a simple feeding arrangement employed in frequency scanning arrays. A common feed point is connected to the radiating elements through transmission lines of varying lengths. Relative to the zeroth element, the path between the common feed point and a radiating element is longer by l for the first element, by $2l$ for the second, and by $3l$ for the third. Thus, the path length for the ith element is

$$l_i = il + l_0, \qquad (9.128)$$

where l_0 is the path length of the zeroth element. Wave propagation at a frequency f on a transmission line of length l_i is characterized by a phase factor $e^{-j\beta l_i}$, where $\beta = 2\pi f / u_p$ is the phase constant of the line and u_p is its propagation velocity. Hence, the *incremental* phase delay of the ith element, relative to the phase of the zeroth

element, is

$$\psi_i(f) = -\beta(l_i - l_0) = -\frac{2\pi}{u_p} f(l_i - l_0)$$

$$= -\frac{2\pi i}{u_p} fl. \qquad (9.129)$$

Suppose that at a given reference frequency f_0 we choose the incremental length l such that

$$l = \frac{n_0 u_p}{f_0}, \qquad (9.130)$$

where n_0 is a *specific* positive integer. In this case, the phase delay $\psi_1(f_0)$ becomes

$$\psi_1(f_0) = -2\pi \left(\frac{f_0 l}{u_p} \right) = -2 n_0 \pi \qquad (9.131)$$

and, similarly, $\psi_2(f_0) = -4 n_0 \pi$ and $\psi_3(f_0) = -6 n_0 \pi$. That is, at f_0 all the elements will have equal phase (within multiples of 2π) and the array radiates in the broadside direction. If f is changed to $f_0 + \Delta f$, the new phase shift of the first element relative to the zeroth element is

$$\psi_1(f_0 + \Delta f) = -\frac{2\pi}{u_p}(f_0 + \Delta f)l$$

$$= -\frac{2\pi f_0 l}{u_p} - \left(\frac{2\pi l}{u_p} \right) \Delta f$$

$$= -2 n_0 \pi - 2 n_0 \pi \left(\frac{\Delta f}{f_0} \right)$$

$$= -2 n_0 \pi - \delta, \qquad (9.132)$$

where use was made of Eq. (9.130) and δ is defined as

$$\delta = 2 n_0 \pi \left(\frac{\Delta f}{f_0} \right). \qquad (9.133)$$

Similarly, $\psi_2(f_0 + \Delta f) = 2\psi_1$ and $\psi_3(f_0 + \Delta f) = 3\psi_1$. Ignoring the factor of 2π and its multiples (since they exercise no influence on the relative phases of the radiated fields), we see that the incremental phase shifts are directly proportional to the fractional frequency deviation $(\Delta f / f_0)$. Thus, in an array with N elements, controlling Δf provides a direct control of δ, which in turn controls the scan angle θ_0 according to Eq. (9.125). Equating Eq. (9.125) to Eq. (9.133) and then solving for $\cos \theta_0$ leads to

$$\cos \theta_0 = \frac{2n_0\pi}{kd}\left(\frac{\Delta f}{f_0}\right). \qquad (9.134)$$

As f is changed from f_0 to $f_0 + \Delta f$, $k = 2\pi/\lambda = 2\pi f/c$ also changes with frequency. However, if $\Delta f / f_0$ is small, we may treat k as a constant equal to $2\pi f_0 / c$; the error in $\cos \theta_0$ resulting from the use of this approximation in Eq. (9.134) is on the order of $\Delta f / f_0$.

Example 9-9   Electronic Steering

Design a steerable six-element array with the following specifications:

1. All elements are excited with equal amplitudes.

2. At $f_0 = 10$ GHz, the array radiates in the broadside direction, and the interelement spacing $d = \lambda_0/2$, where $\lambda_0 = c/f_0 = 3$ cm.

3. The array pattern is to be electronically steerable in the elevation plane over the angular range extending between $\theta_0 = 30°$ and $\theta_0 = 150°$.

4. The antenna array is fed by a voltage-controlled oscillator whose frequency can be varied over the range from 9.5 to 10.5 GHz.

5. The array uses a feeding arrangement of the type shown in Fig. 9-34, and the transmission lines have a phase velocity $u_p = 0.8c$.

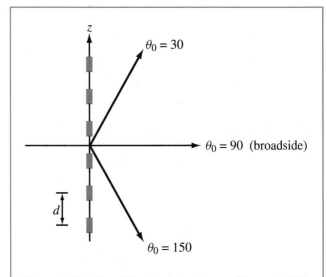

Figure 9-35: Steerable six-element array (Example 9-9).

Solution: The array is to be steerable from $\theta_0 = 30°$ to $\theta_0 = 150°$ (Fig. 9-35). For $\theta_0 = 30°$ and $kd = (2\pi/\lambda_0)(\lambda_0/2) = \pi$, Eq. (9.134) gives

$$\frac{1}{2} = 2n_0\left(\frac{\Delta f}{f_0}\right). \qquad (9.135)$$

We are given that $f_0 = 10$ GHz and the oscillator frequency can be varied between $(f_0 - 0.5$ GHz$)$ and $(f_0 + 0.5$ GHz$)$. Thus, $\Delta f_{\max} = 0.5$ GHz. To satisfy Eq. (9.135), we need to choose n_0 such that Δf is as close as possible to, but not larger than, $\Delta f_{\max}$. Solving Eq. (9.135) for n_0 with $\Delta f = \Delta f_{\max}$ gives

$$n_0 = \frac{f_0}{4\,\Delta f_{\max}} = \frac{10 \text{ GHz}}{4 \times 0.5 \text{ GHz}} = 5.$$

Since n_0 is an integer, we do not need to modify its value; otherwise, we would have had to round it upward to the next whole-integer value.

Application of Eq. (9.130) specifies the magnitude of the incremental length l:

$$l = \frac{n_0 u_p}{f_0} = \frac{5 \times 0.8 \times 3 \times 10^8}{10^{10}} = 0.12 \text{ m} = 12 \text{ cm}.$$

In summary, with $N = 6$ and $kd = \pi$, Eq. (9.127) gives the following expression for the normalized array pattern:

$$F_{\mathrm{an}}(\gamma') = \frac{\sin^2(3\gamma')}{36\sin^2(\gamma'/2)},$$

where, according to Eq. (9.126),

$$\gamma' = kd(\cos\theta - \cos\theta_0) = \pi(\cos\theta - \cos\theta_0),$$

and, from Eq. (9.134),

$$\cos\theta_0 = \frac{2n_0\pi}{kd}\left(\frac{\Delta f}{f_0}\right) = 10\left(\frac{f - 10\ \mathrm{GHz}}{10\ \mathrm{GHz}}\right).$$
(9.136)

The shape of the array pattern is similar to that shown in Fig. 9-30, and its main-beam direction is along $\theta = \theta_0$. For $f = f_0 = 10$ GHz, $\theta_0 = 90°$ (broadside direction); for $f = 10.5$ GHz, $\theta_0 = 30°$; and for $f = 9.5$ GHz, $\theta_0 = 150°$. For any other value of θ_0 between $30°$ and $150°$, Eq. (9.136) provides the means for calculating the required value of the oscillator frequency f. ∎

REVIEW QUESTIONS

Q9.13 Why are antenna arrays useful? Give examples of typical applications.

Q9.14 Explain how the pattern multiplication principle is used to compute the radiation pattern of an antenna array.

Q9.15 For a linear array, what roles do the array amplitudes and phases play?

Q9.16 Explain how electronic beam steering is accomplished.

Q9.17 Why is frequency scanning an attractive technique for steering the beam of an antenna array?

CHAPTER HIGHLIGHTS

- An antenna is a transducer between a guided wave propagating on a transmission line and an EM wave propagating in an unbounded medium, or vice versa.
- Except for some solid-state antennas composed of non-linear semiconductors or ferrite materials, antennas are reciprocal devices; they exhibit the same radiation patterns for transmission as for reception.
- In the far-field region of an antenna, the radiated energy is approximately a plane wave.
- The electric field radiated by current antennas, such as wires, is equal to the sum of the electric fields radiated by all the Hertzian dipoles making up the antenna.
- The radiation resistance R_{rad} of a half-wave dipole is 73 Ω, which can be easily matched to a transmission line.
- The directional properties of an antenna are described by its radiation pattern, directivity, pattern solid angle, and half-power beamwidth.
- The Friis transmission formula relates the power received by an antenna due to that transmitted by another antenna at a specified distance away.
- The far-zone electric field radiated by a large aperture (measured in wavelengths) is related to the field distribution across the aperture by Kirchhoff's scalar diffraction theory. A uniform aperture distribution produces a far-field pattern with the narrowest possible beamwidth.
- By controlling the amplitudes and phases of the individual elements of an antenna array, it is possible to shape the antenna pattern and to steer the direction of the beam electronically.
- The pattern of an array of identical elements is equal to the product of the array factor and the antenna pattern of an individual antenna element.

GLOSSARY OF IMPORTANT TERMS

Provide definitions or explain the meaning of the following terms:

antenna
isotropic antenna
radiation pattern
antenna impedance
far-field (or far-zone) region
short dipole (Hertzian dipole)
spherical propagation factor
power density $\mathbf{S}(\theta, \phi)$
broadside direction
solid angle
radiation intensity (normalized) F
elevation and azimuth planes
radiation lobes
pattern solid angle Ω_p
beamwidth β
antenna directivity D
antenna gain G
radiation efficiency ξ
radiation resistance R_{rad}
loss resistance R_{loss}
effective area (effective aperture) A_e
Friis transmission formula
system noise temperature T_{sys}
signal-to-noise ratio S_n
aperture distribution
antenna array
electronic steering
feeding coefficient
pattern multiplication principle
array factor $F_a(\theta, \phi)$
end-fire direction
frequency scanning

PROBLEMS

Sections 9-1 and 9-2: Short Dipole and Antenna Radiation Characteristics

9.1* A center-fed Hertzian dipole is excited by a current $I_0 = 10$ A. If the dipole is $\lambda/50$ in length, determine the maximum radiated power density at a distance of 1 km.

9.2 A 1-m-long dipole is excited by a 1-MHz current with an amplitude of 12 A. What is the average power density radiated by the dipole at a distance of 5 km in a direction that is 30° from the dipole axis?

9.3* Determine the following:
(a) The direction of maximum radiation.
(b) Directivity.
(c) Beam solid angle.
(d) Half-power beamwidth in the x–z plane.
for an antenna whose normalized radiation intensity is given by

$$F(\theta, \phi) = \begin{cases} 1, & \text{for } 0 \leq \theta \leq 60° \text{ and } 0 \leq \phi \leq 2\pi \\ 0, & \text{elsewhere.} \end{cases}$$

Suggestion: Sketch the pattern prior to calculating the desired quantities.

9.4 Repeat Problem 9.3 for an antenna with

$$F(\theta, \phi) = \begin{cases} \sin^2 \theta \cos^2 \phi & \text{for } 0 \leq \theta \leq \pi \\ & \text{and } -\pi/2 \leq \phi \leq \pi/2 \\ 0 & \text{elsewhere} \end{cases}$$

9.5* A 2-m-long center-fed dipole antenna operates in the AM broadcast band at 1 MHz. The dipole is made of copper wire with a radius of 1 mm.
(a) Determine the radiation efficiency of the antenna.
(b) What is the antenna gain in decibels?
(c) What antenna current is required so that the antenna will radiate 20 W, and how much power will the generator have to supply to the antenna?

---

*Answer(s) available in Appendix D.

Ⓢ Solution available in CD-ROM.

9.6 Repeat Problem 9.5 for a 20-cm-long antenna operating at 5 MHz.

9.7* An antenna with a pattern solid angle of 1.5 (sr) radiates 30 W of power. At a range of 1 km, what is the maximum power density radiated by the antenna?

9.8 An antenna with a radiation efficiency of 90% has a directivity of 6.7 dB. What is its gain in decibels?

9.9* The radiation pattern of a circular parabolic-reflector antenna consists of a circular major lobe with a half-power beamwidth of 2° and a few minor lobes. Ignoring the minor lobes, obtain an estimate for the antenna directivity in dB.

9.10 The normalized radiation intensity of a certain antenna is given by

$$F(\theta) = \exp(-20\theta^2) \quad \text{for } 0 \leq \theta \leq \pi$$

where θ is in radians. Determine:
(a) The half-power beamwidth.
(b) The pattern solid angle.
(c) The antenna directivity.

Sections 9-3 and 9-4: Dipole Antennas

9.11* Repeat Problem 9.5 for a 1-m-long half-wave dipole that operates in the FM/TV broadcast band at 150 MHz.

9.12 Assuming the loss resistance of a half-wave dipole antenna to be negligibly small and ignoring the reactance component of its antenna impedance, calculate the standing-wave ratio on a 60-Ω transmission line connected to the dipole antenna.

9.13* For a short dipole with length l such that $l \ll \lambda$, instead of treating the current $\tilde{I}(z)$ as constant along the dipole, as was done in Section 9-1, a more realistic

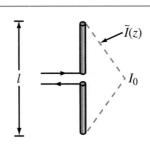

Figure 9-36: Triangular current distribution on a short dipole (Problem 9.13).

approximation that ensures that the current goes to zero at the ends is to describe $\tilde{I}(z)$ by the triangular function

$$\tilde{I}(z) = \begin{cases} I_0(1 - 2z/l), & \text{for } 0 \leq z \leq l/2 \\ I_0(1 + 2z/l), & \text{for } -l/2 \leq z \leq 0 \end{cases}$$

as shown in Fig. 9-36. Use this current distribution to determine the following:
(a) The far-field $\tilde{\mathbf{E}}(R, \theta, \phi)$.
(b) The power density $S(R, \theta, \phi)$.
(c) The directivity D.
(d) The radiation resistance R_{rad}.

9.14 For a dipole antenna of length $l = 3\lambda/2$,
(a) Determine the directions of maximum radiation.
(b) Obtain an expression for S_{max}.
(c) Generate a plot of the normalized radiation pattern $F(\theta)$.
(d) Compare your pattern with that shown in Fig. 9-17(c).

9.15 Repeat parts (a)–(c) of Problem 9.14 for a dipole of length $l = 3\lambda/4$.

9.16 Repeat parts (a)–(c) of Problem 9.14 for a dipole of length $l = \lambda$.

9.17* A car antenna is a vertical monopole over a conducting surface. Repeat Problem 9.5 for a 1-m-long car antenna operating at 1 MHz. The antenna wire is made of aluminum with $\mu_c = \mu_0$ and $\sigma_c = 3.5 \times 10^7$ S/m, and its diameter is 1 cm.

Sections 9-5 and 9-6: Effective Area and Friis Formula

9.18 Determine the effective area of a half-wave dipole antenna at 100 MHz, and compare it to its physical cross-section if the wire diameter is 1 cm.

9.19* A 3-GHz line-of-sight microwave communication link consists of two lossless parabolic dish antennas, each 1 m in diameter. If the receive antenna requires 1 nW of receive power for good reception and the distance between the antennas is 40 km, how much power should be transmitted?

9.20 A half-wave dipole TV broadcast antenna transmits 1 kW at 50 MHz. What is the power received by a home television antenna with 13-dB gain if located at a distance of 30 km?

9.21* A 150-MHz communication link consists of two vertical half-wave dipole antennas separated by 2 km. The antennas are lossless, the signal occupies a bandwidth of 3 MHz, the system noise temperature of the receiver is 600 K, and the desired signal-to-noise ratio is 20 dB. What transmitter power is required?

9.22 Consider the communication system shown in Fig. 9-37, with all components properly matched. If $P_t = 10$ W and $f = 6$ GHz:
(a) What is the power density at the receiving antenna (assuming proper alignment of antennas)?
(b) What is the received power?
(c) If $T_{sys} = 1,000$ K and the receiver bandwidth is 10 MHz, what is the signal-to-noise ratio in decibels?

Sections 9-7 and 9-8: Radiation by Apertures

9.23* A uniformly illuminated aperture is of length $l_x = 20\lambda$. Determine the beamwidth between first nulls in the x–z plane.

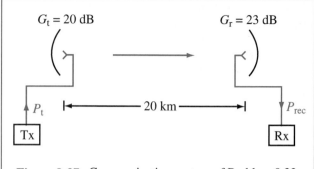

Figure 9-37: Communication system of Problem 9.22.

9.24 The 10-dB beamwidth is the beam size between the angles at which $F(\theta)$ is 10 dB below its peak value. Determine the 10-dB beamwidth in the x–z plane for a uniformly illuminated aperture with length $l_x = 10\lambda$.

9.25* A uniformly illuminated rectangular aperture situated in the x–y plane is 3 m high (along x) and 1 m wide (along y). If $f = 10$ GHz, determine the following:
(a) The beamwidths of the radiation pattern in the elevation plane (x–z plane) and the azimuth plane (y–z plane).
(b) The antenna directivity D in decibels.

9.26 An antenna with a circular aperture has a circular beam with a beamwidth of 1.5° at 20 GHz.
(a) What is the antenna directivity in dB?
(b) If the antenna area is doubled, what will be the new directivity and new beamwidth?
(c) If the aperture is kept the same as in (a), but the frequency is doubled to 40 GHz, what will the directivity and beamwidth become then?

9.27* A 94-GHz automobile collision-avoidance radar uses a rectangular-aperture antenna placed above the car's bumper. If the antenna is 1 m in length and 10 cm in height, determine the following:
(a) Its elevation and azimuth beamwidths.
(b) The horizontal extent of the beam at a distance of 300 m.

9.28 A microwave telescope consisting of a very sensitive receiver connected to a 100-m parabolic-dish antenna is used to measure the energy radiated by astronomical objects at 10 GHz. If the antenna beam is directed toward the moon and the moon extends over a planar angle of 0.5° from Earth, what fraction of the moon's cross-section will be occupied by the beam?

Sections 9-9 and 9-11: Antenna Arrays

9.29* A two-element array consisting of two isotropic antennas separated by a distance d along the z-axis is placed in a coordinate system whose z-axis points eastward and whose x-axis points toward the zenith. If a_0 and a_1 are the amplitudes of the excitations of the antennas at $z = 0$ and at $z = d$, respectively, and if δ is the phase of the excitation of the antenna at $z = d$ relative to that of the other antenna, find the array factor and plot the pattern in the x–z plane for the following:

(a) $a_0 = a_1 = 1$, $\delta = \pi/4$, and $d = \lambda/2$
(b) $a_0 = 1$, $a_1 = 2$, $\delta = 0$, and $d = \lambda$
(c) $a_0 = a_1 = 1$, $\delta = -\pi/2$, and $d = \lambda/2$
(d) $a_0 = 1$, $a_1 = 2$, $\delta = \pi/4$, and $d = \lambda/2$
(e) $a_0 = 1$, $a_1 = 2$, $\delta = \pi/2$, and $d = \lambda/4$

9.30 If the antennas in part (a) of Problem 9.29 are parallel, vertical, Hertzian dipoles with axes along the x-direction, determine the normalized radiation intensity in the x–z plane and plot it.

9.31* Consider the two-element dipole array of Fig. 9-29(a). If the two dipoles are excited with identical feeding coefficients ($a_0 = a_1 = 1$ and $\psi_0 = \psi_1 = 0$), choose (d/λ) such that the array factor has a maximum at $\theta = 45°$.

9.32 Choose (d/λ) so that the array pattern of the array of Problem 9.31 has a null, rather than a maximum, at $\theta = 45°$.

9.33* Find and plot the normalized array factor and determine the half-power beamwidth for a five-element linear array excited with equal phase and a uniform amplitude distribution. The interelement spacing is $3\lambda/4$.

9.34 A three-element linear array of isotropic sources aligned along the z-axis has an interelement spacing of $\lambda/4$ (Fig. 9-38). The amplitude excitation of the center element is twice that of the bottom and top elements, and the phases are $-\pi/2$ for the bottom element and $\pi/2$ for the top element, relative to that of the center element. Determine the array factor and plot it in the elevation plane.

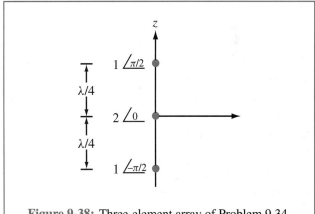

Figure 9-38: Three-element array of Problem 9.34.

9.35* An eight-element linear array with $\lambda/2$ spacing is excited with equal amplitudes. To steer the main beam to a direction 60° below the broadside direction, what should be the incremental phase delay between adjacent elements? Also, give the expression for the array factor and plot the pattern.

9.36 A linear array arranged along the z-axis consists of 12 equally spaced elements with $d = \lambda/2$. Choose an appropriate incremental phase delay δ so as to steer the main beam to a direction 30° above the broadside direction. Provide an expression for the array factor of the steered antenna and plot the pattern. From the pattern, estimate the beamwidth.

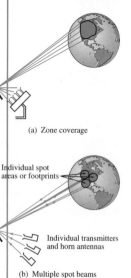

(a) Zone coverage

Individual spot areas or footprints

Individual transmitters and horn antennas

(b) Multiple spot beams

C H A P T E R

10

Satellite Communication Systems and Radar Sensors

APPLICATION EXAMPLES

This concluding chapter presents overviews of the principles of operation of satellite communication systems and radar sensors, with emphasis on electromagnetic-related topics.

10-1 Satellite Communication Systems

Today's world is connected by a vast communication network that provides a wide array of voice, data, and video services to both fixed and mobile terminals [Fig. 10-1]. The viability and effectiveness of the network are attributed in large measure to the use of orbiting satellite systems that function as relay stations with wide area coverage of Earth's surface. From a geostationary orbit position at 35,786 km above the equator, a satellite can view over one-third of Earth's

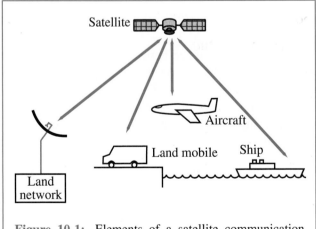

Figure 10-1: Elements of a satellite communication network.

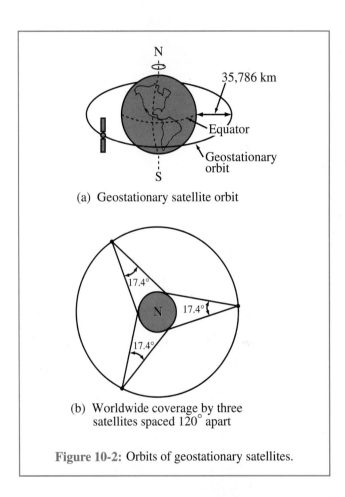

(a) Geostationary satellite orbit

(b) Worldwide coverage by three satellites spaced 120° apart

Figure 10-2: Orbits of geostationary satellites.

surface and can connect any pair of points within its coverage [Fig. 10-2]. The history of communication satellite engineering dates back to the late 1950s when the U.S. navy used the moon as a passive reflector to relay low-data-rate communications between Washington, D.C., and Hawaii. The first major development

391

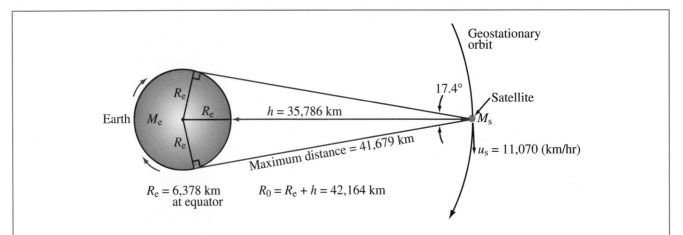

Figure 10-3: Satellite of mass m_s in orbit around Earth. For the orbit to be geostationary, the distance R_0 between the satellite and Earth's center should be 42,164 km. At the equator, this corresponds to an altitude of 35,786 km above Earth's surface.

involving artificial Earth satellites took place in October 1957 when the Soviet Union launched *Sputnik I* and used it for 21 days to transmit (one-way) telemetry information. This was followed by another telemetry satellite, *Explorer I*, which was launched by the United States in January 1958, and in December of that year the satellite *Score* was launched and used to broadcast President Eisenhower's Christmas message, marking the first instance of two-way voice communication via an artificial satellite.

These achievements were followed by a flurry of space activity, leading to the development of operational communication satellites by many countries for both commercial and governmental services. This section provides an overview of the satellite communications link, including transmitter–receiver power calculations, propagation aspects, frequency allocations, and antenna design considerations.

A satellite is said to be in a *geostationary orbit* around Earth when it is in a circular orbit in a plane identical with Earth's equatorial plane and at such an altitude that the orbital period is identical with Earth's rotational period. A satellite of mass M_s in circular orbit around

Earth [Fig. 10-3] is subject to two forces, a gravitational force F_g and a repelling centrifugal force F_c. The magnitudes of these two forces are given by

$$F_g = \frac{G M_s M_e}{R_0^2} \, , \tag{10.1}$$

$$F_c = \frac{M_s u_s^2}{R_0} = M_s \omega^2 R_0, \tag{10.2}$$

where $G = 6.67 \times 10^{-11}$ (N·m^2/kg^2) is the universal gravitational constant, $M_e = 5.98 \times 10^{24}$ kg is Earth's mass, R_0 is the distance between the satellite and the center of Earth, and u_s is the satellite velocity. For a rotating object, $u_s = \omega R_0$, where ω is its angular velocity. In order for the satellite to remain in orbit, the two opposing forces acting upon it have to be equal in magnitude, or

$$G \frac{M_s M_e}{R_0^2} = M_s \omega^2 R_0, \tag{10.3}$$

which yields a solution for R_0 given by

$$R_0 = \left[\frac{G M_e}{\omega^2} \right]^{1/3} . \tag{10.4}$$

To remain stationary with respect to Earth's surface, the satellite's angular velocity has to be the same as that of Earth's own angular velocity around its own axis. Thus,

$$\omega = \frac{2\pi}{T}, \tag{10.5}$$

where T is the period of one sidereal day in seconds. A sidereal day, which takes into account Earth's rotation around the sun, is equal to 23 hours, 56 minutes, and 4.1 seconds. Using Eq. (10.5) in Eq. (10.4) gives

$$R_0 = \left(\frac{GM_e T^2}{4\pi^2}\right)^{1/3}, \tag{10.6}$$

and upon using the numerical values for T, M_e, and G, we obtain the result $R_0 = 42,164$ km. Subtracting 6,378 km for Earth's mean radius at the equator gives an altitude of $h = 35,786$ km above Earth's surface.

From a geostationary orbit, Earth subtends an angle of 17.4°, covering an arc of about 18,000 km along the equator, which corresponds to a longitude angle of about 160°. With three equally spaced satellites in geostationary orbit over Earth's equator, complete global coverage of the entire equatorial plane can be achieved, with significant overlap between the beams of the three satellites. As far as coverage toward the poles, a global beam can reach Earth stations up to 81° of latitude on either side of the equator.

Not all satellite communication systems use spacecraft that are in geostationary orbits; because of transmitter power limitations or other considerations, it is sometimes necessary to operate from much lower altitudes, in which case the satellite is placed in a *highly elliptical orbit* (to satisfy Kepler's law) such that for part of the orbit (near its perigee) it is at a range of only a few hundred kilometers from Earth's surface. Whereas only three geostationary satellites are needed to provide global coverage of Earth's surface, a much larger number is needed when the satellites used are in highly elliptical orbits.

10-2 Satellite Transponders

A communication satellite functions as a distant repeater that receives *uplink* signals from Earth stations, processes the signals, and then retransmits them on the *downlink* to their intended Earth destinations. The International Telecommunication Union has allocated specific bands for satellite communications [Table 10-1]. Of these, the bands used by the majority of U.S. commercial satellites for domestic communications are the *4/6 GHz band* (3.7- to 4.2-GHz downlink and 5.925- to 6.425-GHz uplink) and the *12/14 GHz band* (11.7- to 12.2-GHz downlink and 14.0- to 14.5-GHz uplink). Each uplink and downlink segment has been allocated 500 MHz of bandwidth. By using different frequency bands for Earth-to-satellite uplink segment and for the satellite-to-Earth downlink segments, the same antennas can be used for both functions while simultaneously guarding against interference between the two signals. The downlink segment commonly uses a lower-frequency carrier than the uplink segment, because lower frequencies suffer lower attenuation by Earth's atmosphere, and thus eases the requirement on satellite output power.

We shall use the 4/6 GHz band as a model to facilitate discussion of the satellite-repeater operation, while keeping in mind that the functional structure of the repeater is basically the same regardless of which specific satellite communication band is used.

Figure 10-4 shows a generalized block diagram of a 12-channel repeater as implemented in a typical communications satellite. The path of each channel from the point of its reception by the antenna, followed by its transfer through the repeater up to the point of its retransmission through the antenna, is called a *transponder*. The available 500-MHz bandwidth is allocated to 12 channels (transponders) of 36-MHz bandwidth per channel and 4-MHz separation between channels. The basic functions of each transponder are isolation of neighboring radio frequency (RF) channels, frequency translation,

Table 10-1: Communications satellite frequency allocations.

Use	Downlink Frequency (MHz)	Uplink Frequency (MHz)
Fixed Service		
Commercial (C-band)	3,700–4,200	5,925–6,425
Military (X-band)	7,250–7,750	7,900–8,400
Commercial (K-band)		
Domestic (USA)	11,700–12,200	14,000–14,500
International	10,950–11,200	27,500–31,000
Mobile Service		
Maritime	1,535–1,542.5	1,635–1,644
Aeronautical	1,543.5–1,558.8	1,645–1,660
Broadcast Service		
	2,500–2,535	2,655–2,690
	11,700–12,750	
Telemetry, Tracking, and Command		
137–138, 401–402, 1,525–1,540		

and amplification. With *frequency-division multiple access* (FDMA)—one of the schemes commonly used for information transmission—each transponder can accommodate thousands of individual telephone channels within its 36 MHz of bandwidth (telephone speech signals require a minimum bandwidth of 3 kHz, and the nominal frequency spacing usually used is 4 kHz per telephone channel), several TV channels (each requiring a bandwidth of 6 MHz), millions of bits of digital data, or combinations of all three.

When the same antenna is used for both transmission and reception, a *duplexer* is used to perform the signal separation. Many types of duplexers are available, but among the simplest to understand is the circulator shown in Fig. 10-5. A *circulator* is a three-port device that uses a ferrite material placed in a magnetic field induced by a permanent magnet to achieve power flow from ports 1

to 2, 2 to 3, and 3 to 1, but not in the reverse directions. With the antenna connected to port 1, the received signal is channeled only to port 2; with port 2 being properly matched to the band-pass filter, no part of the received signal will be reflected from port 2 to 3; and with the transmitted signal connected to port 3, the circulator channels it to port 1 for transmission by the antenna.

Following the duplexer shown in Fig. 10-4, the received signal passes through a receiver band-pass filter that ensures isolation of the received signal from the transmitted signal. The receiver filter covers the bandwidth from 5.925 to 6.425 GHz, which encompasses the cumulative bandwidths of all 12 channels; the first received channel extends from 5,927 to 5,963 MHz, the second one from 5,967 to 6,003 MHz, and so on until the twelfth channel, which covers the range from 6,367 to 6,403 MHz. The next subsystem along the received signal path is the

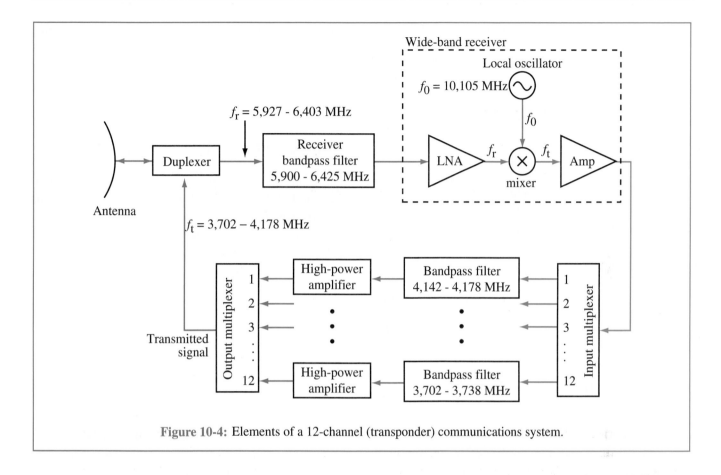

Figure 10-4: Elements of a 12-channel (transponder) communications system.

wideband receiver, which consists of three elements: a low-noise wideband amplifier, a frequency translator, and an output amplifier. The frequency translator consists of a stable local oscillator that generates a signal at a frequency $f_0 = 10,105$ MHz, connected to a nonlinear microwave mixer. The mixer serves to convert the frequency f_r of the received signal (which covers the range from 5,927 to 6,403 MHz) to a lower-frequency signal $f_t = f_0 - f_r$. Thus, the lower end of the received signal frequency gets converted from 5,927 to 4,178 MHz and the upper end gets converted from 6,403 to 3,702 MHz. This translation results in 12 channels with new frequency ranges, but their signals carry the same information (modulation) that was present in the received signals. In principle, the receiver output signal can now be further amplified and then channeled to the antenna through the duplexer for transmission back to Earth. Instead, the receiver output signal is separated into the 12 transponder channels through a *multiplexer* followed by a bank of narrow band-pass filters, each covering the bandwidth of one transponder channel. Each of the 12 channels is amplified by its own *high-power amplifier* (HPA), and then the 12 channels are combined by another multiplexer that feeds the combined spectrum into the duplexer. This channel separation and recombination process is used as a safety measure against losing all 12 channels should a high-power amplifier experience degradation in performance or total failure.

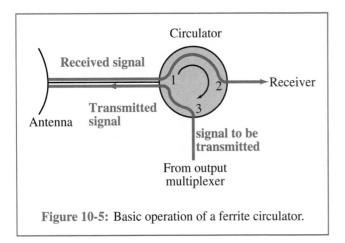

Figure 10-5: Basic operation of a ferrite circulator.

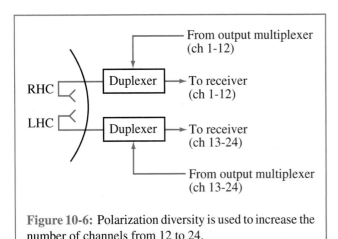

Figure 10-6: Polarization diversity is used to increase the number of channels from 12 to 24.

The information carrying capacity of a satellite repeater can be doubled from 12 to 24 channels over the same 500-MHz bandwidth by using *polarization diversity*. Instead of transmitting one channel of information over channel 1 (5,927 to 5,963 MHz), for example, the ground station transmits to the satellite two signals carrying different information and covering the same frequency band, but with different antenna polarization configurations, such as right-hand circular (RHC) and left-hand circular (LHC) polarizations. The satellite antenna is equipped with a feed arrangement that can receive each of the two

circular polarization signals individually with negligible interference between them. Two duplexers are used in this case, one connected to the RHC polarization feed and another connected to the LHC polarization feed, as illustrated in Fig. 10-6.

10-3 Communication-Link Power Budget

The uplink and downlink segments of a satellite communication link [Fig. 10-7] are each governed by the *Friis transmission formula* [Section 9-6], which states that the power P_r received by an antenna with gain G_r due to the transmission of power P_t by an antenna with gain G_t at a range R is given by

$$P_r = P_t G_t G_r \left(\frac{\lambda}{4\pi R} \right)^2 . \qquad (10.7)$$

The expression given by Eq. (10.7) applies to a lossless medium, such as free space. To account for attenuation

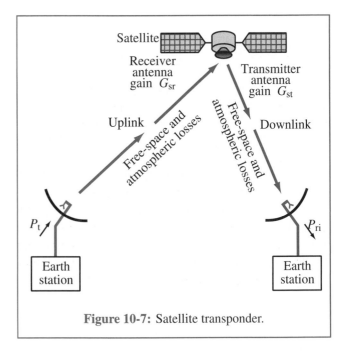

Figure 10-7: Satellite transponder.

by clouds and rain in Earth's atmosphere (when present along the propagation path), as well as absorption by certain atmospheric gases (primarily oxygen and water vapor), we should rewrite Eq. (10.7) in the following form:

$$P_{\rm ri} = \Upsilon(\theta)\, P_{\rm r} = \Upsilon(\theta)\, P_{\rm t} G_{\rm t} G_{\rm r} \left(\frac{\lambda}{4\pi R}\right)^2, \quad (10.8)$$

where now $P_{\rm ri}$ represents the input power at the receiver with atmospheric losses taken into account, and $\Upsilon(\theta)$ is the one-way *transmissivity* of the atmosphere at zenith angle θ. In addition to its dependence on θ, $\Upsilon(\theta)$ is a function of the microwave frequency of the communication link and the rain-rate conditions along the propagation path. At frequencies below 10 GHz, which include the 4/6 GHz band allocated for satellite communications, absorption by atmospheric gases is very small, and cloud and rain attenuation is also relatively small. Consequently, the magnitude of $\Upsilon(\theta)$ is typically on the order of 0.5 to 1 for most conditions. A transmissivity of 0.5 means that twice as much power needs to be transmitted (compared with the free-space case) in order to receive a specified power level. Among the various sources of atmospheric attenuation, the most serious is rainfall, and its attenuation coefficient increases rapidly with increasing frequency. Consequently, atmospheric attenuation assumes greater importance with regard to transmitter power requirements as the communication-system frequency is increased toward higher bands in the microwave region.

The noise appearing at the receiver output, $P_{\rm no}$, consists of three contributions: (1) noise internally generated by the receiver electronics, (2) noise picked up by the antenna due to external sources, including emission by the atmosphere, and (3) noise due to thermal emission by the antenna material. The combination of all noise sources can be represented by an equivalent *system noise temperature*, $T_{\rm sys}$, defined such that

$$P_{\rm no} = G_{\rm rec} K T_{\rm sys} B, \quad (10.9)$$

where K is Boltzmann's constant, $G_{\rm rec}$ is the receiver power gain, and B is its bandwidth. This output noise level is the same as would appear at the output of a noise-free receiver with input noise level

$$P_{\rm ni} = \frac{P_{\rm no}}{G_{\rm rec}} = K T_{\rm sys} B. \quad (10.10)$$

The *signal-to-noise ratio* is defined as the ratio of the signal power to the noise power *at the input of an equivalent noise-free receiver*. Hence,

$$S_{\rm n} = \frac{P_{\rm ri}}{P_{\rm ni}} = \frac{\Upsilon(\theta)\, P_{\rm t} G_{\rm t} G_{\rm r}}{K T_{\rm sys} B} \left(\frac{\lambda}{4\pi R}\right)^2. \quad (10.11)$$

The performance of a communication system is governed by two sets of factors. The first set encompasses the signal-processing techniques used to encode, modulate, combine, and transmit the signal at the transmitter end and to receive, separate, demodulate, and decode the signal at the receiver end. The second set of factors is the gains and losses in the communication link, and they are represented by the signal-to-noise ratio, $S_{\rm n}$. For a given set of signal-processing techniques, $S_{\rm n}$ determines the quality of the received signal, such as the bit error rate in digital data transmission and sound and picture quality in audio and video transmissions. Very high quality signal transmission requires very high values of $S_{\rm n}$; in broadcast-quality television by satellite, some systems are designed to provide values of $S_{\rm n}$ exceeding 50 dB (or a factor of 10^5).

The performance of a satellite link depends on the composite performance of the uplink and downlink segments. If either segment performs poorly, the composite performance will be poor, regardless of how good the performance of the other segment is. It is for this reason that system-power calculations are performed for each segment separately.

10-4 Antenna Beams

Whereas most Earth-station antennas must provide highly directive beams to avoid interference effects, the satellite antenna system is designed to produce beams tailored to match the areas served by the satellite. For global coverage, beamwidths of 17.4° are required. In contrast, for transmission to and reception from a small area, beamwidths on the order of 1° or less may be needed [Fig. 10-8]. An antenna with a beamwidth β

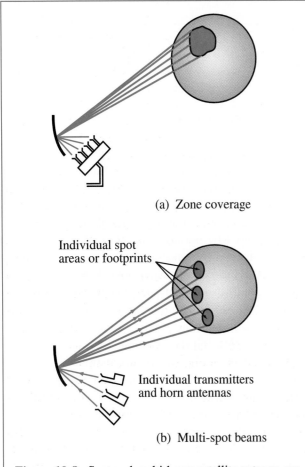

(a) Zone coverage

Individual spot areas or footprints

Individual transmitters and horn antennas

(b) Multi-spot beams

Figure 10-8: Spot and multi-beam satellite antenna systems for coverage of defined areas on Earth's surface.)

of 1° would produce a spot beam on Earth covering an area approximately 630 km in diameter.

Beam size has a direct connection to antenna gain and, in turn, to transmitter power requirements. Antenna gain G is related to the directivity D by $G = \xi D$, where ξ is the radiation efficiency, and D is related to the beamwidth β by the approximate expression given by Eq. (9.26). For a circular beam,

$$G = \xi \, \frac{4\pi}{\beta^2} \,, \qquad (10.12)$$

where β is in radians. For a lossless antenna ($\xi = 1$), a global beam with $\beta = 17.4°$ ($= 0.3$ rad) corresponds to a gain $G = 136$, or 21.3 dB. A narrow 1° beam, on the other hand, corresponds to an antenna gain of 41,253, or 46.2 dB.

To accommodate the various communication functions associated with satellite systems, four main types of antennas are used[*]:

1. Dipoles and helices at VHF and UHF for telemetry, tracking, and command functions;

2. Horns and relatively small parabolic dishes (with diameters on the order of a few centimeters) for producing wide-angled beams for global coverage;

3. Parabolic dishes fed by one or more horns to provide a beam for zone coverage [Fig. 10-8(a)] or multiple spot beams [Fig. 10-8(b)];

4. Antenna arrays consisting of many individual radiating elements for producing multi-spot beams and also for beam steering and scanning.

REVIEW QUESTIONS

Q10.1 What are the advantages and disadvantages of elliptical satellite orbits in comparison to the geostationary orbit?

---

[*]R. G. Meadows and A. J. Parsons, *Satellite Communications*, Hutchinson Publishers, London, 1989.

Q10.2 Why do satellite communication systems use different frequencies for the uplink and downlink segments? Which segment uses the higher frequency and why?

Q10.3 How does the use of antenna polarization increase the number of channels carried by the communication system?

Q10.4 What are the sources of noise that contribute to the total system noise temperature of a receiver?

10-5 Radar Sensors

The term *radar* is a contracted form of the phrase *radio detection and ranging*, which conveys some, but not all, the features of a modern radar system. Historically, radar systems were first developed and used at *radio* frequencies, including the microwave band, but we now also have light radars, or *lidars*, that operate at optical wavelengths. Over the years, the name radar has lost its original meaning and has come to signify any *active* electromagnetic sensor that uses its own source to illuminate a region of space and then measure the echoes generated by reflecting objects contained in the illuminated region. In addition to detecting the presence of a reflecting object and determining its *range* by measuring the time delay of short-duration pulses transmitted by the radar, a radar is also capable of specifying the *position* (*direction*) of the target and its radial velocity. Measurement of the *radial velocity* of a moving object, which is the component of the object's velocity vector along the direction from the object to the radar, is realized by measuring the *Doppler frequency shift* produced by the moving object, which is equal to the difference between the frequencies of the transmitted and received signals. Also, the strength and shape of the reflected pulse carry information about the shape and material properties of the reflecting object.

Radar is used for a wide variety of civilian and military applications, including air traffic control, aircraft navigation, law enforcement, control and guidance of weapon systems, remote sensing of Earth's environment, weather observations, astronomy, and collision avoidance for automobiles. The frequency bands used for the various types of radar applications extend from the megahertz region to frequencies as high as 225 GHz.

10-5.1 Basic Operation of a Radar System

The block diagram shown in Fig. 10-9 contains the basic functional elements of a pulse radar system. The *synchronizer–modulator* unit serves to synchronize the operation of the transmitter and the *videoprocessor–display unit* by generating a train of direct current (d-c) narrow-duration, evenly spaced pulses. These pulses, which are supplied to both the transmitter and the videoprocessor–display unit, specify the times at which radar pulses are transmitted. The transmitter contains a high-power radio-frequency (RF) oscillator with an on/off control voltage actuated by the pulses supplied by the synchronizer–modulator unit. Hence, the transmitter generates pulses of RF energy equal in duration and spacing to the d-c pulses generated by the synchronizer–modulator unit. Each pulse generated by the transmitter is supplied to the antenna through a *duplexer*, which allows the antenna to be shared between the transmitter and the receiver. The duplexer, which often is called the *transmitter/receiver* (T/R) switch, connects the transmitter to the antenna for the duration of the pulse, and then connects the antenna to the receiver for the remaining period until the start of a new pulse. Some duplexers, however, are passive devices that perform the sharing and isolation functions continuously. The circulator shown in Fig. 10-5 is an example of a passive duplexer. After transmission by the antenna, a portion of the transmitted signal is intercepted by a reflecting object (often called a *target*) and is reradiated in many directions. The energy reradiated by the target back toward the radar is collected by the antenna and delivered to the receiver, which processes the signal to detect the presence of the target and to extract information on its location and relative velocity. The receiver converts the reflected RF

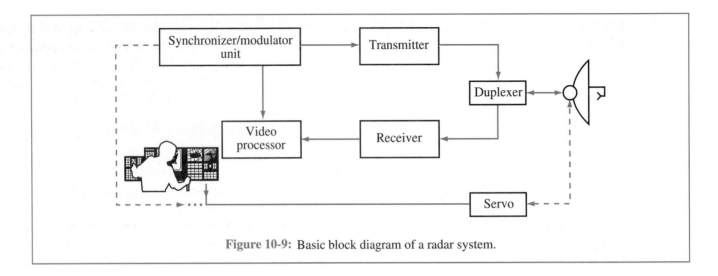

Figure 10-9: Basic block diagram of a radar system.

signals into lower-frequency video-frequency signals and supplies them to the videoprocessor–display unit, which displays the extracted information in a format suitable for the intended application. The servo unit positions the orientation of the antenna beam in response to control signals provided by either an operator, a control unit with preset functions, or a control unit commanded by another system. The control unit of an air traffic control radar, for example, commands the servo to rotate the antenna in azimuth continuously. In contrast, the radar antenna placed in the nose of an aircraft is made to scan back and forth over only a specified angular range.

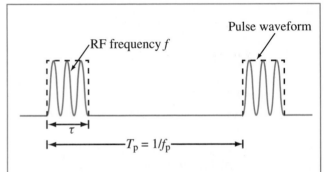

Figure 10-10: A pulse radar transmits a continuous train of RF pulses at a repetition frequency f_p.

10-5.2 Unambiguous Range

The collective features of the energy transmitted by a radar are called the *signal waveform*. For a pulse radar, these features include (1) the carrier frequency f, (2) the *pulse length* τ, (3) the *pulse repetition frequency* f_p (number of pulses per second), or equivalently the *interpulse period* $T_p = 1/f_p$, and (4) the modulation (if any) within the pulses. Three of these features are illustrated in Fig. 10-10. Modulation, which refers to control of the amplitude, frequency, or phase of the

signal, is beyond the level of the present treatment and therefore will not be considered.

The range to the target is determined by measuring the time delay T taken by the pulse to travel to the target and back. For a target at range R,

$$T = \frac{2R}{c}, \qquad (10.13)$$

where $c = 3 \times 10^8$ m/s is the speed of light, and the factor 2 accounts for the two-way propagation. The maximum target range that a radar can measure unambiguously, called the *unambiguous range* R_u, is determined by the interpulse period T_p and is given by

$$R_u = \frac{cT_p}{2} = \frac{c}{2f_p} . \qquad (10.14)$$

The range R_u corresponds to the maximum range that a target can have such that its echo is received before the transmission of the next pulse. If T_p is too short, an echo signal due to a given pulse might arrive after the transmission of the next pulse, in which case the target would appear to be at a much shorter range than it actually is.

According to Eq. (10.14), if a radar is to be used to detect targets that are as far away as 100 km, for example, then f_p should be less than 1.5 kHz, and the higher the pulse repetition frequency (PRF), the shorter is the unambiguous range R_u. Consideration of R_u alone suggests selecting a low PRF, but other considerations suggest selecting a very high PRF. As we will see later in Section 10-6, the signal-to-noise ratio of the radar receiver is directly proportional to f_p, and hence it would be advantageous to select the PRF to be as high as possible. Moreover, in addition to determining the maximum unambiguous range R_u, the PRF also determines the maximum Doppler frequency (and hence the target's maximum radial velocity) that the radar can measure unambiguously. If the requirements on maximum range and velocity cannot be met by the same PRF, then some compromise may be necessary. Alternatively, it is possible to use a *multiple-PRF* radar system that transmits a few pulses at one PRF followed by another series of pulses at another PRF, and then the two sets of received pulses are processed together to remove the ambiguities that would have been present with either PRF alone.

10-5.3 Range and Angular Resolutions

Consider a radar observing two targets located at ranges R_1 and R_2, as shown in Fig. 10-11. Let $t = 0$ denote the

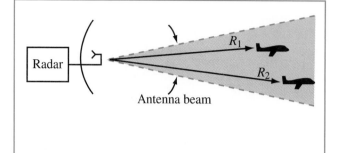

Figure 10-11: Radar beam viewing two targets at ranges R_1 and R_2.

time corresponding to the start of the transmitted pulse. The pulse length is τ. The return due to target 1 will arrive at $T_1 = 2R_1/c$ and will have a length τ (assuming that the pulse length in space is much greater than the radial extent of the target). Similarly, the return due to target 2 will arrive at $T_2 = 2R_2/c$. The two targets will be resolvable as distinct targets as long as $T_2 \geq T_1 + \tau$ or, equivalently,

$$\frac{2R_2}{c} \geq \frac{2R_1}{c} + \tau. \qquad (10.15)$$

The *range resolution* of the radar, ΔR, is defined as the minimum spacing between two targets necessary to avoid overlap between the echoes from the two targets. From Eq. (10.15), this occurs when

$$\Delta R = R_2 - R_1 = c\tau/2. \qquad (10.16)$$

Some radars are capable of transmitting pulses as short as 1 ns in duration or even shorter. For $\tau = 1$ ns, $\Delta R = 15$ cm.

The basic angular resolution of a radar system is determined by its antenna beamwidth β, as shown in

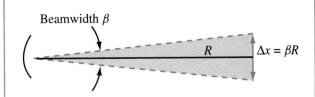

Figure 10-12: The azimuth resolution Δx at a range R is equal to βR.

Fig. 10-12. The corresponding *azimuth resolution* Δx at a range R is given by

$$\Delta x = \beta R, \qquad (10.17)$$

where β is in radians. In some cases, special techniques are used to improve the angular resolution down to a fraction of the beamwidth β. One example is the monopulse tracking radar described in Section 10-8.

10-6 Target Detection

Target detection by radar is governed by two factors: (1) the *signal energy* received by the radar receiver due to reflection of part of the transmitted energy by the target, and (2) the *noise energy* generated by the receiver. Figure 10-13, which depicts the output of a radar receiver as a function of time, shows the signals due to the targets displayed against the noise contributed by external sources as well as by the devices making up the receiver. The random variations exhibited by the noise may at times make it difficult to distinguish the signal reflected by the target from a noise spike. In Fig. 10-13, the mean noise-power level at the receiver output is denoted by $P_{no} = G_{rec}P_{ni}$, where G_{rec} is the receiver gain and P_{ni} is the noise level referred to the receiver's input terminals. The power levels P_{r_1} and P_{r_2} represent the echoes of the two targets observed by the radar. Because of the random nature of the noise, it is necessary to set a threshold level, $P_{r_{min}}$, for detection. For the threshold level 1 indicated in Fig. 10-13, the radar will detect the presence of both targets, but it will also

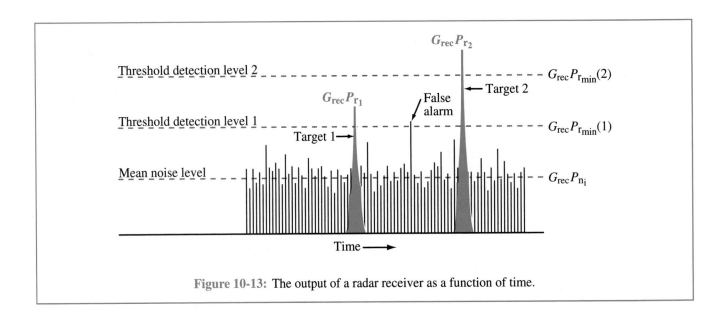

Figure 10-13: The output of a radar receiver as a function of time.

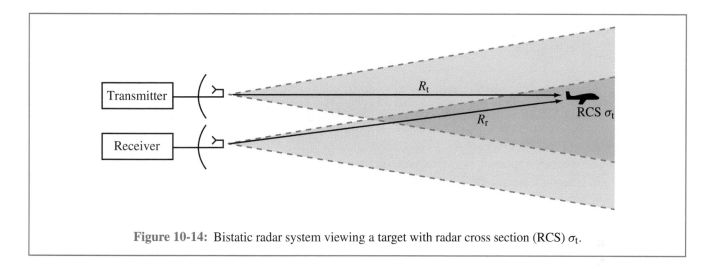

Figure 10-14: Bistatic radar system viewing a target with radar cross section (RCS) σ_t.

detect a *false alarm*. The chance of this occurring is called the *false-alarm probability*. On the other hand, if the threshold level is raised to level 2 to avoid the false alarm, the radar will not detect the presence of the first target. Correct detection of targets is called the *detection probability*. The setting of the threshold signal level relative to the mean noise level is thus made on the basis of a compromise that considers both probabilities.

To keep the noise level at a minimum, the receiver is designed such that its bandwidth B is barely wide enough to pass most of the energy contained in the received pulse. Such a design, called a *matched-filter design*, requires that B be equal to the reciprocal of the pulse length τ (i.e., $B = 1/\tau$). Hence, for a matched-filter receiver, Eq. (10.10) becomes

$$P_{\text{ni}} = K T_{\text{sys}} B = \frac{K T_{\text{sys}}}{\tau} . \qquad (10.18)$$

The signal power received by the radar, P_r, is related to the transmitted power level, P_t, through the radar equation. We will first derive the radar equation for the general case of a *bistatic radar* configuration in which the transmitter and receiver are not necessarily at the same location, and then we will specialize the results to the *monostatic radar* case wherein the transmitter and

receiver are colocated. In Fig. 10-14, the target is at a range R_t from the transmitter and at a range R_r from the receiver. The power density illuminating the target is given by

$$S_t = \frac{P_t}{4\pi R_t^2} G_t \qquad (\text{W/m}^2), \qquad (10.19)$$

where $(P_t/4\pi R_t^2)$ represents the power density that would have been radiated by an isotropic radiator, and G_t is the gain of the transmitting antenna in the direction of the target. The target is characterized by a *radar cross section* (RCS) σ_t (m^2), defined such that the power intercepted and then reradiated by the target is

$$P_{\text{rer}} = S_t \sigma_t = \frac{P_t G_t \sigma_t}{4\pi R_t^2} \qquad (\text{W}). \qquad (10.20)$$

This reradiated power spreads out over a spherical surface, resulting in a power density S_r incident upon the receiving radar antenna and given by

$$S_r = \frac{P_{\text{rer}}}{4\pi R_r^2} = \frac{P_t G_t \sigma_t}{(4\pi R_t R_r)^2} \qquad (\text{W/m}^2). \qquad (10.21)$$

With an effective area A_r and radiation efficiency ξ_r, the receiving radar antenna intercepts and delivers (to the receiver) power P_r given by

$$P_r = \xi_r A_r S_r = \frac{P_t G_t \xi_r A_r \sigma_t}{(4\pi R_t R_r)^2} = \frac{P_t G_t G_r \lambda^2 \sigma_t}{(4\pi)^3 R_t^2 R_r^2}, \quad (10.22)$$

where we have used Eqs. (9.29) and (9.64) to relate the effective area of the receive antenna, A_r, to its gain G_r. For a monostatic antenna that uses the same antenna for the transmit and receive functions, $G_t = G_r \triangleq G$ and $R_t = R_r \triangleq R$. Hence,

$$P_r = \frac{P_t G^2 \lambda^2 \sigma_t}{(4\pi)^3 R^4}. \quad (10.23)$$

Unlike the one-way communication system for which the dependence on R is as $1/R^2$, the range dependence given by the above *radar equation* goes as $1/R^4$, which in effect is the product of two one-way propagation processes.

The detection process may be based on the echo from a single pulse or on the addition (integration) of the echoes from several pulses. We will consider only the single-pulse case here. A target is said to be detectable if its echo signal power P_r exceeds $P_{r_{min}}$, the *threshold detection level* indicated in Fig. 10-13. The *maximum detectable range* R_{max} is the range beyond which the target cannot be detected, and it corresponds to the range at which $P_r = P_{r_{min}}$ in Eq. (10.23). Thus,

$$R_{max} = \left[\frac{P_t G^2 \lambda^2 \sigma_t}{(4\pi)^3 P_{r_{min}}} \right]^{1/4}. \quad (10.24)$$

The signal-to-noise ratio is equal to the ratio of the received signal power P_r to the mean input noise power P_{ni} given by Eq. (10.18):

$$S_n = \frac{P_r}{P_{ni}} = \frac{P_r \tau}{K T_{sys}}, \quad (10.25)$$

and the *minimum signal-to-noise ratio* S_{min} corresponds to when $P_r = P_{r_{min}}$:

$$S_{min} = \frac{P_{r_{min}} \tau}{K T_{sys}}. \quad (10.26)$$

Use of Eq. (10.26) in Eq. (10.24) gives

$$R_{max} = \left[\frac{P_t \tau G^2 \lambda^2 \sigma_t}{(4\pi)^3 K T_{sys} S_{min}} \right]^{1/4}. \quad (10.27)$$

The product $P_t \tau$ is equal to the energy of the transmitted pulse. Hence, according to Eq. (10.27), *it is the energy of the transmitted pulse rather than the transmitter power level alone that determines the maximum detectable range.* A high-power narrow pulse and an equal-energy, low-power long pulse will yield the same radar performance as far as maximum detectable range is concerned. However, the range-resolution capability of the long pulse is much poorer than that of the short pulse [see Eq. (10.16)].

The maximum detectable range R_{max} can also be increased by improving the signal-to-noise ratio. This can be accomplished by integrating the echoes from multiple pulses in order to increase the total amount of energy received from the target. The number of pulses available for integration over a specified integration time is proportional to the PRF. Hence, from the standpoint of maximizing target detection, it is advantageous to use as high a PRF as allowed by other considerations.

10-7 Doppler Radar

The Doppler effect is a shift in the frequency of a wave caused by the motion of the transmitting source, the reflecting object, or the receiving system. As illustrated in Fig. 10-15, a wave radiated by a stationary isotropic point source forms equally spaced concentric circles as a function of time travel from the source. In contrast, a wave radiated by a moving source is compressed in the

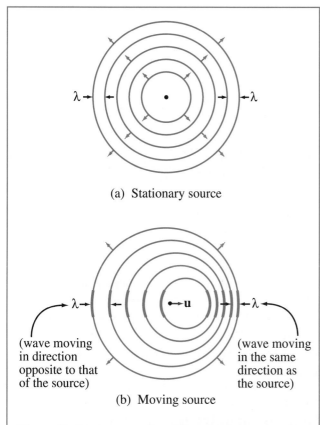

(a) Stationary source

(b) Moving source

(wave moving in direction opposite to that of the source)

(wave moving in the same direction as the source)

Figure 10-15: A wave radiated from a point source when (a) stationary and (b) moving. The wave is compressed in the direction of motion, spread out in the opposite direction, and unaffected in the direction normal to motion.

direction of motion and is spread out in the opposite direction. Compressing a wave shortens its wavelength, which is equivalent to increasing its frequency. Conversely, spreading it out decreases its frequency. The change in frequency is called the *Doppler frequency shift* f_d. That is, if f_t is the frequency of the wave radiated by the moving source, then the frequency f_r of the wave that would be observed by a stationary receiver is

$$f_r = f_t + f_d. \qquad (10.28)$$

The magnitude and sign of f_d depend on the direction of the velocity vector relative to the direction of the range vector connecting the source to the receiver.

Consider a source transmitting an electromagnetic wave with frequency f_t as shown in Fig. 10-16. At a distance R from the source, the electric field of the radiated wave is given by

$$E(R) = E_0 e^{j(\omega_t t - kR)} = E_0 e^{j\phi}, \qquad (10.29)$$

where E_0 is the wave's magnitude, $\omega_t = 2\pi f_t$, and $k = 2\pi/\lambda_t$, where λ_t is the wavelength of the transmitted wave. The magnitude depends on the distance R and the gain of the source antenna, but it is not of concern as far as the Doppler effect is concerned. The quantity

$$\phi = \omega_t t - kR = 2\pi f_t t - \frac{2\pi}{\lambda_t} R \qquad (10.30)$$

is the phase of the radiated wave relative to its phase at $R = 0$ and to a reference time $t = 0$. If the source is moving toward the receiver, as in Fig. 10-16, (or vice versa) at a radial velocity u_r, then

$$R = R_0 - u_r t, \qquad (10.31)$$

where R_0 is the distance between the source and receiver at $t = 0$. Hence,

$$\phi = 2\pi f_t t - \frac{2\pi}{\lambda_t}(R_0 - u_r t). \qquad (10.32)$$

This is the phase of the signal detected by the receiver. The frequency of a wave is by definition equal to the time derivative of the phase ϕ divided by 2π. Thus,

$$f_r = \frac{1}{2\pi} \frac{d\phi}{dt} = f_t + \frac{u_r}{\lambda_t}. \qquad (10.33)$$

Comparison of Eq. (10.33) with Eq. (10.28) leads to $f_d = u_r/\lambda_t$. For radar, the Doppler shift happens twice, once for the wave from the radar to the target and again

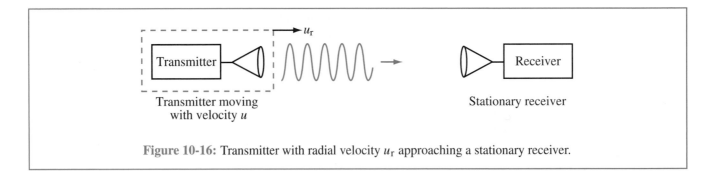

Figure 10-16: Transmitter with radial velocity u_r approaching a stationary receiver.

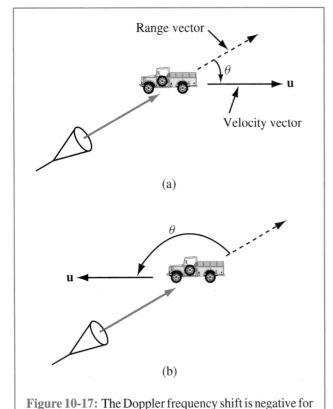

(a)

(b)

Figure 10-17: The Doppler frequency shift is negative for a receding target ($0 \le \theta \le 90°$), as in (a), and positive for an approaching target ($90° \le \theta \le 180°$), as in (b).

for the wave reflected by the target back to the radar. Hence, $f_d = 2u_r/\lambda_t$. The dependence of f_d on direction is given by the dot product of the velocity and range unit

vectors, which leads to

$$f_d = -2\frac{u_r}{\lambda_t} = -\frac{2u}{\lambda_t}\cos\theta, \qquad (10.34)$$

where u_r is the radial velocity component of u and θ is the angle between the range vector and the velocity vector as shown in Fig. 10-17, with the direction of the range vector defined *from* the radar *to* the target. For a receding target (relative to the radar), $0 \le \theta \le 90°$, and for an approaching target, $90 \le \theta \le 180°$.

10-8 Monopulse Radar

On the basis of information extracted from the echo due to a single pulse, a *monopulse radar* can track the direction of a target with an angular accuracy equal to a fraction of its antenna beamwidth. To track a target in both elevation and azimuth, a monopulse radar uses an antenna such as a parabolic dish, with four separate small feeds at its focal point, as illustrated by the horns shown in Fig. 10-18. Monopulse systems are of two types; the first is called *amplitude-comparison monopulse* because the tracking information is extracted from the amplitudes of the echoes received by the four horns, and the second type is called *phase-comparison monopulse* because it relies on the phases of the received signals instead. We shall limit our present discussion to the amplitude-comparison scheme.

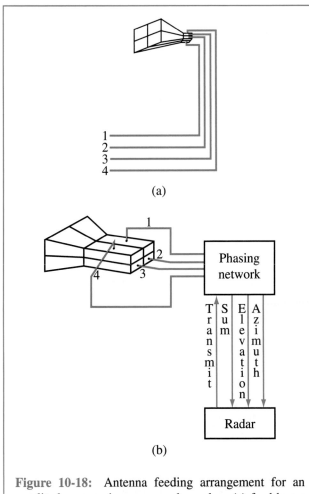

Figure 10-18: Antenna feeding arrangement for an amplitude-comparison monopulse radar: (a) feed horns and (b) connection to phasing network.

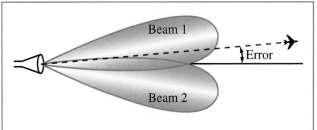

Figure 10-19: A target observed by two overlapping beams of a monopulse radar.

Individually, each horn would produce its own beam, with the four beams pointing in slightly different directions. Figure 10-19 shows the beams of two adjacent horns. The basic principle of the amplitude-comparison monopulse is to measure the amplitudes of the echo signals received through the two beams and then to apply the difference between them to repoint the antenna boresight direction toward the target. Using computer-controlled phase shifters, the *phasing network* shown in Fig. 10-18 can combine the signal delivered to the four-element horn array by the transmitter or by the echo signals received by them in different ways. Upon transmission, the network excites all four feeds in phase, thereby producing a single main beam called the *sum beam*. The phasing network uses special microwave devices that allow it to provide the desired functionality during both the transmit and receive modes. Its equivalent functionality is described by the circuits shown in Fig. 10-20. During the receive period, the phasing network uses power dividers, power combiners, and phase shifters so as to generate three different output channels. One of these is the sum channel, corresponding to adding all four horns in phase, and its radiation pattern is depicted in Fig. 10-21(a). The second channel, called the *elevation-difference channel*, is obtained by first adding the outputs of the top-right and top-left horns [Fig. 10-20(b)], then adding the outputs of the bottom-right and bottom-left horns, and then subtracting the second sum from the first. The subtraction process is accomplished by adding a 180° phase shifter in the path of the second sum before adding it to the first sum. The beam pattern of the elevation-difference channel is shown in Fig. 10-21(b). The third channel (not shown in Fig. 10-20) is the *azimuth-difference channel*, and it is accomplished through a similar process that generates a beam corresponding to the difference between the

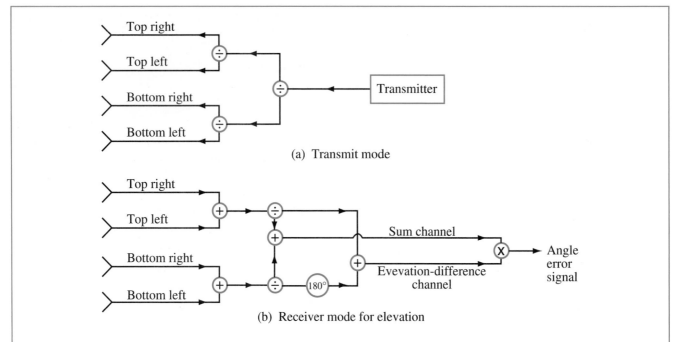

Figure 10-20: Functionality of the phasing network in (a) the transmit mode and (b) the receive mode for the elevation-difference channel.

sum of the two right horns and the sum of the two left horns. If the observed target is centered between the two elevation beams, the receiver echoes will have the same strength for both beams, thereby producing a zero output from the elevation-difference channel. If it is not, the amplitude of the elevation-difference channel will be proportional to the angular deviation of the target from the boresight direction, and its sign will denote the direction of the deviation. In practice, the output of the difference channel is multiplied by the output of the sum channel to increase the strength of the difference signal and to provide a phase reference for extracting the sign of the angle error. This product, called the *angle error signal*, is displayed in Fig. 10-21(c) as a function of the angle error. The error signal activates a servo-control system to reposition the antenna direction. By applying a similar procedure along the azimuth direction using the

product of the azimuth-difference channel and the sum channel, a monopulse radar provides automatic tracking in both directions. The range to the target is obtained by measuring the round-trip delay of the signal.

REVIEW QUESTIONS

Q10.5 How is the PRF related to unambiguous range?

Q10.6 Explain how the false-alarm probability and the detection probability are related to the noise level of the receiver.

Q10.7 In terms of the geometry shown in Fig. 10-17, when is the Doppler shift a maximum?

Q10.8 What is the principle of the monopulse radar?

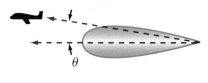

(a) Sum pattern

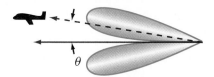

(b) Elevation-difference pattern

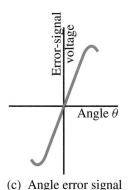

(c) Angle error signal

Figure 10-21: Monopulse antenna (a) sum pattern, (b) elevation-difference pattern, and (c) angle error signal.

CHAPTER HIGHLIGHTS

- Three equally spaced satellites in geostationary orbit can provide coverage of most of Earth's surface.
- The use of polarization diversity makes it possible to double the number of channels per unit bandwidth carried by a satellite repeater.
- A satellite antenna system is designed to produce beams tailored to match the areas served by the satellite. Antenna arrays are particularly suitable for this purpose.

- A radar is an electromagnetic sensor that illuminates a region of space and then measures the echoes due to reflecting objects. From the echoes, information can be extracted about the range of a target, its radial velocity, direction of motion, and other characteristics.
- Due to the random nature of the receiver noise, target detection is a statistical process characterized by false-alarm and detection probabilities.
- A moving object produces a Doppler frequency shift that is proportional to the radial velocity of the object relative to the radar and inversely proportional to λ.
- A monopulse radar uses multiple beams to track the direction of a target, with an angular accuracy equal to a fraction of its antenna beam.

GLOSSARY OF IMPORTANT TERMS

Provide definitions or explain the meaning of the following terms:

geostationary orbit
uplink and downlink
transponder
duplexer
circulator
multiplexer
polarization diversity
atmospheric transmissivity Υ
radial velocity u_r
Doppler frequency shift f_d
pulse length τ
pulse repetition frequency (PRF) f_p
interpulse period T_p
unambiguous range R_u
range resolution
azimuth resolution
false-alarm probability
detection probability
matched filter

bistatic radar
monostatic radar
radar cross section σ_t
threshold detection level
maximum detectable range R_{max}
monopulse radar
sum and difference channels

PROBLEMS

Sections 10-1 to 10-4: Satellite Communication Systems

10.1* A remote sensing satellite is in circular orbit around Earth at an altitude of 1,000 km above Earth's surface. What is its orbital period?

10.2 A transponder with a bandwidth of 500 MHz uses polarization diversity. If the bandwidth allocated to transmit a single telephone channel is 4 kHz, how many telephone channels can be carried by the transponder?

10.3* Repeat Problem 10.2 for TV channels, each requiring a bandwidth of 6 MHz.

10.4 A geostationary satellite is at a distance of 40,000 km from a ground receiving station. The satellite transmitting antenna is a circular aperture with a 1-m diameter, and the ground station uses a parabolic dish antenna with an effective diameter of 30 cm. If the satellite transmits 1 kW of power at 12 GHz and the ground receiver is characterized by a system noise temperature of 1,000 K, what would be the signal-to-noise ratio of a received TV signal with a bandwidth of 6 MHz? The antennas and the atmosphere may be assumed lossless.

Sections 10-5 to 10-8: Radar Sensors

10.5* A collision avoidance automotive radar is designed to detect the presence of vehicles up to a range of 1 km. What is the maximum usable PRF?

10.6 A 10-GHz weather radar uses a 30-cm-diameter lossless antenna. At a distance of 1 km, what are the dimensions of the volume resolvable by the radar if the pulse length is 1 μs?

10.7* A radar system is characterized by the following parameters: $P_t = 1$ kW, $\tau = 0.1$ μs, $G = 30$ dB, $\lambda = 3$ cm, and $T_{sys} = 1,500$ K. The radar cross section of a car is typically 10 m^2. How far away can the car be and remain detectable by the radar with a minimum signal-to-noise ratio of 13 dB?

10.8 A 3-cm-wavelength radar is located at the origin of an x–y coordinate system. A car located at $x = 100$ m and $y = 200$ m is heading east (x-direction) at a speed of 120 km/hr. What is the Doppler frequency measured by the radar?

---

*Answer(s) available in Appendix D.
⦿ Solution available in CD-ROM.

Appendix A:
Symbols, Quantities, and Units

Symbol	Quantity	SI Unit	Abbreviation
A	Magnetic potential (vector)	weber/meter	Wb/m
B	Susceptance	siemens	S
B	Magnetic flux density	tesla or weber/meter^2	T or W/m^2
C	Capacitance	farad	F
D	Directivity (antenna)	(dimensionless)	—
E	Electric-field intensity	volt/meter	V/m
D	Electric flux density	coulomb/meter^2	C/m^2
F	Radiation intensity (normalized)	(dimensionless)	—
F	Force	newton	N
f	Focal length	meter	m
f	Frequency	hertz	Hz
G	Conductance	siemens	S
G	Gain (power)	(dimensionless)	—
H	Magnetic field intensity	ampere/meter	A/m
I	Current	ampere	A
J	Current density (volume)	ampere/meter^2	A/m^2
J$_\mathrm{s}$	Current density (surface)	ampere/meter	A/m
k	Wavenumber	radian/meter	rad/m
L	Inductance	henry	H
l	Length	meter	m
M, m	Mass	kilogram	kg
M	Magnetization vector	ampere/meter	A/m
m	Magnetic dipole moment	ampere-meter^2	A·m^2
n	Index of refraction	(dimensionless)	—
P	Power	watt	W
P	Electric polarization vector	coulomb/meter^2	C/m^2
p	Pressure	newton/meter^2	N/m^2
p	Electric dipole moment	coulomb-meter	C·m
Q, q	Charge	coulomb	C
R	Reflectivity (reflectance)	(dimensionless)	—
R	Resistance	ohm	Ω
S	Standing-wave ratio	(dimensionless)	—
S	Poynting vector	watt/meter^2	W/m^2
S$_\mathrm{av}$	Power density	watt/meter^2	W/m^2
T	Temperature	kelvin	K

Symbol	Quantity	SI Unit	Abbreviation
T	Torque	newton-meter	N·m
T	Transmissivity (transmittance)	(dimensionless)	—
t	Time	second	s
$\mathbf{u}$	Velocity	meter/second	m/s
V	Electric potential	volt	V
V	Voltage	volt	V
V_{emf}	Electromotive force (emf)	volt	V
W	Energy (work)	joule	J
w	Energy density	joule/meter^3	J/m^3
X	Reactance	ohm	Ω
Y	Admittance	siemens	S
Z	Impedance	ohm	Ω
α	Attenuation constant	neper/meter	Np/m
β	Phase constant	radian/meter	rad/m
Γ	Reflection coefficient	(dimensionless)	—
γ	Propagation constant	meter^{-1}	m^{-1}
δ_s	Skin depth	meter	m
$\varepsilon, \varepsilon_0$	Permittivity	farad/meter	F/m
ε_r	Relative permittivity	(dimensionless)	—
η	Impedance	ohm	Ω
λ	Wavelength	meter	m
μ, μ_0	Permeability	henry/meter	H/m
μ_r	Relative permeability	(dimensionless)	—
μ_e, μ_h	Mobility (electron, hole)	meter^2/volt·second	m^2/V·s
ρ_l	Charge density (linear)	coulomb/meter	C/m
ρ_s	Charge density (surface)	coulomb/meter^2	C/m^2
ρ_v	Charge density (volume)	coulomb/meter^3	C/m^3
σ	Conductivity	siemens/meter	S/m
σ_t	Radar cross section	meter^2	m^2
τ	Transmission coefficient	(dimensionless)	—
τ	Pulse length	seconds	s
Φ	Magnetic flux	weber	Wb
$\boldsymbol{\psi}$	Gravitational field	newton/kilogram	N/kg
χ_e	Electric susceptibility	(dimensionless)	—
χ_m	Magnetic susceptibility	(dimensionless)	—
Ω	Solid angle	steradian	sr
ω	Angular frequency	radian/second	rad/s

APPENDIX B:
MATERIAL CONSTANTS
OF SOME COMMON MATERIALS

Table B-1: RELATIVE PERMITTIVITY ε_r OF COMMON MATERIALS[a]

$$\varepsilon = \varepsilon_r \varepsilon_0 \text{ and } \varepsilon_0 = 8.854 \times 10^{-12} \text{ F/m.}$$

Material	Relative Permittivity, ε_r	Material	Relative Permittivity, ε_r
Vacuum	1	Dry soil	2.5–3.5
Air (at sea level)	1.0006	Plexiglass	3.4
Styrofoam	1.03	Glass	4.5–10
Teflon	2.1	Quartz	3.8–5
Petroleum oil	2.1	Bakelite	5
Wood (dry)	1.5–4	Porcelain	5.7
Paraffin	2.2	Formica	6
Polyethylene	2.25	Mica	5.4–6
Polystyrene	2.6	Ammonia	22
Paper	2–4	Seawater	72–80
Rubber	2.2–4.1	Distilled water	81

[a]These are low-frequency values at room temperature (20° C).

Note: For most metals, $\varepsilon_r \simeq 1$.

Table B-2: CONDUCTIVITY σ OF SOME COMMON MATERIALS[a]

Material	Conductivity, σ (S/m)	Material	Conductivity, σ (S/m)
Conductors		*Semiconductors*	
Silver	6.2×10^7	Pure germanium	2.2
Copper	5.8×10^7	Pure silicon	4.4×10^{-4}
Gold	4.1×10^7	*Insulators*	
Aluminum	3.5×10^7	Wet soil	$\sim 10^{-2}$
Tungsten	1.8×10^7	Fresh water	$\sim 10^{-3}$
Zinc	1.7×10^7	Distilled water	$\sim 10^{-4}$
Brass	1.5×10^7	Dry soil	$\sim 10^{-4}$
Iron	10^7	Glass	10^{-12}
Bronze	10^7	Hard rubber	10^{-15}
Tin	9×10^6	Paraffin	10^{-15}
Lead	5×10^6	Mica	10^{-15}
Mercury	10^6	Fused quartz	10^{-17}
Carbon	3×10^4	Wax	10^{-17}
Seawater	4		
Animal body (average)	0.3 (poor cond.)		

[a]These are low-frequency values at room temperature (20° C).

Table B-3: RELATIVE PERMEABILITY μ_r OF SOME COMMON MATERIALS[a]

$\mu = \mu_r\mu_0$ and $\mu_0 = 4\pi \times 10^{-7}$ H/m.

Material	Relative Permeability, μ_r
Diamagnetic	
Bismuth	$0.99983 \simeq 1$
Gold	$0.99996 \simeq 1$
Mercury	$0.99997 \simeq 1$
Silver	$0.99998 \simeq 1$
Copper	$0.99999 \simeq 1$
Water	$0.99999 \simeq 1$
Paramagnetic	
Air	$1.000004 \simeq 1$
Aluminum	$1.00002 \simeq 1$
Tungsten	$1.00008 \simeq 1$
Titanium	$1.0002 \simeq 1$
Platinum	$1.0003 \simeq 1$
Ferromagnetic (nonlinear)	
Cobalt	250
Nickel	600
Mild steel	2,000
Iron (pure)	4,000–5,000
Silicon iron	7,000
Mumetal	$\sim 100,000$
Purified iron	$\sim 200,000$

[a]These are typical values; actual values depend on material variety.

Note: Except for ferromagnetic materials, $\mu_r \simeq 1$ for all dielectrics and conductors.

APPENDIX C:
MATHEMATICAL FORMULAS

Trigonometric Relations

$\sin(x \pm y) = \sin x \cos y \pm \cos x \sin y$

$\cos(x \pm y) = \cos x \cos y \mp \sin x \sin y$

$2 \sin x \sin y = \cos(x - y) - \cos(x + y)$

$2 \sin x \cos y = \sin(x + y) + \sin(x - y)$

$2 \cos x \cos y = \cos(x + y) + \cos(x - y)$

$\sin 2x = 2 \sin x \cos x$

$\cos 2x = 1 - 2 \sin^2 x$

$\sin x + \sin y = 2 \sin \left(\dfrac{x + y}{2} \right) \cos \left(\dfrac{x - y}{2} \right)$

$\sin x - \sin y = 2 \cos \left(\dfrac{x + y}{2} \right) \sin \left(\dfrac{x - y}{2} \right)$

$\cos x + \cos y = 2 \cos \left(\dfrac{x + y}{2} \right) \cos \left(\dfrac{x - y}{2} \right)$

$\cos x - \cos y = -2 \sin \left(\dfrac{x + y}{2} \right) \sin \left(\dfrac{x - y}{2} \right)$

$\cos(x \pm 90°) = \mp \sin x$

$\cos(-x) = \cos x$

$\sin(x \pm 90°) = \pm \cos x$

$\sin(-x) = -\sin x$

$e^{jx} = \cos x + j \sin x$ (Euler's identity)

$\sin x = \dfrac{e^{jx} - e^{-jx}}{2j}$

$\cos x = \dfrac{e^{jx} + e^{-jx}}{2}$

Approximations for Small Quantities

For $|x| \ll 1$,

$(1 \pm x)^n \simeq 1 \pm nx$

$(1 \pm x)^2 \simeq 1 \pm 2x$

$\sqrt{1 \pm x} \simeq 1 \pm \dfrac{x}{2}$

$\dfrac{1}{\sqrt{1 \pm x}} \simeq 1 \mp \dfrac{x}{2}$

$e^x = 1 + x + \dfrac{x^2}{2!} + \cdots \simeq 1 + x$

$\ln(1 + x) \simeq x$

$\sin x = x - \dfrac{x^3}{3!} + \dfrac{x^5}{5!} + \cdots \simeq x$

$\cos x = 1 - \dfrac{x^2}{2!} + \dfrac{x^4}{4!} + \cdots \simeq 1 - \dfrac{x^2}{2}$

$\lim\limits_{x \to 0} \dfrac{\sin x}{x} = 1$

APPENDIX D:
ANSWERS TO ODD-NUMBERED PROBLEMS

Chapter 1

1.1 $p(x, t) = 24 \cos(8\pi \times 10^3 t - 24.24\pi x + 42°)$ (N/m^2)

1.3 $\lambda = 12.5$ cm

1.5 $u_p = 0.83$ (m/s); $\lambda = 10.47$ m

1.7 (a) $y_1(x, t)$ is traveling in positive x-direction, while $y_2(x, t)$ is traveling in negative x-direction.
(b) $x = (\pi/60 + 2n\pi/30)$ cm; $|y_s|_{max} = 1.9$
(c) $x = n\pi/30$; $|y_s|_{min} = 0$

1.9 $T = 1.5$ s; $u_p = 0.56$ m/s; $\lambda = 0.84$ m

1.11 $y_2(t)$ lags $y_1(t)$ by 60°.

1.13 $\alpha = 5 \times 10^{-3}$ (Np/m)

1.15 (a) $z_1 = 3.6e^{-j33.7°}$; $z_2 = 4.5e^{j153.4°}$
(b) $|z_1| = 3.60$
(c) $z_1 z_2 = 16.2e^{j119.7°}$
(d) $z_1/z_2 = 0.80e^{-j187.1°}$
(e) $z_1^3 = 46.66e^{-j101.1°}$

1.17 (a) $t = 3.16 e^{j18.43°}$; $s = 5.10 e^{j78.69°}$
(b) $t = 2.83 e^{-j45°}$; $s = 2.83 e^{j45°}$
(c) $t = 5.2$; $s = 3 e^{j90°}$
(d) $t = 0$; $s = 6 e^{j30°}$

1.19 $\ln(z) = 1.61 - j0.93$

1.21 $v_c(t) = 8.5 \cos(2\pi \times 10^3 t - 62.1°)$ V

1.23 (a) $v(t) = 3 \cos(\omega t - 2\pi/3)$ V
(b) $v(t) = 6 \cos(\omega t + 3\pi/4)$ V
(c) $i(t) = 5 \cos(\omega t + 53.1°)$ A
(d) $i(t) = 3.61 \cos(\omega t + 146.31°)$ A
(e) $i(t) = -\sin \omega t$ A
(f) $i(t) = 2 \cos(\omega t + 3\pi/4)$ A

Chapter 2

2.1 (a) $l/\lambda = 6.67 \times 10^{-6}$; transmission line may be ignored.
(b) $l/\lambda = 0.01$; borderline.
(c) $l/\lambda = 0.2$; transmission line effects should be included.
(d) $l/\lambda = 0.33$; transmission line effects should be included.

2.3 $R' = 1.0$ (Ω/m); $L' = 167$ (nH/m); $G' = 0$; $C' = 172$ (pF/m)

2.5 $\alpha = 0.14$ Np/m; $\beta = 31.5$ rad/m;
$Z_0 = (27.7 + j0.098)$ Ω; $u_p = 2 \times 10^8$ m/s

2.7 $R' = 2$ (Ω/m); $L' = 200$ (nH/m); $G' = 800$ (μS/m);
$C' = 80$ (pF/m); $\lambda = 1$ m

2.9 $R' = 0.8$ Ω/m; $L' = 38.2$ nH/m; $G' = 0.5$ mS/m;
$C' = 23.9$ pF/m

2.11 (a) $b = 3.5$ mm
(b) $u_p = 1.98 \times 10^8$ m/s

2.13 $Z_L = (90 - j120)$ Ω

2.15 $Z_0 = 55.9$ Ω

2.17 $Z_{in} = (60 + j20)$ Ω

2.21 (a) $\Gamma = 0.62e^{-j29.7°}$
(b) $Z_{in} = (12.5 - j12.7)$ Ω
(c) $\tilde{V}_i = 1.40e^{-j34.0°}$ (V); $\tilde{I}_i = 78.4e^{-j11.5°}$ (mA)

2.23 (a) $Z_{in_1} = (35.20 - j8.62)$ Ω
(b) $Z_L' = (17.6 - j4.31)$ Ω
(c) $Z_{in} = (107.57 - j56.7)$ Ω

2.25 $l = \lambda/4 + n\lambda/2$

2.27 $Z_{in} = 240$ Ω

2.29 $l = 0.29\lambda$

2.31 (a) $Z_{in} = (41.25 - j16.35)$ Ω

(b) $\tilde{I}_i = 1.08\, e^{j10.16°}$ A; $\tilde{V}_i = 47.86\, e^{-j11.46°}$ V

(c) $P_{in} = 24$ W

(d) $\tilde{V}_L = 60 e^{-j54°}$ V; $\tilde{I}_L = 0.8\, e^{-j54°}$ A;
$P_L = P_{in} = 24$ W

(e) $P_{Z_g} = 29.15$ W; $P_g = 53.15$ W

2.33 $P_{av}^i = 10.0$ mW; $P_{av}^r = -1.1$ mW; $P_{av}^t = 8.9$ mW

2.35 (a) $\Gamma = 0.5$

(b) $\Gamma = 0.62\underline{/-29.7°}$

(c) $\Gamma = 1.0\underline{/-53.1°}$

(d) $\Gamma = 1.0\underline{/180°}$

2.37 $Z_{01} = 40\ \Omega$; $Z_{02} = 250\ \Omega$

2.39 (a) $Z_{in} = -j154\ \Omega$

(b) $0.074\lambda + (n\lambda/2)$, $n = 0, 1, 2, \ldots$

2.41 (a) $Z_L + j95\ \Omega$

(b) $l = 0.246\lambda$

2.43 $Z_L = (41 - j19.5)\ \Omega$

2.45 $Z_{in} = (95 - j70)\ \Omega$

2.47 First solution: Stub at $d = 0.199\lambda$ from antenna and stub length $l = 0.125\lambda$. Second solution: $d = 0.375\lambda$ from antenna and stub length $l = 0.375\lambda$.

2.49 $Z_{in} = 100\ \Omega$

2.53 $V_g = 19.2$ V; $R_g = 30\ \Omega$; $l = 525$ m

Chapter 3

3.1 $\hat{\mathbf{a}} = 0.45 + \hat{\mathbf{z}}\, 0.89$

3.3 Area $= 9$

3.5 (a) $A = \sqrt{14}$; $\hat{\mathbf{a}}_A = (\hat{\mathbf{x}} + \hat{\mathbf{y}}2 - \hat{\mathbf{z}}3)/\sqrt{14}$

(b) $-12/5$

(c) $\theta_{AC} = 15.8°$

(d) $\mathbf{A} \times \mathbf{C} = \hat{\mathbf{x}} + \hat{\mathbf{y}}4 + \hat{\mathbf{z}}3$

(e) $\mathbf{A} \cdot (\mathbf{B} \times \mathbf{C}) = 13$

(f) $\mathbf{A} \times (\mathbf{B} \times \mathbf{C}) = \hat{\mathbf{x}}54 - \hat{\mathbf{y}}57 - \hat{\mathbf{z}}20$

(g) $\hat{\mathbf{x}} \times \mathbf{B} = -\hat{\mathbf{z}}4$

(h) $(\mathbf{A} \times \hat{\mathbf{y}}) \cdot \hat{\mathbf{z}} = 1$

3.7 $\dfrac{\mathbf{A}(1, -1, 2)}{|\mathbf{A}(1, -1, 2)|} = -\hat{\mathbf{x}}0.17 - \hat{\mathbf{y}}0.70 + \hat{\mathbf{z}}0.70$

3.9 $\hat{\mathbf{a}} = -(+\hat{\mathbf{y}}\, y + \hat{\mathbf{z}}2)/(5 + y^2)^{1/2}$

3.11 $\hat{\mathbf{a}} = (2 + \hat{\mathbf{z}}4)/\sqrt{20}$

3.13 $\mathbf{A} = 0.8 + \hat{\mathbf{y}}1.6$

3.15 $= 0.37 + \hat{\mathbf{y}}0.56 + \hat{\mathbf{z}}0.74$

3.19 (a) $P_1 = (2.24, 63.4°, 0)$ in cylindrical;
$P_1 = (2.24, 90°, 63.4°)$ in spherical

(b) $P_2 = (0, 0°, 3)$ in cylindrical;
$P_2 = (3, 0°, 0°)$ in spherical

(c) $P_3 = (1.41, 45°, 2)$ in cylindrical;
$P_3 = (2.45, 35.3°, 45°)$ in spherical

(d) $P_4 = (4.24, 135°, -3)$ in cylindrical;
$P_4 = (5.2, 125.3°, 135°)$ in spherical

3.21 (a) $P_1(0, 0, 5)$

(b) $P_2(0, \pi, 5)$

(c) $P_3(3, \pi, 0)$

3.23 (a) $V = 21\pi/2$

(b) $V = 125\pi/3$

3.25 (a) $\mathbf{E}_n = -\hat{\mathbf{r}}4$

(b) $\mathbf{E}_t = \hat{\mathbf{z}}4$

3.27 (a) $\theta_{AB} = 90°$

(b) $\pm(\hat{\mathbf{r}}0.487 + \hat{\boldsymbol{\phi}}0.228 + \hat{\mathbf{z}}0.843)$

3.29 (a) $d = \sqrt{2}$

(b) $d = 2.67$

(c) $d = 5$

3.31 (a) $\mathbf{A}(P_1) = \hat{\mathbf{R}}2.856 - \hat{\boldsymbol{\theta}}2.888 + \hat{\boldsymbol{\phi}}2.123$

(b) $\mathbf{B}(P_2) = -\hat{\mathbf{R}}0.896 + \hat{\boldsymbol{\theta}}0.449 - \hat{\boldsymbol{\phi}}5$

(c) $\mathbf{C}(P_3) = \hat{\mathbf{R}}0.854 + \hat{\boldsymbol{\theta}}0.146 - \hat{\boldsymbol{\phi}}0.707$

(d) $\mathbf{D}(P_4) = \hat{\mathbf{R}}3.67 - \hat{\boldsymbol{\theta}}1.73 - \hat{\boldsymbol{\phi}}0.707$

3.33 $T(z) = 10 + (1 - e^{-2z})/2$

3.35 $dV/dl = -3/\sqrt{5}$

3.37 $dU/dl = -3.125 \times 10^{-2}$

3.39 (a) $\oint \mathbf{E} \cdot d\mathbf{s} = -8/3$

(b) $\iiint \nabla \cdot \mathbf{E}\, dv = -8/3$

3.41 (a) $\oint \mathbf{D} \cdot d\mathbf{s} = 150\pi$

(b) $\iiint \nabla \cdot \mathbf{D} \, dv = 150\pi$

3.43 (a) $\oint E \cdot d\mathbf{l} = -1$

(b) $\iint \nabla \times \mathbf{E} \cdot d\mathbf{s} = -1$

3.45 (a) $\oint_C \mathbf{B} \cdot d\mathbf{l} = 8$

(b) $\int_S (\nabla \times \mathbf{B}) \cdot d\mathbf{s} = 8$

3.48 (a) **A** is solenoidal, but not conservative.
(b) **B** is conservative, but not solenoidal.
(c) **C** is neither solenoidal nor conservative.
(d) **D** is conservative, but not solenoidal.
(e) **E** is conservative, but not solenoidal.
(f) **F** is conservative and solenoidal.
(g) **G** is neither conservative nor solenoidal.
(h) **H** is conservative, but not solenoidal.

Chapter 4

4.1 $Q = 2.62$ (mC)

4.3 $Q = 0.173$ C

4.5 (a) $Q = 0$
(b) $Q = \pi a^2 \rho_{s0}/2$
(c) $Q = 2\pi \rho_{s0}[1 - e^{-a}(1 + a)]$
(d) $Q = \pi \rho_{s0}[1 - e^{-a}(1 + a)]$

4.7 $I = 1,570.8$ A

4.9 $\mathbf{E} = \hat{\mathbf{z}} \, 25.61$ kV/m

4.11 $q_2 = -63.13$ (μC)

4.13 $\mathbf{E}(0, 0, z) = \dfrac{706(-\hat{\mathbf{r}}0.02 + \hat{\mathbf{z}}z)}{(4 \times 10^{-4} + z^2)^{3/2}}$ (V/m)

(a) $\mathbf{E}(0, 0, 0) = -\hat{\mathbf{r}}1.76$ (MV/m)
(b) $\mathbf{E}(0, 0, 5 \text{ cm}) = -\hat{\mathbf{r}}90.4 + \hat{\mathbf{z}}226$ (kV/m)
(c) $\mathbf{E}(0, 0, -5 \text{ cm}) = -\hat{\mathbf{r}}90.4 - \hat{\mathbf{z}}226$ (kV/m)

4.15 $\mathbf{E} = \hat{\mathbf{z}} \, (\rho_{s0}h/2\varepsilon_0)[\sqrt{a^2 + h^2} + h^2/\sqrt{a^2 + h^2} - 2h]$

4.17 $\mathbf{E} = 0$

4.21 (a) $\rho_v = y^2 z^3$

(b) $Q = 64/3$ (C)
(c) $Q = 64/3$ (C)

4.23 $Q = 4\pi \rho_0 a^3$ (C)

4.25 $\mathbf{D} = \hat{\mathbf{r}} \rho_{v0}(r^2 - 1)/2r$, for $1 \leq r \leq 3$ m
$\mathbf{D} = \hat{\mathbf{r}} 4\rho_{v0}/r$, for $r \geq 3$ m

4.29 (b) $\mathbf{E} = \hat{\mathbf{z}}(\rho_l a/2\varepsilon_0)[z/(a^2 + z^2)^{3/2}]$ (V/m)

4.31 $V(b) = (\rho_l/4\pi\varepsilon)$
$\times \ln[(l + \sqrt{l^2 + 4b^2})/(-l + \sqrt{l^2 + 4b^2})]$ (V)

4.35 $V_{AB} = -78.06$ V

4.37 (a) $\sigma = 4.32 \times 10^{-4}$ (S/m)
(b) $I = 542.9$ (nA)
(c) $\mathbf{u}_e = -13\mathbf{E}/|\mathbf{E}|$ (m/s); $\mathbf{u}_h = 5\mathbf{E}/|\mathbf{E}|$ (m/s)
(d) $R = 9.21$ (MΩ)
(e) $P = 2.7(\mu\text{W})$

4.41 $R = 2.1$ (mΩ)

4.43 $\theta = 42°$

4.45 $\mathbf{E}_1 = \hat{\mathbf{R}}6\cos\theta - \hat{\boldsymbol{\theta}}3\sin\theta$ (V/m);
$\mathbf{D}_1 = \varepsilon_0(\hat{\mathbf{R}}18\cos\theta - \hat{\boldsymbol{\theta}}9\sin\theta)$ (C/m²)

4.47 $\theta_1 = 71.6°$; $\theta_2 = 78.7°$; $\theta_3 = 81.9°$; $\theta_4 = 45°$

4.49 (a) $|E|$ is maximum at $r = a$.
(b) Breakdown voltage for the capacitor is $V = 1.39$ (MV).

4.51 $W_e = 4.62 \times 10^{-9}$ (J)

4.53 (a) $C = 3.1$ pF
(b) $C = 0.5$ pF
(c) $C = 0.31$ pF

Chapter 5

5.1 $\mathbf{a} = -\hat{\mathbf{y}}2.1 \times 10^{18}$ (m/s²)

5.3 $|\mathbf{B}| = 410$ (mT)

5.5 (a) $\mathbf{F} = -1.41$ N
(b) $W = 0$
(c) $\phi = 0$

5.7 $\mathbf{B} = -\hat{\mathbf{z}}0.3$ (mT)

5.9 $\mathbf{H} = \hat{\mathbf{z}} \dfrac{I\theta(b-a)}{4\pi ab}$, with $\hat{\mathbf{z}}$ being out of the page

5.11 $I_2 = 0.8$ A; direction is clockwise, as seen from above.

5.13 $I = 320$ A

5.15 $\mathbf{F} = -\hat{\mathbf{x}}0.1$ (mN), where $\hat{\mathbf{x}}$ is the direction away from the wire, in the plane of the loop. Thus, the force is pulling the loop toward the wire.

5.17 (a) $\mathbf{H}(0, 0, h) = -\hat{\mathbf{x}}\dfrac{I}{\pi w} \tan^{-1}\left(\dfrac{w}{2h}\right)$ (A/m)

    (b) $\mathbf{F}'_{\mathrm{m}} = \hat{\mathbf{z}}\dfrac{I^2\mu_0}{\pi w} \tan^{-1}\left(\dfrac{w}{2h}\right)$ (N)

    Force is repulsive.

5.19 $\mathbf{F} = \hat{\mathbf{y}} 4 \times 10^{-5}$ N

5.21 (a) $\mathbf{H}_1 = \hat{\boldsymbol{\phi}} J_0$ for $0 \leq r \leq a$

    (b) $\mathbf{H}_2 = \hat{\boldsymbol{\phi}} J_0(a/r)$ for $r \geq a$

5.23 $\mathbf{J} = \hat{\mathbf{z}} 16e^{-2r}$ A/m^2

5.25 (a) $\mathbf{B} = \hat{\mathbf{z}}5\pi \sin \pi y - \hat{\mathbf{y}}\pi \cos \pi x$ (T)

    (b) $\Phi = 0$

    (c) $\Phi = 0$

5.27 (a) $\mathbf{A} = \mu_0 I L/(4\pi R)$

    (b) $\mathbf{H} = (IL/4\pi)[(-y + \hat{\mathbf{y}} x)/(x^2 + y^2 + z^2)^{3/2}]$

5.29 $n_{\mathrm{e}} = 1.5$ electrons/atom

5.31 $\mathbf{H}_2 = \hat{\mathbf{z}} 4$

5.35 $L = (\mu l/\pi) \ln[(d-a)/a]$ (H)

5.37 $W_{\mathrm{m}} = 139I^2$ (nJ)

Chapter 6

6.1 At $t = 0$, current in top loop is momentarily clockwise. At $t = t_1$, current in top loop is momentarily counterclockwise.

6.3 (a) $V_{\mathrm{emf}} = 125e^{-2t}$ (V)

    (b) $V_{\mathrm{emf}} = 62.3 \sin 10^3 t$ (kV)

    (c) $V_{\mathrm{emf}} = 0$

6.5 $B_0 = 1.06$ (nT)

6.7 $I_{\mathrm{ind}} = 37.7 \sin(200\pi t)$ mA

6.9 $V_{12} = -707$ (μV)

6.11 $V_{12} = -3.77$ V

6.13 $V = \omega B_0 a^2/2$

6.15 $I = 1.63 \cos(120\pi t)$ (μA)

6.19 $\rho_{\mathrm{v}} = (6y/\omega) \sin \omega t + C_0$, where C_0 is a constant of integration.

6.23 $k = (4\pi/30)$ rad/m; $\widetilde{\mathbf{E}} = -\hat{\mathbf{z}}941e^{j4\pi y/30}$ (V/m)

6.25 $\mathbf{H} = \hat{\boldsymbol{\phi}}\dfrac{53}{R} \sin\theta \cos(6\pi \times 10^8 t - 2\pi R)$ (μA/m)

Chapter 7

7.1 (a) Positive y-direction

    (b) $u_{\mathrm{p}} = 2 \times 10^8$ m/s

    (c) $\lambda = 1.26$ m

    (d) $\varepsilon_{\mathrm{r}} = 2.25$

    (e) $\widetilde{\mathbf{E}} = -\hat{\mathbf{x}}12.57e^{-j5y}$ (V/m)

7.3 (a) $\lambda = 31.42$ m

    (b) $f = 4.77$ MHz

    (c) $\varepsilon_{\mathrm{r}} = 1.67$

    (d) $\mathbf{H}(z, t) = \hat{\mathbf{x}}22.13 \cos(9.54\pi \times 10^6 t + 0.2z)$ (mA/m)

7.5 $\varepsilon_{\mathrm{r}} = 4$

7.7 $\mathbf{E} = \sqrt{2} \cos(\pi \times 10^{10} t + 104.72\,z)$
    $- \sqrt{2} \sin(\pi \times 10^{10} t + 104.72\,z)$ (V/m)

7.9 $|\mathbf{E}| = 20$; $\psi(t = 0) = 0$; $\psi(t = 5\text{ ns}) = -45°$; $\psi(t = 10\text{ ns}) = -90°$

7.11 (a) $\gamma = 65.5°$ and $\chi = -11.79°$

    (b) Right-hand elliptically polarized

7.19 $u_{\mathrm{p}} = 6.28 \times 10^4$ (m/s)

7.21 $\mathbf{H} = -\hat{\mathbf{y}} 0.16\, e^{-30x} \cos(2\pi \times 10^9 t - 40x - 36.85°)$ (A/m)

7.23 $(R_{\mathrm{ac}}/R_{\mathrm{dc}}) = 143.55$

7.25 $\mathbf{S}_{\mathrm{av}} = \hat{\mathbf{y}}0.12$ (W/m^2)

7.27 (a) $\mathbf{S}_{av} = \hat{\mathbf{z}}500e^{-0.4z}$ (W/m^2)

(b) $A = -1.74z$ (dB)

(c) $z = 23.03$ m

7.29 $u_p = 6 \times 10^7$ (m/s)

Chapter 8

8.1 (a) $\Gamma = -0.67$; $\tau = 0.33$

(b) $S = 5$

(c) $S_{av}^i = 0.13$ (W/m^2); $S_{av}^r = 0.06$ (W/m^2);
$S_{av}^t = 0.07$ (W/m^2)

8.5 (a) $\widetilde{\mathbf{E}}^i = 10(\hat{\mathbf{x}} + j\hat{\mathbf{y}})e^{-j4\pi z/3}$ (V/m)

(b) $\Gamma = -0.2$; $\tau = 0.8$

(c) $\widetilde{\mathbf{E}}^r = -2(\hat{\mathbf{x}} + j\hat{\mathbf{y}})e^{j4\pi z/3}$ (V/m);
$\widetilde{\mathbf{E}}^t = 8(\hat{\mathbf{x}} + j\hat{\mathbf{y}})e^{-1.26 \times 10^{-2}z}e^{-j2\pi z}$ (V/m);
$\widetilde{\mathbf{E}}_1 = 10(\hat{\mathbf{x}} + j\hat{\mathbf{y}})[e^{-j4\pi z/3} - 0.2e^{j4\pi z/3}]$ (V/m)

(d) % of reflected power = 4%;
% of transmitted power = 96%

8.7 $|\widetilde{\mathbf{E}}_1|_{max} = 51.3$ (V/m); $l_{max} = 1.5$ m

8.9 $\varepsilon_{r_2} = \sqrt{\varepsilon_{r_1}\varepsilon_{r_3}}$; $d = c/[4f(\varepsilon_{r_1}\varepsilon_{r_3})^{1/4}]$

8.11 $Z_{in} \simeq (100 - j127)$ Ω; reflected fraction of incident power = 0.24

8.13 $f = 60$ MHz

8.15 $P' = 4.04 \times 10^{-4}$ W/m^2

8.17 $\theta_{min} = 27.92°$

8.19 $\dfrac{S^t}{S^i} = 0.835$

8.21 $d = 15$ cm

8.23 $d = 68.42$ cm

8.25 $f_p = 166.33$ (Mb/s)

8.29 $\theta_t = 18.44°$

8.33 (a) $R = 6.4 \times 10^{-3}$; $T = 0.9936$

(b) $P^i = 0.34$ W; $P^r = 2.2 \times 10^{-3}$ W; $P^t = 0.338$ W

8.35 $s = 0.5$ m; $f = 0.4$ m

8.37 (a) $R = 32$ cm

(b) $M = -4$

8.39 $s' = -25$ cm; $M = 1.25$; image is virtual and erect.

8.43 (a) $\dfrac{1}{f} = \left(\dfrac{n_l - n_m}{n_m}\right)\left(\dfrac{1}{R_1} - \dfrac{1}{R_2}\right)$

(b) $f = -60$ cm in air; $f = -235$ cm in water.

8.45 $s_2' = 8.34$ cm

8.47 $s_2' = \infty$

Chapter 9

9.1 $S_{max} = 1.9$ (μW/m^2)

9.3 (a) Direction of maximum radiation is a circular cone 120° wide, centered around the $+z$-axis.

(b) $D = 4 = 6$ dB

(c) $\Omega_p = \pi$ (sr) = 3.14 (sr)

(d) $\beta = 120°$

9.5 (a) $\xi = 29.7\%$

(b) $G = 0.44 = -3.5$ dB

(c) $I_0 = 33.8$ A; $P_t = 67.3$ W

9.7 $S_{max} = 2 \times 10^{-5}$ (W/m^2)

9.9 $D = 40.13$ dB

9.11 (a) $\xi = 99.3\%$

(b) $G = 1.63 = 2.1$ dB

(c) $I_0 = 0.74$ A; $P_t = 20.1$ W

9.13 (a) $\widetilde{\mathbf{E}}(R, \theta, \phi) = \hat{\boldsymbol{\theta}}\widetilde{E}_\theta = \hat{\boldsymbol{\theta}}j\dfrac{I_0 lk\eta_0}{8\pi}\left(\dfrac{e^{-jkR}}{R}\right)\sin\theta$ (V/m)

(b) $S(R, \theta) = \left(\dfrac{\eta_0 k^2 I_0^2 l^2}{128\pi^2 R^2}\right)\sin^2\theta$ (W/m^2)

(c) $D = 1.5$

(d) $R_{rad} = 20\pi^2(l/\lambda)^2$ (Ω)

9.17 (a) $\xi = 62\%$

(b) $G = 0.93 = -0.3$ dB

(c) $I_0 = 47.5$ A; $P_t = 32.3$ W

9.19 $P_t = 25.9$ (mW)

9.21 $P_t = 0.15$ (mW)

9.23 $\beta_{null} = 5.73°$

9.25 (a) $\beta_{xz} = 0.5°$; $\beta_{yz} = 1.5°$

(b) $D = 5.41 \times 10^4 = 47.3$ dB

9.27 (a) $\beta_e = 1.8°$; $\beta_a = 0.18°$

(b) $\Delta y = \beta_a R = 0.96$ m

9.29 (a) $F_a(\theta) = 4\cos^2\left[\frac{\pi}{8}(4\cos\theta + 1)\right]$

(b) $F_a(\theta) = 5 + 4\cos(2\pi\cos\theta)$

(c) $F_a(\theta) = 4\cos^2(\frac{\pi}{2}\cos\theta - \frac{\pi}{4})$

(d) $F_a(\theta) = 5 + 4\cos(\pi\cos\theta + \frac{\pi}{4})$

(e) $F_a(\theta) = 5 - 4\sin(\frac{\pi}{2}\cos\theta)$

9.31 $d/\lambda = 1.414$

9.33 $F_{an}(\theta) = \dfrac{\sin^2[(15\pi/4)\cos\theta]}{25\sin^2[(3\pi/4)\cos\theta]}$; $\beta = 13.5°$

9.35 $\delta = -2.72$ (rad) $= -155.9°$

Chapter 10

10.1 $T = 105.08$ minutes

10.3 166 channels

10.5 $(f_p)_{max} = 150$ kHz

10.7 $R_{max} = 5.75$ km

BIBLIOGRAPHY

The following list of books, arranged by topic, provides references for further reading.

Electromagnetics

Balanis, C.A., *Advanced Engineering Electromagnetics,* John Wiley & Sons, New York, 1989.

Cheng, D.K., *Fundamentals of Engineering Electromagnetics,* Addison Wesley, Reading, Reading, MA, 1993.

Crowley, J.M., *Fundamentals of Applied Electrostatics,* John Wiley & Sons, New York, 1986.

DuBroff, R.E., S.V. Marshall, and G.G. Skitek, *Electromagnetic Concepts and Applications,* 4th ed., Prentice Hall, Upper Saddle River, New Jersey, 1996.

Hayt, W.H., Jr., *Engineering Electromagnetics,* 5th ed., McGraw-Hill, New York, 1989.

Iskander, M.F., *Electromagnetic Fields & Waves*, Prentice Hall, Upper Saddle River, NJ, 1992.

Johnk, C.T.A., *Engineering Electromagnetic Fields and Waves,* 2nd ed., John Wiley & Sons, New York, 1988.

King, R.W.P. and S. Prasad, *Fundamental Electromagnetic Theory and Applications,* Prentice Hall, Englewood Cliffs, New Jersey, 1986.

Kong, J.A., *Electromagnetic Wave Theory*, 2nd ed., John Wiley & Sons, New York, 1990.

Kraus, J.D., *Electromagnetics,* 4th ed., McGraw-Hill, New York, 1992.

Neff, H.P., Jr., *Introductory Electromagnetics,* JohnWiley & Sons, New York, 1991.

Parton, J.E, S.J.T. Owens, and M.S. Raven, *Applied Electromagnetics,* 2nd ed., Macmillan, London, 1986.

Paul, C.R. and S.A. Nasar, *Introduction to Electromagnetic Fields,* McGraw-Hill, New York, 1987.

Ramo, S., J.R. Whinnery, and T. Van Duzer, *Fields and Waves in Communication Electronics,* 3rd ed., John Wiley & Sons, New York, 1994.

Rao, N.N., *Elements of Engineering Electromagnetics*, 2nd ed., Prentice Hall, Upper Saddle River, New Jersey, 1987.

Shen, L.C. and J.A. Kong, *Applied Electromagnetism,* 2nd ed., PWS Engineering, Boston, Mass., 1987.

Taylor, D.M. and P.E. Secker, *Industrial Electrostatics: Fundamentals and Measurements,* John Wiley & Sons, New York, 1994.

Wait, J.R., *Electromagnetic Theory,* Harper & Row, New York, 1985.

Antennas and Radiowave Propagation

Balanis, C.A., *Antenna Theory: Analysis and Design,* Harper & Row, New York, 1982.

Collin, R.E., *Antennas and Radiowave Propagation,* McGraw-Hill, New York, 1985.

Ishimaru, A., *Electromagnetic Wave Propagation, Radiation, and Scattering,* Prentice Hall, Upper Saddle River, New Jersey, 1991.

Kraus, J.D.,*Antennas,* 2nd ed., McGraw-Hill, New York, 1988.

Sander, K.F. and G.A.L. Reed, *Transmission and Propagation of Electromagnetic Waves,* 2nd ed., Cambridge University Press, Cambridge, England, 1986.

Stutzman, W.L. and G.A. Thiele, *Antenna Theory and Design*, John Wiley & Sons, New York, 1981.

Optical Engineering

Bohren, C.F. and D.R. Huffman, *Absorption and Scattering of Light by Small Particles*, John Wiley & Sons, New York, 1983.

Born, M. and E. Wolf, *Principles of Optics*, 6th ed., Pergamon Press, New York, 1980.

Cho, Z.H., J.P. Jones, and M. Singh, *Foundations of Medical Imaging,* John Wiley & Sons, New York, 1993.

Hecht, E., *Optics,* 2nd ed., Addison-Wesley, Reading, Mass., 1990.

Keiser, G., *Optical Fiber Communications,* 2nd ed., McGraw-Hill, New York, 1991.

Lee, D.L., *Electromagnetic Principles of Integrated Optics,* John Wiley & Sons, New York, 1986.

Smith, W.J., *Modern Optical Engineering: The Design of Optical Systems,* McGraw-Hill, New York, 1991.

Walker, B.H., *Optical Engineering Fundamentals*, McGraw-Hill, New York, 1994.

Microwave Engineering

Freeman, J.C., *Fundamentals of Microwave Transmission Lines,* John Wiley & Sons, New York, 1996.

Pozar, D.M., *Microwave Engineering,* Addison-Wesley, Reading, Mass., 1990.

Richharia, M., *Satellite Communication Systems Design Principles,* McGraw-Hill, New York, 1995.

Scott, A.W., *Understanding Microwaves,* John Wiley & Sons, New York, 1993.

Sinnema, W., *Electronic Transmission Technology,* 2nd ed., Prentice Hall, Englewood Cliffs, New Jersey, 1988.

Skolnik, M.I., *Introduction to Radar Systems,* 2nd ed., McGraw-Hill, New York, 1980.

Stimson, G.W., *Introduction to Airborne Radar,* Hughes Aircraft Company, El Segundo, California, 1983.

INDEX

attenuation constant

$$\alpha = Re\left(\sqrt{(R'+j\omega L')(G'+j\omega C')}\right)$$

phase constant

$$\beta = Im\left(\sqrt{(R'+j\omega L')(G'+j\omega C')}\right)$$

$$Z_0 = \sqrt{\frac{R'+j\omega L'}{G'+j\omega C'}}$$

voltage reflection coefficient

$$\Gamma = \frac{V_0^-}{V_0^+} = \frac{Z_L - Z_0}{Z_L + Z_0}$$

$$Z_L = R + j\omega L$$

$$Z_L = R - \frac{j}{\omega C}$$

voltage standing wave ratio

$$S = \frac{1 + |\Gamma|}{1 - |\Gamma|}$$

matched load $\Gamma = 0$
$$S = 1$$

other $|\Gamma| = 1$
$$S = \infty$$

cylindrical   spherical

$z \quad^r \quad \phi$ $\phi \quad^r \quad \theta$

¼ wave transform

$$Z_{in} = \frac{Z_0^2}{Z_L}$$

$$\beta = \frac{2\pi}{\lambda}$$

$$u = \lambda f$$

Smith Chart

$r=1$ ⊕ $r=1$
sc   oc

vector

$$\theta_{AB} = \cos^{-1}\left[\frac{A \cdot B}{\sqrt{A \cdot A}\sqrt{B \cdot B}}\right]$$

$$= \cos^{-1}\left[\frac{\vec{A} \cdot \vec{B}}{|A||B|}\right]$$

$$Z_{in} = Z_0\left(\frac{Z_L + jZ_0 \tan\beta l}{Z_0 + jZ_L \tan\beta l}\right)$$

$$= Z_0\left[\frac{1 + \Gamma e^{-j2\beta l}}{1 - \Gamma e^{-j2\beta l}}\right]$$

$$L_{eq} = \frac{Z_0 \tan\beta l}{\omega} \qquad \tan\beta l = \sqrt{-\frac{Z^{sc}}{Z^{oc}}}$$

$$C_{eq} = \frac{-1}{Z_0 \omega \tan\beta l}$$

$$Z_0 = \sqrt{Z_{in}^{sc} Z_{in}^{oc}}$$

Torque

$T = NIABo \sin\theta$

$A = area$

$$-1 = e^{-j\pi} = e^{j\pi}$$

$$1 = e^{j2\pi}$$

GRADIENT, DIVERGENCE, CURL, & LAPLACIAN OPERATORS

CARTESIAN (RECTANGULAR) COORDINATES (x, y, z)

$$\nabla V = \hat{\mathbf{x}}\frac{\partial V}{\partial x} + \hat{\mathbf{y}}\frac{\partial V}{\partial y} + \hat{\mathbf{z}}\frac{\partial V}{\partial z}$$

$$\nabla \cdot \mathbf{A} = \frac{\partial A_x}{\partial x} + \frac{\partial A_y}{\partial y} + \frac{\partial A_z}{\partial z}$$

$$\nabla \times \mathbf{A} = \begin{vmatrix} \hat{\mathbf{x}} & \hat{\mathbf{y}} & \hat{\mathbf{z}} \\ \dfrac{\partial}{\partial x} & \dfrac{\partial}{\partial y} & \dfrac{\partial}{\partial z} \\ A_x & A_y & A_z \end{vmatrix} = \hat{\mathbf{x}}\left(\frac{\partial A_z}{\partial y} - \frac{\partial A_y}{\partial z}\right) + \hat{\mathbf{y}}\left(\frac{\partial A_x}{\partial z} - \frac{\partial A_z}{\partial x}\right) + \hat{\mathbf{z}}\left(\frac{\partial A_y}{\partial x} - \frac{\partial A_x}{\partial y}\right)$$

$$\nabla^2 V = \frac{\partial^2 V}{\partial x^2} + \frac{\partial^2 V}{\partial y^2} + \frac{\partial^2 V}{\partial z^2}$$

CYLINDRICAL COORDINATES (r, ϕ, z)

$$\nabla V = \hat{\mathbf{r}}\frac{\partial V}{\partial r} + \hat{\boldsymbol{\phi}}\frac{1}{r}\frac{\partial V}{\partial \phi} + \hat{\mathbf{z}}\frac{\partial V}{\partial z}$$

$$\nabla \cdot \mathbf{A} = \frac{1}{r}\frac{\partial}{\partial r}(rA_r) + \frac{1}{r}\frac{\partial A_\phi}{\partial \phi} + \frac{\partial A_z}{\partial z}$$

$$\nabla \times \mathbf{A} = \frac{1}{r}\begin{vmatrix} \hat{\mathbf{r}} & \hat{\boldsymbol{\phi}}r & \hat{\mathbf{z}} \\ \dfrac{\partial}{\partial r} & \dfrac{\partial}{\partial \phi} & \dfrac{\partial}{\partial z} \\ A_r & rA_\phi & A_z \end{vmatrix} = \hat{\mathbf{r}}\left(\frac{1}{r}\frac{\partial A_z}{\partial \phi} - \frac{\partial A_\phi}{\partial z}\right) + \hat{\boldsymbol{\phi}}\left(\frac{\partial A_r}{\partial z} - \frac{\partial A_z}{\partial r}\right) + \hat{\mathbf{z}}\frac{1}{r}\left[\frac{\partial}{\partial r}(rA_\phi) - \frac{\partial A_r}{\partial \phi}\right]$$

$$\nabla^2 V = \frac{1}{r}\frac{\partial}{\partial r}\left(r\frac{\partial V}{\partial r}\right) + \frac{1}{r^2}\frac{\partial^2 V}{\partial \phi^2} + \frac{\partial^2 V}{\partial z^2}$$

SPHERICAL COORDINATES (R, θ, ϕ)

$$\nabla V = \hat{\mathbf{R}}\frac{\partial V}{\partial R} + \hat{\boldsymbol{\theta}}\frac{1}{R}\frac{\partial V}{\partial \theta} + \hat{\boldsymbol{\phi}}\frac{1}{R\sin\theta}\frac{\partial V}{\partial \phi}$$

$$\nabla \cdot \mathbf{A} = \frac{1}{R^2}\frac{\partial}{\partial R}(R^2 A_R) + \frac{1}{R\sin\theta}\frac{\partial}{\partial \theta}(A_\theta \sin\theta) + \frac{1}{R\sin\theta}\frac{\partial A_\phi}{\partial \phi}$$

$$\nabla \times \mathbf{A} = \frac{1}{R^2\sin\theta}\begin{vmatrix} \hat{\mathbf{R}} & \hat{\boldsymbol{\theta}}R & \hat{\boldsymbol{\phi}}R\sin\theta \\ \dfrac{\partial}{\partial R} & \dfrac{\partial}{\partial \theta} & \dfrac{\partial}{\partial \phi} \\ A_R & RA_\theta & (R\sin\theta)A_\phi \end{vmatrix}$$

$$= \hat{\mathbf{R}}\frac{1}{R\sin\theta}\left[\frac{\partial}{\partial \theta}(A_\phi \sin\theta) - \frac{\partial A_\theta}{\partial \phi}\right] + \hat{\boldsymbol{\theta}}\frac{1}{R}\left[\frac{1}{\sin\theta}\frac{\partial A_R}{\partial \phi} - \frac{\partial}{\partial R}(RA_\phi)\right] + \hat{\boldsymbol{\phi}}\frac{1}{R}\left[\frac{\partial}{\partial R}(RA_\theta) - \frac{\partial A_R}{\partial \theta}\right]$$

$$\nabla^2 V = \frac{1}{R^2}\frac{\partial}{\partial R}\left(R^2\frac{\partial V}{\partial R}\right) + \frac{1}{R^2\sin\theta}\frac{\partial}{\partial \theta}\left(\sin\theta\frac{\partial V}{\partial \theta}\right) + \frac{1}{R^2\sin^2\theta}\frac{\partial^2 V}{\partial \phi^2}$$

$\mathbf{A} \cdot \mathbf{B} = AB \cos \theta_{AB}$ Scalar (or dot) product

$\mathbf{A} \times \mathbf{B} = \hat{\mathbf{n}} AB \sin \theta_{AB}$ Vector (or cross) product, $\hat{\mathbf{n}}$ normal to plane containing $\mathbf{A}$ and $\mathbf{B}$

$\mathbf{A} \cdot (\mathbf{B} \times \mathbf{C}) = \mathbf{B} \cdot (\mathbf{C} \times \mathbf{A}) = \mathbf{C} \cdot (\mathbf{A} \times \mathbf{B})$

$\mathbf{A} \times (\mathbf{B} \times \mathbf{C}) = \mathbf{B}(\mathbf{A} \cdot \mathbf{C}) - \mathbf{C}(\mathbf{A} \times \mathbf{B})$

$\nabla(U + V) = \nabla U + \nabla V$

$\nabla(UV) = U \nabla V + V \nabla U$

$\nabla \cdot (\mathbf{A} + \mathbf{B}) = \nabla \cdot \mathbf{A} + \nabla \cdot \mathbf{B}$

$\nabla \cdot (U \mathbf{A}) = U \nabla \cdot \mathbf{A} + \mathbf{A} \cdot \nabla U$

$\nabla \times (U \mathbf{A}) = U \nabla \times \mathbf{A} + \nabla U \times \mathbf{A}$

$\nabla \times (\mathbf{A} + \mathbf{B}) = \nabla \times \mathbf{A} + \nabla \times \mathbf{B}$

$\nabla \cdot (\mathbf{A} \times \mathbf{B}) = \mathbf{B} \cdot (\nabla \times \mathbf{A}) - \mathbf{A} \cdot (\nabla \times \mathbf{B})$

$\nabla \cdot (\nabla \times \mathbf{A}) = 0$

$\nabla \times \nabla V = 0$

$\nabla \cdot \nabla V = \nabla^2 V$

$\nabla \times \nabla \times \mathbf{A} = \nabla(\nabla \cdot \mathbf{A}) - \nabla^2 \mathbf{A}$

$\int_{v} (\nabla \cdot \mathbf{A}) \, dv = \oint_{S} \mathbf{A} \cdot d\mathbf{s}$ Divergence theorem (S encloses v)

$\int_{S} (\nabla \times \mathbf{A}) \cdot d\mathbf{s} = \oint_{C} \mathbf{A} \cdot d\mathbf{l}$ Stokes's theorem (S bounded by C)